ADVANCE PRAISE FOR
The Future of Nature: Documents of Global Change

"Thi[...] represent[...] and compreh[...] collec[...] of the original publi-
catio[...] no small achievem[...] but what makes the book reall[...] is
the a[...] annotated commentar[...] nd
time[...] and show how collect[...] ay.
The[...] like it." — Thomas E. Lovejoy, Universi[...] ro-
fesso[...] ce and Policy, George Mason Univers[...]

"Thi[...] drawing prim[...] from a 100-year legacy of Western scien[...]fic
litera[...] res related to glob[...] hinking, gives much needed historical context [...] the [...] mass[...] development of human concep[...] of themselves
and [...] wh[...] [...] n relation to exciting [...] habit[...] Cruz Wa[...]en,
New [...] University

"The [...] [...] is a very use[...] [...] [...] as it consists of large[ly]
natu[...] science[...] it is united [...] organized by three humanities scholars. [...] [...] will be enormously useful in bringing together in one v[...]ne
a sel[...] of foundational text for the prevailing thinking about fu[...]re
global [...] " — Fred Horn, Trinity College, Dublin

"Am[...] the greatest challenges for an anthology in the Age of I[...]nt
Dow[...] [...] is to offer a whole that is more than the sum of the book's
dispa[...] selections. With many of these readings easily accessible on-
line, [...] the success of such collections resides in the[...] r's
intro[...] [...] Sörlin, and Warde[...] wonderful job of bringing
toget[...] [...] to conceptual elements under the rubric of global [...] e.'
Thei[...] [...] fler a very appealing way [...] connecting a [...] n-
tal th[...] [...] students in a clear and coherent way. — Ed[...] lo,
Amh[...] [...] ollege

"The [...] of global change [...] [...] [...] an excellent [...] c-
ture a[...] [...] es
as we[...] [...] ntur[...] [...] insightful as well[...] lly
brief. [...] [...] long time span makes this collection particularly valuable."
— Harriet Ritvo, Massachusetts Institute of Technology

"The editors have done a marvelous job of bringing together a fascinating set of primary materials and a superb set of commentaries that provide something we sorely need: more intellectual history of environmental science and thoug[...]

The Future of Nature

The Future
of Nature

Documents of Global Change

EDITED BY

Libby Robin

Sverker Sörlin

Paul Warde

Yale UNIVERSITY PRESS
New Haven and London

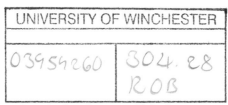
Published with assistance from the foundation established in memory of Calvin Chapin of the Class of 1788, Yale College.

Yale University Press books may be purchased in quantity for educational, business, or promotional use. For information, please e-mail sales.press@yale.edu (U.S. office) or sales@yaleup.co.uk (U.K. office).

Designed by Sonia Shannon.
Set in Electra type by Newgen North America.
Printed in the United States of America.

Library of Congress Cataloging-in-Publication Data

The future of nature : documents of global change / Libby Robin, Sverker Sörlin, Paul Warde, editors.
    p. cm.
Includes bibliographical references and index.
ISBN 978-0-300-18461-7 (pbk : alk. paper)   1. Global environmental change.
2. Overpopulation.   3. Climatic changes.   I. Robin, Libby, 1956–
    GE149.F87 2013
    304.2—dc23

                                                                        2013011467

A catalogue record for this book is available from the British Library.

This paper meets the requirements of ANSI/NISO Z39.48–1992 (Permanence of Paper).

10 9 8 7 6 5 4 3 2 1

Contents

Preface

We live in an age of global change and globalization, when "all that is solid melts into air." The twentieth century was a dark century in many ways, but also one of achievement: of unprecedented improvements in longevity, health care, and income. But now, we suspect, the very foundations of our progress put us at risk: the way we are consuming materials is changing the atmosphere. The ecosystems of the planet are under stress. Mass extinctions threaten. The route to health and wealth for much, if not yet enough, of the world's population may turn out to make us poorer and sicker. We look around and find that the Age of Environment is not all about unqualified progress.

The Age of Environment has been nurtured by the Era of Prediction. As present challenges have arisen, we have learned to expect future challenges of a similar kind. Prediction is important to global imagination too, precisely because we cannot "see" the globe, even if the famous and startling "Only One Earth" image of the lonely blue planet from the Apollo missions has provided us with one, now classic, visual frame. Down on the surface or up in the atmosphere, we perceive only an infinitesimal part of what makes "the globe." Different sciences have taken these snapshots and sought to predict futures for them, and to assess what together they might mean for the whole planet. Environmental activists urge us all to think global and act local. But thinking global is not simple or straightforward, and there is much at stake. The sciences that seek to link the atmosphere and oceans, the terrestrial world and the communities that depend on it (including human societies), and the past world and future world, have had to develop assumptions and ways of thinking. This book gathers together some of the documents that have framed these links over the past three centuries. Thinking about environmental change and global futures has led to particular predictions about the dynamics of our planet. The Earth, perceptions of the planetary, and methods of prediction have histories that are intimately linked.

As citizens and voters, researchers, students and policymakers, we are all interested in how the planet works, what's going to happen next, and what we should be doing about it. How should we understand the predictions of the ever growing army of experts that speak for the planet as a whole? Is this just a matter for experts? Can we also do justice to all the ways that people think about and experience global change? As there is increasing scientific consensus that climate change is real, and that human behaviors are contributing to it, there is an increasing need to understand the political implications of environmental predictions, and their historical context.

In conversations in Copenhagen in 2008, we three editors from different parts of the world developed a sense that the history of "global change" and its associated scholarship could help us reflect on where we think we're going. The history of global change, this extraordinary idea of our time, has not yet been written. It is not easily encapsulated in the history of individual disciplines or in political controversies or social movements. It is a story of the integrative power of certain ideas, techniques and institutions. We wanted to start that. Over the four years since, we have pulled together a wide range of interested researchers at workshops at Norwich, Harvard, Canberra, and Stockholm. Along the way came the idea for an "anthology of documents," each with commentaries. This was a traditional method from literary disciplines, but we apply it here to scientific and other types of writing. Credit goes to our publisher, Jean Thomson Black of Yale University Press, for recognizing that a "source book" was a versatile way to incorporate different voices and documents from a three-hundred-year period into a single volume.

Why an anthology? We want to promote conversations. We don't tell a single story, but rather allow access to the stories that shaped us, and allow readers— students, policymakers, academics, others—to reflect on their own. This is a re-source for reflection and re-visioning. But as historians, we have made selections, interpretations, suggestions. We also chose ten big questions of our time to group the documents. So we offer commentaries by a wide range of people with differ-ing professional interests in "global change": historians, climate scientists, geog-raphers, political scientists, economists . . . and they are all responding to these questions using their own methods.

As we complete our task, we look with gratitude and some surprise at the range, caliber, and generosity of the commentators we have assembled, from emerging researchers to figures of world renown, from a range of places and dif-ferent generations of thinking. A single volume is a small offering in the greater scale of things, but it is a start. We are very grateful to all our contributors, both the original authors of the documents and the commentators on them. Each has brought intellectual excitement and insights that enable us readers to begin to imagine and embrace the enormity of global change and the emerging insight that this change is rapidly becoming a human responsibility, and not just a work of nature.

In this book you will find a little treasure house of ideas, of methods, and visual devices. They include excerpts of both classics and lesser known works;

draw on a range of languages (all translated into English); cover a wide range of fields such as climate science, ecology, history, resilience thinking, demography, economics, forestry, chemistry and future-thinking that sought to integrate across disciplines and sciences. They reflect on lessons learned, unlearned or long forgotten. This is a handbook of how the past saw the future, and how we, their future, might look to ours.

Libby Robin, Sverker Sörlin, and Paul Warde
KTH Environmental Humanities Laboratory, Stockholm
January 2013

How to Use This Book

This is a book you can dip into or read from cover to cover. It is a book of documents that, over three hundred years, have shaped the way we think about our planet and our environment, and the sciences that study global change. Each document is accompanied by a Commentary: a short essay by an expert in the subject that provides context for the document. We have kept outside references in the texts to a minimum, but there is a select bibliography at the back of the book, if you want to read further. Where you find, for example, "(Sörlin and Warde 2009, 16)," you will find the full reference to their book, *Nature's End*, in the bibliography. The relevant point will therefore be found on page 16 of *Nature's End*. Where documents have been adapted or shortened, page numbers from the original are shown in square brackets, with ellipses (. . .) indicating words or sections omitted. When you read a historical document, it is helpful to imagine the audience that it was first intended to reach, as well as to have a sense of its impact today. Each document includes its original date. This is an important clue to the context in which it was written. The commentaries following the documents guide you in your reading. They discuss what the concerns were at the time the document was written, and who was reading the ideas at the time, and since.

The documents are grouped by issues, each with an orienting question. There is an introduction to the book as a whole, "Documenting Global Change," and a short introduction at the beginning of each of the ten parts. References for these are included in the bibliography at the back of the book.

Part Title	Key Question
1. Population	Are we too many, or are we too greedy?
2. Sustainability	Are we limited by resources?
3. Geographies	Are human and natural futures determined or chosen?
4. The "Environment"	How did it emerge?
5. Ecology	How do we understand natural systems?
6. Technology	Does technology create more problems than it solves?
7. Climate	How can we predict change?
8. Diversity	Why do we need it and can we conserve it?
9. Measuring	How do we turn the world into data?
10. The Anthropocene	How can we live in a world where there is no nature without people?

To find a particular document, refer to the contents page. There is also an index at the back of people, places, and topics. Happy reading!

From space, we see a small and fragile ball dominated not by human activity and edifice but by a pattern of clouds, oceans, greenery, and soils. Humanity's inability to fit its doings into that pattern is changing planetary systems fundamentally.

Our Common Future 1987

Introduction

Documenting Global Change

LIBBY ROBIN, SVERKER SÖRLIN,

AND PAUL WARDE

This book is about change. Almost no one denies that the Earth—along with its continents, oceans, and atmosphere, its plants, animals, and diseases—is changing. These collective changes have become familiar enough to be called by a singular expression: *global change*. The concept of global change is a child of the 1980s, fostered by the emergence of such major international research programs as the World Climate Research Programme (established in 1980) and the International Geosphere Biosphere Programme (established in 1987). From this time on, there has been an interdisciplinary "global change science" conducted by a set of specialists who identify themselves as global change scientists.

There have, of course, been stories about the "whole world" for much longer periods: almost every civilization has creation myths. In the cluster of monotheistic religions that emerged in the region known today as the Middle East, there were global catastrophes like "the Fall" and "the Deluge." But in the new modernist turn to global thinking, the leaders have been almost invariably scientists with an intellectual lineage in the ideas of the Enlightenment. It is rare to find social scientists or scholars of the humanities identifying their subject as "global change." Global change science is mostly large scale, exploring natural systems that encompass the entire Earth or significant portions of it (the oceans, the atmosphere, the polar regions). Their work demands large-scale computer models, which have been greatly enhanced by the information technology revolution that has also gathered pace since the 1980s. The emergence of global change science would have been hard to conceive of in earlier periods when calculations had to be worked by hand and when comparable data were scarcer and traveled much slower. However, it would have been perfectly possible to conceive of a contemporary science without global change science. The idea of "the global" is highly historical—and strongly linked to the second half of the twentieth century.

Another factor in the emergence of this particular science was the idea of "the environment," a concept that emerged in its modern form in the 1940s, to describe the human interface with the planet and nature and all its life-sustaining processes. It might seem now that "the environment" has always been with us, but in fact the idea is a relatively recent creation. It draws attention to the fact that our discovery of "global change" has something to do with what we humans are doing. It is a change in our heads, but also a realization that *we* are driving the change: the idea that humans are unsettling everything, to such a degree that we have come to see everything else as a kind of unity, ranging from the very small (a nearby pond, or an entangled bank) to the oceanic, or atmospheric. Humans have both far-reaching impacts and a responsibility to manage at many scales. Thus the emergence of "global change" and its scientific study is the product of recognizing change as both anthropogenic and global. Since such change is partly driven by people, the new scientific thinking must deal with more than just "forces of nature."

"Global" usually applies to spatial concepts. But *time* is also interwoven in the idea of global change—glaciers melt on a different time scale from political decision-making. One of the millennial initiatives that was created to encourage thinking and planning for the next ten thousand years was the Long Now Foundation, established in 01996 (to use their five-digit dates). The concept was popularized by musician Brian Eno and Stewart Brand, publisher of *The Whole Earth Catalog* (1968–1972). The idea of the "Long Now" is to treat not just what is happening today as "now," but also simultaneously what has happened over seven generations (from grandparents through to grandchildren). As people have recognized that humanity is driving global change, groups like the Long Now Foundation have grappled to translate the dynamics of planetary processes into concepts that humans can comprehend and act on. If humanity is to take responsibility for global change and act as stewards of the planet, we need ways of conceptualizing different rates of change, responses, and lagged impacts, as we are attempting to "think like a planet," to paraphrase Aldo Leopold.

The "Anthropocene" is a term coined by atmospheric chemist Paul Crutzen (see Part 10) to describe the beginning of a new geological era in which the actions of people can be traced in all the biophysical systems of Earth and at many scales—from the microscopic to the whole planet. Crutzen and his co-author, ecologist Eugene F. Stoermer, associated these changes with changes in atmospheric chemistry that coincided with the Industrial Revolution and therefore argued that the Anthropocene dated to the 1780s. While the term "Anthropocene" is still under review by the geological community (which usually determines geological epochs), and its chronology and the implicit causal arguments about global change are still evolving as part of these discussions, historians observe that humanity finds itself in an age of conceptual ferment. Living with global change is forcing humans to find ways to comprehend the scope and scale of the change we perceive going on around us. Science is being called on by global organizations to inform decision-making on unprecedented scales (although the relationship between advice and action is rarely straightforward). Environmental issues cannot be understood in isolation from each other. As world historian David Christian commented, "Climate change, acidification of oceans, high rates of extinction, and deforestation are all linked and have to be seen as expressions of a single phenomenon: the astonishing technological creativity of our species that has culminated in the Anthropocene epoch" (Christian 2012, n.p.).

It is of course an empirical issue to establish the extent to which these phenomena are linked and what that really implies. To approach these problems, we must focus on *precisely* what links them and how they interact with each other. We need a *global* approach because they cross national boundaries. Will Steffen, a prominent global change scientist, states that climate change is truly global, because it is about the interactions between Earth's two great fluids, the atmosphere and the ocean, which transport and exchange material and energy all around the planet. We may add that it is global also because each and every human being is affected. Yet this effect is also uneven, both across people alive today and future generations—the new "Other"—through the "climate colonialism" that previous and present generations impose on their descendants for decades, even centuries. This book connects a new planetary understanding and the new human condition under which the current and future generations of human societies will live (Pálsson et al. 2012).

Creating an Archive for the Future

The Anthropocene is a new era to think with. A book like this would not have made sense in the 1780s. Nor would it make sense even in the mid-twentieth century, a time when many of the ideas about global thinking in science and society emerged. Then it was still a case of gathering data, synthesizing, calibrating, beginning to take the measure of *Man's role in changing the face of the earth*, as the influential Princeton conference put it in 1955. Its proceedings were published in a volume of that name in 1956, and we have an excerpt (Ordway) from this in Part 2. It was not until the turn of the millennium, with the definition of the Anthropocene, that a collection of documents from a range of disciplines, published over a total of three hundred years, could be seen as forming a corpus of work belonging to or leading up to "global change thinking." The idea of the global and of anthropogenic change has been compounded by the identification of significant climate change, and aided by new concepts and ideas about collective global impacts such as "planetary boundaries" (Rockström et al., Part 10). It is only in the twenty-first century that we pause to ask: How are we *thinking* about this new type of change as it now operates at an unprecedented scale? How did this key feature of modern thinking emerge? What is its intellectual history?

We editors all trained as historians, but in different places—Australia, Sweden, and the United Kingdom. Each of these places has become conscious of the forces of global change in our lifetimes. Each of us works practically with global change scientists, using our humanities skills in interdisciplinary environmental teams. Thus, we are not independent of global change thinking, but rather we are "participant observers," reflecting on our own practice and that of our colleagues whose training is different. Reflective self-consciousness is one of the key contributions that scholars of the environmental humanities can offer interdisciplinary teams.

We decided that rather than writing a monograph describing the historical emergence of global change ideas, we would map its growth in its own words, using documents with historical integrity. This anthology documents the rise of global change science and enables readers to reflect on what its implications are in different local contexts. The collection of documents constitutes an archive that scopes discussion—for practitioners, scientists, and students alike. Commentaries explain the context and significance of each document for the science at the time and since: some of these are contributed by historians, but many are by people working in the field from other perspectives. Following the development of key ideas historically allows readers to think about the questions they raise and the context in which they were discussed. These are not the only documents we could have chosen, but this selection together captures the history of many crucial ideas in play. The documents too represent a broad spectrum of intellectual traditions.

The collection is not a "canon." We do not seek to define or limit the scope of global change science, but rather to demonstrate its convergence as a field of endeavor, its extent, its deeper history, and its multidisciplinary nature. Taken together, the archive of documents presents contemporary global thinking with its own "long now." It lays bare and reflects on processes that have informed planetary understanding, even before we had a sense of "global." This book, above all, provides access to an archive, and points to a diversity of ways of thinking. We hope it will be useful to readers curious about where this big idea of our times came from. All the documents have been relieved of unnecessary technical detail. Many have been shortened for convenience and easy access. This is not a collection created for political purposes, in any traditional sense: it is part of the global endeavor that must transcend the politics of individual parishes and nations.

Not all disciplines or intellectual traditions are represented in this compact overview, but our choices reflect a breadth of contributions. That these largely belong in a still dominant tradition of Western thinking is, broadly speaking, an effect of how and where this science has been produced up until very recent times. We might wish it were different, but we inherit the history we have, which is not to say that we have fully comprehended that inheritance, or that there is no more to be found in it. Quite the contrary! If this collection were assembled only a generation later, or say in 2063 rather than 2013, we might have seminal documents originating from science carried out in India, China, Brazil, Mexico, Malaysia, Africa—and there might have been papers representing knowledge traditions we do not even know about today. The task we have set ourselves is above all to understand how people early in the twenty-first century "think globally" about the environment and make predictions about its future. At this time, the Western natural sciences dominate the discourse in global change organizations, and so too in the documents we select. We, the editors, hope that by putting the documents together and reflecting on them, we can also begin to ask questions about the human condition—the ethics, justice, and effects of change—which are beyond the methods of natural sciences. Scientific evidence and deterministic processes can provide an excellent start for insight into the plight of humanity today, but without recognizing the political and ethical implications of this plight, the story is incomplete. This is a story that affects us all, not just the Western world, and it affects nonhuman members of our world as well. We humans are social, ethical, spiritual, political, and, of course, biological creatures. Future versions of reading the global may in particular link in social or humanist experiences and methodologies, as Mike Hulme indicates in the most recent of all our documents (pp. 506–519), or perhaps approaches that try in deeper terms to dissolve the divide between the natural and the social, one of the most fundamental of Western ideas, suggesting entirely new frameworks for what is global and what is change (and even what is science). In our selected bibliography at the end of the book we have included a range of recent contributions to the literature that hint on where current thinking may be taking us, although it remains to be seen what will be regarded tomorrow as the defining texts of today.

We ourselves are working in "the future" for the authors of the documents in this book. Our thinking is shaped by the choices made by our predecessors in particular circumstances, for their own very good reasons. It also turns out that, thankfully, here in the future, we can still think for ourselves. And so will our successors.

Concepts for Global Change

What sorts of concepts matter in a natural world irrevocably shaped by the presence of people? We want to highlight some ideas that seem to us to be particularly prominent in the development of global change thinking.

The first is the idea of the *future*. Not all of our documents expressly talk about the future, but most do so at least implicitly. The manifest preoccupation with a secular and nonfictional (that is, also nonmetaphysical and non-utopian) future seems to be largely a phenomenon of the twentieth century, with a gestation period located in the 1920s and 1930s (Bell 1997). Global change science and its precursors seek to establish facts about Earth's past not as a goal in itself, but rather with the explicit ambition to gain knowledge for the future condition of the planet, which they consider possible.

A second concept is *prediction*. Global change scholarship situates itself as a predictive science. It uses models and theory to fulfill the dictum coined by nineteenth-century French philosopher Auguste Comte: *savoir pour pouvoir pour prévoir*, "to know in order to be able to foresee." Global change science has deep roots in the quantitative tradition that uses numbers to describe and calculate what might be called *the state of the earth*. Gradually, and especially since the advent of computers, projections, scenarios, "foresighting," and other concepts with varying levels of precision enabled new scope for prediction. Prediction increasingly became the reason for the study of the environment, and environment was increasingly understood as that phenomenon whose change was measurable and, potentially, worrying. The people who managed prediction became key environmental *experts* and at the same time masters of a particular kind of contemporary angst (for example, Borgström (pp. 40–50), Ehrlich (pp. 54–58), Vogt (pp. 187–190).

This leads to a third concept: *expertise*. Traditionally, many kinds of knowledge (local, traditional, religious, philosophical, scientific) had supported human interaction with natural surroundings. But in the mid-twentieth century, a new hybrid strand of environmental science, with multiple roots, and later global change science crystallized the knowledge that became regarded as environmental expertise, often supported by practical operators with global reach, even the United States military (Doel 2003).

The fourth concept, already alluded to, is *the environment*. It is the glue that binds all the other concepts, that situates diverse sciences (both life and natural sciences) and the subjects of their study into relation with each other. As it is also a *human* environment, we would expect to find space for the social sciences and the humanities too, but these have not figured prominently in environmental expertise until comparatively recently and hardly at all in global change science yet. In the 1940s and 1950s the idea of "the environment," and the crisis about its future, emerged together. Prediction that the whole global system was falling into degradation was co-determined with the very discovery of that system. Strangely, this notion also lived alongside one of the most techno-industrialist and optimistic periods of the history of mankind, a "boy's own" version of manageable dystopia mixed with sanguine Rostowian "stages of economic growth" (Rostow 1960). It emerged amid evergreen engineered suburbia—in a world that is increasingly urban. This is what has been called the Cold War Modern (Crowley et al. 2008). The idea of the environment oriented thinking toward the future and became inextricably linked with crisis, a concept constructed in part by prediction and in part by fear of future catastrophe.

Trust in Numbers: Meaningful Measurement for Predicting Change

Global change science has placed its trust in metrics: forms of measurement and ways of relating the measures to each other. Environmental science was only named as a discrete discipline in 1962, according to the *Oxford English Dictionary*. It immediately defined its purpose as identifying and measuring "environmental variables." Global change science, coming later again, took the idea of measurement further, integrating between measured empirical properties and projecting numerical values into the future. Yet these developments drew on long-established and internationally agreed-upon practices involving standardization of units, the conditions under which phenomena could be measured, the recording of data, the development of institutions, and infrastructure that created global comparability of measures, and defended these through academic conventions in discourse and analysis (see Part 9, Measuring). Inevitably, elements of the environment that lent themselves to measurement, and had been integrated into internationally established disciplinary procedures, were most highly valued in the emerging discourses of environmental science, and later global change. Although novelty and innovation have been characteristic of scientific endeavor for a long time, new ways of thinking are still built upon the foundations (or ruins) of the old, while new institutions frame these developments, not least the global institutions of the postwar era, under the umbrella of UNESCO. It is an essential task of good science and policy-making to reflect on how this history has led practice down certain pathways, and the gains and possibly losses involved.

People who can discover, measure, and interpret metrics have been highly valued since the Enlightenment, and global change has granted a growing prominence to those with particular kinds of expertise: those who can express concepts and evidence numerically and who can stipulate mathematical relationships between variables. This is driven by what Theodore Porter (1995) calls *trust in numbers*. When comparisons or decisions have to be made, it is easier if the possible outcomes have a numerical value: it suggests that the decision is "objective" rather than subjective. Things that are different in kind may be placed on scales that permit comparison and investigation. Since the information technology revolution, vast, intricate, and complex processes can be mapped quickly and the results broadcast globally. Prediction can appear to be something more scientific than prophesy, much more than a simple intuition of future blessings or woe. The predictor is no longer a prophet or a demagogue, ruled by emotion, but a calculator, a purveyor of bald "facts." The documents in this book reflect on facts and predictions that have become the currency of public policy-making.

Another property of numbers is that they offer the capacity to switch scales. We find that those disciplines that facilitate scaling up or scaling down (often from models that originate in studying the very local, but expand to encompass the global) have been very important in the development of environmental thinking. For example, demography and ecology are especially prominent. "Rescaling" can work in time as well as space, allowing very short-term and very long-term phenomena to be related. We argue that the current idea of the environment, what it actually means, was built from the ground (or ocean) up: through a capacity to measure. In measuring, the environment became something viewed as predictable, and indeed, where a primary interest in it was its future. The future, in turn, became less a place for myth and revelation and more a set of scientific observations. Even a probability value attached to the future, such as the percentage change of rainfall, or the number of runs of a computerized climate model predicting a certain outcome, somehow seems more *concrete* than someone telling us that their best guess is that it might rain and it might get hotter. Numerical values of course represent choices, even when they appear objective. But this does not mean we can take any number, that "anything goes." Scientific peer review is an important part of agreeing on ways to take the measure of something, and the potential of such information for further study. Numbers are not independent of the people and institutions that provide them, as has been demonstrated for the predictive climate sciences (Edwards 2010). They sometimes create new institutions and disciplines, and entirely new infrastructures of observation and calculation, and they give authority to certain experts over others.

These reflections are not a critique of "trust in numbers" per se. Rather, we want to ask why societies generate, and trust, certain numbers, and how these stand in relation to other things societies value and approaches they might take. Environmental and social systems are interconnected—but they are measured differently. There is a fundamental incommensurability about social values and environmental facts. They mean different things, and operate on different scales. However, their histories are not so far apart as might sometimes be supposed; as Porter's work (among others') has shown, the modern notions of "society" and its "population" have themselves been the result of advances in statistics, measurement, census taking, and information processing from the seventeenth to the nineteenth centuries. Thus, while modern prediction about environmental futures emerged at a particular historical moment, it did so not in a vacuum, but rather as a product of its long history.

With this perspective, we suggest that our current predicament is not just a matter of human society increasingly steering the environment of the whole planet, or of environmental change having great implications for the future of human society (although we believe these things to be true). We begin to see that society, science, and environment have always been intertwined, and that these connections are inextricable from the history and the future of ideas.

A Great Acceleration of Ideas

A history of the planetary system is already being written, asking how and when people started to overwhelm "normal" change in environmental systems. This also means identifying what could reasonably be called normal, a job that has turned out to be harder than expected. If the decade of the 1780s (or thereabouts) marks the beginning of the Anthropocene, the 1950s marked the beginning of what some scholars have called the Great Acceleration. All those indicators we can measure and graph—carbon levels in the atmosphere, global temperatures, pollution of the air and seas, population, species loss, Gross Domestic Product, and even globalization itself ("measured," for example, through the proxy of the expanding number of McDonald's outlets)—could be graphed in the succeeding time period. The Great Acceleration is reflected in many J-curves (see Steffen figures pp. 488–489), with a flat start in the 1950s and an increasingly alarming vertical angle by the time we reach the twenty-first century.

Some indicators can be plotted on graphs more easily than others. Such graphs in turn often frame understandings of the environment and its "crisis." We cannot count documents in the same way, or enumerate "ideas about global change." Yet, in the mid-twentieth century, there is no doubt that ideas about "the environment" and environmental concepts also started to proliferate exponentially, following the hockey stick curve pattern of other well-known indicators. If we could graph the use of words or expressions such as "the environment" or "global change" as they are found and archived electronically in book titles and text searches, we would undoubtedly find a similar acceleration to those other indicators of global change.

A graph of the uses of "the environment" would not, however, tell us that much about *why* people found it a useful concept to employ. Nor, given the fact that *environment* is an old word with medieval origins, whose meaning has

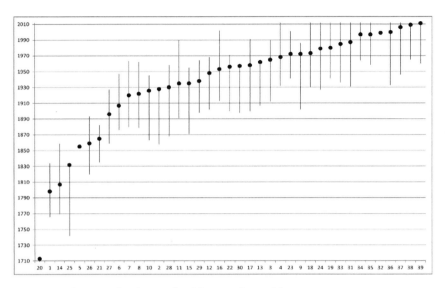

Figure 1. Life spans of authors and publication dates of documents

proliferated over time and of which "the environment" is only one usage, would such a graph tell us much about what people meant when they said it.

What can Figure 1 tell us? It shows the life spans of all of the authors of our documents, and the date at which the document provided in the anthology was published. As a pattern, it shows nothing more than the choices we have made as editors. It might be said simply to *illustrate* our preferences. The graph is not *data* to which we must apply interpretations. The only metric to be found here is our collective judgment, and different authors represent a different case: Svante Arrhenius (pp. 303–312) and Guy Stewart Callendar (pp. 327–334) were not taken very seriously when they first published these documents, but later became recognized as scientific "forerunners." By contrast, forester Hans Carl von Carlowitz (pp. 67–74) or ecologist Eugene Odum (pp. 233–241) were highly influential in the development of their respective disciplines at the time they published these documents; some authors were true originals, while others seem more illustrative of many publications that appeared at much the same time.

Yet the figure also illuminates aspects of what we think is important in the story of the emergence of global change science. It alerts us to the historical context of the story that we present. Given that global change science dates only to the 1980s, we find the period from the 1920s to the 1950s to be surprisingly prominent. The thesis of the collection, if one can identify a singular one, is that the onset of the Great Acceleration coincides with an explosion of ideas about the whole planet. The 1940s are also the decade that touches the life spans of the greatest numbers of our authors. The institutions of the global, including UNESCO, which set up a project to write global history for peace in the wake of World War II, were also mostly born in this era, as is argued in detail by Libby Robin and Will Steffen (2007). This in turn reminds us that patterns in time are also located in space. Global institutions are located mostly in Europe or eastern North America, and global change science largely emerged from this type of institution. European expansion and modernity are entwined, and planetary models and global thinking are, in a sense, their children. Thus we find that many of the older documents were not originally in English, but increasingly, as global institutions adopted English, the conversations about global change science were also conducted in the new lingua franca.

History and Global Change Thinking in the Anthropocene

Change is the chief focus in this book, but the documents themselves are generally about science. What do scientific documents tell about their own history? Stephen Jay Gould picks out 1776 as a big year, the year of "ideas that shaped the flow of human history" (Gould 1997, 13). He cites three significant events, which all took place in Europe or on the eastern seaboard of North America: the publication of Adam Smith's *Wealth of Nations*, the Declaration of Independence by the United States of America on 4 July, and the lecture in Paris by Pierre-Simon Laplace (1749–1827) of a mathematical formulation of the future. As Gould describes, Laplace's laws of probability enabled him to make the bold claim that if he were given the position of every particle in the universe and the forces acting on it, he could specify in detail any future event, even those that are seemingly capricious, inconsequential, or under the influence of human free will. Gould credits Laplace for his pioneering work in the mathematics of the probability for understanding the behavior of random systems and processes, and argues that since this time "human futures cannot be meaningfully predicted . . . [without questioning] the premise of Laplacian determinism" (Gould 1997, 15–16). Gould argues that Laplace's theories overstate the case for "adamantine physics" over other sciences, and over other forms of knowledge and human creativity. Gould's aim in revisiting Laplace is to "learn to avoid the emptiest category of millennial questions—our yearning to know the future and to charge science with accurate prognosis" (Gould 1997, 19).

Paul Crutzen (pp. 483–485) also chooses the late eighteenth century as a significant moment for his Anthropocene concept. Amid the ferment of the Industrial Revolution, not just human but *planetary* history began to change forever. The "onset of the Anthropocene" ushered in a new geological era. While this radical proposal is still officially under review by the geological community, the concept has tremendous traction as a transdisciplinary tool as humanity struggles to understand its enmeshed role in the future of humans and the rest of nature on this planet.

It turns out that the vision of Laplace's "demon" (as it came to be known) has become a *desideratum* for living in the epoch that was inaugurated at the very time when Laplace developed his theories. His formulation gives a privileged standing to science as the ultimate guide in the era since the Industrial Revolution, where, as Gould puts it, "Ability to predict becomes the chief criterion of understanding . . . and science owns the tools of prediction" (Gould 1997, 16).

As Laplace remarked, "We may regard the present state of the universe as the effect of its past and the cause of its future. An intellect which at a certain moment would know all forces that set nature in motion, and all positions of all items of which nature is composed, if this intellect were also vast enough to submit these data to analysis, it would embrace in a single formula the movements of the greatest bodies of the universe and those of the tiniest atom; for such an intellect nothing would be uncertain and the future just like the past would be present before its eyes" (Laplace [1814] 1951, 4).

For all the careful caveats with which global modelers may hedge their results, this sounds something like a vision to which they might aspire; we are certainly much closer to it than Laplace could have imagined possible. But what of those two other events of 1776, the ones that historians have expended considerably more energy writing about? Adam Smith provided a theory of political economy: how people come to mutually identify and satisfy wants through exchange, through patterns of "truck and barter" and conversation. The American Declaration of Independence, abolishing previous claims to political authority, necessarily set those men (they were all men) who were enfranchised to the task of deciding how people belonging to a political community should make decisions as to their own destiny. These, as well as technological change (paradigmatically at that moment, the steam engine), also shaped the Anthropocene.

How does an understanding of wants, barter, conversation, or politics fit into global change? How does global change shape our needs and arguments? These are questions that this collection opens up for discussion. This is partly because they are frequently alluded to, if often somewhat vaguely, as elements *driving* change: something about what people want, and how they might behave. As the millennium turned over, scientists began to recognize increasingly that providing predictions did not ensure appropriate action; or, perhaps we should say, a new generation of scientists came to this recognition again. Even where excellent predictions were available and considerable scientific consensus could be mounted for them, policy-making did not shift as quickly and decisively as many hoped. The twentieth century had seen the rise of global institutions, but national decision-making, itself an agglomeration of interests at different "scales," is still crucial to global outcomes for biodiversity, climate, population, environmental justice—indeed all aspects of global change. History, with its traditional specialist skills in the ways of nations, is thus paradoxically drawn into global change scholarship.

There is further reason for this collection. If we acknowledge that global change thinking has its own history, a collection like this enables us to begin to sketch its contours. A historical perspective helps explain the form that global change scholarship adopts, the practices it employs (and does not), the institutions that govern it, and the types of expertise that are valued by it in different eras. Global change science is itself the product of conversations and institution building; appreciating *how* this has come about helps to explain its successes and failings.

Global Change for Whom, and on Whose Scale?

Sometimes a decade is a long time, and at others, several millennia a mere blip. A chronometer is not the only way to measure time. This is a challenge that "global change" also presents to historians, or the humanities: it was perhaps easier to think about eternity than the Long Now. At the same time, it makes us think differently about the time of human lives. As Rob Nixon has argued, the environment of the world's poor is impoverished in ways that are slower than the eye can see. His concept of "slow violence," processes profound but not newsworthy, unsettles an account of history that is driven by drama, by moments like 1776, or indeed that treat the advance of science of a story of ever-expanding "enlightenment" (Nixon 2011). Thus, while our documents taken together are largely a reflection on the historical and imperial expansion of Europe during the Anthropocene, we are reminded of the importance of ideas that feedback from other places into the dominant Western world of science and technology, and how that dominant vision is having effects we can hardly perceive.

Global change is a momentous idea in our time. Many scientists use this term in preference to climate change or global warming, the terms in the popular press, because they want to emphasize the *human* factors in the change and get beyond "reducing the future to climate," Mike Hulme's apt phrase (pp. 506–519). We should say here that we three editors of this book accept global change and its potential humanity. We agree that there are "global" things happening. We are not critics, but participants. We are conscious that a rhetoric of metrics can fail to include human dimensions. But where technologies or metrics are available—for example, those that measure extinctions, atmospheric carbon, or ocean acidification—human understanding can benefit from the stories unlocked, and such insights must be made available or crucial issues will go unheeded. Technical expertise does not guarantee a capacity to communicate findings with nonspecialists or the ability to understand how changes unfold at a human scale. Yet increasingly we define our understanding of phenomena like climate change, and limit possible futures, because technical expertise is favored over plain language, and people lacking the technical expertise no longer follow the debate. If we are talking about society's future, the expert might be an economist. In terms of the nonhuman world, it is usually the natural or life scientist. All frequently employ quantitative methods, and work toward "predictions" that often exclude contingent, imaginative, and humanistic conceptions for possible futures.

In an era where sustainable development is increasingly defined as a journey rather than a destination, we need our humanity as we travel. The future itself is no longer somewhere we go, it is something we *create*, according to social commentator and former Australian Commissioner for the Future Ian Lowe (2012, 187). This is the experience of the Anthropocene. It makes all of us who can exercise some choice in the matter experts in creating a future where we would want to live (Adams 1996). If so, we need all the humanity we can engage in making that future: science, economics, the full range of humanities and social disciplines, and the human imagination. No one can fully *predict* the future, but we do need to bring our imagination to understanding its relations to our present condition.

PART 1

Population

Are We Too Many, or Are We Too Greedy?

The relationship between the number of people and the resources available to them has long been a matter of human concern, at first understood where pressures were most immediately felt, by the family and the local community. For many centuries this equation was mostly considered only on a basic level: is there enough food to go around? From the seventeenth century, what mattered was simply how many people there were, and the early accounting of people—what we now call demography—was an important stimulus to the development of data collection, quantitative methods of analysis, and statistics. Yet with the dramatic population increases of the nineteenth century and especially the twentieth, eventually associated with the "demographic transition" that saw large falls in mortality and then fertility rates, the question began to look more complex. The question of inequality now loomed larger: should countries that had gone through rapid population growth in the past deny the same to societies that developed later? And in a hugely unequal world, was pressure on resources really generated by excess numbers of people, or rather the lifestyles of those who consumed most?

We often think that demography began with Thomas Malthus. This is not really true, but he provided such a striking and clear argument that it set the terms of the "problem of population" down to our own day. For Malthus, human destiny was determined by the power of unchecked population growth to overwhelm the availability of resources, no matter how productive the earth became. Thus, the fundamental problem for him, and his successors, was how to check population. Because this was essentially an argument from design, it was also an argument about the whole Earth, not just any one part of it. Despite the fact that populations were measured only in countries or regions, demography could treat humanity as a whole, and arguments about local destinies were arguments about the whole species—as we see in the excerpts from both Malthus and George Knibbs.

By the time Knibbs wrote in the early twentieth century, governments were beginning to produce detailed data on population; they considered keeping track of it one of their basic tasks. This allowed the development of the "global audit," a balancing-out of populations and available resources, that has expanded out as our notion of global change has become more inclusive. Much of the ground-work in the development of data collection and methods of analysis was done in the 1910s and 1920s, before the increasing emergence of international institutions and efforts at global governance with the founding of the United Nations in the 1940s.

From Malthus onward, population was seen as unruly. The "constancy of the passion between the sexes" overrode wiser counsel. Once it was established, after much debate during the eighteenth century, that populations were indeed growing, the question was posed of how far that growth would go; the future was "the problem of problems," as Knibbs wrote. People had both an intellectual and a moral incapacity to understand what they were getting themselves into, and needed guidance from population experts: "Humanity has both to be instructed and governed" (p. 34).

Solutions to the "population problem" were thus also often seen as moral, whether to be adopted voluntarily and encouraged through education, or im-posed coercively. Governments and campaigners used both methods during the twentieth century. In particular, the poor, supposedly "lesser" races, and the less able were to be discouraged from having children. These ideas were at the fore-front of the eugenics movement in the first half of the twentieth century. The "problem" was global and necessitated international cooperation, but it did not give everyone an equal voice.

Already in Knibbs, we see the suggestion that the likely future of a much larger population could also lead to an argument for a different kind of moral "elevation": the choice to be poorer and share resources more equally. From the 1920s, there was increased sensitivity to differences in diet and expectation; could humanity be treated as a whole, or did policy have to reflect local differences? Older methods of defining food intake by the production of an acre, already ap-plied in calculating maximum population by William Petty in the seventeenth century, were inadequate to this task. By the 1950s, we see the work of Borgström (pp. 40–50) attempting to take account of both differences in diet and the already heavy dependence of population on trade with his idea of "ghost acreage": the local acreage that would be required to produce the food the country actually consumed. With this kind of work we see the emergence of the idea of "carrying capacity," whether of a region or the whole planet.

The 1960s saw a new wave of publicly prominent population theorists, who looked with dismay at the fact that global population growth was now more rapid than ever. In fact, this was in part due to the success of public health policy and the work of earlier generations of demographers that led to significant falls in mortality rates, especially infant mortality. These new theorists—Paul Ehrlich (pp. 54–58) most prominent among them—often came from the natural sciences, and in some ways reverted to seeing humanity as a single "population." They failed to perceive that what they were observing was itself the result of public policy, and instead developed a moral critique, whether of basic human selfishness or the excessive demand on resources of the developed world, that would inevitably be followed by the developing world. The horizon of overpopulation became much closer, reduced even to just a few years; the world was about to be overwhelmed. In much of this thinking, greed was the problem for the rich world, where numbers had now stabilized, but the problem of "population"— merely existing—was the excess propensity of the poor to reproduce. It was indeed the governments of developing nations such as China and India that were most alarmed. Their often coercive family planning policies and controversy over the ethics of "neo-Malthusians" have made population control a much less acceptable and prominent policy issue in the past two decades. Meanwhile, population continues to grow to over twice the level of 1960, and soon to surpass the "limit" of 7.8 billion that Knibbs imagined in 1926.

An Essay on the Principle of Population

THOMAS MALTHUS

Chapter 1

. . .

I think I may fairly make two postulata.

First, That food is necessary to the existence of man.

Secondly, That the passion between the sexes is necessary and will remain nearly in its present state.

These two laws, ever since we have had any knowledge of mankind, appear to have been fixed laws of our nature, and, as we have not hitherto seen any alteration in them, we have no right to conclude that they will ever cease to be what they now are, without an immediate act of power in that Being who first arranged the system of the universe, and for the advantage of his creatures, still executes, according to fixed laws, all its various operations.

I do not know that any writer has supposed that on this earth man will ultimately be able to live without food. But Mr Godwin has conjectured that the passion between the sexes may in time be extinguished. As, however, he calls this part of his work a deviation into the land of conjecture, I will not dwell longer upon it at present than to say that the best arguments for the perfectibility of man are drawn from a contemplation of the great progress that he has already made from the savage state and the difficulty of saying where he is to stop. But towards the extinction of the passion between the sexes, no progress whatever has hitherto been made. It appears to exist in as much force at present as it did two thousand or four thousand years ago. There are individual exceptions now as there always have been. But, as these exceptions do not appear to increase in number, it would surely be a very unphilosophical mode of arguing to infer, merely from the existence of an exception, that the exception would, in time, become the rule, and the rule the exception.

Thomas Malthus. 1798. Excerpts from Chapters 1 and 2 in *An essay on population*. First printed for J. Johnson, in St. Paul's Church-Yard. London.

Assuming then my postulata as granted, I say, that the power of population is indefinitely greater than the power in the earth to produce subsistence for man.

Population, when unchecked, increases in a geometrical ratio. Subsistence increases only in an arithmetical ratio. A slight acquaintance with numbers will shew the immensity of the first power in comparison of the second.

By that law of our nature which makes food necessary to the life of man, the effects of these two unequal powers must be kept equal.

This implies a strong and constantly operating check on population from the difficulty of subsistence. This difficulty must fall somewhere and must necessarily be severely felt by a large portion of mankind.

Through the animal and vegetable kingdoms, nature has scattered the seeds of life abroad with the most profuse and liberal hand. She has been comparatively sparing in the room and the nourishment necessary to rear them. The germs of existence contained in this spot of earth, with ample food, and ample room to expand in, would fill millions of worlds in the course of a few thousand years. Necessity, that imperious all pervading law of nature, restrains them within the prescribed bounds. The race of plants and the race of animals shrink under this great restrictive law. And the race of man cannot, by any efforts of reason, escape from it. Among plants and animals its effects are waste of seed, sickness, and premature death. Among mankind, misery and vice. The former, misery, is an absolutely necessary consequence of it. Vice is a highly probable consequence, and we therefore see it abundantly prevail, but it ought not, perhaps, to be called an absolutely necessary consequence. The ordeal of virtue is to resist all temptation to evil.

This natural inequality of the two powers of population and of production in the earth, and that great law of our nature which must constantly keep their effects equal, form the great difficulty that to me appears insurmountable in the way to the perfectibility of society. All other arguments are of slight and subordinate consideration in comparison of this. I see no way by which man can escape from the weight of this law which pervades all animated nature. No fancied equality, no agrarian regulations in their utmost extent, could remove the pressure of it even for a single century. And it appears, therefore, to be decisive against the possible existence of a society, all the members of which should live in ease, happiness, and comparative leisure; and feel no anxiety about providing the means of subsistence for themselves and families.

Consequently, if the premises are just, the argument is conclusive against the perfectibility of the mass of mankind.

I have thus sketched the general outline of the argument, but I will examine it more particularly, and I think it will be found that experience, the true source and foundation of all knowledge, invariably confirms its truth.

Chapter 2

The different ratio in which population and food increase—The necessary effects of these different ratios of increase—Oscillation produced by them in the condition of the lower classes of society—Reasons why this oscillation has not been so much observed as might be expected—Three propositions on which the general argument of the Essay depends—The different states in which mankind have been known to exist proposed to be examined with reference to these three propositions.

I said that population, when unchecked, increased in a geometrical ratio, and subsistence for man in an arithmetical ratio. Let us examine whether this position be just. I think it will be allowed, that no state has hitherto existed (at least that we have any account of) where the manners were so pure and simple, and the means of subsistence so abundant, that no check whatever has existed to early marriages, among the lower classes, from a fear of not providing well for their families, or among the higher classes, from a fear of lowering their condition in life. Consequently in no state that we have yet known has the power of population been left to exert itself with perfect freedom.

Whether the law of marriage be instituted or not, the dictate of nature and virtue seems to be an early attachment to one woman. Supposing a liberty of changing in the case of an unfortunate choice, this liberty would not affect population till it arose to a height greatly vicious; and we are now supposing the existence of a society where vice is scarcely known.

In a state therefore of great equality and virtue, where pure and simple manners prevailed, and where the means of subsistence were so abundant that no part of the society could have any fears about providing amply for a family, the power of population being left to exert itself unchecked, the increase of the human species would evidently be much greater than any increase that has been hitherto known.

In the United States of America, where the means of subsistence have been more ample, the manners of the people more pure, and consequently the checks to early marriages fewer, than in any of the modern states of Europe, the population has been found to double itself in twenty-five years.

This ratio of increase, though short of the utmost power of population, yet as the result of actual experience, we will take as our rule, and say, that population, when unchecked, goes on doubling itself every twenty-five years or increases in a geometrical ratio.

Let us now take any spot of earth, this Island for instance, and see in what ratio the subsistence it affords can be supposed to increase. We will begin with it under its present state of cultivation.

If I allow that by the best possible policy, by breaking up more land and by great encouragements to agriculture, the produce of this Island may be doubled in the first twenty-five years, I think it will be allowing as much as any person can well demand.

In the next twenty-five years, it is impossible to suppose that the produce could be quadrupled. It would be contrary to all our knowledge of the qualities of land. The very utmost that we can conceive, is, that the increase in the second twenty-five years might equal the present produce. Let us then take this for our rule, though certainly far beyond the truth, and allow that, by great exertion, the whole produce of the Island might be increased every twenty-five years, by a quantity of subsistence equal to what it at present produces. The most enthusiastic speculator cannot suppose a greater increase than this. In a few centuries it would make every acre of land in the Island like a garden. Yet this ratio of increase is evidently arithmetical.

It may be fairly said, therefore, that the means of subsistence increase in an arithmetical ratio. Let us now bring the effects of these two ratios together.

The population of the Island is computed to be about seven millions, and we will suppose the present produce equal to the support of such a number. In the first twenty-five years the population would be fourteen millions, and the food being also doubled, the means of subsistence would be equal to this increase. In the next twenty-five years the population would be twenty-eight millions, and the means of subsistence only equal to the support of twenty-one millions. In the next period, the population would be fifty-six millions, and the means of subsistence just sufficient for half that number. And at the conclusion of the first century the population would be one hundred and twelve millions and the means of subsistence only equal to the support of thirty-five millions, which would leave a population of seventy-seven millions totally unprovided for.

A great emigration necessarily implies unhappiness of some kind or other in the country that is deserted. For few persons will leave their families, connections, friends, and native land, to seek a settlement in untried foreign climes, without some strong subsisting causes of uneasiness where they are, or the hope of some great advantages in the place to which they are going.

But to make the argument more general and less interrupted by the partial views of emigration, let us take the whole earth, instead of one spot, and suppose that the restraints to population were universally removed. If the subsistence for man that the earth affords was to be increased every twenty-five years by a quantity equal to what the whole world at present produces, this would allow the power of production in the earth to be absolutely unlimited, and its ratio of increase much greater than we can conceive that any possible exertions of mankind could make it.

Taking the population of the world at any number, a thousand millions, for instance, the human species would increase in the ratio of –1, 2, 4, 8, 16, 32, 64, 128, 256, 512, etc. and subsistence as –1, 2, 3, 4, 5, 6, 7, 8, 9, 10, etc. In two centuries and a quarter, the population would be to the means of subsistence as 512 to 10: in three centuries as 4096 to 13, and in two thousand years the difference would be almost incalculable, though the produce in that time would have increased to an immense extent.

No limits whatever are placed to the productions of the earth; they may increase for ever and be greater than any assignable quantity. Yet still the power of population being a power of a superior order, the increase of the human species can only be kept commensurate to the increase of the means of subsistence by the constant operation of the strong law of necessity acting as a check upon the greater power.

The effects of this check remain now to be considered.

Among plants and animals the view of the subject is simple. They are all impelled by a powerful instinct to the increase of their species, and this instinct is interrupted by no reasoning or doubts about providing for their offspring. Wherever therefore there is liberty, the power of increase is exerted, and the superabundant effects are repressed afterwards by want of room and nourishment, which is common to animals and plants, and among animals by becoming the prey of others.

The effects of this check on man are more complicated. Impelled to the increase of his species by an equally powerful instinct, reason interrupts his career and asks him whether he may not bring beings into the world for whom he cannot provide the means of subsistence. In a state of equality, this would be the simple question. In the present state of society, other considerations occur. Will he not lower his rank in life? Will he not subject himself to greater difficulties than he at present feels? Will he not be obliged to labour harder? And if he has a large family, will his utmost exertions enable him to support them? May he not see his offspring in rags and misery, and clamouring for bread that he cannot give them? And may he not be reduced to the grating necessity of forfeiting his independence, and of being obliged to the sparing hand of charity for support?

These considerations are calculated to prevent, and certainly do prevent, a very great number in all civilized nations from pursuing the dictate of nature in an early attachment to one woman. And this restraint almost necessarily, though not absolutely so, produces vice. Yet in all societies, even those that are most vicious, the tendency to a virtuous attachment is so strong that there is a constant effort towards an increase of population. This constant effort as constantly tends to subject the lower classes of the society to distress and to prevent any great permanent amelioration of their condition.

The way in which these effects are produced seems to be this. We will suppose the means of subsistence in any country just equal to the easy support of its inhabitants. The constant effort towards population, which is found to act even in the most vicious societies, increases the number of people before the means of subsistence are increased. The food therefore which before supported seven millions must now be divided among seven millions and a half or eight millions. The poor consequently must live much worse, and many of them be reduced to severe distress. The number of labourers also being above the proportion of the work in the market, the price of labour must tend toward a decrease, while the price of provisions would at the same time tend to rise. The labourer therefore must work harder to earn the same as he did before. During this season of distress, the discouragements to marriage, and the difficulty of rearing a family are so great that population is at a stand. In the mean time the cheapness of labour, the plenty of labourers, and the necessity of an increased industry amongst them, encourage cultivators to employ more labour upon their land, to turn up fresh soil, and to manure and improve more completely what is already in tillage, till ultimately the means of subsistence become in the same proportion to the population as at the period from which we set out. The situation of the labourer being then again tolerably comfortable, the restraints to population are in some degree loosened, and the same retrograde and progressive movements with respect to happiness are repeated.

This sort of oscillation will not be remarked by superficial observers, and it may be difficult even for the most penetrating mind to calculate its periods. Yet that in all old states some such vibration does exist, though from various transverse causes, in a much less marked, and in a much more irregular manner than I have described it, no reflecting man who considers the subject deeply can well doubt.

Many reasons occur why this oscillation has been less obvious, and less decidedly confirmed by experience, than might naturally be expected.

One principal reason is that the histories of mankind that we possess are histories only of the higher classes. We have but few accounts that can be depended upon of the manners and customs of that part of mankind where these retrograde and progressive movements chiefly take place. A satisfactory history of this kind, on one people, and of one period, would require the constant and minute attention of an observing mind during a long life. Some of the objects of inquiry would be, in what proportion to the number of adults was the number of marriages, to what extent vicious customs prevailed in consequence of the restraints upon matrimony, what was the comparative mortality among the children of the most distressed part of the community and those who lived rather more at their ease, what were the variations in the real price of labour, and what were the observable differences in the state of the lower classes of society with respect to ease and happiness, at different times during a certain period.

Such a history would tend greatly to elucidate the manner in which the constant check upon population acts and would probably prove the existence of the retrograde and progressive movements that have been mentioned, though the times of their vibrations must necessarily be rendered irregular from the operation of many interrupting causes, such as the introduction or failure of certain manufactures, a greater or less prevalent spirit of agricultural enterprise, years of plenty, or years of scarcity, wars and pestilence, poor laws, the invention of processes for shortening labour without the proportional extension of the market for the commodity, and, particularly, the difference between the nominal and real price of labour, a circumstance which has perhaps more than any other contributed to conceal this oscillation from common view.

It very rarely happens that the nominal price of labour universally falls, but we well know that it frequently remains the same, while the nominal price of provisions has been gradually increasing. This is, in effect, a real fall in the price of labour, and during this period the condition of the lower orders of the community must gradually grow worse and worse. But the farmers and capitalists are growing rich from the real cheapness of labour. Their increased capitals enable them to employ a greater number of men. Work therefore may be plentiful, and the price of labour would consequently rise. But the want of freedom in the market of labour, which occurs more or less in all communities, either from parish laws, or the more general cause of the facility of combination among the rich, and its difficulty among the poor, operates to prevent the price of labour from rising at the natural period, and keeps it down some time longer; perhaps till a year of scarcity, when the clamour is too loud and the necessity too apparent to be resisted.

The true cause of the advance in the price of labour is thus concealed, and the rich affect to grant it as an act of compassion and favour to the poor, in consideration of a year of scarcity, and, when plenty returns, indulge themselves in the most unreasonable of all complaints, that the price does not again fall, when a little rejection would shew them that it must have risen long before but from an unjust conspiracy of their own.

But though the rich by unfair combinations contribute frequently to prolong a season of distress among the poor, yet no possible form of society could prevent the almost constant action of misery upon a great part of mankind, if in a state of inequality, and upon all, if all were equal.

The theory on which the truth of this position depends appears to me so extremely clear that I feel at a loss to conjecture what part of it can be denied.

That population cannot increase without the means of subsistence is a proposition so evident that it needs no illustration.

That population does invariably increase where there are the means of subsistence, the history of every people that have ever existed will abundantly prove.

And that the superior power of population cannot be checked without producing misery or vice, the ample portion of these too bitter ingredients in the cup of human life and the continuance of the physical causes that seem to have produced them bear too convincing a testimony.

But, in order more fully to ascertain the validity of these three propositions, let us examine the different states in which mankind have been known to exist. Even a cursory review will, I think, be sufficient to convince us that these propositions are incontrovertible truths.

Commentary

Thomas R. Malthus, *An Essay on the Principle of Population* (1798)

BJÖRN-OLA LINNÉR

The icon of population pessimism, Thomas Robert Malthus (1766–1834), presented his famous population theory in *An Essay on the Principle of Population* (1798). It postulates that population will inevitably grow faster than the supply of nutrition: "the power of population is indefinitely greater than the power in the earth to produce subsistence for man."

Malthus described this relationship as a natural law. He noted that the population grows exponentially, so that human population, if not checked, would double every twenty-five years. Food production, in contrast, increased only linearly. Crop yields could not keep pace with the constant doubling of unrestrained population growth, and humankind could not put its hope in unlimited agricultural advances. More recently, foreboding titles such as *Limits to Growth* (1972) and "planetary boundaries" (pp. 491–501) have recalled Malthus's observation that "a careful distinction should be made between an unlimited progress and a progress where the limit is merely undefined" (Malthus 1798, 52). History had proved to Malthus that future improvements in the human condition could merely be marginal. Even in more prosperous regions, population would expand until these regions' abundant resources were also exhausted.

Two types of constraints could restrain population growth, according to Malthus. First, unplanned checks such as war, epidemics, and famines maintained a balance between populations and their means of subsistence:

> Famine seems to be the last, the most dreadful resource of nature. The power of population is so superior to the power in the earth to produce subsistence for man, that premature death must in some shape or other visit the human race. The vices of mankind are active and able ministers of depopulation. . . . Should success be still incomplete, gigantic inevitable famine stalks in the rear, and with one mighty blow levels the population with the food of the world. (Malthus 1798, 51–52)

Humanity was not confined to relying on unplanned population controls. Societies could also choose a second set of controls, in which people deliberately limited the number of children born. Morally chosen continence by means of late marriage and restrained sexual activity was, for Malthus, the way Christian morals could avert catastrophe.

Malthus doubted whether poor people had the capacity to restrain their reproduction by themselves. Improved living conditions would instead inevitably lead to even bigger families when constraints such as starvation or ill health were eliminated. If preventive measures were not taken, these aforementioned unplanned checks would do the job, bringing even more severe miseries on a greater number of people. Poverty was part of the law of nature. Malthus was accordingly skeptical of relief measures for the poor. He argued vehemently for gradually abolishing the English poor laws. Handouts would only result in more children, more mouths to feed.

In later editions, Malthus provided his essay on population with a new subtitle: A *view of its past and present effects on human happiness: with an inquiry into our prospects respecting the future removal or mitigation of the evils which it occasions*. He published six editions of his treatise between 1798 and 1826; each edition added new data, addressed his critics, and slightly changed his view of population checks. Most notorious was his addition of the parable of the "feast" in the second edition.

> A man who is born into a world already possessed, if he cannot get subsistence from his parents on whom he has a just demand, and if the society do not want his labour, has no claim of *right* to the smallest portion of food, and, in fact, has no business to be where he is. At *Nature's mighty feast* there is no vacant cover for him. *She* tells him to be gone, and will quickly execute her own orders, if he does not work upon the compassion of some of her guests. (Malthus 1798, 531)

It would lessen the pleasure of those already at the feast if more were constantly let into the feast out of compassion. Faced with heavy criticism for lack of compassion, Malthus excluded the passage from later editions.

Malthus's theory has been widely debated ever since. Both nineteenth-century liberals and socialists criticized him for not recognizing the human capacity to increase the food supply. He was foremost denounced for using putatively scientific laws to promote an unjust social order. Friedrich Engels called it "the crudest, most barbarous theory that ever existed." Influential thinkers, such as John Stuart Mill and Charles Darwin, drew inspiration from his population theory, leaving his social conclusions aside.

Late nineteenth-century Malthusianism developed into a social and political movement that emphasized the relationship between baby booms and poverty. It encompassed a wide range of political views, appealing to conservatives, socialists in favor of family planning, and women's rights advocates. Most notable of the neo-Malthusians was the Malthusian League, founded in 1877. Overpopulation was seen as a principal source of poverty, which should be countered by promoting contraceptives and family planning.

With the population surge after World War II, Malthus's theory attracted renewed interest. Two prominent streams characterize postwar neo-Malthusianism. One is socially focused, emphasizing family planning, contraception, and women's rights in developing countries with high fertility rates. The other is ecologically motivated, focusing on the impact of high population densities on scarce resources.

Postwar neo-Malthusianism held that the exponential growth in the number of people consuming and polluting already scarce resources would inevitably result in famine spreading around the world. Another mechanism was added to the unplanned population checks predicted by Malthus: environmental degradation caused by industrialization would also severely limit population growth. The earth simply could not support additional billions of people living at Western material standards.

Despite the basic similarities, there is a clear difference between Malthus's socially and politically conservative conclusions and mainstream environmentalist neo-Malthusianism, which calls for nutritional redistribution and a fair world order. However, a few environmentalists have drawn another conclusion in line with Malthus's "feast" argument: if it is already too late to save all humankind, it would be better to save a select few than let all perish. Accordingly, "lifeboat ethics" emerged as a triage concept in the late 1960s.

The postwar era saw an outpouring of treatments of Malthusian thought, including Paul Ehrlich's bestseller *The Population Bomb* and the Club of Rome's 1972 report *Limits to Growth*. The constant reappearance of the Malthusian population principle was described as follows in 1977 by economist Herman E. Daly: "Malthus has been buried many times, and Malthusian scarcity with him. But as Garrett Hardin remarked, anyone who has been reburied so often cannot be entirely dead" (Daly 1991, 43).

Today, Malthusian theory reverberates in international organizations such as the United Nations Food and Agriculture Organization and in the food industry. Echoing Malthus, the multinational plant-breeding corporation Monsanto claims: "A fast-growing world population . . . will inevitably outstrip our capacity to produce enough food to meet our needs." In the climate change debate, the "people pollute" theme of environmental neo-Malthusianism is once again being invoked. Malthusian population theory is indeed not only barely buried, but is alive—and breeding.

Further Reading

Connelly, M. 2008. *Fatal misconception: The struggle to control world population.* Cambridge, MA: Belknap Press.

Daly, Herman E. 1991. Steady state economics. Washington, DC: Island Press.

James, P. 2006. *Population Malthus: His life and times.* London: Routledge.

Linnér, B.-O. 2003. *The return of Malthus: Environmentalism and post-war population-resource crises.* Isle of Harris: White Horse Press.

Monsanto. 2001. *Conservation tillage: How new farming practices promist a better world.* St. Louis, MO: Monsanto.

Petersen, W. 1999. *Malthus*, 2nd ed. London: Heinemann.

Winch, Donald (ed.). 1992. *Malthus: An essay on population.* Cambridge: Cambridge University Press.

The Shadow of the World's Future

GEORGE KNIBBS

Chapter XI: Conclusions as to Population Increase

We have already, in Chapter VI, given some slight indication of the significance of the population-question. Owing to the imperfections and inadequacies
of existing statistics, we cannot fix the population-limits with any precision, and
we have shown that it is dependent largely upon factors at our disposal, viz., our
economic and ethical advance, and the standard-of-living which we are prepared
to accept. What has appeared in regard to the significance of rates has shown us
that, even if the "unspecified" area of the world's surface should turn out to be
"productive," the issues are not materially altered. The shadow is not lifted. We
may now revert to what has been established in the preceding pages, and ask,
"What are the conclusions to be drawn in regard to the problems of the world's
future?" "Is there really a population menace, constituting the Shadow of the
World's future?"

The rate, at which Man has increased for more than a century, informs us
that we have unquestionably entered upon a new era. That rate will probably
not diminish except through the arrival of unforeseen troublous times. Of itself,
the rate will create enormous difficulties, for mankind has not yet become an
economic unity, nor has it yet learnt to regard issues from the standpoint of the
good of the whole. The time available for all necessary adjustments is so short
that Man's immediate task is indeed a very heavy one, and it is inescapable. History reveals, however, that the building up of the character of a people is a slow
process and one which involves centuries of experience and effort.

Existing conditions are such that, if they continue, mankind could perhaps
attain to 3800 millions, double its present numbers. This would involve no more
variation of its organisation than would seem to be easily possible with any sincere

George Knibbs. 1928. "Conclusions as to population increase." Chapter 11 in *The shadow
of the world's future, or, the Earth's population possibilities and the consequences of the
present rate of increase of the Earth's inhabitants.* London: Ernest Benn.

and well-directed effort. But to reach even this population without world-wide calamities supervening, quite special efforts will be essential, as anyone will readily perceive who has taken account of the movements in the East, in Africa, and in America. Man is face to face with issues which demand attention, and which call for an incisive inquiry into the position of the inferior and the so-called coloured races. A new liberalism, and a less egoistic regard for the well-being of all races, is being called into existence.

For the world to attain to thrice its existing numbers, that is to 5850 millions, fundamental changes in the existing characters of human civilisations will not necessarily be involved; but it will involve great improvements in respect of international economics, and in respect of the moral aspects of national and international life. It will involve also many further advances in science and technology, advances greatly surpassing those of the past and present century. Doubtless, too, it will involve the cultivation of areas now neglected.

For the earth to quadruple its numbers, that is to attain to 7800 millions, is a huger task, involving not only a much more efficient use of its surface, but also a deeper study of the climatological factors which can aid in the enormous improvement of its food-supplies that will be required. But that is not all. The chemico-physical factors are relatively simple as compared with what is also essential, viz., the virtual elimination of all forms of unscrupulous egoism in the life of nations and in the relations of races. This means that thorough and sympathetic studies of those things in international life which reveal that Man is subject to moral law will have to be undertaken. (Is there any real expression of such law in the inter-relations of mankind?) Moreover, the financial and economic systems, and the different productivities of the various peoples of the earth will have to be co-ordinated with the greatest possible equity and goodwill.

This limit of 7800 millions will be passed only with the greatest difficulty and probably very slowly. It is, however, quite possible that still further increase can take place, to the order of say five times the present population, viz., to 9750 millions, and ultimately it might reach even six times, say 11,700 millions. It seems certain that, under any conditions whatsoever, the numbers of the human race can never surpass this. Even to attain to 9750 millions, the perfection of all human organisation would have to be so high on the moral as well as on the physical plane, that it is very difficult even to imagine how this can transpire in the limits of time which are probably available. The history of the human race appears to indicate that only very slow changes, if any at all, in the fundamental elements of Man's character are possible. Unless the changes arrive through intelligent reproductive controls, taking every advantage of appropriate methods of reproduction, it would seem unlikely that the 7800 limit will ever be passed.

Although the history of Japan has been a revelation of how rapidly a people, with devoted and mentally capable leaders, may develop in a particular direction, viz., in that which has characterised Western civilisation, and although the rapid rise of various other peoples has been almost equally surprising, there is no doubt that, to attain to a high population-density, the prevailing aims of human lives will have to be less concerned with complications in the mere standard-of-living. A more humble life, physically, with a deeper regard for the higher issues, is a *sine qua non*, if it be really desired to see the earth covered with contented peoples, whose well-being is assured and whose living is a disclosure of generous attitude and noble purpose. When such propositions are really examined, it is at once apparent, that for mankind to multiply greatly is not merely a physical difficulty: it is one involving his higher powers, and that view lies behind much of what Sir Rabindranath Tagore has had to say in regard to the limitations of nationalism. We do not accept his view as an unassailable verdict as to the essence of the whole position, but it is one which it is desirable to analyse carefully. And, reverting again to Japan, the Bushido ideals of that nation, exemplified in the readiness of the former Daimios to forgo their privileges for the nation's well-being, shows that ethical elements can play a very real part in the numerical and dynamic development of a people.

Finally, one may say that although the dream of a densely peopled earth, living in relative contentment, is not an impossible one, it is, of course, a dream of the impossible, *as things are*. The earth numbers to-day only 1950 millions, after its long life-history; an amazing fact even after only its 10,000 years of *recent* development. This has been because of its great intellectual, and also its great moral, limitations. The cynic may well say "these dreams of the world's possible future are but idle phantasies," and the sneer would be well-founded. Nevertheless, it is suggested that the world's future *can* be vastly better than its present, and the future is worthy of sympathetic consideration, as to ways and means of advance, by the finest minds and the noblest characters.

At the present time the mere increase of population, coupled with the fact that Man's moral development has not kept pace with scientific knowledge, is threatening trouble. With the collisions of interest that are now in existence, the future looks not merely threatening but very ominous indeed. If that future is to be better than appears, it will depend largely upon the attitude of its inhabitants to the era that is dawning. The matter even of its growth in numbers is truly momentous, and, with its assertive and unscrupulous greeds, is no less alarming to any one who has any vision, and who realises to what past history is pointing. The frightful indifference to ghastly miseries and unspeakable sufferings which made the last war possible, reveal the spirit which is governing so large a part of mankind even now. *That spirit is a limiting factor* to the growth of the human race and to material and spiritual advances in its future. Virtually we are told it will never change; if that be true, then the Shadow of the Future will be very dark.

The World's Future is, then, *the problem* of problems. That we should at once face it, is revealed by the fact that the rapidity of the increases in population-numbers is already threatening us with apparently almost insoluble difficulties: we are rapidly approaching numbers that make the problem a stupendous, aye, even an appalling, one. At the present time one country, at least, must make provision for the emigration of some of its inhabitants. We may elect to ignore these matters, but if we do we only accentuate our future difficulties. It is here that we see that the way of humility is needed, for the ablest are intellectually incompetent, and the noblest fall short of the splendour of purpose, demanded for its solution.

Anyone who has read Dean Inge's *England* attentively will realise something of the magnitude, not merely of England's problems, but those of the world. His epilogue sums up the situation. The issues for all great nations do not differ materially. What Dean Inge has to say, in his most able review of the problems pressing upon the British nation for solution, reveals, either directly or incidentally, what the world situation is. It discloses also how enormous the work to be done is if, with an ignorant and selfish humanity, ill-consequences are to be *minimised*. Humanity has both to be instructed and governed. Naturally enough the situation appears to be well-nigh hopeless, not because it is essentially insoluble, not because it is beyond the reach of the intelligent, but because humanity is mentally and morally what it is at the present time. Nevertheless, the World's Future calls for consideration by all who are not wholly wrapped in the garment of utter indifference, for the auspices are not favourable, and the population-pressures, so rapidly developing, are inescapable.

When one thinks of the periods which have been necessary for the development of all the highly civilised peoples among mankind, and of the crude stages only now reached by the backward peoples, it would seem that no possible effort during the remaining three-fourths of the present century can materially alter the conditions existing, at any rate for the greater portion of the human race. The complexity of modern life with the more advanced nations, the range and excellence of their comforts, the elaboration of their methods, of their customs and their enjoyments, the luxury and ostentation of their appointments, the enormous expenditures of money, or its equivalent in labour, of those who control the social and political world, all imply an accentuation of the elements of human nature which constitute the main promptings of modern Man. It is these things that make the future difficult.

When one knows something of the world's surface and of its peoples, and finds it possible for a country like Switzerland to carry a population of 247 to the square mile, while a country like the United States of America is carrying only 39, being told also by certain special and able students that it can never carry more than 66 to the square mile, one realises how superficial are some of the studies of the world's possibilities.

The data do not yet exist by means of which a really exhaustive estimate can be made of the world's population-limits, as things are at present, nor as they are likely to be. But we do know enough to affirm with confidence, that the fear that a country with immense resources can carry only 66 to the square mile is created by too narrow a view of the problem in hand. No sufficient account has been taken of the standard-of-living assumed to be essential, nor of the fact that the theory leading to this estimate is based upon merely temporary, undeveloped and unessential conditions. It may of course be true that the easy state of things in any new country must pass as the world's peoples multiply, and that the standards existing must perforce change. If they do change in the direction of less luxury, then the estimate of 66 people to the square mile goes by the board.

Even should our estimates of the limits of population be too modest, it still remains true that mankind is profligate in the use of such of Nature's materials as are immediately at his disposal, and he is applying them, and the food-stuffs likely to be available, recklessly. For this reason Man will certainly be pulled up in the near future, and the Shadow of his future remains in being. What we said in our report on the Australian Census of 1911 remains true. Our words were:—

"The limits of human expansion are much nearer than popular opinion imagines; the difficulty of future food supplies will soon be of the gravest character; the exhaustion of sources of energy necessary for any notable increase of population or advance in the standards of living, or both combined, is perilously near. Within periods of time, insignificant compared with geologic ages, the multiplying force of living things, man included, must receive a tremendous check."

And we went on to add the following:—

"The present rate of increase in the world's population cannot continue. . . . The extraordinary increase in the standard of living, which has characterised the last few decades, must quickly be brought to a standstill, or be determined by the destructive forces of human extravagance. Very soon the world-politic will have to face the question, whether it is better that there should be larger numbers and more modest living, or fewer numbers and lavish living; whether world-morality should aim at the enjoyment of life by a great multitude, or aim at the restriction of life-experience to a few, that they may live in relative opulence."[1]

We pointed out that the student of the future would "utilise all discovery of the mysterious play, and no less cryptic limitation, of life-force to make prediction sure." And further that with "co-ordinated international effort, there would be no difficulty in so directing future statistical technique" that a more perfect study could be made of the drift of mankind "in the more important relations of civic, national, and international life" (p. 454).

Certainly, in so far as Man is ignorant he is both the puppet of fortune and the victim of desire. He knows but little of the driving forces in the world of life. He sees but the surface of things and his science is far from being a perfectly co-ordinated system of concepts, representing the world as he beholds it. It may be that his future was written upon the Tablets of Destiny, æons before his world came into being. And thus one cannot learn whether the present tendencies of Man's life on earth are but a hint of the passing of this civilisation, or are remediable through the labours of the leaders of humanity. Thus attention to the menaces of the future may presage the lifting of the World's Shadow, or may be merely the informing of human intellect of the dark promise of the immediate Future.

Note

1. *Report of Census*, 1911, Appendix A, Vol. I, p. 453.

Commentary

George Knibbs, "Conclusions as to Population Increase" (1928)

ALISON BASHFORD

George Knibbs (1858–1929) wrote his short book *The Shadow of the World's Future* in the light of the first World Population Conference, held in Geneva in 1927. As a distinguished statistician, he had the mathematical capacity to assess carefully trends in growth rates, and the status to have his projections and concerns about the future read seriously by auspicious commentators. As an internationalist, Knibbs considered population far from a matter of mere numbers, however. His was a generation for whom population growth was automatically a question of food security: it therefore raised directly the politics and economics of land, war, and peace.

The Shadow of the World's Future was Knibbs's attempt to apply Malthus's ideas not just to a new century, but to a new epoch. He considered his own an entirely new age, because of unprecedented population growth over the nineteenth century, mainly European and American, because of a world war, and because of a geopolitically closed world. As he and others would continually state, there were no new hemispheres or continents to colonize and inhabit. One response was to shore up national lands for nationalist interests, and many commentators on population growth pursued this agenda. Another response—George Knibbs's—was deeply anti-nationalist. He argued an endangered common future for "the whole of the human race."

By 1928, Knibbs had been troubled for many years by acceleration of population growth. The book was an attempt to popularize a warning that he had first developed in an unlikely text, the appendix to the 1911 Australian census report. In it, he strayed well beyond the genre of a census report, leaving the national(ist) project behind in the wake of a catastrophic world warning: "only 450 years to exhaust the food required." For Knibbs, the future of humans on earth had come down to the impact of population growth "of the last ten decades" and the effect of this "within the next ten" (Knibbs 1917). The fallout from this world problem was comprehended as urgently imminent for that generation, much as climate change is now.

This appendix was itself abstracted and republished many times as an independent document, sought by European and U.S. government statisticians and extensively quoted in South African census reports and by Indian economists. Czech mathematician Emanuel Czuber translated and repackaged Knibbs's work into a widely cited German edition. In part, Knibbs's work became well known internationally because of his innovations in statistical method, but it was his concluding set of remarks about the imperiled future of humanity that resonated strongly in an international public sphere already primed to consider global-level issues of land, food security, and war.

It was common for that generation to imagine human futures in planetary terms: neither Sputnik nor earthrise photographs were necessary for this to be so. In part, this planetary vision derived from Malthus's *Essay* itself, and in part from late Victorian Malthusians. At issue here were not only nations and empires in war and peace, or even a supranational comprehension of social or geological space: it was about the bounded space of a singular earth, as one planet among many. It is not incidental that Knibbs lectured in astronomy, or that his initial education was in geodesy, a discipline that required both technical knowledge of, and capacity to imagine, the shape and positioning of the globe in space. Trumping even Malthus's reputation for gloom, he wrote: "From the world point of view . . . human beings are relatively but the merest specks on the earth's surface." And even worse: "Doubtless, from the widest point of view, the obliteration of the whole solar system is insignificant" (19–20). The problem was occasionally for him not even about humans: "Man's view of the world is frankly anthropocentric" (124).

There was, though, a window of opportunity. The "New Malthusianism," he called it. Humans did have the capacity to manage fertility rates as a population, and reproductive conduct as individuals and families. New technologies of contraception helped, and responsible states, he argued, should remove barriers to the dissemination of birth control information or technologies. The late nineteenth and early twentieth centuries were a period when sudden local declines in fertility growth were clearly evident and much remarked upon. Historians have focused strongly on this so-called race suicide, in a way that has problematically overshadowed the concerns of the likes of Knibbs about overall (that is, world) population growth. Knibbs, not least as government statistician, was more expert than most on the facts of fertility decline, and he was aware that this posed national conundrums. But in the end he considered this a hopeful sign, not a problem, a trend that should become worldwide.

Fertility control was not the only solution, however. For Knibbs and others in this period, "population control" was less the reproductive or health issue it would become later in the twentieth century, and more a political economy question received, as it were, from Malthus: land use and improvement, food, and the value of labor were the issues at hand. Though they were often characterized as crude "population controllers," it was in fact standard for the likes of Knibbs always to address (at least) two sides of the population/resource coin. In short, humans were both too many and too greedy. Internationally cooperative economic policy, equity in standards of living, managed resource use, especially soil conservation, and management of reproduction were jointly the way forward. That population was an energy problematic did not in the least escape him, and he saw a fuel (petroleum) crisis looming, especially, he said, if mechanical power came to replace organic power in non-Western agricultural economies.

This led Knibbs to undertake global audits of world lands, soils, and agricultural practices, and he determined that the limit of human numbers was 7800 million. But this depended significantly on the nature of social and economic organization, and on expected and accepted standards of living. It was not just a political but an ethical problem, and a world able to sustain 7800 million people required general and cross-civilizational (as he would put it) acceptance of a "more humble life." This would be possible, he thought, only if national antagonisms and competitions were replaced by international agendas, and even by systems of world governance.

This was a characteristically 1920s agenda that became less viable over the 1930s as the world's experiment in internationalism—the League of Nations— started to come undone. Yet this strand of analysis about planetary carrying capacity, already long established, reemerged after World War II, in equal measure as the implementation of demographic transition theories, as feminist argument for birth control, as a politics of soil conservation, and as a freshly urgent global political ecology.

Further Reading

Bashford, A. 2013. *Global population: History, geopolitics and life on Earth.* New York: Columbia University Press.

Connelly, M. 2008. *Fatal misconception: The struggle to control world population.* Cambridge, MA: Belknap Press.

Mitman, G. 1992. *The state of nature: Ecology, community, and American social thought, 1900–1950.* Chicago: University of Chicago Press.

Knibbs, G. H. 1917. "Mathematical theory of population, of its character and fluctuations, and of the factors which influence them," Appendix A, *Census of the Commonwealth of Australia, 1911.* Vol. 1: *Statistician's Report* (Melbourne: Government Printer, 1917), 455–456.

Perkins, J. H. 1997. *Geopolitics and the green revolution: Wheat, genes, and the cold war.* New York: Oxford University Press.

Rothschild, E. 1994. "Population and common security." In Laurie Ann Mazur (ed.), *Beyond the numbers: A reader on population, consumption, and the environment.* Washington, DC: Island Press.

Robertson, T. 2012. *The Malthusian moment: Global population growth and the birth of American environmentalism.* New Brunswick, NJ: Rutgers University Press.

Ghost Acreage

GEORG BORGSTRÖM

It is a well-known fact that most countries do not subsist merely on agriculture but depend upon the importation of food and feed and in addition obtain essential protein from oceans, rivers, and lakes. This should be self-evident, but private and national economies with their vested interests enter as disrupting or blurring factors into these simple relationships. Nevertheless, economic and nutritional measures are commonly discussed and analyzed as though they concerned agriculture alone. There are many factors which might explain this state of affairs. One important reason for the lack of understanding of these fundamental forces is the fact that there have so far been no data, readily understandable, whereby one could on a commensurate basis compare trade, fisheries, and agriculture. In short, there has been no common denominator, aside from the misleading standard of the dollar, pound, franc, ruble, etc.

A method was presented in chapter 2 whereby it became possible to evaluate the true role of fisheries in individual countries, based on the prime function of fish as supplying animal protein. By calculating the acreage necessary to produce in the most acreage-saving way for each country an amount of animal protein equivalent to what presently is obtained through fisheries and with present techniques in the agricultural production of this very same country, we arrived at what I have termed the *fish acreage*. This constitutes one basic element in what this author has chosen to call the *ghost acreage* of a country. This is the computed, non-visible acreage which a country would require as a supplement to its present visible agricultural acreage in the form of tilled land in order to be able to feed itself. If part of this acreage is taken up by grazing lands or pastures, the calculated ghost acreage would be correspondingly larger.

Georg Borgström. (1962) 1965. "Ghost acreage." Chapter 5 in *The hungry planet: The modern world at the edge of famine*. Originally published in Swedish in 1953. English translation, New York: Macmillan.

The second element of the ghost acreage is represented by *trade* acreage—calculated as the acreage, in terms of tilled land, required to produce, also with present techniques, the agricultural products constituting the *net importation*. The traditional trade balance sheets are based upon either metric ton figures or monetary values. Both concepts are, however, less suited for a realistic appraisal of the feeding capacity and overall food balance of a country. As is well known, weight varies considerably with the water content, which renders the traditional adding-up of various kinds of food on a weight basis as nutritionally absurd. Unfortunately this is still frequently done with industrially manufactured foods of various kinds. Such data may be of interest in relation to transportation costs and when it comes to procuring tonnage for shipment, but such total sums poorly reflect the nutritional significance of shipment loads.

For more than thirty years the food prices both on the world market and in individual countries have rarely reflected the true production costs. Production regulations, subsidies, tariffs, taxes, and subvention purchases have long ago replaced free competition in this area—fortunately, it may be fair to add. At the same time, however, monetary appraisals of food and feed have become far removed from what would be acceptable from a nutritional point of view. As a

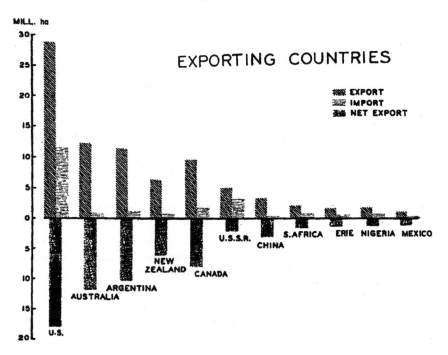

Figure 19. Trade acreages based on 1958–59 figures.

contribution to a discussion along new and more meaningful guidelines, I have therefore introduced this acreage concept. In the first place, this was done to place protein in its key role in human feeding. Protein raised through soils is in general the most acreage-demanding constituent. Besides, protein content and value have hardly ever been the prime yardstick in determining prices, although in so many cases it holds the first line in determining the nutritional value to man and to livestock.

In order to make these calculations more reliable, and with the aim of offering real guidance, trade acreages are computed separately for each individual country. Even in this connection it is true that when it comes to huge countries such as the United States and the U.S.S.R. these figures become less relevant and need to be supplemented with corresponding regional analyses, such as for example the Soviet Far East, European U.S.S.R., the Rocky Mountain States, the New England States, or the Pacific Region of the United States. Only when referring to a region reasonably uniform in topographic and climatic respects do such analyses take on full meaning.

Another complication arises from agricultural products which cannot be produced in a given country (such as coffee and bananas in the United States) and are imported. In trade statistics food importation is sometimes split up between supplementing and substituting products, often given the designation supplementary and complementary—the latter not normally grown in the country. This distinction is important.

It should be underlined that the trade acreages comprise all categories of agricultural products (the nutritive values as well as their demand on acreage), also non-food items such as fibers and tobacco, as they all dispose tilled land and pastures. In other words, they affect the acreage available for the raising of food in the exporting and importing countries. On the whole, however, acreages for nonfood items are minor compared to those used for food and feed. Another complication refers to commodities that are complementary and not supplementary. Also in this case the total acreage is minor, but no reasonable yield figures would be valid for the importing country. In this particular detail the world averages were employed in my computations.

From the above it is evident that one acre in Scandinavia does not mean the same as one acre in the United States. Acre figures are directly comparable in only a very few countries. But this objection does in effect apply to most acreage data which have so far been used in economic geography and agriculture. Even in two neighboring countries, an acre of tilled land may not, in terms of production, mean the same thing. Such discrepancies do not imply, however, that in regions with a fairly uniform topography and comparable soil structure, acre figures constitute a reasonably satisfactory basis for comparisons and can

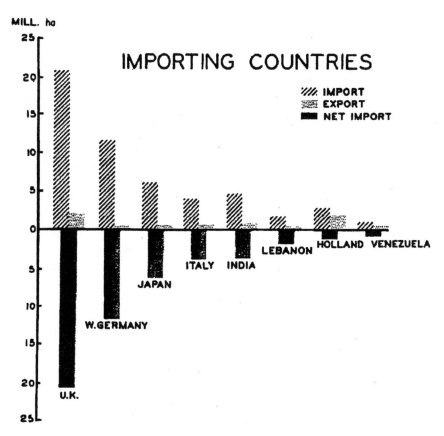

Figure 20. Trade acreages based on 1958–59 figures.

be used as a kind of commensurate gauge. At any rate, the computed data offer clear indications of fundamental differences. Further detailed research will be needed, however, in order to evaluate fully the true significance of these data on ghost acreages.

Let us take a closer look at a few illustrative results from an analysis of the kind outlined above. These studies show that Scandinavia has a ghost acreage of almost 6.7 million acres—almost an entire additional agricultural Sweden in terms of tilled land—this is the case in spite of the fact that Denmark through its export deliveries, primarily to Western Europe and the United States, places at the disposal of the world household a net acreage of 1.1 million acres. This means, among other things, that the total food export of Denmark would not suffice to cover more than one-third of these net import needs of Scandinavia as measured in acreages. It seems appropriate again to stress the fact that such acreage figures far better measure the real significance of this trade in food and feed, as these calculations pay attention to and directly reflect the nutritive values as well as the demand on acreage for their production.

The fish acreage of Scandinavia, for example, amounts to almost three million acres, or half the tilled land of Sweden. The total ghost acreage of Sweden exceeds 36 per cent of the tilled land of the country, an almost identical land area. Importation accounts for 24 per cent. It is worth noting, that in this case these figures come close to the estimates made in the conventional way on the money value of the trade. This might, therefore, even provide argument for questioning the need of approaching this field from the nutritional angle along the principles advocated here. For a number of countries, however, the discrepancies between the two approaches becomes considerable.

Presently this author is devising methods whereby the use of commercial fertilizers and the energy inputs are computed in corresponding terms and added to the ghost acreages. Furthermore, it has proved most useful to present the unfilled nutritional needs of man in similar, unequivocal terms. The question is then posed: What acreages would in respective countries be required to fill the gap between minimal nutritional needs and the present conditions of undernutrition or malnutrition? In countries where comprehensive dietary surveys are lacking, one can by this method readily establish what acreage with the present state of the techniques would be needed to improve the nutritional standard to a defined level such as that of Japan, Brazil, or Italy.

Contrary to all analyses so far presented, the ghost acreage of Japan has increased continuously since 1952, primarily owing to the rapid increase in the importation of wheat, soybeans, and feeding-stuffs. In principle these imports constitute a large-scale influx of proteins. Japan's balance sheet at the turn of the latest decade looked this way; the fish acreage was equal to 1.54 of the tilled land of Japan; the net importation accounted for a *ghost acreage* equaling the entire tilled acreage of the country. Both these figures have grown since 1958–59. The *ghost acreage* of Japan consequently zooms to 2.55 Japan in terms of the present tilled land. In addition to the tilled acreage there are available for food production within the borders improved pastures, which constitute merely 0.27 of the tilled land. . . .

This means that Japan obtains less than one-third of its food from its own soils (28 per cent). Approximately the same quantity is taken from abroad and a little less than half (44 per cent) from the sea through fisheries. Anyone starting to add up calories in the conventional manner would soon find that agriculture is accounting for more than half of the calorie intake. If we consider protein and recognize the relatively large acreage required to produce this precious commodity, we obtain a strongly modified and more realistic picture of Japan: (1) the

Table 12

Ghost Acreage In Relation To Tilled Land

One (1) Is the relative figure given to tilled land in every country listed.

Country	Fisheries	Net Import	Ghost Acreage
Switzerland	0.33	3.22	3.55
England	0.38	2.90	3.28
Belgium	0.33	2.58	2.91
Israel	1.31	1.38	2.69
Japan	1.54	1.01	2.55
West Germany	0.28	1.33	1.61
Holland	0.47	0.99	1.46
Norway	0.66	0.64	1.30
Egypt	0.48	0.14	0.62
Ceylon	0.23	0.30	0.53
Portugal	0.29	0.08	0.37
China	0.23	0.03	0.26

enormous degree of dependence upon territory other than its own; (2) the key position of its fisheries; and finally, in relative terms, i.e., tilled land, (3) the modest role of its agriculture.

Holland is another illuminating example. It is looked upon as a food-exporting country, which it is. But this is often interpreted to mean the country has a food surplus from which it generously delivers to other less fortunate nations. The Dutch are pictured as magicians of the soils who feed more people per acre than any other nation. These notions are misconceptions.

Another illustrative example is Israel, which feeds only one-third of its population from its agricultural acreage despite all impressive accomplishments. The very narrow margin for new soils, and their critical dependence on water, set obvious limitations to the productive capacity of the country. Irrigation is, even when expanded and superefficient (depending on the rainfall and condensation in the watershed area), mainly the river Jordan systems. Fisheries contribute one-third of the feeding acreage and the net importation another third.

In these tables only a few examples have been chosen of particular interest from one viewpoint or another. But they all serve to demonstrate the role of the ghost acreages. They also bring out the hollowness of conventional self-sufficiency concepts. Figure 21 shows the degree to which a number of listed countries depend on their agriculture for feeding their peoples. Fishery and net

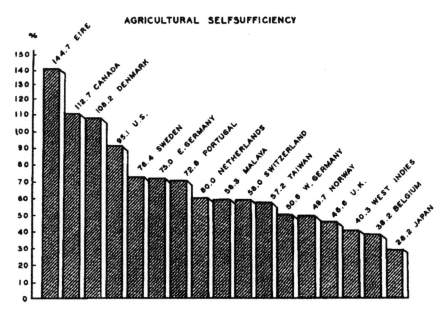

Figure. 21. The feeding of a country rests on a tripod—agriculture, fisheries, and trade. In this study all three have been computed in terms of acreages referring to each country respectively. Net exportation reduces the agricultural acreages. This chart shows in these terms to what degree agriculture provides the feeding basis of a country.

trade fill in, or in the case of net export, reduce the operational basis of the country.

Population density figures may also be modified to take into account these ghost acreages. There is in this particular respect a great deal of glib talk and confused reasoning when discussing "miracles," such as Japan, Holland, Ceylon, and others—how they manage to "feed" so many from so little acreage. The graphs (Figs. 22 and 23) refute the validity of these reasonings, as in all these cases substantial contributions are made through net importation and fishing. Without fisheries it would in most cases not be feasible for any of the Southeast Asian countries to maintain their trade deliveries of such items as rice, copra, tea, and rubber.

Nutritional Acreages

All the computing done so far and discussed above has been based on the assumption that each country is meeting the needs of its inhabitants. This is a fallacy. Dietary surveys are available for a number of countries. Several of these

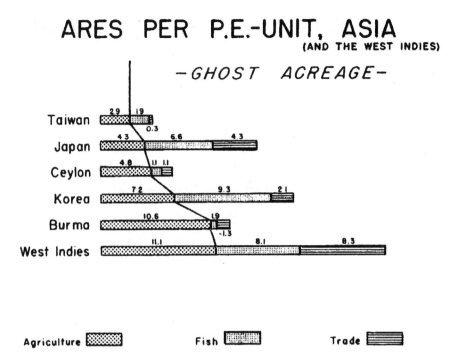

Figure. 22. Ghost acreages calculated as acreage per population equivalent. The most densely populated countries of Asia and Europe have been included.

point to the presence of extensive undernourishment and nutritional deficiencies. In those cases, where the results of such surveys present comprehensive computations primarily of protein shortage, these findings can be related to the food production of each individual country. In this way it becomes feasible to add to the ghost acreage what could be termed a *nutritional acreage*. By this would be meant the additional acreage required to satisfy nutritional minimum needs but still taking into consideration the dietary habits prevailing in each country. Calculations are not based on such unrealistic questions as: What would be required to give the Brazilians a United States diet? Instead, they aim at answering more realistic queries such as: How many more acres of beans or how much additional pasture would be needed to give the undernourished in the country a minimal diet without changing its relative composition?

The extensive malnutrition among the peoples of the islands of the Caribbean area will be discussed in the chapter on Latin America. It was also briefly touched upon in the chapter on the protein crisis. Take Haiti as one concrete example. The estimates show that the additional acreage needed to fill the protein

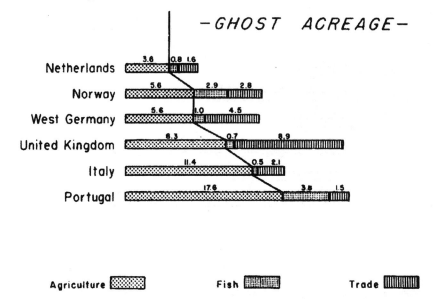

ARES PER P.E.-UNIT, EUROPE

– GHOST ACREAGE –

Figure. 23. Ghost acreages calculated as acreage per population equivalent. The most densely populated countries of Asia and Europe have been included.

gap and satisfy minimal nutrition requirements would amount to one extra Haiti in terms of tilled land and pastures. Once more it should be pointed out that there is of course the alternate proposition of doubling the yields per acre. This does not appear, however, to be a very realistic likelihood for Haiti within the foreseeable future. Undoubtedly Norwegian dried cod or Peruvian fish meal would constitute a cheaper and more reliable expedient in filling their present protein gap.

Another method which elucidates fairly well the relationship between agricultural production and nutritional standard is to calculate how large an acreage, and alternately how large an increase in yield, would be required, under present production conditions, to attain a raised nutritional level. Such estimates give a rather clear indication of the degree of feasibility for such programs.

Most experts would agree that Indonesia, for instance, is seriously overpopulated. Calculations of the suggested kind bear this out still more clearly. In order to raise the nutritional standard of Indonesia up to Soviet level, an acreage four times as large as the present would be needed if it were to be done chiefly through rice, and a doubling of the acreage if done through peanuts.

Table 13

Thousand Hectares	Agriculture		Ghost Acreage		Total Feeding Acreage	Agricultural Self-sufficiency %
	Tilled Land	Pastures	Fisheries	Trade		
Denmark	2,736	381	255.5	491.5	2,881	—
%	95	13.5	9	27.5	—	108.5
Finland	2,586	283	110	648	3,627	—
%	71.5	8	3	17.5	—	79.5
Norway	829	203	546	525	2,103	—
%	39	10	26	25	—	49
Sweden	3,712	724	434	924	5,794	—
%	64	12.5	7.5	16	—	76.5
Scandinavia	9,863	1,591	1,345.5	1,605.5	14,405	—
%	68.5	11	9.5	11	—	79.5
England (U.K.)	7,126	12,238	2,711	20,629	42,704	—
%	16.5	29	6.5	48	—	45.5
W. Germany	8,727	5,689	2,475	11,607	28,498	—
%	30.5	20	9	40.5	—	50
Netherlands	1,054	1,251	491	1,045	3,841	—
%	27	33	13	27	—	60
Belgium	990	725	331	2,549	4,595	—
%	21.5	15.5	7.5	55.5	—	37
Portugal	4,130		1,208	334	5,672	—
%	73		21	6	—	73
Israel	378	222	494	523	1,617	—
%	23	14	30	33	—	37
Switzerland	445	1,727	135	1,427	3,734	—
%	12	46	3.5	38.5	—	58
Italy	15,809	5,127	982	3,808	25,726	—
%	62	20	4	8	—	82
China	109,354	177,996	25,667	3,232	309,785	—
%	35	58	8	1	—	93
India	160,006	12,207	6,685	3,619	182,517	—
%	88	7	3.5	1.5	—	95
Indonesia	17,681		2,773	518	20,972	—
%	84		13	3	—	84
Ceylon	1,523		347	458	2,328	—
%	65		15	20	—	65
Egypt	2,618		1,250	355	4,223	—
%	62		30	8	—	62

Brazil would need either a 60 per cent acreage increase or a 60 per cent higher yield per acre of wheat in order to be able, from its own soils, to give the nation an Italian standard. Various combinations are obviously possible, such as 30 per cent greater yield and a 30 per cent increase in acreage; in both cases the result would be a rather modest improvement and yet would require exceptional measures, besides what normally would be needed to feed the more than one million extra now added to this country every year.

England's Recurrent Crises

It is not surprising that England off and on has payment difficulties and is goaded by its ministers of finance to new and ever increasing export achievements. No one in charge of the British treasury has yet suggested, however, that the nation needs to adjust to reality and its true resources by cutting its coat according to its cloth. The crises now persistently hampering this country seem in the long run chronic, when nearly half of the nation's food and feed has to be paid for from hard-earned foreign exchange. Yet no one would suggest curing the financial ills by reducing the conventional intakes of meat, butter, and eggs, which now constitute more than half of the imported acreage. The remainder is represented by food oils and grain for food and feed.

West Germany allows itself the same extravagance but to a lesser degree. Relatively speaking, less of its imported food consists of animal products, and more of bread, feeding grain, and oils. But even in this case an adjustment to the nutritional realities of our day's world would constitute a policy economically more sound in the long run. In the case of both these countries, maintaining the present dietary habits implies the earmarking of a considerable amount of the export incomes for regular food consumption. The immense global food shortage is a reminder of the world's huge unfilled needs and markets, which far more desperately require this food. Signs of adjustment to these realities are seen in the opening up of China, Japan, and other Asiatic markets to grains and other agricultural commodities from Australia, Canada, and New Zealand.

Commentary

Georg Borgström, "Ghost Acreage" (1962)

SVERKER SÖRLIN

Georg Borgström (1912–1990), a plant physiologist who earned a Ph.D. in botany from Lund University in 1939, made a name as a neo-Malthusian expert on the geography of food with a solid emphasis on shortages and food crises. He was a prolific author and a tireless public speaker and media personality, in the United States and elsewhere, not least in his native Sweden, where he became a household name in the 1960s and 1970s. His pessimism about population growth became central to the understanding of the environmental issues for some time. In 1967 he was named "the international Swede of the year" by the tabloid newspaper *Expressen*.

Borgström started his career as a food chemist and got his second status as a docent in food technology at Chalmers School of Technology in 1951. He was hired by the package and canning industry as the director of their institute for applied food and conservation research. From this position he pursued some of the neo-Malthusian work for which he was later famous, but which was regarded unfavorably by the industry. The situation was further aggravated when Borgström argued against the use cans of and throwaway glass packages in the packaging industry, recommending instead vacuum storage and recyclable glass containers, which later became the norm. The conflict with his employer and its funding companies, especially the glass container giant PLM, grew to such proportions that he had to quit his position in 1951. In 1956 he chose to leave Sweden and settle as a professor of food technology and, from 1960, economic geography at Michigan State University. At that time he was, interestingly, held in low regard by the scientific establishment in Sweden, although after his later fame in 1980 he was elected a member of the Royal Swedish Academy of Sciences, by which time he had returned to his mother country.

Borgström wrote about the emerging, already grave, and in his view drastically worsening problems of feeding the growing world population. Virtually all his books, most of them from the 1960s and 1970s, deal with this issue one way or the other, to the extent that his books are quite repetitive. Borgström thought that the world was in a deep crisis because of the growing population, and that most politicians and other elites had not yet faced that fact or its implications. His message was always quantitative; not only did he use a lot of figures and statistics, he would also present his data and ideas in visual form and draw maps and elegant, often dramatic, diagrams that spoke as much as the dithyrambic prose of his many volumes.

The concept "ghost acreage" (*spökareal* in Swedish) that he first used in the late 1950s, was the title of a chapter in his book *The Hungry Planet* (1965). It is an utterly pedagogical presentation of the concept and at the same time a rehearsal of his basic credo: that national trade and economic statistics were not acceptable because they were expressed in monetary units. Such statistics failed to capture adequately the real values, as expressed in calories, nutrition units, or especially protein, which Borgström regarded as the most valuable food component. Such units could much better serve to present the real economies of nations and households.

The panorama of his ideas was classically Malthusian. He advocated contraception and was pessimistic about the possibility of matching growing population with limited resources. He made his debut as an environmental pessimist with a radio series in 1952 in the wake of the 1948 books by William Vogt, *The Road to Survival* (Part 4) and Fairfield Osborn, *Our Plundered Planet*, and followed their tendency to catalogue interrelated problems as "environmental," problems ranging from soil erosion and industrial pollution to poverty, underdevelopment, and above all, overpopulation, which he regarded as the driver of most of the others. His series of radio talks was collected and published as a book (in Swedish) in 1953. Borgström held that the world as he knew it, with selfish and greedy individuals organized into selfish and greedy nation-states, would need to be reorganized in order to face the grave challenges. He wanted to see a world government organizing what he called "the world household." He also advocated the building of global grain reserves.

Borgström belongs in the first generation of postwar prophets and alarmists of global change. He clearly saw humans as the cause of environmental decay and also as threatening violence and global warfare through fighting over limited resources. He can also be seen in lineage of the earlier resource determinists like Griffith Taylor (Part 3) and Vilhjalmur Stefansson (Part 3). Like them, he saw resources as the defining element of world development and of the global distribution of wealth and power. He was, one might say, a "protein determinist," since economic powers such as states could increase their share of protein through trade and, if need be, war. He is typical of the postwar alarmist prophets also in the sense that he was a scientist but commented widely on politics, culture, history, and other social and human subjects. He had a planetary perspective and was innovative in translating scientific observations into larger trends and rates of change. He was a masterful presenter, eloquent and convincing; it has been said that it was no coincidence that he was a clergyman's son.

Further Reading

Borgström, Georg. 1969. *Too many: A study of the Earth's biological limitations*. New York: MacMillan.

Linnér, B.-O. 2003. *The return of Malthus: Environmentalism and the post-war population-resource crisis*. Isle of Harris: White Horse Press.

The Population Bomb

PAUL EHRLICH

Americans are beginning to realize that the underdeveloped countries of the world face an inevitable population-food crisis. Each year food production in these countries falls a bit further behind burgeoning population growth, and people go to bed a little bit hungrier. While there are temporary or local reversals of this trend, it now seems inevitable that it will continue to its logical conclusion: mass starvation. The rich may continue to get richer, but the more numerous poor are going to get poorer. Of these poor, a *minimum* of ten million people, most of them children, will starve to death during each year of the 1970s. But this is a mere handful compared to the numbers that will be starving before the end of the century. And it is now too late to take action to save many of those people.

However, most Americans are not aware that the U.S. and other developed countries also have a problem with overpopulation. Rather than suffering from food shortages, these countries show symptoms in the form of environmental deterioration and increased difficulty in obtaining resources to support their affluence.

In a book about population there is a temptation to stun the reader with an avalanche of statistics. I'll spare you most, but not all, of that. After all, no matter how you slice it, population is a numbers game. Perhaps the best way to impress you with numbers is to tell you about the "doubling time"—the time necessary for the population to double in size.

It has been estimated that the human population of 8000 B.C. was about five million people, taking perhaps one million years to get there from two and a half million. The population did not reach 500 million until almost 10,000 years later—about 1650 A.D. This means it doubled roughly once every thousand years or so. It reached a billion people around 1850, doubling in some 200 years. It took only 80 years or so for the next doubling, as the population reached two billion

Paul Ehrlich. (1968) 1971. Excerpt from *The population bomb*, revised and expanded ed., 3–6, 9–12, and 16–17. New York: Ballantine Books.

around 1930. We have not completed the next doubling to four billion yet, but we now have well over three and a half billion people. The doubling time at present seems to be about 35 years. Quite a reduction in doubling times: 1,000,000 years, 1,000 years, 200 years, 80 years, 35 years. Perhaps the meaning of a doubling time of around 35 years is best brought home by a theoretical exercise. Let's examine what might happen on the absurd assumption that the population continued to double every 35 years into the indefinite future.

If growth continued at that rate for about 900 years, there would be some 60,000,000,000,000,000 people on the face of the earth. Sixty million billion people. This is about 100 persons for each square yard of the Earth's surface, land and sea. A British physicist, J. H. Fremlin, guessed that such a multitude might be housed in a continuous 2,000-story building covering our entire planet. The upper 1,000 stories would contain only the apparatus for running this gigantic warren. Ducts, pipes, wires, elevator shafts, etc., would occupy about half of the space in the bottom 1,000 stories. This would leave three or four yards of floor space for each person. I will leave to your imagination the physical details of existence in this ant heap, except to point out that all would not be black. Probably each person would be limited in his travel. Perhaps he could take elevators through all 1,000 residential stories but could travel only within a circle of a few hundred yards' radius on any floor. This would permit, however, each person to choose his friends from among some ten million people! And, as Fremlin points out, entertainment on the worldwide TV should be excellent, for at any time "one could expect some ten million Shakespeares and rather more Beatles to be alive."

Could growth of the human population of the Earth continue beyond that point? Not according to Fremlin. We would have reached a "heat limit." People themselves, as well as their activities, convert other forms of energy into heat which must be dissipated. In order to permit this excess heat to radiate directly from the top of the "world building" directly into space, the atmosphere would have been pumped into flasks under the sea well before the limiting population size was reached. The precise limit would depend on the technology of the day. At a population size of one billion billion people, the temperature of the "world roof" would be kept around the melting point of iron to radiate away the human heat generated.

But, you say, surely Science (with a capital "S") will find a way for us to occupy the other planets of our solar system and eventually of other stars before we get all that crowded. Skip for a moment the virtual certainty that those planets are uninhabitable. Forget also the insurmountable logistic problems of moving billions of people off the Earth. Fremlin has made some interesting calculations on how much time we could buy by occupying the planets of the solar system.

For instance, at any given time it would take only about 50 years to populate Venus, Mercury, Mars, the moon, and the moons of Jupiter and Saturn to the same population density as Earth.

What if the fantastic problems of reaching and colonizing the other planets of the solar system, such as Jupiter and Uranus, can be solved? It would take only 200 years to fill them "Earth-full." So we could perhaps gain 250 years of time for population growth in the solar system after we had reached an absolute limit on Earth. What then? We can't ship our surplus to the stars. Professor Garrett Hardin of the University of California at Santa Barbara has dealt effectively with this fantasy. Using extremely optimistic assumptions, he has calculated that Americans, by cutting their standard of living down to 18% of its present level, could in *one year* set aside enough capital to finance the exportation to the stars of *one day's* increase in the population of the world.

Interstellar transport for surplus people presents an amusing prospect. Since the ships would take generations to reach most stars, the only people who could be transported would be those willing to exercise strict birth control. Population explosions on space ships would be disastrous. Thus we would have to export our responsible people, leaving the irresponsible at home on Earth to breed.

Enough of fantasy. Hopefully, you are convinced that the population will have to stop growing sooner or later and that the extremely remote possibility of expanding into outer space offers no escape from the laws of population growth.

. . .

Demographically, the whole problem is quite simple. A population will continue to grow as long as the birth rate exceeds the death rate—if immigration and emigration are not occurring. It is, of course, the balance between birth rate and death rate that is critical. The birth rate is the number of births per thousand people per year in the population. The death rate is the number of deaths per thousand people per year. Subtracting the death rate from the birth rate, ignoring migration, gives the rate of increase. If the birth rate is 30 per thousand per year, and the death rate is 10 per thousand per year, then the rate of increase is 20 per thousand per year ($30 - 10 = 20$). Expressed as a percent (rate per hundred people), the rate of 20 per thousand becomes 2%. If the rate of increase is 2%, then the doubling time will be 35 years. Note that if you simply added 20 people per thousand per year to the population, it would take 50 years to add a second thousand people ($20 \times 50 = 1,000$). But the doubling time is actually much less because populations grow at compound interest rates. Just as interest dollars themselves earn interest, so people added to population produce more people. It's growing at compound interest that makes populations double so much more rapidly than seems possible. Look at the relationship between the annual percent increase (interest rate) and the doubling time of the population (time for your money to double):

Annual percent increase	Doubling time
1.0	70
2.0	35
3.0	24
4.0	17

Those are all the calculations—I promise. If you are interested in more details on how demographic figuring is done, you may enjoy reading Thompson and Lewis's excellent book, *Population Problems*, or my book, *Population, Resources, Environment*.

There are some professional optimists around who like to greet every sign of dropping birth rates with wild pronouncements about the end of the population explosion. They are a little like a person who, after a low temperature of five below zero on December 21, interprets a low of only three below zero on December 22 as a cheery sign of approaching spring. First of all, birth rates, along with all demographic statistics, show short-term fluctuations caused by many factors. For instance, the birth rate depends rather heavily on the number of women at reproductive age. In the United States the low birth rates of the late 1960's are being replaced by higher rates as more post World War II "baby boom" children move into their reproductive years. In Japan, 1966, the Year of the Fire Horse, was a year of very low birth rates. There is widespread belief that girls born in the Year of the Fire Horse make poor wives, and Japanese couples try to avoid giving birth in that year because they are afraid of having daughters.

But, I repeat, it is the relationship between birth rate and death rate that is most critical. Indonesia, Laos, and Haiti all had birth rates around 46 per thousand in 1966. Costa Rica's birth rate was 41 per thousand. Good for Costa Rica? Unfortunately, not very. Costa Rica's death rate was less than nine per thousand, while the other countries all had death rates above 20 per thousand. The population of Costa Rica in 1966 was doubling every 17 years, while the doubling times of Indonesia, Laos, and Haiti were all above 30 years. Ah, but, you say, it was good for Costa Rica—fewer people per thousand were dying each year. Fine for a few years perhaps, but what then? Some 50% of the people in Costa Rica are under 15 years old. As they get older, they will need more and more food in a world with less and less. In 1983 they will have twice as many mouths to feed as they had in 1966, if the 1966 trend continues. Where will the food come from? Today the death rate in Costa Rica is low in part because they have a large number of physicians in proportion to their population. How do you suppose those physicians will keep the death rate down when there's not enough food to keep people alive?

One of the most ominous facts of the current situation is that over 40% of the population of the underdeveloped world is made up of people under 15 years old. As that mass of young people moves into its reproductive years during the next decade, we're going to see the greatest baby boom of all time. Those youngsters are the reason for all the ominous predictions for the year 2000. They are the gunpowder for the population explosion.

. . .

It is, of course, socially very acceptable to reduce the death rate. Billions of years of evolution have given us all a powerful will to live. Intervening in the birth rate goes against our evolutionary values. During all those centuries of our evolutionary past, the individuals who had the most children passed on their genetic endowment in greater quantities than those who reproduced less. Their genes dominate our heredity today. All our biological urges are for more reproduction, and they are all too often reinforced by our culture. In brief, death control goes with the grain, birth control against it.

In summary, the world's population will continue to grow as long as the birth rate exceeds the death rate; it's as simple as that. When it stops growing or starts to shrink, it will mean either the birth rate has gone down or the death rate has gone up or a combination of the two. Basically, then, there are only two kinds of solutions to the population problem. One is a "birth rate solution," in which we find ways to lower the birth rate. The other is a "death rate solution," in which ways to raise the death rate—war, famine, pestilence—find us. The problem could have been avoided by population control, in which mankind consciously adjusted the birth rate so that a "death rate solution" did not have to occur.

Commentary

Paul Ehrlich, *The Population Bomb* (1968)

MICHAEL EGAN

Doomsday prophecies have been a staple of American rhetoric since its inception. Puritan foretellings of disaster are deeply rooted in the American ethos.
Warning against environmental cataclysm—and situating that disaster on the
imminent horizon—has been a prominent feature of American environmental
rhetoric, especially since World War II. This form of dire Doomsday prediction
reached its climax among a group of politically engaged biologists by the end
of the 1960s. George Wald, for example, offered a decidedly bleak picture of
the future; he estimated that "civilization will end within fifteen or thirty years
unless immediate action is taken against problems facing mankind" (Bailey
2000). At a talk at Swarthmore College just before the first Earth Day (22 April
1970), Kenneth Watt warned, "We have about five more years at the outside to
do something" (Bailey 2000). A couple of months earlier, Barry Commoner had
appeared on the cover of *TIME* magazine, hailed as the "Paul Revere of ecology," the revolutionary hero who had warned of imminent danger. In the same
article, Commoner and his scientific colleagues were also labeled the "new
Jeremiahs."

The new jeremiad married accessible science with a new rhetoric reminiscent of the biblical jeremiad. Like its predecessor, this new jeremiad anticipated
doom and destruction, but urged ecological—rather than spiritual—awakening.
Just as the original Jeremiah's dire predictions warned of the destruction of Je-
rusalem, the new Jeremiahs prophesied the destruction of the Earth's ability to
sustain life; both lamented the human fall from grace and saw the human condition and attempts at redemption as almost hopeless. Almost. In the jeremiad's
rhetoric, there lingered a glimmer of hope to which audiences were meant to
cling. The environmental jeremiad aimed to lead their audiences to despair,
but then redeem them through the narrowest of hopes. Biologist Paul Ehrlich
was among the most adept practitioners of this style, and certainly one of the
most prominent leaders of American environmentalism. During the years immediately before and after the first Earth Day, Ehrlich emerged as a veritable
eco-celebrity, with an effective message about the hazards of uncontrolled human population growth and a massive audience. His 1968 book *The Population
Bomb* became a bestseller, selling more than three million copies over the next
decade. In addition, Ehrlich was also a regular guest on *The Tonight Show* with
Johnny Carson. He was witty, charismatic, and handsome; possessed a deep and

strong voice; and spoke clearly and convincingly to a popular audience about the dangers of human population growth. According to journalist Stephen Fox, "Ehrlich—with his thundercloud visage and deeply resonant voice—seemed the very personification of the Voice of Doom" (Fox 1985, 311).

Ehrlich was the most popular and articulate of a new brand of neo-Malthusians, who worried that overpopulation posed an ecological strain on the Earth's carrying capacity and food production limits (Malthus and Borgström, discussed in Part 1 of this book, and Vogt, discussed in Part 4). The crux of Malthus's treatise was the difference in scale between population growth and the growth of food production. The postwar iteration of Malthus's treatise linked population growth not just to resource scarcity but also to increased pollution. Indeed, Ehrlich and his colleagues believed the technological revolution that followed World War II was a response to population pressures. Furthermore, the neo-Malthusianism of the 1960s was laced with foreboding. In one particularly famous interview that appeared in *Look* magazine in April 1970, Ehrlich said: "When you reach a point where you realize further efforts will be futile, you may as well look after yourself and your friends and enjoy what little time you have left." He concluded by warning: "that point for me is 1972."

Ehrlich's prognostications about population growth and inevitable ecological decline were not guesswork wrapped in ominous warnings. He grounded his analysis in a formula he introduced with his collaborator, physicist John Holdren, designed to measure the various factors of pollution. They argued that environmental impact (pollution) was the product of population, affluence, and technology; the number of people, the quantity of goods people consume, and the technologies people employ to produce the goods. They presented the compact formula as:

$$I = P \times A \times T$$

Essentially, this simple equation dictated that an increase in population or in consumption or in polluting technologies would result in an increase in environmental impact. Growing population—especially in the developing world—would generate enormous ecological pressures on the planet's carrying capacity as global affluence increased. As population continued to grow, Ehrlich worried about a planet bursting with too many people. And while it might be worth noting that Ehrlich's predictions were not especially accurate, it might also warrant noting that the praxis of his jeremiad contributed markedly to the growing public interest and literacy in ecology.

Of course, the jeremiad constituted something of a devil's bargain. While the jeremiad unquestionably drew attention to and stressed the urgency of environmental issues, its tone seemed to move debate away from the rational scientific claims that grounded their discourse. Moreover, many jeremiads overreached in their rhetoric, forecasting ecological doom before the end of the decade. The jeremiad itself began to wear thin. As their predictions proved to be off the mark, they hurt their professional standing. Scientists were tuned out as popular culture adapted to or resisted the alarmist rhetoric. Frederick Buell described the transition as "from apocalypse to way of life" (Buell 2004).

Indeed, Ehrlich himself received some negative publicity regarding a subsequent wager with the economist Julian Simon. Critical of Ehrlich's anticipation of ecological catastrophe, Simon challenged the notion that Ehrlich's scarcity data was well founded. In 1980, Simon invited Ehrlich to select five commodity metals. According to Ehrlich's population hypothesis, increased numbers of humans would create greater demand and stress the supply of the resources, thereby leading to an increase in the metals' market prices. Simon bet that their prices would decline. Ehrlich chose copper, chromium, nickel, tin, and tungsten; by 1990—in spite of the largest population increase in a single decade in all of human history—all of the commodities' prices had dropped. While measuring metals rather than biophysical limits might not have been the most accurate way to identify the nature of population growth and its environmental impact, the bet reduced the credibility of much of Ehrlich's population thesis, particularly in mainstream fora.

Nevertheless, *The Population Bomb* and Ehrlich's Malthusian concerns warrant careful attention—both in their historical moment as a vital chapter at the dawn of the "Age of Ecology" and as vital examples of the future of nature. Ehrlich's view of the environmental future was decidedly bleak. He had begun *The Population Bomb* with the stark statement that "the battle to feed all of humanity is over" (Ehrlich 1968, xi). According to Ehrlich, famine and devastation were inevitable. The human death rate—galvanized by famine, war, and disease—would increase with pollution while resources dwindled. For Ehrlich, population was the heart of this postwar environmental crisis: "Too many cars, too many factories, too much detergent, too much pesticide, multiplying contrails, inadequate sewage treatment plants, too little water, too much carbon dioxide—all can be traced to *too many people*" (Ehrlich 1968, 66–67). The lone grain of hope was immediate and rigid population control: "There are only two kinds of solutions to the population problem," he argued. "One is a 'birth rate solution,' in which we find ways to lower the birth rate. The other is a 'death rate solution,' in which ways to raise the death rate—war, famine, pestilence—*find us*" (Ehrlich 1968, 34).

Further Reading

Bailey, R. 2000. "Earth Day, then and now." *Reason* (May), http://reason.com/archives; original.

Buell, F. 2004. *From apocalypse to way of life*. New York: Routledge.

Fox, S. R. 1985. *The American conservation movement: John Muir and his legacy*. Madison: University of Wisconsin Press.

Watt, K. 1970. "Earth Day organizers hope to halt pollution." *New York Times*, 23 April.

PART 2

Sustainability

Are We Limited by Knowledge or Resources?

Wanting more and having to get by with less is not just a human story, but a description of the life of many species. Fear of scarcity and management of risk means that regulation and control of resource use have ancient roots and are to be found in most human societies. But during the seventeenth century, the development of theories on the political order of Europe and its new colonies, and on relative economic performance, suggested concrete questions more orientated toward the future. Was the destiny of states determined by the resources at their disposal? Were states successful because they controlled a lot of resources? Or was success built upon social institutions or individual virtues—for example, industriousness or inventiveness—that drew resources to them? Was there a limit to such success?

From the 1640s, an increasingly prevalent idea in the English language was that social life should be a process of "improvement," and steered by "improvers." Human virtues, capacities, and ingenuity were the wellsprings of success, and were equally a demonstration of godliness in this Christian society. In this way of thinking, the idea of a limit to resources was a challenge to innovation, a divinely ordained challenge to call forth virtue. So we see that the emergence of "sustainability" as a problem stood in close relation to an expectation and desire for further social and economic development. It was not so much a question of husbanding what you have, as unleashing potential. Knowledge, rather than resources, was limited. Oddly enough to our eyes, this way of thinking was most prominent in a society almost entirely dependent on what we think of as "renewables," and limited by ecology. But "sustainability" was not a steady state, an equilibrium, but rather the challenge of keeping supply in step with ambition. This required expertise, especially among foresters. Forestry is the world of our first excerpt, from Hans Carl von Carlowitz, and was repeated in debates about "timber famine" in the late nineteenth and early twentieth centuries.

The "problem" of resource limits seemed more acute—ironically—when society moved to dependence on what were then very large stocks of fossil fuels and metals. Consumption expanded rapidly in the modern industrial economy, far ahead of the rate of population growth. Already in the late eighteenth century writers expressed worries about what we might call "peak coal." By the time W. S. Jevons wrote in 1865, the development of statistical data on consumption, and statistical techniques for establishing rates of change, facilitated comparison between the past, present, and possible futures. In this new world of massive extraction of nonrenewables, knowledge might not suffice to secure future well-being, if modern society was dependent on a resource for which there was no substitute.

Resource problems were "big" problems, connected to the destiny of whole societies. Yet they tended to be studied in isolation, and the new economics that Jevons among others developed accentuated this trend: you had to separate out what you can measure, and compare what can be calibrated; nature and society as a laboratory, everything else *ceteris paribus*.

Thus by the time of the "environmental" moment after World War II, studies had identified pressure on a huge range of resources: metals, water, soil fertility, energy . . . all of which could be better understood because of problems faced during the huge logistical efforts of the war. In the 1930s, they were also linked to a surprisingly recent notion—"the economy," a national system as imagined most fully by John Maynard Keynes, and given a number to measure it (Gross Domestic Product) by Simon Kuznets. The question asked by men like Samuel Ordway was essentially the same as that of Carlowitz and Jevons: how long can economic growth continue? But now it was posed in terms of decades rather than centuries, and was linked to the complex interaction of demands for a host of materials. Modern economic growth had come to be understood as an unusual historical event and problem that itself had to be "sustained." Given that we now look back from a time beyond many of the predicted moments of resource exhaustion, these postwar thinkers seemed pessimistic. In many cases, resource exploitation had hardly begun—and yet we are still here.

Yet ideas of interdependence of society and nature remained intuitions, with individual elements viewed largely in isolation, until the means emerged to put them together analytically. It was the computer that offered this possibility, another product of the war. The most famous of the new predictive models was produced by the Club of Rome's *Limits to Growth* report in 1972, something envisaged as only the first stage in modeling what we might now call an entire socio-ecological system. This was the tool of the moment: the jeremiahs shouting in the wilderness could now make their case with algorithms and datasets and plot the future in lines shimmering across screens. The emergence of the "world system" was followed by the 1970s energy crisis. Sustainability was no longer a question of resource extraction keeping up with growth, or cutting one's cloth to likely shortages, but the management of a huge range of interacting variables, now including pollutants. This was a global agenda like nothing before—but the new approaches also offered more sophisticated policy instruments. The issue was not so much a moral change or the challenge to invent (although these certainly did not go away) as a tweaking of the feedback systems and efforts toward technological improvement—or what we might call transition.

Sustainability emerged in the English language in the 1970s. The most famous manifestation of the idea was as "sustainable development" in the World Commission on Environment and Development report, *Our Common Future*, headed by Gro Harlem Brundtland, where it was defined as meeting our own needs without undermining the capacity of future generations to meet their needs. In fact, this definition combined very old ideas about inter-generational justice found in previous centuries with the relatively modern dilemma of how economic growth should be "sustained," although with less certainty than in the past about how the future might define its own needs. This idea became widely written into governmental and corporate policy following the United Nations Conference on Environment and Development in Rio de Janeiro, 3–14 June 1992 (informally known as the Earth Summit). Sustainability offered many more paths to the future than just "running out" of resources or professional knowledge saving the day. It also endorsed a new kind of expert, the systems analyst, the manager of information, the envisioner of global dynamics. Ecology and climate science became influential parts of the system. But in this world, the idea of an end to growth, of the necessity for a more modest and moral living, has perhaps less purchase on policy makers than in previous decades. The legacy of the 1960s has been, surprisingly, to temper many of the expectations of limits that existed before; here our collective memory fails us. The reply to a long history of jeremiads is the fact that we are richer (in a material sense) and healthier than ever, and in truth the gloomy prophecy has always been left in the shadow of the gospel of improvement. The transition to sustainability is more than ever a problem of knowledge.

Sylvicultura oeconomica

HANS CARL VON CARLOWITZ

Most serene and all-powerful King, most gracious Lord!

Just as your royal Majesty's great and glorious *Actiones* in Imperial affairs, and the unending *Inclinationes*, that You graciously direct to the raising up of your diverse territories and people, to your majesty's most high and undying fame throughout the whole world, and ringing out to every man's great amazement, that are granted to all, and therefore can nevermore be sufficiently extolled and venerated; and shining out especially among other things, from your royal Majesty's incomparable care consoling your poor subjects, the raising of trade and commerce, and through this *conserving* ample sustenance and shift for them, under which the mines of your royal Majesty's world-renowned Ore Mountains are above all to be esteemed, as a great factor for the best of the common weal, through which many magnificent cities, towns and villages are founded, many thousands of people maintained, great sums of money brought into motion in this and neighboring territories, and by means of those metals and minerals and the *manufactures* made from them, the *commerce* of merchants drawn into the land, yes ever more strengthened and maintained, and following your royal Majesty's interest, promoted as far as is possible.

Beholding this, and especially how the mines, as the noble jewel und inestimable holy means of sustenance in your royal Majesty's Electoral Saxon territory, might not in future fall into decline through apparent wood shortage, and through this hinder the flourishing commerce, therefore have I here taken it upon myself, according to my duty as your royal Majesty's faithful vasall and mining officer, or Observer of your high mining *Regalis*, not only to sketch my few thoughts in writing, but also out of faithful patriotic feelings, to have it printed, and in this little work draw attention to how the state of the wood in your royal

H. C. von Carlowitz. (1713) 2009. *Sylvicultura oeconomica*, 2nd ed., f. 3v–4v, 28–29, 32–33, 68–69. Trans. Paul Warde. Originally published in Leipzig: Johann Friedrich Braun, 1713. Facsimile of 2nd ed., Remagen: Verlag-Kassel, 1732.

Majesty's electoral Saxon territories is to be maintained, and the feared wood shortage replaced little by little by the growth and rejuvenation of young wood, on those great bare patches, and areas of stumps of those many thousands of acres of cleared and felled woodlands, for the best for our descendents, and thereby the dear mines (which by God's blessing are inexhaustible in your royal Majesty's territories, but that without ample wood, may not be run) may come to be increased and expanded both now and into the future, especially because land is sufficiently available, and with a careful plantation of wood on it can never henceforth become in short supply.

If, most gracious King, I might have doubts, to most obediently present such a trivial work to your royal Majesty; nevertheless in consideration of the above-mentioned your royal Majesty's most gracious *Intention* as father of this land, that you lead with undying fame, to bring all possible help and grace to your subjects in the raising up and promotion of their sustenance and *conservation* of their commerce and especially the beloved mines, I have out of most loyal devotion fixed myself absolutely upon the resolution, now sufficiently established, that through the mines from which so much precious metal is annually brought out of the Earth, the territory will always become richer in money, and that the precious coins, alongside the sales of those many metals and minerals, and finished wares and *manufactures* made from them, will continually increase trade. Therefore I herewith in deepest *submission* lay this little script at Your Royal Majesty's feet, with the most obedient request, that Your Royal Majesty will most graciously be moved to allow your glance to fall upon it, and have it subjected to your Royal protection. I hope also that God will bless this work, that today and in the future the necessary contributions arise to both the mining and the state of woods in your royal Majesty's kingdom and territories, by which the poor subjects and dear *posterity* may look forward to the maintenance of their sustenance, a harvest we beg sincerely from God, and abandon to all of God's merciful blessings, and remain unceasingly in most faithful *devotion* with this most heartfelt desire all our life long.

> Your Royal Majesty's
> Most obedient faithful and dutiful servant
> Hannß Carl von Carlowitz

. . .

1. Although Germany, as has been repeatedly detailed above, was previously for the most part covered with woodlands, it has however not remained so in the succeeding ages; instead these dreadul woodlands and wastes have been turned into the most beautiful fields, meadows, vineyards, pastures and ponds, yeah into many thousands of those most perfect cities, castles and villages, to the extent that

as a result of that, and the great consumption, a wood shortage threatens to occur almost continuously.

2. Similar great woodlands may also be found today in America in many locations, different European nations employ great efforts towards their eradication, because owing to these same immense quantities, or trees standing thick to one another, one can almost not traverse them, or with great difficulty and only on foot, as many travel accounts prove, and confirm beyond argument. Indeed, if Europeans, especially in those provinces possessed by the English and French, lay out gardens, meadows or fields, or want to cultivate the land, they must with great effort and costs fell the wood and clear it from the site beforehand, and eradicate the great stumps. One can read of a memorable example of this that took place at the island of Newfoundland, namely that in order to make the island fertile, and to introduce cultivation and other economic activities, the foreign arrivals alongside the inhabitants set fire to the great woodlands, and burned the wood in indescribable quantities, so that much of the sap, resin, ash and pitch of the trees was washed into the sea by the rain and flowing waters, and the salts, acids and pitch spoiled the water, and caused the renowned and delicious catch of fish to be banned for seven years from parts of the region around this island, and the fish could not tolerate the taste, but moved away, from which is to conclude, how great an amount of wood must have been available, and burned.

3. Indeed it is not to be wondered, that such thick woodlands are to be encountered in locations, in which beforehand the Indians that lived within them had lacked in blades and axes to fell them, and those same nations also kept a poor economy, from indolence and lack of knowledge neglecting cultivation and livestock-raising, and hence raising no valuable buildings, and therefore having no need of much wood or large timbers. Equivalently great immense woodlands can yet be found in Siberia, in Samojita, on the borders of the great Tartar lands, in Lappland, Norway, where, because of the great cold, there are few dwellings, and the timber of the consequently uninhabited territories is not consumed, but they are able to help themselves to firewood.

4. In contrast there is in our local fully cultivated and inhabited northern lands a far different situation, in that already in times long past, and above all since that hard age of war, the woodlands have been devastated, those fields and meadows that had become covered with wood, have where possible been assarted, burned over, and the soil again brought to cultivation and other uses; and from the wood many thousands of castles, palaces, houses, barns, stalls, bridges and such like built in many cities great and small, in market towns, villages, manors and so forth; from which it is easy to conclude, that through such a great mass of inhabitants, the consumption of firewood and timber will little by little, as it were unnoticed, make for a wood shortage, as was complained of already in the

old wood-ordinances from some centuries past down to us, a decline that now is
felt much more severely. And through this not a little damage is caused, and in
the future is yet far more to be feared, as no person, yes no economy, no matter
how small it may be, can dispense with the use of the fire and of wood. Therefore
without doubt every man, high and low, must be concerned of how this is to be
remedied; otherwise we will hear the groans of poverty, and we will suffer distress,
trouble and loss of sustenance, of health and finally life itself.

. . .

14. Among these and other causes of the wood shortage, which will be detailed in the following, the primary one is the practice up until now of careless
and wasteful cutting down and clearing away of the wood. Because one could not
think that the great amounts of the same, that were to be encountered on occasion after the general peace of the year 1648, could have an end, that might have
given cause to take proper care of it, so that one did not come to emulate those in
other territories who continually cut down the wood, and persecuted it as its bit-
terest enemy, burning the felled wood in great heaps, or leaving it to decay, flung
into holloways, mires and pools and left to rot. Then in many regions only the
best wood was taken, and the lesser simply not made use of or gathered up, and
was thus entirely relinquished to decay; that which every man properly, to the
extent that the welfare of his fatherland is dear to him, should have stood guard
over. But if the clearances are arranged with regard to the annual regrowth of
the woods with sowing and planting; if it comes to pass namely that those can be
sufficiently replaced and the woods maintained with care, then a wood shortage
will easily be avoided. In this regard the following question arises: whether one
may, according to the composition of a woodland, justifiably fell an individual
tree here and there, an entire wood or at least a piece thereof? Hopefully the decision will be made according to the proportions of the woodlands and the cleared
areas within it, but also to promote the regrowth therein and to maintain this
for the best. The woodcutters go by the saying *Copia fastidium parit*, that excess
of a thing makes it tiresome, and are become so choosy, that if a felled trunk is
so ill-formed, knotty, and not suited to splitting, they will leave it lying, and fell
another. Not to speak of the fact that with wood that is so inexpensive and easy
to obtain, one must go begging the buyers to purchase; also the hammer-works,
furnaces and lime-kilns are continually increasing, but in contrast the woodlands
are greatly reduced, therefore in those locations and territories, where those
wood-devouring works are present, it becomes very necessary to yield something,
otherwise as a consequence a shortage will arise, as one work consumes annually
1, 2 or up to 3 000 *Schragen*, that may in the future only attain a few hundred and
those at expense, which would bring unspeakable damage to many lands and
be very *fatal*, that these lucrative works could not be fully furnished with wood

and kept in operation, through which many thousands of people among miners, smelters, hammer-smiths, pitch-makers, artisans, charcoalers, woodcutters, rafters, cartmen and other workers would have their sustenance taken away from them, and indeed would have to vacate this land. By such reasons the great woodland reserves are going to be made sparse, felled and cleared in a few years, to a degree that will seem almost beyond belief. Certainly if one recalls those great reserves of wood that were still available in places half a century ago, one proves thereby that an indescribable and unbelievable sum of wood has been cleared. And no-one could have been persuaded of or imagined this, otherwise without doubt they would have not assaulted the woods so fiercely, that so great a store of wood could have been wasted away so soon.

15. One of the most principal causes of the wood shortage is also that with this great decline that already has affected us, we did not attend in timely fashion to how those woods yet standing might through planting and sowing be preserved, the decline arrested, and the damage so far as possible repaired, especially as in such places over time a greater part of the welfare, flourishing and glory of a land depends on wood supply.

But no less importantly wood is consumed in great amounts, and if now every person must sense and see, that wood can utterly not be dispensed with, and yet certain means offer themselves, to address most securely and speedily this disaster, and to arrest this insufferable wood shortage, then at the very least should be considered, how one can bring back growth, or might sow, by which to supply themselves and their descendants with wood. And as is well known, our forefathers had *prognosticated* such long ago and had taken the best possible measures that the regrowth was again coming forth, ere and before the remaining woods were felled. However the all too-powerful consumption interrupted that care of our forefathers in the conserving, sowing and planting of wood which would, if it had occurred, with the sowing and planting continuing, have seen this land import many tons of gold. Because it is unfortunately a general affliction, that most of the woods have been cleared, and on many clearings no growth is to be seen. We certainly have to consider this a particular punishment of God in that we have not addressed this evil with the proper sowing and planting of the wild trees.

. . .

17. If one considers and counts, how many trees are to be found on a devastated and not fully regrown and covered acre of wood, and sets against this how many more trunks could stand there than are now available, then will one find and see, that there are many lacking, and that this arises only from the negligence of the owner. If one then made a reckoning of so many 1000 acres, e.g. if on one acre there are 20 trees lacking, then on 1000 acres some 20000, and in the reserves of the whole territory many 100000. As every man would concede, this is

no matter of small *importance*, if such could become increased through sowing
and planting. And if at the same time one wanted to say, if so many 1000 trees
more as otherwise would be set and planted on a certain reserve, the land could
not bear it, or give them sufficient sustenance, or it would at the very least wear
out the ground and soil and make it lean; so will prudent Nature herself answer,
namely that every tree bears its annual manure on itself and casts it down, namely
the leaves, twigs, shoots, blossom, bark, needles and fatness, that is washed from
it by rain and meltwaters and through which the ground and soil becomes improved again.

18. If one now considers further, what advantages are to be hoped for, if so
many clearings, bare places and reserves grazed bare from domestic and wild
beasts, became sown and planted and well-stocked with wood, or happily regrown, and how much more highly the same would be valued, that now are of
no worth. It is also easy to *judge*, how a territory well stocked with wood is to be
esteemed, against such wastelands, and how the valuation of one against the
other with so many of those bare places and clearings, must bring disadvantage
as much to to the land-lords as to the inhabitants in their sustenance. If consequently the woodlands of our land, so soon as they were cleared, were again
brought back to growth, how many tons more gold would such be worth, that
they now are not? By this consideration will it without doubt be incumbent on
each and every householder, whom God has endowed with woods and forests,
that they take the due care and measures, whether Father of the land or patriarch
of a household, by which a constant and continual usage is maintained, that the
best kind of wood from every species grows and is available, and in particular that
no unsown or unplanted place or expanse is to be found.

19. It is not necessary that one refers at length to instructions and ordinances
alone. The holy text gives us sufficient commands. Because Gen II.v.5 & 15 reveals that Almighty God has instructed mankind to cultivate the land, and therefore propagate plants, therefore also the wild trees. And in particular after the
Fall it has been pleasing to his most holy will that he does not give mankind
sustenance and comfort through indirect means, but when the work is taken
on through these our hands. This was fulfilled by Abraham himself, in that in
Gen. XXI he planted trees, or after the common saying, a woodland or copse.
Thus according to this the sowing and planting of wood is to be respected as a
praiseworthy, honorable, and most necessary care, undertaking and labor, yeah
that cannot sufficiently be praised, lauded or fully honored with a proper title,
because it will give us the most certain means to avert wood shortage in the future in this and other lands; especially in many extensively cleared, and entirely
bare wood reserves, which however may in a short time become again bedecked

with growth, increased and continually maintained, and the gaps filled with so many million tree-trunks, *in infinitum* and to an unbelievable extent, not solely through the woodland itself, but also through the resulting *commerce*. Thereby we also recognize with consideration, amazement and contemplation of the gift of God's Creation, God's beneficence, and should praise him, and use the same to our advantage and not squander it, or neglect its regrowth, rather all the more spare no effort, labor, care or reflection, until it is brought into a happy state.

20. There is however no time to lose with these good intentions, *natura progrediens simper multiplicatur per media*. That is, Nature increases only through certain means. Because for every year that passes, in which nothing is planted or sown, the longer must one wait for the advantages, and the more that so many thousands suffer over time, then all the more devastation happens, that finally the woods yet remaining are attacked, entirely *consumed*, and must be more and more diminished. *Cum labor in damno est, crescit mortalium egestas*. That is, Where damage comes from labors unperformed, there grows poverty and wretchedness. The cultivation of the wood can not so promptly be handled as with the tillage; because although with the latter bad harvests may happen two, three or more years in succession, so can one single blessed and fruitful year, just as with the vineyards, make up for everything; in contrast, if the wood is just once devastated, the damage, especially where it concerns the coarse and strong building-timber, is not to be remedied in many years, yeah even in a century; especially as in betwixt all kinds of *vicissituines Rerum* and changes may occur.

In case one house-holder promotes and cultivates, another in contrast neglects and indeed lays waste, what had for some years been improved; and speaking in general, where as a result of delay in dealing with some danger the resulting damage is irretrievable, then one must lose no time, and therefore one must take the sowing and planting of trees most urgently in hand, as all the while a long period will be required, ere the wild trees can be brought to the proper height, strength and uses, especially given, as already mentioned, there is absolutely no doubt that the wonderful and beautiful woods have been hitherto the greatest treasure of many lands, and held for inexhaustible. Indeed it has undoubtedly been viewed as a store of wealth, in which endured the welfare and prosperity of this land, and had been so to speak the *Oraculum*, that good fortune could never be lacking, in that thereby so many treasures and all kinds of metals could be got hold of. But when the deepest parts of the Earth have been exposed through so much effort and trouble, then a shortage of wood and charcoal occurs, to smelt the same metal. Therefore the great Arts, Science, Industriousness and Order of our Land will be founded on how a *conservation* and cultivation of wood is set in place, so that it gives a *continual*, constant and sustainable use, because it is an indispensible matter, without which the Land may not remain in good heart [*Esse*]. Because just as other lands and kingdoms are blessed by God with cereals, beasts, fisheries, ship-faring and other things; so is it here the wood, by which the noble jewel of this land, namely the mines, are maintained, and the ores processed and used for other necessities.

Commentary

Hans Carl von Carlowitz, *Sylvicultura oeconomica* (1713)

PAUL WARDE

The *Sylvicultura oeconomica* by Hans Carl von Carlowitz (1645–1714) has become renowned for introducing the term "sustainability" to the world (in its German form, *Nachhaltigkeit*). Carlowitz did not actually use the term "sustainability" (*Nachhaltigkeit*) but only "sustainable" (*nach haltend*). However, subsequent eighteenth-century German foresters turned the adjective into a noun, making "sustained yield theory" a cornerstone of their practice, and leading to Carlowitz being viewed as the originator of the idea.

Carlowitz was certainly not the first to write about what we now call sustainable yield, or the need of current generations to ensure that resources (especially wood) are available to their descendants. There was a long heritage stretching back to medieval times of making this argument, related to often hysterically expressed fears of wood shortage and the crisis and disorder that would result from "timber famine." Such fears provided justification throughout Europe from the fifteenth century for passing laws to ensure future wood supplies, regulating the use of forests and wood consumption, and stimulated writings in a range of countries. Carlowitz placed himself in this heritage, providing references to numerous examples of predecessors from ancient and early modern times; he also came himself from a long line of foresters and mining officials in Saxony in eastern Germany that can be traced back a century before his birth. *Sylvicultura oeconomica*, written in his final infirmity in his late sixties and published just a year before he died, is a distillation of a lifetime of personal experience working as a forestry and mining official after an education at the University of Jena and travels across Europe. No one had previously written on such a scale, and with the authority of such long direct experience in woodland management.

His originality lay in the absolutely central place he gave to "artificial" (*Kunstliche*) techniques in forestry, where "artificial" should be understood in its older meaning of requiring human artifice and interference. This placed the authority and expertise of the forester at the center of resource management. He transferred the active techniques of garden horticulture into forests, rather than relying, as was more usual at the time, on encouraging and protecting natural regrowth of felled woodland. Through these techniques, foresters became engineers of nature, rather than mere policemen of the woods. The book came to be seen as the precursor to the explosion of literature leading to "scientific forestry" in Germany. In contrast, the earlier work of Englishman John Evelyn on arboriculture, *Sylva* (1664), remained famous in Britain because it had no successors, remaining the only anglophone work of its type. Evelyn's *Sylva* had relatively little influence on forests in Britain, which lacked a professional cadre of foresters, while Carlowitz's work is considered a canonical, even founding text by the forestry profession.

For Carlowitz and his contemporaries, predictions of wood shortage were based on a mixture of observation and assumption. They assumed that the growth of human populations and cultivation had led relatively recently to wholesale deforestation, but for the most part the blame for shortages was placed squarely on a conviction that most people (especially peasants) were poor and wasteful managers of the woods who gave little consideration to future generations. Current and future shortages were created by ignorance and sloth, rather than population growth or absolute limits to resource availability. The remedy was thus better management and care, the dissemination of the knowledge and power held by men such as Carlowitz himself. Particularly given the slow rate of growth of trees, they had to be managed by people with a vision of the future and the capacity for long-term planning. He was explicit in arguing that God would not have provided a world where resources were insufficient to support people, and indeed that his generosity should not be squandered. This stood clearly in the tradition of "improving" European thought, often associated with Francis Bacon (1561–1626), which saw God's design of Creation as a challenge to the ingenuity of godly men, to understand and master it.

Much of Carlowitz's own forestry work was related to the supply of timber and fuel to the mining and metallurgical industries of the Kingdom of Saxony, and this is the context in which he developed his thinking on "sustainable" practice. He was a state employee, and his work is couched as an appeal to the ruler of the state who was interested in both the welfare of his people and increased tax revenue. The development of the metallurgical industries was seen as the best way to compete with other commercial centers in Europe and reduce dependence on imports. The task of forestry was to keep up the supply of wood to enable commercial expansion, to permit a "continual and sustainable use" of timber (as the very first use of the term "sustainable" reads); lassitude in this matter would lead to poverty. Thus the fate of society was linked to the principles that economic growth required an enhanced supply of resources, and that those resources (trees) should be replaced when consumed.

After Carlowitz, writing on forestry, which previously had usually been subsumed within texts on agriculture, rapidly developed its own disciplinary literature, most prominently within the numerous forestry administrations and universities that dotted competing German states in the period before its unification in 1871. Prognostications of imminent wood shortage and economic disadvantage remained the central justification for the development of a more "scientific" forestry that would supplant wasteful or low-yielding traditional practices. The framework of producing a "sustainable" supply of which Carlowitz provided such a systematic and detailed account increasingly became a route by which the Enlightenment fascination with geometry and mathematics was employed to predict future wood supplies and organize the forest in such a way that the harvesting and regrowth of wood, along with costs and revenue, could be, in theory at least, planned and predicted. Such techniques were not at all present in Carlowitz's writing but by the end of the eighteenth century had become central to the education of foresters, who also developed a preference for monocultures of trees and strong control over woodland space, excluding grazing and other uses. This in turn would provide a blueprint for forestry practice all over the world, nearly always connected to the idea that scientific forestry was a necessity called into existence by imminent shortages, and the savior of society from scarcity. The widespread transformation of woodland space even in Germany did not occur for nearly a century and half after Carlowitz's death, but his legacy has been as enduring and global as German forestry practice itself.

Further Reading

Lowood, H. E. 1990. "The calculating forester: Quantification, cameral science and the emergence of scientific forestry management in Germany." In T. Frangsmyr, J. L. Heilbron, and R. E. Rider (eds.), *The quantifying spirit in the eighteenth century*. Berkeley: University of California Press.

Radkau, Joachim. 2011. *Wood: A history*: Cambridge: Polity Press.

Warde, Paul. 2006. "The fear of wood shortage and the reality of the woodlands in Europe, c. 1450–1850." *History Workshop Journal* 62:29–57.

Warde, Paul. 2011. "The invention of sustainability." *Modern Intellectual History* 8:153–170.

Williams, Michael. 2003. *Deforesting the earth: From prehistory to global crisis*. Chicago: Chicago University Press.

The Coal Question

WILLIAM STANLEY JEVONS

Chapter I: Introduction and Outline

Day by day it becomes more evident that the Coal we happily possess in excellent quality and abundance is the mainspring of modern material civilization. As the source of fire, it is the source at once of mechanical motion and of chemical change. Accordingly it is the chief agent in almost every improvement or discovery in the arts which the present age brings forth. It is to us indispensable for domestic purposes, and it has of late years been found to yield a series of organic substances, which puzzle us by their complexity, please us by their beautiful colours, and serve us by their various utility.

And as the source especially of steam and iron, coal is all powerful. This age has been called the Iron Age, and it is true that iron is the material of most great novelties. By its strength, endurance, and wide range of qualities, this metal is fitted to be the fulcrum and lever of great works, while steam is the motive power. But coal alone can command in sufficient abundance either the iron or the steam; and coal, therefore, commands this age—the Age of Coal.

Coal in truth stands not beside but entirely above all other commodities. It is the material energy of the country—the universal aid—the factor in everything we do.

With coal almost any feat is possible or easy; without it we are thrown back into the laborious poverty of early times.

With such facts familiarly before us, it can be no matter of surprise that year by year we make larger draughts upon a material of such myriad qualities—of such miraculous powers. But it is at the same time impossible that men of foresight should not turn to compare with some anxiety the masses yearly drawn with the quantities known or supposed to lie within these islands.

W. S. Jevons. (1865) 1866. Chapters 1 and 2 in *The coal question: An enquiry concerning the progress of the nation, and the probable exhaustion of our coal-mines*, 2nd ed., 1–6, 15–17, 28–35. London: Macmillan.

Geologists of eminence, acquainted with the contents of our strata, and accustomed, in the study of their great science, to look over long periods of time with judgment and enlightenment, were long ago painfully struck by the essentially limited nature of our main wealth. And though others have been found to reassure the public, roundly asserting that all anticipations of exhaustion are groundless and absurd, and "may be deferred for an indefinite period," yet misgivings have constantly recurred to those really examining the question. Not long since the subject acquired new weight when prominently brought forward by Sir W. Armstrong in his Address to the British Association, at Newcastle, the very birthplace of the coal trade.

This question concerning the duration of our present cheap supplies of coal cannot but excite deep interest and anxiety wherever or whenever it is mentioned: for a little reflection will show that coal is almost the sole necessary basis of our material power, and is that, consequently, which gives efficiency to our moral and intellectual capabilities. England's manufacturing and commercial greatness, at least, is at stake in this question, nor can we be sure that material decay may not involve us in moral and intellectual retrogression. And as there is no part of the civilized world where the life of our true and beneficent Commonwealth can be a matter of indifference, so, above all, to an Englishman who knows the grand and steadfast course his country has pursued to its present point, its future must be a matter of almost personal solicitude and affection.

The thoughtless and selfish, indeed, who fear any interference with the enjoyment of the present, will be apt to stigmatise all reasoning about the future as absurd and chimerical. But the opinions of such are closely guided by their wishes. It is true that at the best we see dimly into the future, but those who acknowledge their duty to posterity will feel impelled to use their foresight upon what facts and guiding principles we do possess. Though many data are at present wanting or doubtful, our conclusions may be rendered so far probable as to lead to further inquiries upon a subject of such overwhelming importance. And we ought not at least to delay dispersing a set of plausible fallacies about the economy of fuel, and the discovery of substitutes for coal, which at present obscure the critical nature of the question, and are eagerly passed about among those who like to believe that we have an indefinite period of prosperity before us.

The writers who have hitherto discussed this question, being chiefly geologists, have of necessity treated it casually, and in a one-sided manner. There are several reasons why it should now receive fuller consideration. In the first place, the accomplishment of a Free Trade policy, the repeal of many laws that tended to restrain our industrial progress, and the very unusual clause in the French Treaty which secures a free export of coals for some years to come, are all events tending to an indefinite increase of the consumption of coal. On the other hand, two most useful systems of Government inquiry have lately furnished us with new and accurate information bearing upon the question; the Geological Survey now gives some degree of certainty to our estimates of the coal existing within our reach, while the returns of mineral statistics inform us very exactly of the amount of coal consumed.

Taking advantage of such information, I venture to try and shape out a first rough approximation to the probable progress of our industry and consumption of coal in a system of free industry. We of course deal only with what is probable. It is the duty of a careful writer not to reject facts or circumstances because they are only probable, but to state everything with its due weight of probability. It will be my foremost desire to discriminate certainty and doubt, knowledge and igno-rance—to state those data we want, as well as those we have. But I must also draw attention to principles governing this subject, which have rather the certainty of natural laws than the fickleness of statistical numbers.

Chapter II: Opinions of Previous Writers

One of the earliest writers who conceived it was possible to exhaust our coal mines was John Williams, a mineral surveyor. In his "Natural History of the Min-eral Kingdom," first published in 1789, he gave a chapter to the consideration of *"The Limited Quantity of Coal of Britain."* His remarks are highly intelligent, and prove him to be one of the first to appreciate the value of coal, and to foresee the consequences which must some time result from its failure. This event he rather prematurely apprehended; but in those days, when no statistics had been collected, and a geological map was unthought of, accurate notions were not to be expected. Still, his views on this subject may be read with profit, even at the present day.

Sir John Sinclair, in his great Statistical Account of Scotland, took a most enlightened view of the importance of coal; and, in noticing the Fifeshire coal-field, expressed considerable fears as to a future exhaustion of our mines. He correctly contrasted the fixed extent of a coalfield with the ever-growing nature of the consumption of coal.

In 1812 Robert Bald, another Scotch writer, in his very intelligent "General View of the Coal Trade of Scotland," showed most clearly how surely and rapidly a consumption, growing in a "quick, increasing series," must overcome a fixed store, however large. Even if the Grampian mountains, he said, were composed of coal, we would ultimately bring down their summits, and make them level with the vales.

In later years, the esteemed geologist, Dr. Buckland, most prominently and earnestly brought this subject before the public, both in his evidence before the

Parliamentary Committees of 1830 and 1835, and in his celebrated "Bridgewater Treatise." On every suitable occasion he implored the country to allow no waste of an article so invaluable as coal.

Many geologists, and other writers, without fully comprehending the subject, have made so-called estimates of the duration of the Newcastle coal-field. Half a century ago, this field was so much the most important and well known, that it took the whole attention of English writers. The great fields of South Wales and Scotland, in fact, were scarcely opened. But those who did not dream of the whole coal-fields of Great Britain being capable of exhaustion, were early struck by the progressive failure of the celebrated Newcastle seams. Those concerned in the coal trade know for how many years each colliery is considered good; and perhaps, like George Stephenson in early youth, have had their homes more than once moved and broken up by the working out of a colliery. It is not possible for such men to shut their eyes altogether to the facts.

. . .

We ought, therefore, to compare the total supply within the kingdom with the total probable demand, paying little or no regard to local circumstances.

Mr. Hull has made such a comparison. He compared the 79,843 millions of tons of his first estimate with the 72 million tons of coal consumed in 1859, and deduced that, at the same rate of consumption, the supply would last 1100 years.

"Yet we have no right," he very truly remarked, "to assume that such will be the actual duration; for the history of coal mining during the last half century has been one of rapid advance." Our consumption, in short, had about doubled itself since 1840; and, supposing it to continue doubling every twenty years, our "total available supply would be exhausted before the lapse of the year 2034."

"If we had reason," he continues, "to expect that the increase of future years was to progress in the same ratio, we might well tremble for the result; for that would be nothing less than the utter exhaustion of our coal-fields, with its concomitant influence upon our population, our commerce, and national prosperity, in the short period of 172 years!"

No sooner has Mr. Hull reached this truly alarming result than he recoils from it. "But are we," he says, "really to expect so rapid a drain in future years? I think not." Economy will reduce our consumption; the burning waste-heaps of coal will be stopped; America will relieve us from the world-wide demand for our coal, and will eventually furnish even this country with as much as we want. Such are some of the fallacious notions with which Mr. Hull, in common with many others, seeks to avoid an unwelcome conclusion. More lately, he has said: "Notwithstanding these facts, however, it would be rash to assume that the experience of the past is to be a criterion of the future. We neither wish for, nor expect, an increase during the remainder of the second half of this century, at all proportionate to that of the earlier half; and this view is borne out by some of the later returns. Some of our coal-fields, as has been shown, have passed their meridian, and, having expended their strength, are verging to decay. Others have attained their maximum, or nearly so; this, indeed, is the case with the majority. The younger coal-fields will have much of their strength absorbed in compensating for the falling off of the older; so that, in a few years, the whole of our coal-producing districts will reach a stage of activity beyond which they cannot advance, but around which they may oscillate. Entertaining these views, I am inclined to place the possible maximum of production at 100 millions of tons a year; and yet it has been shown that, even with this enormous 'output,' there is enough coal to last for eight centuries."

The reader will easily see, in the course of our inquiry, how mistaken is Mr. Hull, in supposing our production of coal to be limited to 100 millions. It has already exceeded 92 millions without counting the waste of slack coal, and is yet advancing by great strides. And the public seems unaware that *a sudden check to the expansion of our supply would be the very manifestation of exhaustion we dread*. It would at once bring on us the rising price, the transference of industry, and the general reverse of prosperity, which we may hope not to witness in our days. And the eight centuries of stationary existence he promises us would be little set off against a nearer prospect so critical and alarming.

Facts, however, prove the hastiness of these views. The number of collieries is rapidly increasing up to the very last accounts (1864); and new collieries being mostly larger works than the old ones laid in, we may conclude that coal owners are confident of pushing the production for many years to come.

The remarks of Sir W. Armstrong on this subject, in his Address to the British Association at Newcastle, in 1863, are so excellent that I quote them at length:—
"The phase of the earth's existence, suitable for the extensive formation of coal, appears to have passed away for ever; but the quantity of that invaluable mineral which has been stored up throughout the globe for our benefit is sufficient (if used discreetly) to serve the purposes of the human race for many thousands of years. In fact, the entire quantity of coal may be considered as practically inexhaustible.

"Turning, however, to our own particular country, and contemplating the rate at which we are expending those seams of coal which yield the best quality of fuel and can be worked at the least expense, we shall find much cause for anxiety. The greatness of England much depends upon the superiority of her coal, in cheapness and quality, over that of other nations; but we have already drawn, from our choicest mines, a far larger quantity of coal than has been raised in all other parts of the world put together; and the time is not remote when we shall have to encounter the disadvantages of increased cost of working and diminished value of produce.

"Estimates have been made at various periods of the time which would be required to produce complete exhaustion of all the accessible coal in the British Islands. The estimates are certainly discordant; but the discrepancies arise, not from any important disagreement as to the available quantity of coal, but from the enormous difference in the rate of consumption at the various dates when the estimates were made, and also from the different views which have been entertained as to the probable increase of consumption in future years. The quantity of coal yearly worked from British mines has been almost trebled during the last twenty years, and has probably increased tenfold since the commencement of the present century; but as this increase has taken place pending the introduction of steam navigation and railway transit, and under exceptional conditions of manufacturing development, it would be too much to assume that it will continue to advance with equal rapidity.

"The statistics collected by Mr. Hunt, of the Mining Record Office, show that, at the end of 1861, the quantity of coal raised in the United Kingdom had reached the enormous total of 86 millions of tons, and that *the average annual increase in the eight* preceding years amounted to 2¾ millions of tons.

"Let us inquire, then, what will be the duration of our coal-fields if this more moderate rate of increase be maintained. By combining the known thickness of the various workable seams of coal, and computing the area of the surface under which they lie, it is easy to arrive at an estimate of the total quantity comprised in our coal-bearing strata. Assuming 4,000 feet as the greatest depth at which it will ever be possible to carry on mining operations, and rejecting all seams of less than two feet in thickness, the entire quantity of available coal existing in these islands has been calculated to amount to about 80,000 millions of tons, which, at the present rate of consumption, would be exhausted in 930 years; but with a continued yearly increase of 2¾ millions of tons would only last 212 years.

"It is clear that, long before complete exhaustion takes place, England will have ceased to be a coal-producing country on an extensive scale. Other nations, and especially the United States of America, which possess coal-fields thirty-seven times more extensive than ours, will then be working more accessible beds at a smaller cost, and will be able to displace the English coal from every market. The question is, not how long our coal will endure before absolute exhaustion is effected, but how long will those particular coal-seams last which yield coal of a quality and at a price to enable this country to maintain her present supremacy in manufacturing industry. So far as this particular district is concerned, it is generally admitted that 200 years will be sufficient to exhaust the principal seams, even at the present rate of working. If the production should continue to increase as it is now doing, the duration of those seams will not reach half that period. How the case may stand in other coal mining districts, I have not the means of ascertaining; but, as the best and most accessible coal will always be worked in preference to any other, I fear the same rapid exhaustion of our most valuable seams is everywhere taking place."

With almost every part of this statement I can concur, except the calculation by a fixed annual increase of consumption, which I shall show to be contrary to the principles of the subject, and not to reach the whole truth.

Commentary
W. S. Jevons, *The Coal Question* (1865)
PAUL WARDE

William Stanley Jevons wrote and published *The Coal Question* to thunderous
acclaim in 1865, only the thirtieth year of his life. John Stuart Mill, the towering
political economist of the day, and himself later a frequent target of criticism
from Jevons, was prompted to raise questions in Parliament at the grim scenario
it outlined. Britain was at that time unquestionably the economic powerhouse of
the world, and it was generally accepted that economic supremacy was founded
on a combination of the best technology, above all steam power, with immense
reserves of cheap coal to fuel it. But coal was a finite, nonrenewable stock. How
much was left, asked Jevons, and what would happen when supplies began to
dwindle? Could economic growth be sustained at all? The young man who had
completed his university studies only two years earlier found himself discussing
his work with Prime Minister William Gladstone in Downing Street, and only a
matter of months later, Parliament was prompted to set up a commission to examine the future of coal use in Britain. This body assembled the greatest ever survey
of coal consumption and mineral reserves in a report delivered in 1871. The conclusion was that Jevons's predictions were too pessimistic, but the "coal question"
has remained. How long will the sources of national prosperity endure?

Yet Jevons was not doing anything novel. At the beginning of the book, he
provides a sober, chronological literature review of previous estimates of how
long coal reserves would last, all the way back to the writing of Welsh geologist
John Williams in 1789. It is equally apparent that Jevons was indebted to the
population theory of Thomas Malthus. So what was new? Previous writers had attempted to assess the size of national coal stocks and set this limit against annual
consumption, and quibbled with each other's figures. Jevons was, in contrast,
sensitive to the dynamic interrelationships that governed both supply and demand, understanding above all how what we would call *marginal* rates of change
interacted and changed over time. Both the rate of coal consumption, and the
rate of coal extraction, could be expected to vary in a manner mediated by price.
He also argued for what we now call the "Jevons Paradox": that improving the
efficiency of fuel consumption by steam engines and fireplaces would not reduce
overall coal consumption, but have quite the opposite effect by reducing the real
price of the *benefits* or *services* from energy use and stimulating demand for them.
No less important, perhaps, he was a fine, urgent writer.

The Coal Question thus made clear that the future of coal use was determined by several mutually interacting trends, raising costs as mining went deeper,

reducing demand by efficiency gains, expanding demand as the population and industry grew. Although for simplicity Jevons tended to use constant rates of expansion of demand, his overall schema was a departure from the Malthusian model that had tended to see its key variables as fixed, above all in the "constancy of the passion between the sexes."

The work was written in a period of extraordinary intellectual fecundity. Born in 1835 to a family of wealthy Liverpool iron merchants, his mother coming from a family of financiers and enthusiastic art collectors, he had undertaken intermittent university studies largely focused on chemistry and mathematics. He would go on to publish authoritative works in a very broad range of fields, embracing political economy (which he would help turn into "economics"), meteorology, logic, chemistry, and even physiology. He enthusiastically conducted scientific experimentation and constructed machines, and stood as a national figure of renown and professor of political economy when he drowned in a swimming accident in 1882.

Economics for Jevons was based on logic, and the essence of his argument that British prosperity would be undermined by the coming scarcity of coal reserves was logical. Although he calculated coal consumption and likely trends in demand for coal from a rising population and industry, in the end Jevons thought this data immaterial for the basic accuracy of his argument. The procedures of economics were, for him, the same as in any of the other sciences he wrote on, the logical working through of hypotheses. One could obtain "Exact and useful knowledge of the universe" through number, and any phenomena that could be conceived of in terms of more or less, of quantity, could usefully be grasped by mathematics. But this did not make for an *exact science*, which was a misnomer in all cases. Science could not replicate the super-complexity of the universe, and could proceed only through hypothesis testing that gave useful approximations.

If reality was super-complex, how could it be predicted? Jevons's view was that the universe was entirely mechanistic, but that we are incapable of grasping the full functioning of the mechanism. Nevertheless, in most cases it did not matter that we could not grasp or measure most causes acting on a phenomenon. As with much of his thinking Jevons drew directly on mechanics, and especially the balance, an instrument that had fascinated him as a child and that provided him with his livelihood during the late 1850s when he worked as a gold assayer for the Australian mint, interrupting his university studies to support his wider family that had gone bankrupt in the mid-century railway boom and bust. The key to his balancing act was that one did not need to know anything about the total forces acting on either side of a balance to produce equilibrium, but only the marginal change that brought everything to a level. To comprehend a phenomena one did not have to address the totality of forces, but only that which made the crucial difference. A procedure was apparently opened up by which long-term trends could be rigorously analyzed.

It was logic and mechanical reasoning that in the end permitted prediction: "to reap where we have never sown." Yet this still relied on history, and the historical data that Jevons collected assiduously. "All predictions . . . proceed on the assumption that new events will conform to the conditions detected in our observations of past events." To make comparison of related trends possible, Jevons had also pioneered the use of index numbers in his earliest acclaimed work, on the influence of the gold rush on prices. But the trends that emerged from such data simply assisted in assigning a date to a likely occurrence. It was the logical framework that implied the possibility of prediction, and that already determined outcome. The system was closed but timing was open.

Although sometimes treated as a nineteenth-century equivalent of "peak oil," the primary concern expressed in *The Coal Question* was not the exhaustion of coal leading to economic collapse (although viewed as a possibility) but the surrendering of economic supremacy to the United States because of its greater mineral reserves. He thought that coal consumption was likely to rise greatly for a considerable period of time—a trend it turns out that he underestimated—and he refused to speculate on technological advances that we have seen lead to oil and varied forms of electricity generation. The logic was inescapable and the fate set. His policy recommendation was that current income should be used to pay off the national debt, to ensure that future generations were not encumbered with that burden as well as the disappearance of the original source of their wealth.

William Stanley Jevons's legacy was broad and varied, but above all rested on the possibility of a unified method of scientific endeavor, by which "history" and thus the future could be envisaged and studied as an interacting set of functions. It was not necessary to grasp all the particularities of any given event to fit it into an explanatory narrative, and indeed events in isolation had significance insofar that they related to trends. The complexity of history and economics did not demand a focus only on particulars, but was analogous to problems faced in mechanics and no obstacle to developing a new science of mathematical principles and applied statistics.

Further Reading

Great Britain. Coal Commission. *Report of the Commissioners appointed to inquire into the several matters relating to coal in the United Kingdom*, Vol. 3. London: Printed by G. E. Eyre and W. Spottiswoode for Her Majesty's Stationery Office, 1871.

Maas, Harro. 2005. *William Stanley Jevons and the making of modern economics*. Cambridge: Cambridge University Press.

Madureira, Nuno Luis. 2012. "The anxiety of abundance: William Stanley Jevons and coal scarcity in the nineteenth century." *Environment and History* 18 (3): 395–421.

Shabbas, Margaret. 1990. *A world ruled by number: William Stanley Jevons and the rise of mathematical economics*. Princeton, NJ: Princeton University Press.

Possible Limits of Raw-Material Consumption

SAMUEL H. ORDWAY JR.

Whether there may be any unbreakable upper limits to the continuing growth of our economy, we do not pretend to know, but it must be part of our task to examine as present themselves.

—President's Materials Policy Commission, *Resources for Freedom* (1952)

This paper does not seek to analyze the availability, present or prospective, of particular raw materials. This already has been done effectively by the President's Materials Policy Commission, and some of its findings, summarized below, constitute basic information from which the arguments herein stem. This information leads to presentation of a theory of the limit of growth. The possibility of avoiding, in the United States, arrival at such a limit by imports and technological discovery is minimized. The remainder of the paper is addressed to possible ways by which our prosperity may be sustained. It is suggested that, for the next few decades at least, there will be increased prosperity and plenty. In the same period there will continue to be decreases in the hours men have to work, accompanied by an abundance of leisure for many. These factors of plenty and leisure can lead to individual and group activity which can develop new national awareness and discipline and a needed ethic for an age of conservation.

Samuel H. Ordway Jr. 1956. "Possible limits of raw-material consumption." In *Man's role in changing the face of the Earth*, W. L. Thomas (ed.), 987–992. Chicago: University of Chicago Press.

Basic Information

British Political and Economic Planning (PEP), in a draft report dated January, 1955, excerpted and summarized relevant findings from the President's Materials Policy Commission's report (1952) most succinctly as follows:

There is a Materials problem of considerable severity affecting the United States and the industrialized nations of Western Europe. Unless the problem is effectively met, the long range security and economic growth of this and other free nations will be seriously impaired. The Commission's report is primarily concerned with the United States problem, which cannot, however, be isolated from the rest of the free world problem.

The basic reason for the problem is soaring demand. This country took out of the ground two-and-a-half times more bituminous coal in 1950 than in 1900; three times more copper, four times more zinc, thirty times more crude oil. *The quantity of most metals and mineral fuels used in the United States since the first World War exceeds the total used throughout the entire world in all of history preceding 1914.* Although almost all materials are in heavily increasing demand, the hard core of the materials problem is minerals.

In 1950, the United States consumed 2.7 billion tons of materials of all kinds—metallic ores, non-metallic minerals, agricultural materials, construction materials, and fuels—or about 36,000 pounds for every man, woman and child in the country. With less than 10 per cent of the free world population, and only 8 per cent of its area, the United States consumed more than half of 1950's supply of such fundamental materials as petroleum, rubber, iron ore, manganese and zinc.

War would alter the patterns of materials demand and supply in swift and drastic ways; yet if a permanent peace should prevail, and *all the nations of the world should acquire the same standard of living as our own, the resulting world need for materials would be six times present consumption.* In considering materials at long range, therefore, we have roughly the same problems to face and actions to pursue, war or no war.

For the last hundred years, the United States' total output of all goods and services (the Gross National Product, or GNP) has increased at the average rate of 3 per cent a year, compounded. Such a rate means an approximate doubling every twenty-five years (which would mean a nineteen-fold increase in a full century). As of 1950, the GNP was approximately $283 billion. In considering the next quarter century the Commission has made no assumption more radical than that the GNP will continue to increase at the same 3 per cent rate compounded every year, which is the average of the last century, all booms and depressions included. This would mean a GNP in the middle of the 1970's of about $566 billion, measured in dollars of 1950 purchasing power. The Commission has also assumed, after consultation with the Bureau of the Census, that population will increase to 193 million by 1975, and the working force to 82 million, compared to the 1950 figures of 151 million and 62 million. It has also assumed a shortening working week, but that man-hour productivity will continue to rise somewhat more than in the recent past. But even these conservative assumptions bring the United States up against some very hard problems of maintaining materials supply, for natural resources, whatever else they may be doing, are not expanding at compound rates.

Absolute shortages are not the threat in the materials problem. We need not expect we will some day wake up to discover we have run out of materials and that economic activity has come to an end. *The threat of the materials problem lies in insidiously rising costs* which can undermine our rising standard of living, impair the dynamic quality of American capitalism, and weaken the economic foundations of national security. These costs are not just dollar costs, but what economists refer to as *real costs*—meaning the hours of human work and the amounts of capital required to bring a pound of industrial material or a unit of energy into useful form. Over most of the 20th century these real costs of materials have been declining, and this decline has helped our living standards to rise. But there is now reason to suspect that this decline has been slowed, that in some cases it has been stopped, and in others reversed. The central challenge of the materials problem is therefore to meet our expanding demands with expanding supplies while averting a rise in real costs per unit.

In materials, there is always a tendency for real costs to rise because invariably people use their richest resources first and turn to the leaner supplies only when they have to. What is of concern today is that the combination of soaring demand and shrinking resources creates a set of upwards costs pressures much more difficult to overcome than any in the past. In the United States there are no longer large mineral deposits in the West waiting to be stumbled upon and scooped up with picks and shovels; nor are there any longer vast forest tracts to be discovered. We can always scratch harder and harder for materials, but declining or even lagging productivity in the raw materials industries will rob economic gains made elsewhere. The ailment of rising real costs is all the more serious because it does not give dramatic warning of its onset; it creeps upon its victim so slowly that it is hard to tell when the attack began.

In recent years, the general inflation has struck with special force at many materials, causing their price to rise more than the price structure as a whole. Some materials prices are high today because demand has temporarily outrun supply; here we can expect the situation to adjust itself. But in other cases the problem is more enduring than this, and reflects a basic change of supply conditions and costs. It would be wishful, for example, to expect lumber prices to settle back to their pre-1940 price relationships. We are running up against a physical limitation in the supply of timber, set by the size and growth rates of our forests, and cost relief through easy expansion is not to be expected. For such metals as copper, lead and zinc, United States discovery is falling in relation to demand, and prices reflect the increasing pressure against limited resources.

Although the GNP can be expected to double between now and 1975, the total materials input necessary for this will not double, but perhaps rise only 50 to 60 per cent. Demand for materials will rise most unevenly, sometimes increasing one-quarter or less, sometimes rising fourfold or more. Among the major classifications something like this might be expected in the United States (1975 compared to 1950):

Demand for minerals as a whole, including metals, fuels and non-metallics, will rise *most*—about 90 per cent, or almost double.

Demand for all agricultural products will rise about 40 per cent.

Demand for industrial water will increase roughly 170 per cent.

Taking these classifications one by one shows wide ranges and various problems within each. There can be plenty of room for argument as to how high demand for this or that will really rise, but the central fact is: *demand for everything can be expected to rise substantially*. These projections, which look high today, may look low tomorrow.

The above figures apply to the United States alone. For the rest of the free world the projections made by the commission are necessarily much rougher, but they suggest that demand in other free nations, building as it will on a smaller base, will be even larger in its percentage increase than United States demand, and that the United States, although its total of materials consumption will increase greatly, will probably consume a somewhat smaller share of the free world's total supply.

Demand for iron, copper, lead and zinc might rise only 40 to 50 per cent over the next quarter century, but other increases might be: fluorspar, threefold; bauxite for aluminium, fourfold; magnesium, eighteen to twenty fold (the largest projected increase for any material).

Industrial water, which used to be had for the taking in most of the country except the arid parts of the West, now shows a growing shortage problem, because modern industry uses it in such vast quantities. About 18 barrels of water are needed in refining a barrel of oil, and more than 250 tons of water must go to make a ton of steel or a ton of sulphate wood pulp. During World War II plans for at least 300 industrial or military establishments had to be abandoned or modified because of inadequate water supplies, and an already serious problem may be much sharper by 1975.

All of the above facts are symptoms of the same condition: the United States is outgrowing its present usable domestic resource base. This condition has been a long time in the making, but it was not until the 1940's that we completed the change from being a raw materials *surplus* nation to being a raw materials *deficit* nation. Whereas at the start of the century we produced some 15 per cent more raw materials than we consumed (excluding food), by mid-century we were consuming 10 per cent more materials than we produced. This is a peacetime situation, and the trend seems firmly established.

With the nation facing such situations the Commission draws a sharp distinction between being alarmist, which it is not, and seriously concerned, which it is. The nation cannot assume that "everything will be all right if we just leave things alone"; the forces causing the materials problem will increase, not diminish. We must become conscious of the existence of the materials problem and guide ourselves by its seriousness in every way possible [Political and Economic Planning 1955, 2–6].

Theory of the Limit of Growth

It is our opinion that the President's Materials Policy Commission, while concerned, is not sufficiently concerned by the implications of its findings. It is obvious that we shall not come to the end of any of our raw materials suddenly and without warning, but it is submitted that continuing consumption each year of more raw materials essential to industrial expansion than the earth and man together re-create will bring us some day to a limit of growth. We have been living, and we are still living, on resource capital as well as income to make possible continuing industrial expansion and higher levels of living for ever more people.

This part of the thesis has been developed by the author of this paper in the following terms:

None of us can doubt that population densities have sharply increased in many parts of the world, that many people are undernourished and that there are today growing shortages of strategic and necessary materials in some places much of the time. None of us can doubt that technology has found new sources of raw material and new products to substitute for old, to our great betterment as well as to our woe. No one seems to have examined carefully enough the causes responsible for our current fantastic consumption of raw materials—causes which are more significant than population growth or preparation for war. No one has assumed to analyze the basic philosophy, indeed religion, of modern man, that makes us what we are: a race working, struggling, inventing, fighting, living *to create an ever higher level of living for all mankind*. That is our great inspiration, our almost universal goal, and it may turn out to be our great illusion.

This aspiration for an ever higher level of living has become the obsession of mankind. It is an expression of the democratic aim towards greater equality; it is the dictator's justification for a five-year plan; it is a tangible fulfilment of the spiritual aim of the church to better the poor as well as the rich; it becomes the internationalist's formula for relieving the pressures which "have not" people exert on wealthier nations, and a major means of preventing war. If we keep on raising the standard of living, want will be satisfied. To most of us today an ever higher level of living is the very meaning of human progress.

To laymen, human progress must have tangible expression. It means more and better food, clothing, and housing, better health and longer life, greater leisure (often confused with the idea of freedom) and more security, accompanied, of course, by less physical effort.

In the United States we have steadily moved to attain these things. Despite occasional setbacks and depressions, production has increased miraculously in all areas almost year by year; wages have risen, working hours decreased, investment and income and corporate profits have mounted. Industry has been able to plow back millions and millions of dollars into new plants and new equipment. This is expansion, economic growth, the realization of a dream.

Economists substantiate these obvious evidences of economic growth. Industrialists boast of industrial expansion. Despite Malthus and the fact of population growth, our aim, our goal, our aspiration, our way of life is being fulfilled. We are achieving an ever higher level of living.

For this end, and this ideology, Americans have worked ever harder than they have worked for any other article of faith. Our modern religion is growth. But at what cost, material and spiritual, to the nation? The President's Materials Policy Commission affirms, as right, our American faith in the principle of growth, because "it seems preferable to any opposite, which to us implies stagnation and decay" [Ordway 1953, 5–8].

An ever higher level of living for an ever expanding population has been brought about primarily by increased industrial and agricultural production. Expanding industry has converted raw materials into more useful products. It has produced in the last thirty years the most remarkable tools that man has ever known—tools which help find and extract more raw materials and refine them more economically into time and labor-saving devices. Industry also produces machines and tools and fertilizers which help increase the growth of living things. This enables us now to feed and clothe more people at lower cost. Our laboratories, directly and indirectly supported by industry, have found substitutes

for scarce materials, and new chemicals and drugs that cure ills heretofore in-
curable, and save and prolong life. Industry thus increases wealth, health, and
leisure.

At the same time the growth of industrial production has increased consumption of raw materials out of all proportion to our increase in population—
although it is population growth which continues to arouse concern in neo-
Malthusians. The population of the United States has exactly doubled in the
last fifty years and neo-Malthusians still insist that mankind will eat itself out of
house and home within foreseeable time, particularly if the birth rate not only
here but in underdeveloped countries in the rest of the world follow their current upwards patterns and death rates, due to improved medical care and longer
life, continue to decline. Vogt [1948] says; "It is obvious that fifty years hence the
world cannot support three billion people at any but coolie standards—for most
of them. One third of an acre cannot decently feed a man, let alone clothe him
and make possible control of the hydrologic cycle." It is this threat which the
cornucopian scientists believe new discovery and new technology can offset; but
what has not been discussed at length in the literature, or adequately analyzed in
conservation forums, is the fact that continuing and increasing consumption of
raw materials due to economic expansion is proportionally far greater than the
population rise.

While the number of persons in the United States doubled in fifty years,
the production of all minerals increased 8 times; the consumption of power increased 11 times; the consumption of paper and paperboard increased 14 times
over the same period. The use of some raw materials more and more exceeds
domestic production with the result that we are increasingly dependent upon
foreign imports to supply our higher level of living.

At the same time, despite technological advances in farming, our gross farm
product is not increasing currently as fast as our population. This is a rich land.
Nevertheless, since 1945 our food production has increased 50 per cent less than
our population. And manufacture each year is using more and more organic
products from the land.

While food in the United States is not scarce, many of our inorganic resources are scarce. Thirty-three separate minerals are presently on the critical
list. Millions of dollars a year are being spent to speed the search for new deposits
and to find substitutes (other raw materials) for them. No one doubts that there is
a limit to the supply of these raw materials we are consuming so fast in the earth,
sea, and air. There is doubt as to the extent of undiscovered supplies . . . there
is doubt as to the extent of discoverable substitutes . . . and there is doubt about
the time in which we shall reduce the supply to a point where rising costs will
curtail use. Yet, to the satisfaction of all who have faith in an ever higher level of

living, increasing use of our natural resources goes steadily on. Both industry and population continue to grow. The resource base grows less.

Much of the wealth produced by industry today is reinvested in expansion—to the extent of more than twenty billion dollars per year! One company alone last year paid out dividends of 34 million dollars, but withheld 623 millions of its earnings for new equipment and expansion. Few of us appreciate the extent to which continuing expansion makes further inroads, quantitatively and qualitatively, on the productivity of the earth [Ordway 1953, 10–15].

Increasing pressure on renewable resources also affects adversely productivity of the land itself, unless extraordinary and expensive techniques are used to restore and increase its productivity—techniques which are not known by, and frequently beyond the means of, smaller owners who eke out their living from the earth. One *Yearbook of Agriculture* [1938] stated: "Fifty million acres of farm land have already been abandoned by farmers because they are no longer productive, and 30.000.000 acres more are in the process of abandonment." Year by year, productive acreage per person is declining.

There is little unexplored land left in the United States. As lands are exhausted and abandoned (further increasing the new acreage needed) more and more farmers enter industry. Urban and industrial pressures continue to grow.

Not alone in terms of food, but also in terms of industrial production, shortage of agricultural land is a limiting factor. Approximately one half of the raw material used by business and industry is organic in origin. Automobiles are made of animal and vegetable, as well as of mineral products.

Reports of technicians to the Materials Policy Commission emphasize the growing industrial drain on agricultural products and the fact that the problem is worldwide. They predict a 17 per cent increases in *industrial* consumption of cotton and wool by 1975. They also anticipate a 34 per cent increase in *industrial* requirements for wood, which exceeds the output considered probable at that time by 39 per cent. The shortage of wood predicted for the United States will not be prevented, the report states, by imports from other parts of the world. Even with free trade, and full Soviet participation in meeting estimated 1979 import needs of the present free world, the gap in 1979 between such requirements and available supply would still remain far from closed [Ordway 1953, 21–23].

The theory of the limit of growth is based on two premises:

Levels of human living are constantly rising with mounting use of natural resources.

Despite technological progress we are spending each year more resource capital than is created.

The theory follows: If this cycle continues long enough, basic resources will come into such short supply that rising costs will make their use in additional production unprofitable, industrial expansion will cease, and we shall have reached the limit of growth.

Despite reductions in the prices of many finished products caused by increased production, raw materials—food, wood, water, and minerals—are becoming dearer. The limit of expansion will not be reached until raw materials have become so scarce that the industrial product can no longer be sold at a profit.

This is not a matter of temporary boom and depression, or artificially stimulated high or low prices, or overstocking which commonly causes ups and downs in industrial production—sometimes with drastic temporary effects on the economy. This is a matter of expanding industry, approaching maximum profitable use of its resource base and finally overreaching that maximum at a time when (unless we are prepared) it is too late to alter values voluntarily, willingly abandon the dream of higher levels of living, and peacefully adapt our thinking and our ideals to another very different way of life. This kind of enforced, unexpected reversal of a faith, this end of an expanding industrial civilization, could be the end of the culture we know.

That kind of end may well overtake us, despite all our apparent wealth, unless a new philosophy of conservation, by which we reorient our views of the Good Life and many of our values, becomes generally accepted within a reasonable time [Ordway 1953, 31–35].

Commentary

Samuel H. Ordway Jr., "Possible Limits of Raw-Material Consumption"
(1956)

PAUL WARDE

Samuel Ordway's short essay on the future of the world's resources was prepared for what, with hindsight, has become one of the most famous events in the development of environmental concern during the twentieth century: the symposium *Man's Role in Changing the Face of the Earth* held in Princeton, New Jersey, in the middle of June 1955. The conference was the brainchild of William L. Thomas, director of the Wenner-Gren foundation on anthropological research that provided the funding, who drafted in three extraordinarily influential professors to preside over the mix of academics and figures from policy making and industry: Americans geographer Carl Sauer, entomologist Marston Bates, and polymathic historian, philosopher, and planning theorist Lewis Mumford. Of the seventy-three participants in this early attempt to convene global environmental expertise there was only one woman, Indian plant geneticist Edavaleth Jamaki Ammal. Forty percent of the attendees came from the earth sciences, 28 percent from the biological sciences, 12 percent from the social sciences and humanities, and 20 percent from applied fields such as planning.

The papers gathered together were intended more as a study of process than a definitive audit, and they aimed to provide accessible versions of research already published elsewhere. Ordway's paper was no exception, drawing heavily on his 1953 book *Resources and the American Dream*. After a decade of proliferation of international agencies and the rapid expansion of the United Nations, the ethos behind *Man's Role* was that problems were global and had to be studied in an interdisciplinary fashion. Although debate at the event was extensive and reported in some detail, there was no serious attempt at a synthesis, and like some more polemical works of the time, it contained relatively little hard data. "We did not want it to run into shoals of statistics," commented Thomas. The resulting volume was dedicated to American conservationist George Perkins Marsh, who published the pathbreaking *Man and Nature* in 1864, and much of the ethos of the American conservation movement pervaded the contributions: the contributors wanted to explain how humans had shaped their planet in a way that highlighted risks but also expressed optimism at the prospects for more enlightened attitudes being able to avert catastrophic ecological degradation. With perhaps a nod to Cold War balance, Thomas also praised the pioneering work of Russian geographer Alexander Woeikof (1842–1914) and his 1901 essay "De l'influence de l'homme sur la terre."

The "prospects" section edited by Mumford that looked to the future, and where we find Ordway's text, was small, containing only five from dozens of papers. Ordway's work is representative of a widespread concern in the postwar era that very rapid population and economic growth would lead to shortages in metals, energy, and water, and soil erosion. There was little methodological innovation in these studies: the argument that "the essentials of the position are the balance of rates: the rate of production of alternatives or substitutes for each mineral must equal or exceed the rate of consumption of available reserves" was no advance on the work of Jevons in the 1860s. What had changed was the vast expansion of data available on human population, agricultural output, and resource consumption, and the invention of new metrics to marshal that data, such as Simon Kuznets's idea of "Gross Domestic Product" from 1934. Ordway used sources that were themselves compilations of this great wave of measurement, produced by the report of the American President's Materials Policy Commission, *Resources for Freedom* (1952), and the work of a British think-tank, *Political and Economic Planning* (1955). These works reflected the alarm at rates of growth, the fact that in the 1940s the United States had become a net importer of raw materials, and extensive knowledge built up from the experience of wartime planning.

Ordway had trained as a lawyer, been a government official and naval officer, and become a leading conservationist (in the American sense, preoccupied above all with resource supplies). He argued for a "theory of the limit to growth," where the cause of economic stagnation would be "insidiously rising costs" driven by a demand for increasingly scarce resources to satisfy increasingly high expectations of material wealth. In this he opposed those such as William Vogt who stressed Malthusian population problems. In common with official predictions, Ordway thought that the world would be able to adequately supply itself for around sixty years, but with a risk that metals and water could become dangerously scarce. He assumed that scarcity would be reflected in price rises prompting attempts at technical fixes; supply would not suddenly fall off the edge of a cliff. But above all, the increased leisure afforded by a growing economy would allow an ethical change, a collective reorientation toward conservation that would revive the spirit of an older, frontier America that had enjoyed "an ethic for survival," built around its barn-raising, quilt-making, and husking bees.

Although the presentation and papers were relatively simple, we find in the contents of the *Man's Role* symposium, and Ordway's own contribution, much of the terms of debate with which we are still familiar today. Could resource problems be resolved by better management, or did they require a fundamental reorientation of lifestyle? Was it worth pursuing policies that ran counter to the American dream? Is prediction from current trends a "statistical illusion," as key socio-cultural changes are contingent and unknowable (Lewis Mumford imagined here a confident Roman just before the fall predicting an upward trend in the number of public baths over the next few centuries)? There was dissent and argument between techno-optimism and gloomier prognoses. Ordway's argument that the world was living off its "resource capital" is a precursor to attempts to calculate "natural capital" today, and the modern theory of "genuine savings" that could as well be described by his words written in 1955: "The only way to avoid reaching that limit some day is eventually to cease to consume more resources each year than nature and man together create."

Further Reading

Demeny, Paul. 1988. "Demography and the limits to growth." In "Population and resources in western intellectual traditions." *Population and Development Review* 14 (suppl.): 213–244.

Ordway, Samuel. 1953. *Resources and the American dream*. New York: Ronald Press.

Political and Economic Planning. 1955. *World population and resources*. London: Allen & Unwin.

Putnam, P. C. 1953. *Energy in the future*. New York: Van Nostrand.

Raushenbush, Stephen (ed.). 1952. "The future of our natural resources." Special issue, *Annals of the American Academy of Political and Social Science* 281 (May).

The Limits to Growth

DONELLA H. MEADOWS, JORGEN RANDERS, AND

DENNIS L. MEADOWS FOR THE CLUB OF ROME

The overwhelming growth in world population caused by the positive birth-rate loop is a recent phenomenon, a result of mankind's very successful reduction of worldwide mortality. The controlling negative feedback loop has been weakened, allowing the positive loop to operate virtually without constraint. There are only two ways to restore the resulting imbalance. Either the birth rate must be brought down to equal the new, lower death rate, or the death rate must rise again. All of the "natural" constraints to population growth operate in the second way—they raise the death rate. Any society wishing to avoid that result must take deliberate action to control the positive feedback loop—to reduce the birth rate.

In a dynamic model it is a simple matter to counteract runaway positive feedback loops. For the moment let us suspend the requirement of political feasibility and use the model to test the physical, if not the social, implications of limiting population growth. We need only add to the model one more causal loop, connecting the birth rate and the death rate. In other words, we require that the number of babies born each year be equal to the expected number of deaths in the population that year. Thus the positive and negative feedback loops are exactly balanced. As the death rate decreases, because of better food and medical care, the birth rate will decrease simultaneously.

Such a requirement, which is as mathematically simple as it is socially complicated, is for our purposes an experimental device, not necessarily a political recommendation.[1] The result of inserting this policy into the model in 1975 is shown in figure 44.

In figure 44 the positive feedback loop of population growth is effectively balanced, and population remains constant. At first the birth and death rates

Donella H. Meadows, Jorgen Randers, and Dennis L. Meadows for the Club of Rome. 1972. *The limits to growth: A report for the Club of Rome's project on the predicament of mankind*, 158–175. New York: Universe Books.

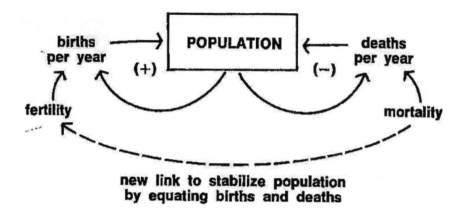

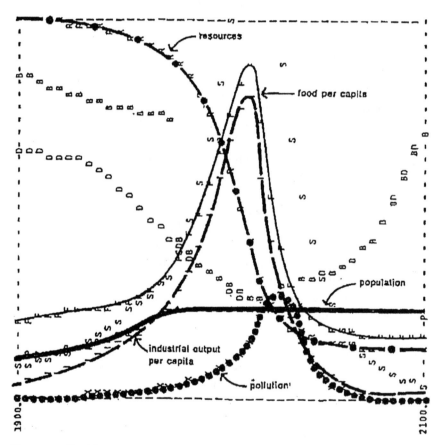

Figure 44. World Model with Stabilized Population. In this computer run conditions in the model system are identical to those in the standard run (figure 35), except that population is held constant after 1975 by equating the birth rate with the death rate. The remaining unrestricted positive feedback loop in the system, involving industrial capital, continues to generate exponential growth of industrial output, food, and services per capita. Eventual depletion of nonrenewable resources brings a sudden collapse of the industrial system.

are low. But there is still one unchecked positive feedback loop operating in the model—the one governing the growth of industrial capital. The gain around that loop increases when population is stabilized, resulting in a very rapid growth of income, food, and services per capita. That growth is soon stopped, however, by depletion of nonrenewable resources. The death rate then rises, but total population does not decline because of our requirement that birth rate equal death rate (clearly unrealistic here).

Apparently, if we want a stable system, it is not desirable to let even one of the two critical positive feedback loops generate uncontrolled growth. Stabilizing population alone is not sufficient to prevent overshoot and collapse; a similar run with constant capital and rising population shows that stabilizing capital alone is also not sufficient. What happens if we bring *both* positive feedback loops under control simultaneously? We can stabilize the capital stock in the model by requiring that the investment rate equal the depreciation rate, with an additional model link exactly analogous to the population-stabilizing one.

The result of stopping population growth in 1975 and industrial capital growth in 1985 with no other changes is shown in figure 45. (Capital was allowed to grow until 1985 to raise slightly the average material standard of living.) In this run the severe overshoot and collapse of figure 44 are prevented. Population and capital reach constant values at a relatively high level of food, industrial output, and services per person. Eventually, however, resource shortages reduce industrial output and the temporarily stable state degenerates.

What model assumptions will give us a combination of a decent living standard with somewhat greater stability than that attained in figure 45? We can improve the model behavior greatly by combining technological changes with value changes that reduce the growth tendencies of the system. Different combinations of such policies give us a series of computer outputs that represent a system with

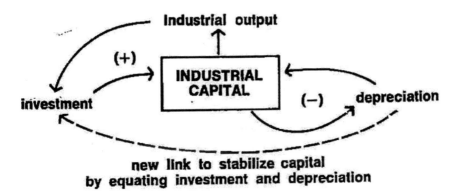

**new link to stabilize capital
by equating investment and depreciation**

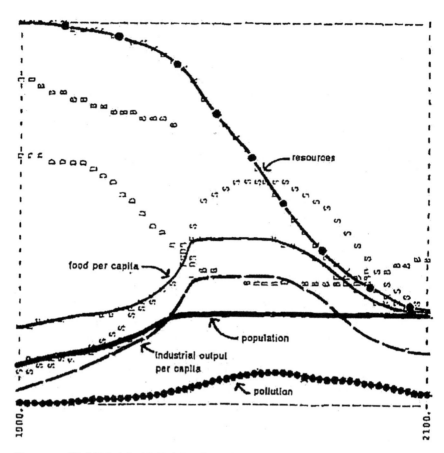

Figure 45. World Model with Stabilized Population and Capital. Restriction of capital growth, by requiring that capital investment equal depreciation, is added to the population stabilization policy of figure 44. Now that exponential growth is halted, a temporary stable state is attained. Levels of population and capital in this state are sufficiently high to deplete resources rapidly, however, since no resource-conserving technologies have been assumed. As the resource base declines, industrial output decreases. Although the capital base is maintained at the same level, efficiency of capital goes down since more capital must be devoted to obtaining resources than to producing usable output.

reasonably high values of industrial output per capita and with long-term stability. One example of such an output is shown in figure 46.

The policies that produced the behavior shown in figure 46 are:

1. Population is stabilized by setting the birth rate equal to the death rate in 1975. Industrial capital is allowed to increase naturally until

1990, after which it, too, is stabilized, by setting the investment rate equal to the depreciation rate.

2. To avoid a nonrenewable resource shortage such as that shown in figure 45, resource consumption per unit of industrial output is reduced to one-fourth of its 1970 value. (This and the following five policies are introduced in 1975.)

3. To further reduce resource depletion and pollution, the economic preferences of society are shifted more toward services such as education and health facilities and less toward factory-produced material goods. (This change is made through the relationship giving "indicated" or "desired" services per capita as a function of rising income.)

4. Pollution generation per unit of industrial and agricultural output is reduced to one-fourth of its 1970 value.

5. Since the above policies alone would result in a rather low value of food per capita, some people would still be malnourished if the traditional inequalities of distribution persist. To avoid this situation, high value is placed on producing sufficient food for *all* people. Capital is therefore diverted to food production even if such an investment would be considered "uneconomic." (This change is carried out through the "indicated" food per capita relationship.)

6. This emphasis on highly capitalized agriculture, while necessary to produce enough food, would lead to rapid soil erosion and depletion of soil fertility, destroying long-term stability in the agricultural sector. Therefore the use of agricultural capital has been altered to make soil enrichment and preservation a high priority. This policy implies, for example, use of capital to compost urban organic wastes and return them to the land (a practice that also reduces pollution.)

7. The drains on industrial capital for higher services and food production and for resource recycling and pollution control under the above six conditions would lead to a low final level of industrial capital stock. To counteract this effect, the average lifetime of industrial capital is increased, implying better design for durability and repair and less discarding because of obsolescence. This policy also tends to reduce resource depletion and pollution.

In figure 46 the stable world population is only slightly larger than the population today. There is more than twice as much food per person as the average value in 1970, and world average lifetime is nearly 70 years. The average industrial output per capita is well above today's level, and services per capita have tripled. Total average income per capita (industrial output, food, and services combined)

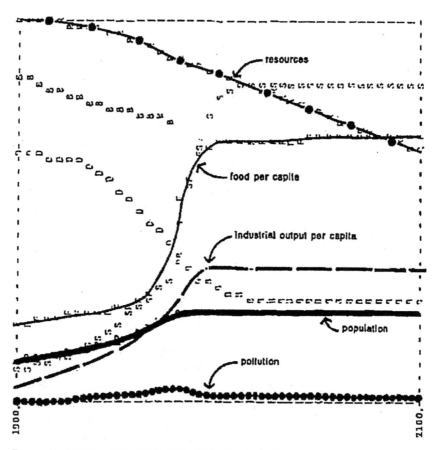

Figure 46. Stabilized World Model I. Technological policies are added to the growth-regulating policies of the previous run to produce an equilibrium state sustainable far into the future. Technological policies include resource recycling, pollution control devices, increased lifetime of all forms of capital, and methods to restore eroded and infertile soil. Value changes include increased emphasis on food and services rather than on industrial production. As in figure 45, births are set equal to deaths and industrial capital investment equal to capital depreciation. Equilibrium value of industrial output per capita is three times the 1970 world average.

is about $1,800. This value is about half the present average US income, equal to the present average European income, and three times the present average world income. Resources are still being gradually depleted, as they must be under any realistic assumption, but the rate of depletion is so slow that there is time for technology and industry to adjust to changes in resource availability.

The numerical constants that characterize this model run are not the only ones that would produce a stable system. Other people or societies might resolve the various trade-offs differently, putting more or less emphasis on services or food or pollution or material income. This example is included merely as an illustration of the levels of population and capital that are *physically maintainable* on the earth, under the most optimistic assumptions. The model cannot tell us how to attain these levels. It can only indicate a set of mutually consistent goals that are attainable.

Now let us go back at least in the general direction of the real world and relax our most unrealistic assumptions — that we can suddenly and absolutely stabilize population and capital. Suppose we retain the last six of the seven policy changes that produced figure 46, but replace the first policy, beginning in 1975, with the following:

1. The population has access to 100 percent effective birth control.
2. The average desired family size is two children.
3. The economic system endeavors to maintain average industrial output per capita at about the 1975 level. Excess industrial capability is employed for producing consumption goods rather than increasing the industrial capital investment rate above the depreciation rate.

The model behavior that results from this change is shown in figure 47. Now the delays in the system allow population to grow much larger than it did in figure 46. As a consequence, material goods, food, and services per capita remain lower than in previous runs (but still higher than they are on a world average today).

We do not suppose that any single one of the policies necessary to attain system stability in the model can or should be suddenly introduced in the world by 1975. A society choosing stability as a goal certainly must approach that goal gradually. It is important to realize, however, that the longer exponential growth is allowed to continue, the fewer possibilities remain for the final stable state. Figure 48 shows the result of waiting until the year 2000 to institute the same policies that were instituted in 1975 in figure 47.

In figure 48 both population and industrial output per capita reach much higher values than in figure 47. As a result pollution builds to a higher level and resources are severely depleted, in spite of the resource-saving policies finally introduced. In fact, during the 25-year delay (from 1975 to 2000) in instituting the stabilizing policies, resource consumption is about equal to the total 125-year consumption from 1975 to 2100 of figure 47.

Many people will think that the changes we have introduced into the model to avoid the growth-and-collapse behavior mode are not only impossible, but un-

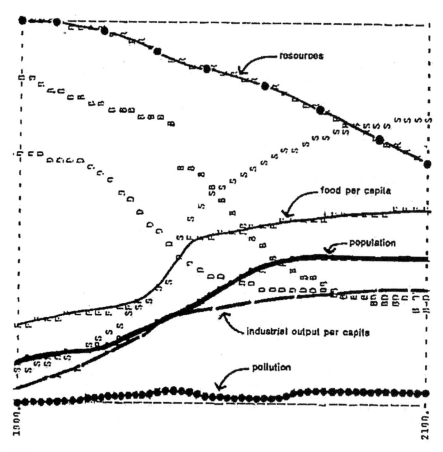

Figure 47. Stabilized World Model II. If the strict restrictions on growth of the previous run are removed, and population and capital are regulated within the natural delays of the system, the equilibrium level of population is higher and the level of industrial output per capita is lower than in figure 46. Here it is assume that perfectly effective birth control and an average desired family size of two children are achieved by 1975. The birth rate only slowly approaches the death rate because of delays inherent in the age structure of the population.

pleasant, dangerous, even disastrous in themselves. Such policies as reducing the birth rate and diverting capital from production of material goods, by whatever means they might be implemented, seem unnatural and unimaginable, because they have not, in most people's experience, been tried, or even seriously suggested. Indeed there would be little point even in discussing such fundamental changes in the functioning of modern society if we felt that the present pattern of unrestricted growth were sustainable into the future. All the evidence available

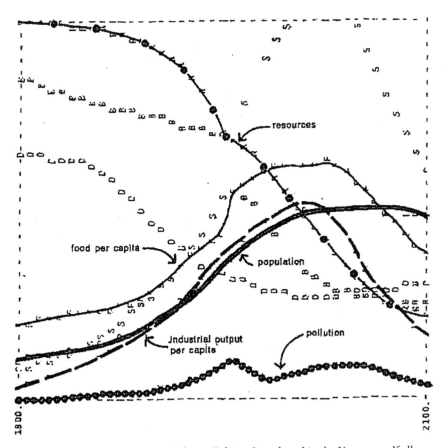

Figure 48. World Model with Stabilizing Policies Introduced in the Year 2000. If all the policies instituted in 1975 in the previous figure are delayed until the year 2000, the equilibrium state is no longer sustainable. Population and industrial capital reach levels high enough to create food and resource shortages before the year 2100.

to us, however, suggests that of the three alternatives—unrestricted growth, a self-imposed limitation to growth, or a nature-imposed limitation to growth—only the last two are actually possible.

Accepting the nature-imposed limits to growth requires no more effort than letting things take their course and waiting to see what will happen. The most probable result of that decision, as we have tried to show here, will be an uncontrollable decrease in population and capital. The real meaning of such a collapse is difficult to imagine because it might take so many different forms. It might occur at different times in different parts of the world, or it might be worldwide. It could be sudden or gradual. If the limit first reached were that of food production, the nonindustrialized countries would suffer the major population decrease. If the first limit were imposed by exhaustion of nonrenewable resources, the industrialized countries would be most affected. It might be that the collapse would leave the earth with its carrying capacity for animal and plant life undiminished, or it might be that the carrying capacity would be reduced or destroyed. Certainly whatever fraction of the human population remained at the end of the process would have very little left with which to build a new society in any form we can now envision.

Achieving a self-imposed limitation to growth would require much effort. It would involve learning to do many things in new ways. It would tax the ingenuity, the flexibility, and the self-discipline of the human race. Bringing a deliberate, controlled end to growth is a tremendous challenge, not easily met. Would the final result be worth the effort? What would humanity gain by such a transition, and what would it lose?

. . .

The Equilibrium State

We are by no means the first people in man's written history to propose some sort of nongrowing state for human society. A number of philosophers, economists, and biologists have discussed such a state and called it by many different names, with as many different meanings.

We have, after much discussion, decided to call the state of constant population and capital, shown in figures 46 and 47, by the term "equilibrium." Equilibrium means a state of balance or equality between opposing forces. In the dynamic terms of the world model, the opposing forces are those causing population and capital stock to increase (high desired family size, low birth control effectiveness, high rate of capital investment) and those causing population and capital stock to decrease (lack of food, pollution, high rate of depreciation or obsolescence). The word "capital" should be understood to mean service, industrial, and agricultural capital combined. *Thus the most basic definition of the state of global equilibrium is that population and capital are essentially stable, with the forces tending to increase or decrease them in a carefully controlled balance.*

There is much room for variation within that definition. We have only specified that the stocks of capital and population remain constant, but they might theoretically be constant at a high level or a low level—or one might be high and the other low. A tank of water can be maintained at a given level with a fast inflow and outflow of water or with a slow trickle in and out. If the flow is fast, the average drop of water will spend less time in the tank than if the flow is slow. Similarly, a stable population of any size can be achieved with either high, equal birth and death rates (short average lifetime) or low, equal birth and death rates (long average lifetime). A stock of capital can be maintained with high investment and depreciation rates or low investment and depreciation rates. Any combination of these possibilities would fit into our basic definition of global equilibrium.

What criteria can be used to choose among the many options available in the equilibrium state? The dynamic interactions in the world system indicate that the first decision that must be made concerns time. *How long should the equilibrium state exist?* If society is only interested in a time span of 6 months or a year, the world model indicates that almost any level of population and capital could be maintained. If the time horizon is extended to 20 or 50 years, the options are greatly reduced, since the rates and levels must be adjusted to ensure that the capital investment rate will not be limited by resource availability during that time span, or that the death rate will not be uncontrollably influenced by pollution or food shortage. The longer a society prefers to maintain the state of equilibrium, the lower the rates and levels must be.

At the limit, of course, no population or capital level can be maintained forever, but that limit is very far away in time if resources are managed wisely and if there is a sufficiently long time horizon in planning. Let us take as a reasonable time horizon the expected lifetime of a child born into the world tomorrow— 70 years if proper food and medical care are supplied. Since most people spend a large part of their time and energy raising children, they might choose as a minimum goal that the society left to those children can be maintained for the full span of the children's lives.

If society's time horizon is as long as 70 years, the permissible population and capital levels may not be too different from those existing today, as indicated by the equilibrium run in figure 47 (which is, of course, only one of several possibilities). The rates would be considerably different from those of today, however. Any society would undoubtedly prefer that the death rate be low rather than high, since a long, healthy life seems to be a universal human desire. To maintain equilibrium with long life expectancy, the birth rate then must also be low. It would be best, too, if the capital investment and depreciation rates were low, because the lower they are, the less resource depletion and pollution there will be. Keeping depletion and pollution to a minimum could either increase the maximum size of the population and capital levels or increase the length of time the equilibrium state could be maintained, depending on which goal the society as a whole has preferred.

By choosing a fairly long time horizon for its existence, and a long average lifetime as a desirable goal, we have now arrived at a minimum set of requirements for the state of global equilibrium. They are:

1. *The capital plant and the population are constant in size.* The birth rate equals the death rate and the capital investment rate equals the depreciation rate.
2. *All input and output rates—births, deaths, investments, and depreciation are kept to a minimum.*
3. *The levels of capital and population and the ratio of the two are set in accordance with the values of the society.* They may be deliberately revised and slowly adjusted as the advance of technology creates new options.

An equilibrium defined in this way does not mean stagnation. Within the first two guidelines above, corporations could expand or fail, local populations could increase or decrease, income could be more or less evenly distributed. Technological advance would permit the services provided by a constant stock of capital to increase slowly. Within the third guideline, any country could change its average standard of living by altering the balance between its population and its capital. Furthermore, a society could adjust to changing internal or external factors by raising or lowering the population or capital stocks, or both, slowly and in a controlled fashion, with a predetermined goal in mind. The three points above define a *dynamic* equilibrium, which need not and probably would not "freeze" the world into the population-capital configuration that happens to exist at the present time. The object in accepting the above three statements is to create freedom for society, not to impose a straitjacket.

What would life be like in such an equilibrium state? Would innovation be stifled? Would society be locked into the patterns of inequality and injustice we see in the world today? Discussion of these questions must proceed on the basis of mental models, for there is no formal model of social conditions in the equilibrium state. No one can predict what sort of institutions mankind might develop under these new conditions. There is, of course, no guarantee that the new society would be much better or even much different from that which exists today. It seems possible, however, that a society released from struggling with the many problems caused by growth may have more energy and ingenuity available for solving other problems. In fact, we believe . . . that the evolution of a society that favors innovation and technological development, a society based on equality and justice, is far more likely to evolve in a state of global equilibrium than it is in the state of growth we are experiencing today.

Note

1. This suggestion for stabilizing population was originally proposed by Kenneth E. Boulding in *The meaning of the 20th century* (New York: Harper and Row, 1964).

Commentary

Donella H. Meadows, Jorgen Randers, and Dennis L. Meadows for the
Club of Rome, *The Limits to Growth* (1972)

MICHAEL EGAN

The Limits to Growth holds an important place both in environmental prediction
and in global environmental history. It constitutes a methodological sea change
in the history of engaging with future natures. Through the use of computer
modeling and the adoption of the then recently conceived field of system dynamics, the book forecasts a variety of economic collapse scenarios should production
and consumption patterns continue to grow exponentially. During a period in
which gloomy predictions of environmental futures were fairly commonplace,
the study—a report produced for the Club of Rome by a team headed by Donella
Meadows—offered a unique approach that they interpreted as a global *prob-
lematique.* For its authors, system dynamics represented a portal through which
society could better understand the origins, significance, and interrelationships
between its myriad components. The group's use of computers to process data
and model futures heralded a new era of large-scale and complex environmental
prediction. *The Limits to Growth* transformed the nature of the debate surrounding the earth's carrying capacity and anticipating global trends in population,
economics, and the environment.

The central message of *The Limits to Growth* was that the human ecological
footprint could not continue to grow at the same pace experienced in the twentieth century. Within the next century, humanity's ecological footprint would
overshoot the Earth's carrying capacity. The book's introduction boldly stated the
group's findings:

1. If the present growth trends in world population, industrialization,
   pollution, food production, and resource depletion continue unchanged, the limits to growth on this planet will be reached some-
   time within the next one hundred years. The most probable result
   will be a rather sudden and uncontrollable decline in both population and industrial capacity.
2. It is possible to alter these growth trends and to establish a condition
   of ecological and economic stability that is sustainable far into the
   future. The state of global equilibrium could be designed so that the
   basic material needs of each person on earth are satisfied and each
   person has an equal opportunity to realize their individual human
   potential.

3. If the world's people decide to strive for this second outcome rather than the first, the sooner they begin working to attain it, the greater will be their chances of success. (Meadows et al. 1972, 23–24)

While *The Limits to Growth* was not unique in its bleak forecast of humanity's environmental future, the amount of data it gathered for the analysis, its international collaborative nature, and its novel approach with computer modeling brought considerable attention to its findings. It was published in thirty-seven languages and sold twelve million copies worldwide.

The Club of Rome was formed in 1968 at the behest of Dr. Aurelio Peccei. Its task was to develop a thorough analysis of "the present and future predicament of man" (ibid., 9). The Club of Rome was not a formal organization, but rather an independent, informal, and international body. It comprised an "invisible college" of experts in policy, economics, and the natural and social sciences. By 1970, it had seventy-five members from twenty-five countries and had identified sixty-six "Continuous Critical Problems" that together constituted a focus for the predicament they sought to resolve. Confronted with a seemingly inchoate list of social, economic, political, and environmental issues—poverty, war, terrorism, pollution, crime, racism, resource depletion, economic instability, drug addiction— American computer engineer Jay Wright Forrester, at a Club meeting in Bern, Switzerland, suggested that *growth* could be a unifying thread among the disparate problems.

Forrester had worked on aircraft flight simulators immediately after World War II and played an important role in developing the Whirlwind digital computer. Forrester was an inveterate problem-solver; by 1956, he left computer development and turned his attention to building the Sloan School of Management at the Massachusetts Institute of Technology (MIT), where he began to think more explicitly about systems. Whereas simple models frequently consisted of closely related causes and effects, more complex models put greater distance and time between cause and effect, making it much more difficult to identify or predict the defining relationships and realize sound solutions to specific problems. Forrester's efforts to demystify the inherent complexity in multifaceted systems led him to devise a mathematical modeling technique, data for which could be fed into early computer systems. In essence, this modeling technique was the heart of system dynamics, a method for understanding the dynamic behavior of complex systems. It used a holistic analysis of the system and the interactions between its elements, rather than a concentration on the workings of its individual components.

Spurred by Forrester's assertions that system dynamics was a useful tool for uncovering the root causes of global problems, the Club of Rome initiated a first phase of the Project on the Predicament of Mankind to delve into resource and environmental problems. The group was headed by Dennis Meadows, a graduate student of Forrester's (Dennis's wife, Donella Meadows, was the lead author of the study). The Project on the Predicament of Mankind set the research parameters for *The Limits to Growth* and was designed to address a series of interrelated global problems—namely, "poverty in the midst of plenty; degradation of the environment; loss of faith in institutions; uncontrolled urban spread; insecurity of employment; alienation of youth; rejection of traditional values; and inflation and other monetary and economic disruptions" (ibid., 10). The first (research) phase—the basis of *The Limits to Growth*—involved a concerted examination of five specific factors that dictate the boundaries of life on the planet, and thereby determine and limit growth: population, agricultural production, natural resources, industrial production, and pollution. Each of these factors was universal, and all possessed interacting social, technical, economic, and political components ideal for analysis in terms of system dynamics.

Building on Forrester's earlier efforts, the MIT team produced the global computer model World3 in order to chart the dynamic implications of growth. Assembling indexes to predict population growth, resource consumption, and pollution, the authors recognized that the rates of change were growing exponentially but at different rates, and that they would need to factor in the implications of their changing interrelationships over time in order to determine long-term trends. At the heart of this problem—and, indeed, at the heart of system dynamics—lies the importance of the feedback loop, or "vicious circle," which is a closed path that connects an action to its disturbance on surrounding conditions. This in turn sets the tone for subsequent action. In this mode of analysis, which allowed for the computation of vastly complex arrays of data, systems dynamics adopted the (almost ecological) notion that the whole system was greater than the sum of its parts. Several decades later, *The Limits to Growth* remains one of the best-known examples of a study employing system dynamics.

Where method and message warned against continued growth, *The Limits to Growth* emphasized equilibrium as a desirable end stage. This served as one of many catalysts at the beginning of the 1970s for the new global movement for environmental sustainability. While sustainability is often linked to the 1972 United Nations Conference on Humans and the Environment in Stockholm—and the subsequent Brundtland Commission report, *Our Common Future*, of 1987, seeds of its genesis are present in *The Limits to Growth*'s concluding observations, reproduced here. In effect, *The Limits to Growth* began a conversation that evolved into a complex debate about the merits of steady-state economics and the biophysical limits of economic growth. In 1972, this was fresh, provocative, and controversial—its dependence on computer modeling, even more so.

Further Reading

Blanchard, E. T. Vieille. 2007. "Croissance ou stabilité? L'entreprise du Club de Rome et le débat autour des modèles." In A. Dahan (ed.), *Les modèles du future*, 19–43. Paris: La Découverte.

Meadows, D., J. Richardson, and G. Bruckmann. 1982. *Groping in the dark: The first decade of global modeling*. Chichester: John Wiley.

Meadows, D. H., J. Randers, and D. L. Meadows. 2004. *The limits to growth: The thirty year update*. London: Routledge.

Simon, J. L., and H. Kahn, eds. 1984. *The resourceful earth*. Oxford: Blackwell.

Smil, V. 2005. "Limits to growth: A review essay." *Population and Development Review* 31(1): 157–164.

PART 3

Geographies

Are Human and Natural Futures Determined or Chosen?

Natural surroundings—"airs, waters, places"—have been used as an explanation for the state of civilizations and cultures since the writings of Hippocrates in classical times. After the great colonial expansion of Europe beginning in the fifteenth century, not only were the new empires confronted with a need to categorize, understand, and find uses for the great number of new plants, animals, and minerals that they encountered, they also wondered what might explain the great variety of cultures scattered around the globe, and why some appeared to be more successful than others. Was the diversity of society and fortune they found something to do with the diversity of geography, and if so, which shaped the other? Their investigations favored the development of systematic knowledge on the global scale fueled by close observation and evidence over divine revelation—the Enlightenment learning that has dominated the whole of the three-hundred-year period considered in this book, but that also drew on ancient sources.

The "progress" or "improvement" of Europe, as it was seen (or increasing "civilization"), had significant effects on nature, most of them considered benign, but with observations of serious decline already in the seventeenth and eighteenth centuries. But equally important, the physical features of nature and climate—in the nineteenth century largely the domain of geographers—also became thought of as a determining factor that over time shaped cultures and civilizations, even "races," a concept in respectable scientific use until the 1940s. Geography, climate, and environment were used to explain the nature of society, but at the same time—rather paradoxically—had prompted Europeans to transcend their environment, to understand it, to tame and control it. Geographers like Griffith Taylor were called on to advise governments, as "experts" on the potential for natural environments to support progressive agendas. Science, progress, and environment thus became intimately linked.

Geographies defined spaces for development until the major shift to the idea of "the environment" in the middle of the twentieth century, which drew on ecological understandings of the biotic and abiotic world together (see Parts 4 and 5), when environment turned from a concept primarily to denote the influence of natural, climatic, and geographical features on human societies into one focusing on the human impacts on nature, now often perceived as malign. The earlier use of the concept was comprehensive in geography, but also in disciplines such as archaeology and anthropology, and with some presence in early sociology (notably the human ecology of the Chicago school in the early decades of the twentieth century). But since the environment was also thought to have a determining influence on individuals, it became influential in the behavioral sciences, health, and medicine as well. "Environmentalism" in older textbooks and encyclopedias therefore commonly denotes this strand of environmental determinism, rather than the later meaning of organized work to protect the environment, which came into wider circulation only from the 1960s.

The explanatory power of geography had its intellectual roots in the political theories of society of French Enlightenment philosopher Montesquieu (1689–1755). Geographers rose to the peak of their influence in the second half of the nineteenth century at the time of the expansion of European imperialism, and their discipline became closely linked both to the idea of progress and to imperialism. Since this kind of progress was considered as having its origins in European civilization, civilization and its "natural context" became the object of intense interest, especially in geography. Under what environmental conditions would modern civilization flourish? Was it possible to measure the level of civilization? Did civilization relate to race, and if so, how? Was European civilization possible everywhere, or were there climatic or environmental constraints? Would it flourish even better in new environmental conditions—for example, in the far north? Was there perhaps a predictable pattern of large-scale history, so that civilizations flourished and declined as a consequence of climatic conditions? Was environment itself cyclical—for example, so that climates changed according to principles that science could reveal?

The papers in this part all deal with manifestations of environmental determinism in the early twentieth century, thus providing a fundamental piece of the background thinking about physical geographies elevated to the role of historical actor affecting the fate of humanity and the future of the planet, at a time when European imperialism was in decline and new understandings of "the global" were emerging. Overall they provide determinism with a spatial, even geopolitical, dimension.

Ellsworth Huntington and Vilhjalmur Stefansson were progressive optimists, and (Thomas) Griffith Taylor a pessimist, but all were commenting on the expansion of empires, or perhaps even more important, on parts of former empires in a period when the empires had started to break down into large but independent parts that were looking for their own destiny in the world—in particular, the United States, Australia, and Canada. Of Britain's superiority there had never been any doubt, as the workshop of the world with an empire where the sun never set. In the distant stretches of the former imperial states, economic geography was an absolutely key discipline with the purpose of finding out the precise conditions of progress in the home country, an enterprise predicated on the idea that the roots of success were in the natural resources of the territory. Progress also demanded much of the people: their hardiness and ingenuity, again seen as factors primarily of race and breed rather than a product of, for example, education. Thus Griffith Taylor would argue that the only way to make the hot heart of Australia productive was to use non-British immigrants, adapted by racial characteristics to heat and the climate of the north and inland.

This line of environmental determinism was decisively, even fanatically predictive: knowing the physical, racial, and economic geographies was the same as knowing the future prowess and productivity of the country, thus limiting choice. This was both an important and slightly ominous knowledge, lending a particular kind of prophetic expertise to its practitioners, who shrouded themselves in an aura of quantitative language and visuals that would shine as beacons illuminating lofty statements about what was good, what was ahead, and what would become the future and who would lead it.

The paradox was that key concepts were shifting even as deterministic ideas were voiced; population was seen as an *impact* rather than as a resource, for example, and the exploitation of natural resources and burning of fossil fuels had become a source of "pollution and climate change" rather than the signatures of wealth and grandeur. These contradictions, together with constantly failing predictions and shifting attitudes to racism, made this strand of economic and geographic determinism obsolete after World War II.

A major feature of these documents and their authors, implicit in their obsession with prediction, is the preoccupation with *the future* as an object of geographical and environmental concern. After all, most of their earth science contemporaries were preoccupied with getting the record of the past straight, leading to increased understanding of ice ages and climate change, even in the recent past. The strand of geography reflected in the papers here was much bolder and eagerly used the knowledge at hand, however scattered, to build meta-historical narrative patterns of change, thus providing an explanatory power to the future that claimed not to be rooted in eschatology or ideology but to stand firm on the ground of science, high above the abyss of metaphysics. This again pointed forward to a major feature of later global change science, although the morals of the determinist geographers were totally different; the documents in this section rarely display caution or regret for the expansive human enterprise.

In the 1920s and 1930s, just as Huntington, Stefansson, and Taylor launched their grandiose schemes of global geopolitical upheavals, there were increasing hints of this new alternative vision of constraint concerning "environment" and global change, as something suffering human expansion rather than facilitating it, which would become hegemonic in the environmental discourse during the decades following World War II. In these first few decades of the twentieth century, while determinism dominated, progress was firmly anchored in Darwinian competition between states where the most richly endowed—in population, industry, and resources—had unquestioned advantages. Again, while progress was still marching on, this limited choice and challenged governance. Constraint was thus built into even the most triumphalist notion of Western progressivism. Was progress possible everywhere? Would it be attainable for all?

The Pulse of Asia

ELLSWORTH HUNTINGTON

In the progress of human knowledge the marked advances in each science have been made under the stimulus of a great fundamental principle. Astronomy could proceed but little beyond astrology until Newton discovered the law of gravitation; physics remained empirical until the conservation of energy was recognized; chemistry was merely alchemy until its pioneers worked out the unfailing law of the replacement of atom by atom; and geology would still be miner's lore, if scientists had not seen that in the course of ages the earth as we know it has been slowly evolved by processes identical with those still in action. So too, in the biological sciences, botany, zoology, and physiology, all was confusion until Darwin touched the key of evolution and a vast number of apparently unrelated facts fell into their appointed places, and the way was open for the wonderful advances of the last half century.

The anthropological sciences are also bound together by the unifying principle of evolution. Geography, anthropology, history, and sociology form an anthropological group possessing a unity as great as that of the biological sciences, although this has been perceived only within a few years. The average man thinks of geography, the oldest of all the sciences, as a schoolboy study of maps and of empirical descriptions of places and people. He forgets that the leaders of geographic thought have gone far beyond this, and are beginning to see that their science deals not only with the *distribution* of organic and inorganic forms in space, but also with the *relation*, both direct and indirect, of the entire group of organic forms inhabiting any part of the earth's surface to the inorganic forms in the same region. Geography, according to the new view, tells us not only what forms of plants and animals live together in mutual dependence, but also why the human inhabitants of a given region possess certain habits, occupations, and mental

Ellsworth Huntington. 1907. "Introduction: The significance of Central Asia." In *The pulse of Asia: A journey in Central Asia illustrating the geographic basis of history*, 1–17. Boston: Houghton, Mifflin and Company.

and moral characteristics, and why they have adopted a certain form of social organization. Among primitive people the relation of inorganic causes to organic results is universally recognized. Literature is full of references to the nearness of the Red Man to Nature, and to the complete dependence of primeval man upon her. Among highly civilized people, the relation is lost sight of because of the mixture of races, the growing control which man exercises over nature, and the rise of great religious or ethical ideas, racial hatreds, and dominant personalities. Nevertheless, it is there, and a patient untangling of the snarled threads of history will bring it to light.

In searching out the foundations upon which to build the new sciences of anthropology and sociology, students are turning more and more to geography in its broader sense. The anthropologist finds that the development of civilized man from the savage state is inextricably bound up with the various types of physical environment in which successive generations have lived. The sociologist discovers that the conditions of human society to-day are in part the result of racial characteristics due to past environments, and in part the result of present geographic conditions. Climate, the relation of land and sea, the presence of mountains, the location of trade routes, and the suitability of a region for agriculture, mining, or manufacturing are all potent factors in determining sociological conditions. The dependence of history upon geography is equally great. In recent years there has arisen the so-called "bread and butter school" of historians, who hold that the deepest cause of historical events is the necessity of mankind to subsist. The ambition of kings, the hatred of race for race, the antagonisms of religion, may agitate the surface and cause the waves which seem to us so portentous; but far down below all these there is the unending struggle for bread. It is this primarily which makes men work. It manifests itself in the discontent of the poor peasants of Russia, in the disputes between labor and capital in America, and in the bitter cry of the famine-stricken millions of India and China against the foreigners who seem to rob them of bread. An increasing supply of food has made Egypt contented and prosperous during the last few decades. Scarcity of food, present or prospective, for its increasing population has brought Japan into conflict with Russia, and is bringing it face to face with the United States in California, where the Japanese coolie is said to take the bread from the mouth of the native-born American laborer. According to this view, geography is clearly the basis of history, for the productivity of a country depends upon geographic facts, especially upon climate.

In saying that geography is the basis of the anthropological sciences, it is not meant to imply that physical processes explain all the qualities of man. They do not explain life, or mind, or ideals. At present we can only confess that we do not understand what these are, or how we came to possess them. We can only ascribe their origin to the same great Intelligence which framed the material universe and gave it immutable laws. We know, however, that they are the greatest forces in the world, the motive power which moves mankind. In the past, men have supposed that the human race either progressed blindly, or was led onward by the direct interposition of some unseen divine power. Now, we begin to see that man's course has been guided by his physical surroundings, just as a railroad winds here and there at the command of river, hill, or lake. To carry the analogy farther, the living mind of man, with its ideals, its love, and its pain, is the motive force to which is due the progress of human institutions; and history is the track along which man has advanced. Sometimes his course has been straight, sometimes devious, and at times it has doubled back on itself; but on the whole, it has led toward a dimly seen goal of uprightness, freedom, justice, and love.

We have studied the energizing mind, and know something of how it acts, though not of what it is. We have examined the human institutions of the home, the church, the state, and the social organization of industry; and our knowledge of them is large. The track, too, has been scrutinized minutely by historians; and we know its curves and grades, both up and down. One thing alone has been neglected: we have not looked at the country through which we have passed. To-day we are beginning to study our surroundings, and to see that we have reached our present position because of certain geographic facts. Historians have been slow to accept this view. When they found a piece of downgrade in the track, they looked at the cars and the engine to find its cause. They have failed to see that the swift descent of the United States into a financial panic and the hard pull out of it may be due to the fact that the train is crossing a valley, and not to overloading of the cars in the shape of over-production, or to poor couplings in the shape of a weak financial system, though these may precipitate disaster. It may be, as we shall see, that panics are due to the regular recurrence of periods of deficient rainfall, causing poor crops and fluctuating prices. If this is so, we must not only look to our couplings and our load; we must bridge the next valley, or cut and fill the road-bed so as to diminish the grades.

Again, as we look at the past, we see the track of history double far back on itself at the time of the fall of Rome before barbarian invaders. At present we are facing a similar, albeit peaceful invasion on the part of the starving millions of China: the fear that our track may again turn back is before us. The relapse of Europe in the Dark Ages, as future chapters will show, was due apparently to a rapid change of climate in Asia and probably all over the world—a change which caused vast areas which were habitable at the time of Christ to become uninhabitable a few centuries later. The barbarian inhabitants were obliged to migrate, and their migrations were the dominant fact in the history of the known world for centuries. We of to-day shall do well to ascertain whether we too are not facing the problem which faced the Romans. Parts of China have been growing drier and less habitable during recent centuries, and if the process continues, we are in danger of being overrun by hungry Chinese in search of bread. We cannot, perhaps, prevent their migration; but if we understand the cause, we can profit by the lessons of the past and avoid the danger, as a railroad engineer avoids turning back by choosing a place where he can tunnel through the mountains to the broad uplands on the other side.

The importance of climate and of changes of climate in history and the allied sciences has never been fully realized. It is climate which causes the Eskimo to differ so widely from the East Indian; it is climate which almost irresistibly tempts the Arab to be a plunderer as well as a nomad, and allows the Italian to be an easy-going tiller of the soil. And, if Percival Lowell is right, it is the dry climate of Mars which has caused the inhabitants of that planet to adopt an advanced form of social organization, where war is unknown, and each man must be keenly conscious of the interdependence of himself and the universal state.

Four years of life in Asiatic Turkey and three years of travel in Central Asia have impressed upon me the importance of the geographic basis in the study of the anthropological sciences. Hence this book. It is an attempt to describe Central Asia in such a way as to show the relation of geography to history and the related sciences, and to show the immense influence which changes of climate have exerted upon history.

From the Caspian Sea on the west to Manchuria on the east, Central Asia is largely a country of deserts. It is politically divided into the countries of Persia, Afghanistan, Baluchistan, northern India, Tibet, China, and Asiatic Russia. It varies in elevation from the low depression of the Caspian Sea and the small basin of Turfan, lying three hundred feet below sea-level in the very heart of Asia, to the plateaus of Tian Shan, Tibet, and the Pamirs at an elevation of from 10,000 to 20,000 feet above the sea. Although usually the mountainous parts are comparatively rainy and are often well covered with vegetation, the lowlands, which comprise most of the country, are intensely dry and almost absolutely desert. The people are equally varied, the fierce Afghan being as different from the sycophant Persian, as the truculent Mongol from the mild Chanto of Chinese Turkestan. Yet in spite of all this, not only the physical features of the country, but the habits and character of its inhabitants, possess a distinct unity; for all alike bear the impress of an arid climate.

Central Asia, more fully perhaps than any other part of the world, exemplifies the great geographic type in which the topography, vegetation, animal life, and human civilization have developed along the lines characteristic of prolonged aridity. We all know something of arid countries, empirically or from observation. We need, however, a more general concept, so that the term "arid" shall bring to mind the essential features of a definite geographic type, just as the term "bovine" brings to mind the spreading horns, large eyes, heavy body, cloven hoofs, cud, and other essential features of a definite zoölogic type. If once the geographic type is well understood in its highly developed form in Central Asia, it will be easy to comprehend how similar conditions of climate in other parts of the world give rise to similar topographic features, and how the two combine to determine the distribution and nature of life of all forms.

The rainfall of Central Asia is so small that the rivers fail to reach the sea. Hence the whole of a vast region, stretching three thousand miles east and west, and having an area nearly equal to that of the United States, is made up of enclosed basins, from which there is no outlet. Each consists essentially of a peripheral ring of higher land,—usually mountainous, but sometimes merely a broad, gentle arch,—within which a desert plain of gravel, sand, and clay, brought from the mountains by rivers, surrounds a salt lake, or the saline beds whence the waters of an ancient lake have evaporated. Where the peripheral ring of higher land is at all mountainous, it is flanked, and often half buried, by vast slopes of barren rock-waste—typical piedmont deposits of gravel, washed out from the uplands by floods. Because of the aridity, vegetable life is scanty except along the courses of streams and in the rainy plateaus. Far less than a tenth of the country is permanently habitable: the rest is either absolute desert, or mitigated desert which supports vegetation part of the year, but is too dry among the plains, and too cold among the mountains, to allow permanent occupation. Hence the inhabitants must either live in irrigated oases along the rivers, or wander from place to place in search of pasture for their flocks. There are no manufacturing communities, either large or small; no commercial centres except local bazaars, and no continuous agricultural population, such as that of the Mississippi Valley, dependent on rain for its water supply. Two main types of civilization prevail: the condition of nomadism with its independent mode of life, due to the scattered state of the sparse population, and the condition of intensive agriculture in irrigated oases with its centralized mode of life, due to the crowding together of population in communities whose size is directly proportional to that of the streams. Because of the arid climate and the consequent physical characteristics of Central Asia, its types of civilization have been, are, and probably must continue to be fundamentally different from those of well-watered regions such as most of America and Europe.

My acquaintance with Central Asia began in 1903, when I was appointed by the Carnegie Institution of Washington to assist Professor William M. Davis of Harvard University in the physiographic work of an expedition to Russian Turkestan, under the lead of Mr. Raphael Pumpelly. I remained in Central Asia from May, 1903, to July, 1904, spending most of the time in Russian Turkestan.

I crossed into Chinese Turkestan for a month, however, the first summer, and spent four months in eastern Persia during the winter. The results of the expedition are recorded in "Explorations in Turkestan," a volume published by the Carnegie Institution. The following year I had the good fortune to be invited by Mr. Robert L. Barrett to accompany him on an expedition to Chinese Turkestan. Arriving in India in February, 1905, we proceeded north to the Vale of Kashmir among the Himalayas, crossed them in May, and reached Chinese Turkestan in June. There we worked together for two months, and then undertook independent expeditions, Mr. Barrett going east from Khotan to China Proper, while I went east by another route to Lop-Nor, and then, turning north, arrived at Turfan in March 1906, and reached home the following May via Siberia and Russia.

The journey through Chinese Turkestan from India to Siberia forms the main theme of this book, but I shall devote a few chapters to other parts of Central Asia. This volume, like the majority of so-called "geographical" books, is a description of a journey; but, as I have already said, is also an attempt to describe certain parts of Asia as illustrations of the great principles of geography. My conception of that science, as stated above, is the one which has been spread abroad in the world at large, and especially in America, during the last few years by the persistent labors of Professor Davis. According to his definition, geography is primarily the study of the various natural divisions or provinces of the earth's surface as illustrations of the relations between the inorganic physical facts of the earth, air, and water on the one hand, and the organic facts of the vegetable, animal, and human world on the other. To illustrate: The investigation of the structure, origin, form, and climate of a lofty plateau and a neighboring arid plain is not geography, but geology, physiography, or meteorology. Neither can the study of the methods of plant growth and animal nutrition rightly be called geography, but botany or zoology. When, however, we consider the fact that because of the elevation of the plateau its climate is such that grass grows abundantly in summer; while the plain, being lower, has less rainfall, and bears only a sparse growth of grass in the early spring, we at once bring in the element of *relation* between the organic and the inorganic, and the study becomes geography. For the purposes of geography it is only necessary to understand enough of the plateau, the plain, and the grass to gain a clear conception of how the one acts on the other. If animals inhabit the country, they must be such as can live on grass, or can prey on their grass-eating companions. Further, if the plain is waterless in summer, and the plateau is deeply buried in snow in winter, the animals must perforce migrate, and a new geographic factor is introduced. When man enters the region, he finds it too dry in one part and too cold in another for agriculture. Hence he must live upon animals, either as a hunter, or, when the population becomes a little denser and the wild animals diminish in number, as a shepherd. In either case he must wander from place to place. Such a nomadic life induces certain habits as to cleanliness, eating, traveling, sleeping, working, resting, and the like. The habits in turn develop certain moral qualities, such as gluttony alternating with abstemiousness, hardihood under physical difficulties, laziness, hospitality, and others. Thus the physical features of the region give rise to certain kinds of vegetation, which in turn determine the species and movements of animals, and so cause man to adopt the nomadic life. And man, because he happens to be a pastoral nomad, develops certain habits, physical, mental, and moral, which, taken together, constitute character. Geography, it seems to me, cannot logically be content, as many geographers would have it, with the mere description of physical features and of their influence on the distribution of living species. It must deal with a given region, or natural province, as a whole; and must describe the entire assemblage of organic forms which result from a specified group of inorganic controlling features. The description is not complete unless it includes the highest and most interesting realm of geography,—the influence of physical environment, directly or through other forms of life, upon the mental and moral condition of man.

In accordance with this view of geography, I shall describe some of the chief and most typical physical features of Central Asia, not for their own sake, but as a preparation for the study of their relation to life. Then I shall set forth certain events, conversations and scenes which fell within my own experience, and shall show how they illustrate the influence of the physical environment already described upon the habits, thought, and character of the people. The descriptions centre in five basins located in northern India, western China, eastern Persia, and Asiatic Russia. The first basin, that of Kashmir, lies among the Himalaya mountains. Unlike the others, it has sufficient rainfall, so that it is not self-contained, but is drained by the Jhelum River, which flows out through a gorge in the surrounding mountains and reaches the sea. Hence the conditions of life are different from those of Central Asia in general, and resemble those of moister countries, such as Italy. The next three basins, those of Lop and Turfan in China, and Seyistan (Sistan, or Seistan) in Persia, are so arid that their rivers either dwindle to nothing in the desert, or end in shallow salt lakes. They closely resemble one another, and when the main features of one have been comprehended, but little need be added as to the others. The last basin, the so-called Aralo-Caspian depression, possesses many of the characteristics of its more arid neighbors, but its great size and the absence of mountains to the north give it a diversity of climate unknown to the others. I shall not consider it except in relation to the problem stated in the next paragraph.

In the study of the five basins along the lines of the definition of geography given by Professor Davis, I discovered a number of facts which lead to a new

application of the geographic principle of cause and effect. In order to understand the present condition, that is the geography, of Central Asia, we must look
upon it not as the result of the long-continued action of *fixed* physical conditions,
but as the result of *changing* conditions. During the recorded occupation of the
country by man there appear to have been widespread changes of climate. It has
long been surmised by historians that certain parts of Asia have been growing
more arid, but the surmise has lacked scientific confirmation. Indeed, meteorological data seem to stand directly opposed to it, for they show that there is no
evidence of any appreciable change since records have been kept instrumentally.
The oldest records, however, date back little more than a hundred years, and
hence cannot be considered as proving anything in regard to antiquity. The data
which I obtained in Central Asia, on the other hand, confirm the surmise of the
historians. There is strong reason to believe that during the last two thousand
years there has been a widespread and pronounced tendency toward aridity. In
drier regions the extent of land available for pasturage and cultivation has been
seriously curtailed; and the habitability of the country has decreased. In certain
moister districts among the mountains, on the other hand, the change has been
beneficial: they have become less damp and snowy, and hence more habitable.
Moreover, in both the drier and the moister regions the change of climate does
not appear to have been all in one direction. After a period of rapidly decreasing
rainfall and rising temperature during the early centuries of the Christian era,
there is evidence of a slight reversal, and of a tendency toward more abundant
rainfall and lower temperature during the Middle Ages.

In relatively dry regions increasing aridity is a dire calamity, giving rise to
famine and distress. These, in turn, are fruitful causes of wars and migrations,
which engender the fall of dynasties and empires, the rise of new nations, and
the growth of new civilizations. If, on the contrary, a country becomes steadily
less arid, and the conditions of life improve, prosperity and contentment are the
rule. There is less temptation to war, and men's attention is left more free for the
gentler arts and sciences which make for higher civilization.

The main outlines of the history of Central Asia agree with what would be
expected from a knowledge of the changes of climate through which the country
has passed. The favorable changes coincide with periods of prosperity and progress; the unfavorable with depression and depopulation. My own investigations
show that the parallelism between climatic changes and history applies to an area
extending at least three thousand miles, from Turkey of the west to China Proper
on the east. Other evidence, which has not as yet been investigated in detail, indicates that the parallelism applies to all the historic lands of the Old World and
possibly to the New. As we look back into the past, we are forced to the conclusion that whatever the motive power of history may be, one of the chief factors

in determining its course has been geography; and among geographic forces, changes of climate have been the most potent for both good and bad.

In the last chapter of this book I shall consider this conclusion its broader outlines as part of the philosophy of history. For the most part, however, I shall confine myself to a statement of the phenomena which have led to its adoption. Briefly restated, the fundamental idea of this volume is that geography is the basis of history. The physical features of the earth's surface limit the organic inhabitants of a given region to certain species of plants and animals, including man, which live together in mutual dependence. The world is naturally divided into geographic provinces characterized by definite organic and inorganic forms. Among primitive men the nature of the province which a tribe happens to inhabit determines its mode of life, industries, and habits; and these in turn give rise to various moral and mental traits, both good and bad. Thus definite characteristics are acquired, and are passed on by inheritance or training to future generations. If it be proved that the climate of any region has changed during historic times, it follows that the nature of the geographic provinces concerned must have been altered more or less. For example, among the human inhabitants of Central Asia widespread poverty, want, and depression have been substituted for comparative competence, prosperity, and contentment. Disorder, wars, and migrations have arisen. Race has been caused to mix with race under new physical conditions, which have given rise to new habits and character. The impulse toward change and migration received in the vast arid regions of Central Asia has spread outward, and involved all Europe in the confusion of the Dark Ages. And more than this, the changes of climate which affected Central Asia were not confined to that region, apparently, but extended over a large part of the inhabited earth. Everywhere they were the most potent of geographic influences, working sometimes for progress and sometimes for destruction. Such in brief is the broad conclusion to which we are led by a study of Central Asia as an example of the influence of geography upon history. Before accepting it, it behooves us to examine with the closest scrutiny all the evidence in relation to climatic changes which may have been so momentous in the world's history.

Commentary

Ellsworth Huntington, *The Pulse of Asia* (1907)

CAROLE CRUMLEY

Determinism may be thought of as a search of history for unique causation and a particular way of laying putatively scientific claims on the future. There are many types of determinism, the most familiar of which are racial or genetic (scientific racism), economic or class-based (social Darwinism), sex/gender-based (Freudian theory), and climatic or environmental. Beginning in the late eighteenth and early nineteenth centuries, geological, paleontological, and archaeological evidence grew for a significantly earlier appearance of the human species than indicated in biblical texts. By the late nineteenth and early twentieth centuries, science-based authority had dramatically expanded, and researchers endeavored to connect the emerging social sciences with similarly expanding work in evolutionary biology. While today their conclusions are recognized as severely flawed, it is worth remembering that in the latter half of the nineteenth century no major figure in the social and biological sciences entirely escaped the influence of the new scientific field of evolution and its hasty and flawed application to society.

The ideas of progress and determinism were fundamental assumptions guiding the policies of late nineteenth- and early twentieth-century nation-states. Scientific racism and nationalism proved particularly compatible, offering a solid justification for nineteenth-century colonialism, class privilege, and the persecution of minorities. Equally attractive was the (environmental) deterministic argument that Europeans and certain areas of North America were doubly blessed with the world's most intellectually invigorating climate and its most enlightened population.

In Germany, National Socialism adopted a suite of ideas that combined geographical determinism (drawing on Tacitus's *Germania*), cultural determinism (funding and promoting the work of linguist and archaeologist Gustaf Kossinna), and genetic determinism (the idea that human social and behavioral qualities are manifest in the form of "racial character"). By 1933 the Nazis had embraced the work of several prominent American scholars, among them physical anthropologists Aleš Hrdlička and Charles Davenport and geographers Walter Christaller and Ellsworth Huntington.

The German and American eugenics movements' origins paralleled one another, both beginning just after 1900 and establishing societies, journals, and a strong following in only a few years. One of the most influential converts to what was termed "race hygiene" was Ellsworth Huntington. Huntington (1876–1947) was born into the family of a Congregational minister and grew up just outside Boston. He earned a B.A. at Beloit College in 1897 in geology and rhetoric, then joined the Congregational foreign missions and was posted to Turkey. On his return he attended Harvard (M.A. 1902), but he was refused the doctorate due to a "deficiency in climatology" and began teaching part-time at Yale, which awarded him the Ph.D. in 1909, while he was on the teaching staff. Yale never properly hired him as a regular member of the faculty, thinking him an explorer rather than an academic. Huntington was an enormously prolific writer, publishing twenty-eight books and hundreds of articles in both scientific and popular publications between 1907 and 1945. Among them was *The Pulse of Asia* (1907), perhaps the most-reviewed book of the early twentieth century, which was translated into dozens of languages. Always in financial straits, Huntington wrote constantly—articles, textbooks, brochures, workbooks—to maintain his large family.

He was already controversial by 1914; a *Nature* review in 1921 remarks that his "tendency to lack . . . due care in generalizing [has] too many evil consequences." Even his Harvard mentor, physical geographer and geologist William Davis, was critical of his sloppiness; yet by 1923 he was president of the American Association of University Geographers. Those most kindly disposed to Huntington thought he applied his ideas to history "with premature enthusiasm," while others considered him racist. However, substantive criticism in print of his ideas was not mounted until 1930; this was despite comments Huntington made, as early as 1917, that "America is seriously endangering her future by making fetishes of equality, democracy, and universal education."

Huntington was a founding member (1916) and president (1917) of the Ecological Society of America. He joined the American Genetic Association in 1926; his Committee on Eugenics (practices intended to improve a human population's genetic composition) strongly influenced U.S. immigration policy in the 1920s. Huntington helped found the Population Association of America in 1931, was treasurer of the American Eugenics Society in 1932, and was its president from 1934 to 1938. The U.S. National Research Council, under the Society's influence, helped set immigration policy and ushered in an era of eugenic sterilization. Between 1907 and 1963 more than 65,000 individuals were forcibly sterilized under eugenic legislation in the United States.

Famous on both sides of the Atlantic, Huntington wrote most of two generations' geography and social science textbooks; at first he was simply an embarrassment to academic geographers, who considered him an explorer and only later began to realize what a huge ethical liability he had become. His insistence on the genetic inferiority of the mentally ill, those suffering from epilepsy, and non-Aryans became increasingly untenable to the scholarly community. Undaunted, by 1935 when *Tomorrow's Children: The Goal of Eugenics* was published, Huntington had laid out for the United States a program of racist genetics that did not differ substantially from that of the Nazis.

Although Huntington's name is today linked with environmental determinism, at the height of his career it was his genetic determinism that colleagues found particularly distasteful. By 1930, most anthropologists had abandoned the premise that culture and environment co-vary (environmental determinism). As a result of anthropologist Franz Boas's influence, they had also distanced anthropology from biological determinism by forcefully arguing for the importance of particular historical circumstances and the social transmission of cultural practices (AAA 2011).

Huntington and Franz Boas embody the struggle in American academic circles to delineate the relative influence of "nature" (biology, especially genetics) over "nurture" (culture, history). Boas (1858–1942) was a German Jew who emigrated to the United States in 1887. In his scholarship he moved from physics to geography to ethnography and cultural anthropology; his neo-Kantian philosophy offered a compromise between idealism (mind) and materialism (reality). He took a position against all-encompassing evolutionary schemes (that is, the various determinisms) and gave equal weight (depending on history) to diffusion and independent origins. By the time his book *The Mind of Primitive Man* was published in 1911, he had become the foremost American spokesman in refutation of scientific racism.

The differences between these intellectual contemporaries delineate the edges of an argument that seems never quite resolved, however clear the evidence. Its essential elements are geography, nation, class, and race. In the preface of *The Pulse of Asia*, the roots of Huntington's later eugenics argument can be discerned. Huntington's prose style is calm and confident; he speaks with what appears to be scientific authority. Some of his assertions, such as historic changes in climate, have been confirmed. Yet scarcely two decades later he had snapped these elements together into the intellectual framework of the Third Reich. While the plasticity of the relationship between geography and society, the polyvalence of intelligence, the key role of education, and the fallacy of racial classifications have been systematically demonstrated, the popularity of national, cultural, economic, and genetic superiority endures, always finding new ears in new eras.

Further Reading

American Anthropological Association (AAA). 2011. *Race: Are we so different?* http://understandingrace.org.

Martin, G. J. 1973. *Ellsworth Huntington: His life and thought.* Hamden, CT: Archon Books.

Kühl, Stefan. 2002. *The Nazi connection: Eugenics, American racism, and German National Socialism.* 2nd ed. New York: Oxford University Press.

"Nature *Versus* The Australian"

GRIFFITH TAYLOR

All Australians are anxious to see their homeland develop speedily, so that it may come as soon as possible into its proper place in the comity of nations. There are two distinct methods of helping toward this happy goal. One of these follows what we may call commercial lines, and the other (not necessarily antagonistic) travels over the accepted routes of scientific research.

The former plan might be likened to putting all the best goods into the window, and, having arrived attracted the customers, to see that they purchase something. The other method is to give as much attention to the less valuable assets as to the most attractive, in the full belief that thorough knowledge will pay best in the end. The only antagonism arises when the exponent of the former method complains that the scientist spoils his chances by a too-open discussion of disabilities which were better discreetly hidden!

However, there is much less of this false patriotism nowadays. The authorities are encouraging the investigation of the physical controls which govern conditions in our less populous areas, believing that only by so doing can we make the best of our heritage. It is entirely with the object of helping this good work that the following article is written—for, to continue our parable, nothing can be such a bad advertisement as a misled or discontented purchaser.

One may misquote a well-known saying—"They little know of this lone land, who only this land know"; and we can only hope to arrive at the possibilities of Australia and can only estimate its resources by comparing them with similar aspects in other countries. Isolation has its advantages in some respects. The "tight little island" [Britain] no doubt kept free from medieval wars, but its marvellous progress was chiefly due to its natural resources in climate and coal. Here in Australia isolation is also advantageous in times of trouble, but in times of peace it tends to react adversely on the national character, fostering, perhaps, too optimistic a trust in Nature's endowment of the continent.

Thomas Griffith Taylor. 1920. Excerpts from "Nature *versus* The Australian." *Science and Industry* 11(8):1–14.

This article will therefore be largely concerned with comparisons based on recent physiographic research, and we shall find that we have reason to be proud of the future of Australia, even though it is not so well endowed as the older centres of white settlement.

Australia is particularly well suited for climatic studies, since it has the least diverse topography of all the continents and the most uniform outline. Hence it is free from variations due to elevated plateaux, high ranges, or deep gulfs and inland seas.

Unfortunately, these very characteristics are a distinct handicap as a dwelling place for man, and we may well devote some time to a consideration of how such physical controls affect Australia.

If we glance at a world-map representing either population, or vegetation, or rainfall, we shall find that the most striking feature in all three maps is the belt of empty arid lands which lies along the tropics in both hemispheres. (See Fig. A.) These are the regions where the trade winds are supreme; and where they blow from the continents to the ocean they are desiccating winds. Their realm covers half the surface of the globe, and where a broad belt of land is affected (as in Australia and Northern Africa) the result from an economic point of view is disastrous.

[3] Here, then is the first and the chief burden which Nature has laid on the Australian. Nothing can make up for the large extent of our continent which lies below the constant sweep of desiccating trade winds. It is no help to know that in fairly late geological times the continent extended into more clement regions

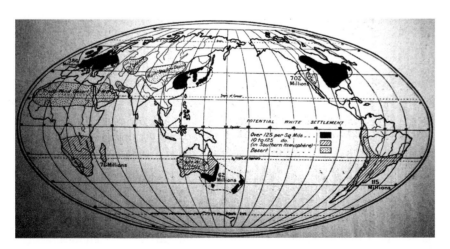

Figure. A. The Status of Australia Among the Continents. (Only potential WHITE Settlement is considered.)

to the south and east where the climate was undoubtedly better suited for closer settlement. . . .

[6] Let us now glance briefly at the climatic controls which affect the development of Australia. We are concerned primarily with temperature and rainfall, but we shall find that the humidity is of great importance in connexion with tropical settlement.

Since the southern hemisphere consists so largely of water, it is natural that on the whole it should have a cooler temperature than the northern. But where large masses of land are involved this difference practically vanishes. Tropical seas have a temperature of 80 degrees F [26.7°C]. fairly generally, and the effect of a large ocean does not extend indefinitely inland. If we consider actual temperature records, we find that four regions in the world exceed an actual average temperature of 84 degrees F [28.9°C.]. These are around Timbuktu, Massowah (on the Red Sea), Tinnevelli (at the southern tip of India) and Wyndham, in north-west Australia. The two former are arid. The two latter are very wet in the hot months. Hence Wyndham (with 84.6 degrees F [29.2°C]) is undoubtedly one of the least favoured regions in the world so far as temperature is concerned.

[7] The "heat equator" must be drawn through Wyndham, Darwin and Thursday Island, for these places are much hotter than any to the north or south. We are unable, therefore to say that tropical Australia is cooler than other similarly placed regions with the possible exception of North Africa. (The current temperature maps have fostered the error by recording temperature reduced to sea-level, instead of the actual temperatures of the places concerned. Consider what this means for regions like Mexico.)

Heat as such is not of paramount importance, provided it be dry heat. In this respect Australian is fortunate in her arid interior for the humidity is never very high except in the coastlands. Throughout the summer months only the region north of a line through Broome, Daly Waters, and Rockhampton has a high humidity. It is very hot inland, certainly with many days in succession over 90 degrees [32.2°C]. The human organism can bear this without much discomfort or loss of health, provided the air be dry—but it is otherwise with plant life. Hence from a hygienic point of view, a low rainfall is desirable in our tropics, but from an economic point of view this is disastrous.

[8] It is a truism that if the rainfall were only reliable it would not matter so much if it were comparatively small in amount. By long experience farmers have learnt that in the Western Australian wheat belt a lower rainfall will suffice than in western New South Wales. Now that our records are fairly complete in many parts of the Commonwealth it is possible to study this aspect scientifically.

Consider the following table where the same months in *consecutive* years have been taken at the places named:

Unreliable Rainfall.

Onslow (Western Australia) ..	April, 1900　..　11·0″ „　1901　..　o (April Average, o·9″)	May, 1900　..　10·5″ „　1901　..　o·5″ (May Average, 1·5″)
Borroloola (Northern Territory) ..	March, 1899　..　29·0″ „　1900　..　o·5″ (March Average, 6·0″)	Feb. 1896　..　21·4″ „　1897　..　4·7″ (Feb. Average, 7·4″)
Charlotte Waters (South Australia)　..	March, 1908　..　5·0″ „　1909　..　o (March Average, o·7″)	Jan. 1877　..　9·7″ „　1878　..　o (Jan. Average, o·8″)

[9] One often hears that agriculture may be found profitable in the interior where the *average rainfall* is 10 inches [254 mm], because wheat can be grown with such a rainfall in the Mallee [northwest Victoria] or in the northern region of Western Australia. As a matter of fact, none of the Mallee is below 10 inches — and I have never been able to find a region in Western Australia below 10 inches where wheat is grown at all regularly.

If the rainfall at the right season exceeds 11 or 12 inches [280–305 mm], and is reliable, well and good. But, as I shall now proceed to show, the greater part of arid Australia is cursed, not only by a low rainfall, but by a very uncertain one. By a method of reckoning which I have explained elsewhere, I have drawn up the reliability isopleth shown in Fig C.

The most variable region in Australia is around Onslow (W.A.). The best is around Perth, in the same State. The reason for the success in wheat in the south-west corner (Swanland) is shown at a glance. All the arid country (under 10 inches per year), except in Swanland and along the Trans-Australian Railway, has a variability exceeding 10 per cent. of its average total. Moreover, a great deal of the country, with better rainfall — especially to the south of the Gulf of Carpentaria — also has very erratic rainfall. Thus the Barkly Tableland seems to be a very unpromising agricultural region, in spite of its average of 15 or 20 inches [380–500 mm] per year.

As regards future prospects of settlement as based on rainfall, we may subdivide Australia into seven regions. In the following table the regions are arranged approximately in order of value. (See Fig. C.)

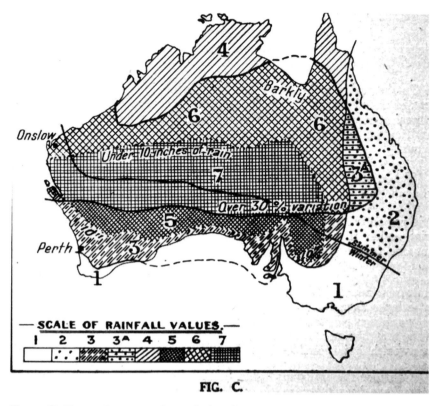

Figure. C. Economic regions of Australia based on variations in the amount, season, uniformity, and reliability of the Rainfall. (For regions see Text.)

Class	Sub-class	Chief Localities
Uniform	With winter max. (1)	Riverina, Victoria, Tasmania
	With summer max. (2)	Eastern Queensland, NE NSW
Seasonal	Moderate winter rain (3)	Swanland (WA)
but reliable	Summer rain (4)	Kimberley (WA); Northern Territory
	Arid winter rain (5)	Coolgardie (WA) to Broken Hill (NSW)
Erratic	Summer only (6)	Pilbara (WA); Macdonnells (NT)
	Arid only (7)	Central Australia

. . .

[12] If 70 degrees wet bulb is the limit of comfortable living for white families, then the following table is not encouraging as regards our northern coast lands:

However, in this, as in other controversial points, the last word is not yet said. But obviously our tropical problem is a climatological one, and it would seem obvious that money might well be spent on an investigation along these lines.

Scale of Discomfort, based on Average Wet Bulb.

Based on the Graphical Study of Climate in Meteorological Bulletin No. 14.

	45°–55° F. most comfortable.	55°–65° occasionally uncomfortable	65°–75° often uncomfortable.	Over 75° F. almost continuously uncomfortable.
	Months.	Months.	Months.	Months.
Melbourne ..	6	6	o	o
Sydney ..	5	7	o	o
Hobart* ..	6	3	o	o
Coolgardie ..	7	5	o	o
Perth ..	5	7	o	o
Alice Springs ..	5	6	1	o
Brisbane ..	3	4	5	o
Townsville ..	o	3	9	o
Nullagine ..	o	5	7	o
Wyndham ..	o	3	3	6
Darwin ..	o	o	6	6
Thursday Island ..	o	o	6	6
Wellington (New Zealand) ..	8	4	o	o
London* ..	5	3	o	o
New York* ..	4	3	2	o
Batavia ..	o	o	10	2
Madras ..	o	o	6	6
Sierra Leone ..	o	o	o	12

* Some months below 45° *i.e.*, cool, but comfortable.

In no other portion of the world, so far as I am aware, is there a settlement of northern Europeans resembling our sugar-growers in the Cairns district [North Queensland]. In Brazil, near Rio, are some Spanish and Portuguese—but they belong to a much warmer clime originally. All honour, therefore, to the Australians, who are very practically fighting Nature in their attack on the luxuriant wealth of the tropics.

I must not neglect to point out that we have been favoured with an almost complete absence of the worst tropical diseases. Malaria is certainly diminishing under adequate sanitary treatment. Yellow fever and beri-beri have never been of any great importance. These diseases have been the greatest enemies in other tropical regions, so that the paucity of our [A]boriginal population would seem to be a distinct advantage as lessening the risk from such contagious diseases.

I have now traversed rapidly most of the ground indicated by the title of the paper. In conclusion, it has seemed to me possible to make a first approximation to a map showing the *habitability* of the earth.

[13] The study of future white settlement gives very definite conclusions at to the status of Australia. By dividing the world into 74 natural regions, each assessed quantitatively for (1) temperature, (2) rainfall, (3) coal, (4) location, it was possible to draw up a map which showed the areas of potential white population. Western Europe, Eastern North America, and North China (though this does not concern white folk) showed the highest figures—with considerable areas where over 200 people to the square mile would settle.

The United States is well entitled to call itself (in the material sense at least) "God's Own Country." The Japanese realize the value of Northern China, and hence their desire to obtain control of it. In the southern hemisphere, New Zealand and South-east Australia are the most favoured, being able, perhaps, to take ultimately 120 people to the square mile. There are, however, much larger areas of moderate habitability in South Africa and South America. (See Fig. A)

Grouped in nations, we may show that the British Empire (when it is "saturated" to the extent of Europe today) may reach a potential white population of 377,000,000.

The following table gives my conclusions for a future date, which is by no means outside the province of present-day politics:—

Table Indicating Future White Settlement

	British Control.	U.S.A. Control.	Other States.	Total.
	Millions.	Millions.	Millions.	Millions.
1. North America ..	179	513	10	702
2. Europe ..	60	..	326	386
3. Argentine, &c. ..	..	..	115	115
4. South Africa ..	76	..	6	82
5. Australia and New Zealand ..	62	..	..	62
Total	377	513	457	1,347

Commentary

Griffith Taylor, "Nature *Versus* The Australian" (1920)

CAROLYN STRANGE

Griffith Taylor (1880–1963) was a scientist who lived life at an exclamation point. Only one appears in this article, written a decade into Taylor's career as a climate scientist and public policy commentator. However, the title provides a sense of his provocative answer to one of modernity's key questions: did nature dictate progress, or did progress depend on the manipulation of nature and the transcendence of environmental limits? For Taylor, "Nature" invariably won, particularly in Australia.

Was Taylor an old-fashioned thinker? Many of his contemporaries thought so. And worse: any Australian who stressed Nature's power over "man's" ingenuity was unpatriotic. Politicians, landowners, and economic developers, who believed in the first flush of nationalism (Australian colonies federated in 1901) that nothing could stop Australia from becoming the United States of the southern hemisphere, were like shopkeepers who fill their display windows with the most appealing items on offer, keeping the shabbier goods from view. In this article Taylor claims, as he did throughout his career, that he was neither old-fashioned nor unpatriotic. Quite the contrary: he was a modern scientist, committed to expose and explain "the physical controls which govern conditions."

For Australia the question of its capacity to support closer settlement and a population comparable to Europe's or North America's came down to the viability of its "less populous areas"—namely, the arid interior and the tropical north. Taylor, who had worked from the mid-1910s for the federal Bureau of Meteorology after having studied geology at the University of Sydney, believed that "too much optimistic thinking" threatened to warp population policy. To counter baseless faith, he offered science, in the form of comparative meteorological data.

By the time Taylor published this article he was employed by the University of Sydney to establish its geography department. Using his cartographic and propagandistic talents, he composed maps to support his arguments. In Figure A, "The Status of Australia among the Continents," he projects the growth potential of Australia and New Zealand (along with that of other climatic regions of the world) and pegged it at 62 million, less than one-tenth of his estimate for the United States. Whether or not Australians accepted the messenger's gloomy news, the facts were plain to see: the country was "burdened" by Nature.

To prove his point, it made sense, from a climatological perspective, for Taylor to compare Australian locales to such "foreign regions" as North Africa and the Congo. Yet, his arguments about the limited prospects of tropical settlement and the unlikelihood of economic development in arid regions unsettled most Australians' belief that the nation merely awaited modern development. His data confirmed that regions with uniform and reliable rainfall were the only places suited to sustain agriculture. No amount of wishful thinking would contradict his conclusions.

In other respects, Taylor's projections of Nature's limits conformed to the race-based notions of "civilization" that prevailed in this period. Cautious population projections were warranted for Australia's tropics because he thought solely in terms of future "white" settlement (which he states explicitly in the subtitle to his world map). It was commonly understood that people of northern European stock and "Britishers" fared best in temperate climates, meaning regions with moderate rainfalls and no extremes of heat. First Australians, along with peoples who might migrate from hotter and more humid regions of the world, could endure the extremes of the Australian climate, but that was not the point. Like his political and economic opponents Taylor envisioned Australia as part of the "comity" of predominately "white" nations. The closest he came to challenging the "White Australia Policy" was to advocate, hesitantly, that Australia consider introducing "limited Asian migration," on the grounds of their "fitness" to adapt to tropical living. In the meantime it was unwise, even unchivalrous, to "settle white women and children in our northern tropical coasts." Here, again, he relied upon a conventional understanding of domesticity as the sine qua non of civilization.

So why did Australians listen to foreign blow-ins, rather than the scientist who had spent a decade studying rainfall patterns, prevailing winds, and seasonal temperatures across the country? This is what occurred in 1923, when the Commonwealth funded the trip of Vilhjalmur Stefansson (1879–1962), "famous Arctic explorer" and expert on scarce resources, to investigate the productive potential of inland Australia. The Canadian delivered an upbeat report, which boosters in government and industry were quick to interpret as the vindication of their vision of "Australia Unlimited." Taylor's patience was growing thin by this point. Privately, he complained to Stefansson that "explorers" were ill suited to provide settlement and development advice: that was up to scientists like himself. But Stefansson's optimistic assessment of man's potential for ingenuity, even though based on a breezy tour, found more supporters than Taylor's emphasis on limits.

Taylor's petulant dispute with Stefansson flags wider debates, from the late nineteenth to the early twentieth centuries, between "determinists" (most notably, in addition to Taylor, Americans Ellen Churchill Semple and Elsworth Huntington) and "possibilists," the founders being Paul Vidal de la Blache and Carl Sauer. By the 1920s possibilists rejected the determinists' affection for patterns and laws in favor of locally based research, which appeared to affirm humans' capacity to shape the natural environment. Citing evidence ranging from the Hoover Dam to air conditioning, possibilists captured the "can-do" spirit of the age, leaving Taylor, with his global climatic regions and naturally determined patterns of development, sounding like an intellectual relic.

Thus it is surprising that Taylor began to earn respect from scientists in the early twenty-first century. Headlong exploitation of natural resources is quite evidently unsustainable, and the use of arid land for crops such as cotton and rice, requiring massive irrigation schemes, appears far less sensible than it did in the 1920s. Tim Flannery paints Taylor as a brave truth-teller, willing to withstand public pressure and stick by his data. Indeed, as a scientist, Taylor was at his best when compiling meteorological data and determining homoclimes, not humans or their racial characteristics!

Further Reading

Powell, J. M. 1997. "The pulse of citizenship: Reflections on Griffith Taylor and 'nation-planning.'" *Australian Geographer* 28(1):39–52.

Strange, C. 2010. "The personality of environmental prediction: Griffith Taylor as 'latter-day prophet.'" *Historical Records of Australian Science* 21(2):133–48.

Strange, C., and Bashford, A. 2008. *Griffith Taylor: Visionary, environmentalist, explorer,* Canberra and Toronto: National Library of Australia/University of Toronto Press.

Taylor, Griffith. 1958. *Journeyman Taylor: The education of a scientist.* London: Hale.

The Northward Course of Empire

VILHJALMUR STEFANSSON

Chapter 1

Man, as an animal, is indeed, a tropical animal. But man, as distinguished from animals, is not at his best in the tropics or very near them. His fight upward in civilization has coincided in part at least with his march northward over the earth into a cooler, clearer, more bracing air.

For the last few centuries, and especially in America, our attention has been centered upon the proposition that "Westward the course of empire takes its way." It has indisputably taken a westerly course during the last few centuries. But it is equally indisputable and more significant (because it rests upon broader natural causes) that northward the course of civilization has been taking its way, not only through the long period of written history and of tradition, but also through that far longer period, the records of which are the skeletons of the forerunners of men and of near-men, and of indubitable men who developed a civilization through millenniums of crude stone tools and polished stone and copper and bronze and iron down to Egypt and China as our histories show them.

There are but two commonly held theories of the origin of man. Each places the spot of origin in or near the tropics, the one because the skeletons of the anthropoids or pre-anthropoids, from which they think man descended, have been found chiefly in the tropics, and the other because tradition says the Garden of Eden was in tropical lands. With many divergences, both fundamental and superficial, the two theories agree on the geographic origin of man.

Man as an animal is not only tropical in origin but is also by the nature of his body unfit to flourish in any other sort of climate. Even those who assert he was once hairy refrain from contending that he had fur. Hairy as he was he would

Stefansson, V. 1922. Excerpts from *The northward course of empire*, 1–19. New York: Harcourt, Brace and Company have shivered in Italy and could not have prospered at all in the winter climate of North Dakota or of Russia. Nor would the most thoroughgoing advocate of a meat diet pretend he could flourish through hunting until after the invention of weapons and traps. He must have lived in a country not too cold for an unclad, furless animal where vegetables and fruits could be found at all times of year to constitute either his main diet or at least the bridges over necessary gaps in the meat supply.

Then came the inventions of fire and clothing for combating the cold, and of weapons for killing the grass-eating animals upon which man could subsist though he could not directly upon the grass. With these inventions commenced the northward march of civilization, and we do not yet know how far north it will continue. At least that contention can be made, though it has to be made in the face of an overwhelming public opinion to the effect that the northward limit has already been reached.

Men at every period of history have been generally of the opinion that the ultimate limit of the northward spread of civilization had then at length been reached.

It is a reasonable assumption, deduced from what we know of later history, that even the thoughtful men of Memphis and of Babylon failed to see potentialities for much beyond barbarism in the Greece and Italy of their time. We know as a matter of recorded opinion that the Greeks and Romans not only considered the people to the north of them inferior, but believed that that inferiority must continue, largely because of a supposedly hostile climate of the lands to the north. Tacitus probably knew as much as any of his contemporaries about the lands beyond the Alps, and was merely voicing the general opinion of his time and countrymen when he said that nobody could conceive that any one, unless forced by the stern necessity of war, would willingly leave the fertile shores of Africa or the plains of Italy for the country north of the Alps, where the climate is as disagreeable as the soil is sterile. This was undoubtedly a truism of his time; but it is a fact of our time that many people live in Paris and other parts of France by choice. . . .

It is human nature that we undervalue the distant and exaggerate the difficulties of the unknown. My friend, Professor Ellsworth Huntington of Yale, once sent out a letter of inquiry to about two hundred geographers, ethnographers, and other men of wide information in various lands, asking them among other things to give their opinion on the degree of civilization of the people of Iceland. The classification was to be made on the basis of a scale of 10, the people of high civilization being Group No. 10, and those of the darkest savagery in Group No. 1. The civilization of Iceland was graded as follows:

The Asiatics..put Iceland in group 3	
" Latin Europeans............................. " " " " 4	
" Americans...................................... " " " " 5	
" British .... " " " " 6	
" Germans and Scandinavians " " " " 8	

Each in his own country these authorities were of approximately equal culture, rank and native intelligence, yet the Asiatics because they were geographically and culturally remote, put the Icelanders near the bottom of the intellectual scale, while the nearest neighbors of Iceland placed it not far from the top. From this classification we accordingly learn nothing reliable or valuable about Iceland; but we get instead further confirmation of the principle that we tend to undervalue whatever is remote.

To the peoples of the centers of civilization the un-colonized North has been more or less remote geographically and almost infinitely remote from a cultural and historical point of view, for the information about it was in considerable part misinformation and its history and problems lay in the future.

On the basis of distance and misinformation the North has always been supposed to be dreadful and devoid of resources. These judgments have always been wrong and this we could prove by dozens of further instances although we shall adduce only two or three. . . .

In 1867 in America a great war had come to a close. During that war the side which eventually triumphed had not been supported so consistently by any major European power as by Russia. The country was grateful to Russia and it became necessary to translate that gratitude into substantial terms. To put it in modern parlance, they wanted to "slip some coin" to Russia as a reward for kindness received, and they carried out what was for that time an extremely large but otherwise quite ordinary political transaction by purchasing Alaska for $7,200,000. Such are the views of many historians as to the reason for the Alaska Purchase. Woodrow Wilson's history seems to consider as the chief motive, the extension of the Monroe Doctrine to still another part of the American continent, while others think the United States bought Alaska for some ready money partly to show European nations, which doubted America's solvency and power to recuperate after a devastating war, that the country was not really broke.

There are still other explanations of why Alaska was purchased but none of them rests on the assumption that the territory was intrinsically worth the price. It may have been that Secretary Seward and a few others realized that the money was not an actual gift and that Alaska had a great future, although, if that was so, Seward must have been a good deal wiser in his generation than Benjamin Franklin had been with reference to Canada in an earlier one. However that be, the Republican Party and Secretary Seward were attacked in the next presidential campaign for having spent several millions of public money for a lump of ice.

If you want to make up your mind what people really thought of Alaska at the time of its purchase and for many years after, turn to the files of the newspapers for the next presidential campaign (which resulted in the election of Grant) and you will find the Democrats attacking the Republicans on the score of the Alaska purchase. They put up a bitter fight on this issue. That in itself does not mean much, for such are the tactics of politics. But turn to the defense made by the Republicans and the lameness of it will convince you that they had no pride in what they had done, nor even faith in the future to exculpate, let alone justify, them. They felt themselves to be the pot and the best they could do was to call the kettle black. They drew a herring across the trail by calling the Democrats traitors and slaveholders; they shifted the battle to the old reliable issue of the tariff.

I am not a profound historical scholar and my memory does not go back to Grant's time, but this is history as I have read it. It was not till about 1900, when gold was discovered in Alaska, that politicians began to "point with pride" to Seward on the score of his purchase and I believe it was Franklin K. Lane (perhaps because he was born in Canada and had therefore a better understanding of the potentialities of the North) who first among cabinet officials had a vision of Alaska's coming greatness.

When it began to dawn on the United States that Alaska was of value, it was her mineral resources they saw. This again is a common historical phenomenon. When Columbus sailed west from Spain he was ostensibly in search of a short route to the Indies. He probably did not expect to find America. At least the popular view was that he had been searching for Asia and when he returned his was one of the many exploring expeditions that have been called failures because they discovered something quite different from that which had been expected. By way of making the best of the unfortunate fact that America blocked the direct sea route to China, those who went there, unless they were searching for a fountain of youth, were commonly looking for gold and precious stones. None of them were looking for the potato, although its unheralded discovery has proved of greater value to the world than all the gold dug out of the two continents. So it was and will be with Alaska—the first things to be looked for were precious metals and furs, but the greatest things to come out of it will not be those originally looked for.

Alaska had its turn as a gold seeker's paradise, and since 1900 has been much in men's minds on that score. Later it was realized that in portions fairly accessible from the Pacific there were huge deposits of copper more valuable than the gold, and coal mines of no less promise, and unless the present industrial trend is altered, the forests are likely to become more valuable than either. . . .

More than two centuries ago the Dutch discovered Spitsbergen, the south tip of which is about 300 miles farther north than the north tip of Alaska (a fact that must, however, be interpreted in the light of the unsymmetrical nature of the polar regions as explained, for instance, in Chapter II of *The Friendly Arctic*, 1921). Whale and seal oil were of far greater commercial importance then than now, and this group of islands soon became an important focus of the whale "fishery." All of it was claimed by Great Britain and all was claimed by Holland, and other countries made various claims, but as a matter of fact most of the country was for a long time controlled by the British and a small part by the Dutch. Later these "fisheries" declined in value and disappeared when Standard Oil began to furnish the light of the world. No British or other sailors made any regular visits for years; and Gladstone, as Prime Minister of Great Britain, renounced any claims that Britain might have had, saying and apparently believing that the islands could only be a bill of expense if possession were maintained.

Some years later the Hamburg-American Line and other steamship lines cultivated Spitsbergen as one of the interesting outposts of the tourist trade, exploiting that most commonplace-looking of marvels, the "midnight sun," which no one can tell from any other sun by anything but reference to a watch carrying local time.

About the beginning of our century there were in Sweden some men of foresight who proposed in the Parliament that Sweden should take possession of Spitsbergen. This proposal was promptly turned down on the ground that Sweden had no claims to Spitsbergen and did not want to have, as the country was not worth claiming.

And then it happened that some Americans visited the place as tourists and came upon some coal on the beach and some iron. On the strength of this and other evidence, engineers were sent there and reported that the islands contained fabulous quantities of easily accessible coal and iron of high grade. An American company was organized for the promotion of these mines, and a Norwegian and an English company were also organized.

Several countries then simultaneously awoke to the realization of the value of Spitsbergen. Holland began to claim it because she had discovered it, Great Britain because she had for a long time held possession of it, and Russia and the Scandinavian countries because they had explored it and had other possessions not so very many hundreds of miles away from it. Even the Germans claimed it. Each country was a dog in the manger so far as all the other countries were concerned, and anarchy was a consequence. Though huge commercial enterprises were being undertaken, there was on the islands no police officer or judge or any vestige of recognized government, and no way of legally obtaining title to any property.

In 1913, on a visit to England I met one of the large coal mine owners of Wales, who told me that it was already then clearly foreseen by himself and all the other coal men whom he knew that Spitsbergen was soon to become one of the chief competitors, if not the chief competitor, of Wales in the coal markets of the world.

The representations of the various commercial concerns finally led to an international convention of the countries involved. This convention had met in Norway and was in session when it was suddenly and automatically dissolved by the conflagration of the World War. Later the American capitalists, doubtless partly because they failed to secure support from their government, sold their holdings to the Norwegians, and Great Britain and Norway remained the two countries most vitally interested.

Now comes a chapter in the story of Spitsbergen that is humorous or tragic or pathetic according to one's attitude toward the statesmen and industrial pioneers of Britain. In the spring of 1920 the newspapers carried an announcement that the British had surrendered to Norway their political claims to Spitsbergen. I was in New York when this news was published and was interviewed on the subject by some enterprising reporters. As it seemed to me clear that Britain had a stronger claim to the islands than any other nation, and certainly a much stronger one than Norway, I gave it as my opinion that there must be behind the transaction some secret political bargain, possibly made at a time of war stress and uncertainty, and that Norway was being rewarded now by Britain for having kept her agreement. Knowing the large investment of English and Scotch capital in the Spitsbergen coal mines I did not conceive it possible that English diplomats had now succeeded in doing in the case of Norway what they had failed in 1763 to do with France, when they tried to give away (or refuse to receive) Canada.[1]

---

1. When strongly impressed with their value, the British have occasionally "grabbed" remote lands, their titles to which were by no means clear. This makes the more curious their propensity as a government to give away, under a short-sighted impression of their worthlessness, lands to which their title was clear. Apropos of certain news despatches about Wrangel Island, this propensity was recently summarized by an Editorial in the New York *Evening Post*: "The occupation of Wrangel Island, nominally for Great Britain and actually for Canada, is designed to make good a claim which could never rest on the mere fact of British discovery. It adds another to the many striking instances of the extension of British sovereignty by energetic subjects without the knowledge and even against the will of the colonial office. London cares nothing about Wrangel Island, and Ottawa

Soon after the publication of this interview I went to England and found that so far as my friends knew, who were interested in the Spitsbergen mines, the unbelievable was true. Their statements may have been colored by the heat of their feelings, but they told me that the substance of the story was this: The Norwegians had said to the British diplomats at Paris that if Britain didn't mind very much they would like, please, to be given Spitsbergen. To this the British had replied in substance that they didn't see why anybody wanted those isolated, frozen islands, but if anybody did want them badly enough to ask for them they didn't see why they shouldn't have them.

If this be a true statement, these British diplomats can at least quote an excellent precedent from Tacitus—they were repeating about Spitsbergen what he would have said about Britain nearly two thousand years earlier. . . .

Meantime the Journal of the Royal Geographical Society of Great Britain tells us that in 1918, in spite of the extraordinary difficulties due to the unsettled condition of Europe (and not to the climate or latitude of Spitsbergen) one hundred thousand tons of coal was exported. The Journal also says that the Admiralty of Great Britain has published a table of the comparative steam values of various kinds of coal, which places that of Spitsbergen higher than the best Welsh coal. It says further that while the rich iron ore of Spitsbergen is at present being exported to smelters in Great Britain this is but a transient phenomenon, for in the course of a few years local smelters are certain to be built. Spitsbergen is one of a very few known places in the world where a large quantity of easily accessible hard coal is found in close proximity to large quantities of easily accessible iron ore of high grade.

It has always been easy for people of that type of mind known as "practical," "sound," and "conservative" to prove that lands as yet of no value cannot

---

little. New Zealand was saved to Great Britain in 1839 by the spirited colonizer, Edward Gibbon Wakefield, only when the French were on the point of taking possession. In Hawaii in 1794 a council of chiefs asked for British protection, and a naval lieutenant hoisted the British flag, while in 1822 Kamechameha II. confirmed the protectorate, but the British government would never acknowledge the acquisition. Tahiti was discovered by the British in 1767 and a British flag hoisted, while in 1825 Queen Pomare was eager for a British protectorate; but Louis Philippe of France was allowed to take over the island. The Fiji islands all but slipped from the careless British grasp—W. T. Pritchard saved them, to the intense indignation of the colonial office. Australia for years after 1867 labored to effect British annexation of eastern New Guinea, but the British resisted until they found that the Germans had gobbled half.

"Islands, as Carlyle said of tools, belong to those who can use them. Under international law the mere hoisting of a flag, without continuing occupation, avails nothing."

possibly ever be of value. In striking contrast to this type of mind is that of the born explorer, who must, above all things, be a man of imagination. Henry Hudson, the second navigator to reach those islands, noted in his journal in the year 1607 that he had no doubt Spitsbergen "would be profitable to whoever should adventure it."

Chief of the arguments against the value of Spitsbergen fifteen years ago was that it was located in an arctic sea which, although it could be navigated at certain seasons, could not be profitably navigated because interrupted navigation was said to be never profitable. This same argument is at present being advanced most convincingly against the feasibility of the Hudson Bay route which the Canadian Government is developing as a means of contact between the prairie provinces and Europe, by way of Hudson Straits. Although the argument sounds convincing when pronounced with conviction, actual trial has failed to confirm it. I have talked with an able mining engineer who at one time was in charge of the mines of the American firm, Ayer and Longyear, in Spitsbergen, and he has told me that he believes coal can be so cheaply mined and transferred from Spitsbergen to Europe that Spitsbergen will drive Newcastle and Wales out of the continental coal markets north of their latitude, which means among others those of the White Sea and the Murman Coast and the northern half of the Scandinavian countries.

"All very interesting," the critics may say, "but it is a long lane that has no turning. Tacitus was wrong when he said people would never by choice live as far north as France; the Moors of the Middle Ages were short-sighted when they undervalued the possibilities of Britain; it is strange that as astute a man as Franklin thought a small tropic isle like Guadeloupe commercially more valuable than Canada; Seward was wise in buying Alaska and Gladstone a simpleton to want to renounce Spitsbergen. But surely there must be somewhere the limit to Northward progress. Have we not come to that limit now?"

We have not come to the northward limit of commercial progress. There was many a pause but no stop to the westward course of empire until we came to the place where East is West. In that sense only is there a northward limit to progress. Corner lots in Rome were precious when the banks of the Thames had no value; the products of Canada were little beyond furs and fish when the British and French agreed in preferring Guadeloupe. But values have shifted north since then and times have changed. Time will continue to change. There is no northern boundary beyond which productive enterprise cannot go till North meets North on the opposite shores of the Arctic Ocean as East has met West on the Pacific.

Commentary

Vilhjalmur Stefansson, *The Northward Course of Empire* (1922)

SVERKER SÖRLIN

In the year 1922 anthropologist and polar explorer Vilhjalmur Stefansson published *The Northward Course of Empire*. It was a book in the style of the day, speculative, spatial in its cognitive structure, and with a suggestive deterministic tendency. The core idea was straightforward: the level of "civilization" rose with time as its center moved from hotter places to colder. In fact, the level of civilization stood in an inverse relation to annual mean temperatures.

In Stefansson's version of the world's development, the cradle of modern civilization was in the Upper Nile Valley and Sumer five thousand years ago. It had then moved via Babylon, ancient Greece, and Rome toward western Europe. In the nineteenth century it reached its peak in London, only to move across the Atlantic to New York. The bold descendant of Icelandic parents who had immigrated to the prairies of Manitoba, Stefansson declared with prophetic self-assurance that the shining star of mankind would in the future continue toward the northwest. It would reach its culmination in ice-cold Winnipeg, the place where Stefansson had been born forty years earlier. As a child, the Icelandic-Canadian moved to North Dakota in the United States, later studying at Harvard. From there, Stefansson spent a field season in Iceland, where he, under dubious circumstances, collected skulls for Harvard's Peabody Museum.

After his *rite d'initiation* as an anthropologist of Iceland, he spent almost two decades on long expeditions in Alaska and northern Canada, building his reputation as an authority on the north, and also as a controversial, somewhat eccentric, visionary with more ideas than judgment. Several men lost their lives in the Bering Straits when his expedition ship, the *Karluk*, sank in 1912. Stefansson himself abandoned the ship and was rescued. During his long stays among the Inuit, and the "Copper Inuit," which he claimed to have discovered, he met Fanny Pannigabluk and fathered a son, Alex. He provided for Fanny and Alex, but he never publicly acknowledged them when he later settled in New York as a writer, collector, and polar expert.

Cold and civilization were irretrievably bound together, Stefansson argued. In *The Northward Course of Empire* he wrote that nothing would stop intrepid modern man in the quest to reach ever further north. Humanity would conquer even this last remaining part of the planet, and eventually ships would cross the Arctic Ocean from bustling harbors; airplanes would fly high above the North Pole, uniting continents on short routes; cities would emerge out of the tundra;

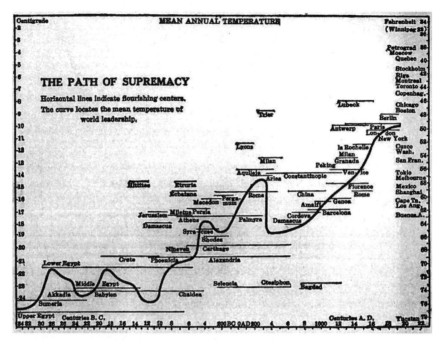

Original frontispiece from *The Northward Course of Empire*.

and industry would resound mightily from the farthest, coldest ends of the earth. Life in the Arctic was in no way as hard and dangerous as rumor had it. In *The Friendly Arctic* (1921), he disparaged adventurous travel in the north as unnecessary and exaggerated. With the right equipment and a little bit of local knowledge you could live well, if not luxuriously, on the tundra.

For some time there was demand for Stefansson's ideas. They underpinned a certain form of identity politics in a Canada that took successive steps toward independence from Great Britain, a process that extended well into the postwar period. By this time, however, Stefansson had fallen into disgrace and been declared a security risk. He lost his funding from the U.S. Office of Naval Research to edit a ten-volume *Encyclopedia of the Arctic* because the Central Intelligence Agency found that he had made economic contributions to radical immigrant groups in New York and to organizations labeled "communist" during the McCarthy years. At that time Stefansson's popularity had gravitated eastward, and his books sold in large numbers in the Soviet Union, where the Arctic remained of major political interest, as evidenced by the establishment of the Arctic Institute in Leningrad in the 1920s and the Agency for the Naval Route to Asia (*Glavsev-*

morput). Stefansson ended his checkered career as a professor at Dartmouth College in Hanover, New Hampshire, where photographs from the 1950s and 1960s show him building igloos in the snow with his students.

Almost everything that Stefansson wrote was about the Arctic, but the wider context of his thinking included contemporary ideas of geography and especially natural resources, and to some extent race, as determining factors of development. In that respect he had many similarities with contemporaries such as Ellsworth Huntington and Griffith Taylor, although his outlook was fundamentally more optimistic than either of these famous determinists. Stefansson was essentially uninterested in more theoretical economic geography, but he was considered important enough to be asked to write a report for the Australian government about the possible use of the central parts of that country. For his ideas about the northward march of civilization he was directly inspired by S. Columb GilFillan, who had published an article in 1920 with almost identical ideas. GilFillan was later to appear in U.S. federal committees on futures studies, and what attracted Stefansson to GilFillan's idea, apart from its bright prospects for the Arctic, was its speculative and predictive element. Geography was regarded by Stefansson as the subject of the future.

There was an element of racism and racial determinism in *The Northward Course of Empire*. But if white superiority was commonly explained as a result of a mild and temperate climate, as opposed to extreme climates in the tropical and Arctic zones that produced only inferior races, Stefansson would claim that it was the cold itself that refined the best breed, and that the pinnacle of civilization would therefore be found further north. Global change was thus, for him, not so much a matter of humans impacting on the environment, but rather the environment providing the tools for a (white and southern) humanity to improve themselves and the extremes of the planet together. Local Arctic residents would contribute important experience, but essentially the new empires of the north would be fully modern and would harness nature's resources for industry. Countries that realized the future value of northern natural resources had an advantage. Such a racially inflected resource optimism was not uncommon in the first third of the twentieth century, but Stefansson's was an extreme case.

Further Reading

Bell, Wendell. (1997) 2003. *Foundations of futures studies*, Vol. 1–2. New Brunswick, NJ: Transaction.

GilFillan, S. Columb. 1920. "The coldward course of progress." *Political Science Quarterly* 35(3):393–410.

Pálsson, Gísli. 2005. *Travelling passions: The hidden life of Vilhjalmur Stefansson*, trans. Keneva Kunz. Winnipeg: University of Manitoba Press.

Price, David H. 2004. *Threatening anthropology: McCarthyism and the FBI's surveillance of activist anthropologists*. Durham, NC: Duke University Press.

Stefansson, Vilhjalmur. 1921. *The friendly Arctic: The story of five years in polar regions*. New York: The Macmillan Company.

PART 4

"The Environment"

How Did the Idea Emerge?

What is "the environment"? After at least half a century of environmental concern, with Rachel Carson's best seller *Silent Spring* (1962) often used as the canonical scripture of the foundation myth, we still know surprisingly little about the concept, its origins, usage, functions, let alone its deeper relations with global change science. This is despite the fact that a lot of that science has the self-understanding of being environmental, or at least potentially beneficial for the environment, a term generally understood as denoting something out there that is important for our well-being but is affected (negatively) by human action.

The latter decades of the twentieth century saw a growth of the environment as a significant social phenomenon. Social movements defended it, and green political parties "spoke" for it. Most of all, it was the focus of vast new integrated management programs, with global environmental institutions established around the world, particularly in northern Europe and North America. These suggest deeper cultural or even religious dimensions to environmental ideas and concerns.

Historical research has accumulated ever more solid and convincing evidence that there is more to the story of the environment. First of all, the concept itself is older and has a more varied use over time than has previously been assumed. A moment of change seems to have occurred in the years following World War II. "The environment," precisely in that form, as a specific noun, started being used then among a few authors—for example, American ecologist William Vogt in his *Road to Survival* (1948)—suggesting that there was also a deeper change in the thinking and social understanding of human-nature relationships going on.

Vogt's book was a success, read by many millions; it was apparently able to captivate an audience around the world that was already disposed toward the alarming message of rising populations, eroding soils, threatened species, and polluted waters. Each and every one of these items on Vogt's comprehensive list of grave problems had their own history, shorter or longer, and their geography as well, regional or even global. "The environment" became the word with which it was from now on possible to talk about many, if not all, these issues, or at least their effects on nature. This word, this idea, is now so entrenched in our language that it seems hard to believe that it ever could have been new, and certainly not so recently.

Thus "the environment" became a global singularity already taken as a fact, a catchall through which human destiny could be expressed and mapped. Something that, like us, had a fate. "Environment" itself was not a new word by any means; it had medieval origins in English, had been popularized by Thomas Carlyle, and it appears with increasing frequency from the 1890s. But in each case it was local and relational, and nearly always qualified: the closest to the modern sense is found in the idea of a "physical environment." Even in the 1960s, some time after the "tipping point" in which the modern environmentalist narrative was formed, the word remained opaque to many, and needed qualification. Rachel Carson in *Silent Spring* in 1962 (the same year that we, according to the *Oxford English Dictionary*, coined the term "environmental sciences") spoke of "the total environment." Vogt linked "*the* environment" not just to the local and its influence on the individual organism, but also to the global. On a broad canvas he painted the world as an increasingly unified place, starting with the age of exploration and culminating in the modern integrated understanding: "that we live in one world in an ecological—an environmental—sense."

The idea of "the environment," and the prediction of a whole global system falling into degradation, emerged together: co-determined, or bound in a kind of double helix that may be impossible to unravel. "The environment" seems to have borne an orientation toward the future in it from its beginnings, just as, from the same time (if not before), one could hardly say "nature" without evoking a certain nostalgia. There were other concepts and ideas around in the preceding decades that can be seen as attempts to do a similar job. Russian scientist Vladimir Vernadsky's "biosphere" from 1926, which only gradually spread in the Western world, brought out the global dimension and underscored its fragility and thinness. Paul Sears in *Deserts on the March* (1935) talks about ecology as the discipline that could be entrusted with the charge of understanding the whole of the ecosystem, the discipline that could also predict changes and thus also be called on for prevention. The common denominator here is that humans and societies, if not economies, were starting to affect various features of the physical world in ways that created a need for concepts that could describe in a neat and practical way that particular new, severe, and potentially, if not actually, global pressure on various features of the natural. It took a few decades, but by the mid-1960s it is clear that the codifying concept was "the environment." Evidence for this was the choice of title for the U.S. Environmental Protection Agency created in 1967. Even then the term was not yet hegemonic, and so it is not surprising that the otherwise progressive Swedish equivalent of the EPA, also formed that year, should chose the name *Nature* Protection Agency. The formative period lasted a whole generation, from the 1930s through to the 1960s.

The environmental outlook was at its "tipping point" when Vogt published in 1948. The environment was first and foremost a work of integrative imagination, of combining a set of already existing issues and problems on various scales into a meaningful whole. Vogt and his fellow American Fairfield Osborn, who published *Our Plundered Planet* (also in 1948) established what we could call the "modern environmental problem catalogue." It included, but was not limited to, population growth (by far the number-one issue at the time), water scarcity, soil erosion, overconsumption, overgrazing, overfishing, pests, industrial wastes, the retarding productivity of soils, and species loss. Only later would these concepts be joined by "climate change," although climate had played a greater or lesser role in discourses about the nature and fate of civilizations since antiquity.

None of the "catalogue" was entirely new, although some concepts were more acknowledged than others. Yet, the integration of "the environment" brought a new quality to the understanding of the earth, to humanity as a species, and to their inextricably bound futures. It was an act of imagination, but one that required and prompted new metrics and perhaps even a genre of its own. There was also a fundamental narrative structure built into the concept of the environment—something degrading, as in declensionist narratives, or something that was returning to a sustainable state, as in recovery narratives. New interpretations would occur later—for example, emphasizing the creative dimensions of environment, and environment as a result of human labor and as extension of the societal into the natural.

The Biosphere

VLADIMIR I. VERNADSKY

The Biosphere as a Region of Transformation of Cosmic Energy

The biosphere may be regarded as a region of transformers that convert cosmic radiations into active energy in electrical, chemical, mechanical, thermal, and other forms. Radiations from all stars enter the biosphere, but we catch and perceive only an insignificant part of the total; this comes almost exclusively from the sun. The existence of radiation originating in the most distant regions of the cosmos cannot be doubted. Stars and nebulae are constantly emitting specific radiations, and everything suggests that the penetrating radiation discovered in the upper regions of the atmosphere by Hess originates beyond the limits of the solar system, perhaps in the Milky Way, in nebulae, or in stars of the Mira Ceti type. The importance of this will not be clear for some time, but this penetrating cosmic radiation determines the character and mechanism of the biosphere.

The action of solar radiation on earth-processes provides a precise basis for viewing the biosphere as both a terrestrial and a cosmic mechanism. The sun has completely transformed the face of the Earth by penetrating the biosphere, which has changed the history and destiny of our planet by converting rays from the sun into new and varied forms of energy. At the same time, the biosphere is largely a product of this radiation.

The important roles played by ultraviolet, infrared, and visible wavelengths are now well recognized. We can identify the parts of the biosphere that transform these three systems of solar vibration, but the mechanism of this transformation presents a challenge which our minds have only begun to comprehend. The mechanism is disguised in an infinite variety of natural colors, forms and

V. Vernadsky. (1926/1929) 1998. Excerpts from *La Biosphère* (Russian original 1926, French translation 1929 [Paris: Félix Alcan], English translation 1998). David B. Langmuir and Mark A. S. McMenamin (eds.). New York: Copernicus.

movements, of which we, ourselves, form an integral part. It has taken thousands of centuries for human thought to discern the outlines of a single and complete mechanism in the apparently chaotic appearance of nature.

In some parts of the biosphere, all three systems of solar radiation are transformed simultaneously; in other parts, the process may lie predominantly in a single spectral region. The transforming apparatuses, which are always natural bodies, are absolutely different in the cases of ultraviolet, visible, and thermal rays.

Some of the ultraviolet solar radiation is entirely absorbed, and some partly absorbed, in the rarefied upper regions of the atmosphere; i.e., in the stratosphere, and perhaps in the "free atmosphere," which is still higher and poorer in atoms. The stoppage or "absorption" of short waves by the atmosphere is related to the transformation of their energy. Ultraviolet radiation in these regions causes changes in electromagnetic fields, the decomposition of molecules, various ionization phenomena, and the creation of new molecules and compounds. Radiant energy is transformed, on the one hand, into various magnetic and electrical effects; and on the other, into remarkable chemical, molecular, and atomic processes. We observe these in the form of the aurora borealis, lightning, zodiacal light, the luminosity that provides the principal illumination of the sky on dark nights, luminous clouds, and other upper-atmospheric phenomena. This mysterious world of radioactive, electric, magnetic, chemical, and spectroscopic phenomena is constantly moving and is unimaginably diverse.

These phenomena are not the result of solar ultraviolet radiation alone. More complicated processes are also involved. All forms of radiant solar energy outside of the four and one-half octaves that penetrate the biosphere are "retained"; i.e., transformed into new terrestrial phenomena. In all probability this is also true of new sources of energy, such as the powerful torrents of particles (including electrons) emitted by the sun, and of the material particles, cosmic dust, and gaseous bodies attracted to the Earth by gravity. The role of these phenomena in the Earth's history is beginning to be recognized.

They are also important for another form of energy transformation—living matter. Wavelengths of 180–200 nanometers are fatal to all forms of life, destroying every organism, though shorter or longer waves do no damage. The stratosphere retains all of these destructive waves, and in so doing protects the lower layers of the Earth's surface, the region of life.

The characteristic absorption of this radiation is related to the presence of ozone (the ozone screen, formed from free oxygen—itself a product of life).

While recognition of the importance of ultraviolet radiation is just beginning, the role of radiant solar heat or infrared radiation has long been known, and calls for special attention in studies of the influence of the sun on geologic and geochemical processes. The importance of solar radiant heat for the existence of life is incontestable; so, too, is the transformation of the sun's thermal radiation into mechanical, molecular (evaporation, plant transpiration, etc.), and chemical energy. The effects are apparent everywhere—in the life of organisms, the movement and activity of winds and ocean currents, the waves and surf of the sea, the destruction of rock and the action of glaciers, the formation and movements of rivers, and the colossal work of snow and rainfall.

Less fully appreciated is the role that the liquid and gaseous portions of the biosphere play as accumulators and distributors of heat. The atmosphere, the sea, lakes, rivers, rain and snow actively participate in these processes. The world's ocean acts as a heat regulator, making itself felt in the ceaseless change of climate and seasons, living processes, and countless surface phenomena. The special thermal properties of water, as determined by its molecular character, enable the ocean to play such an important role in the heat budget of the planet.

The ocean takes up warmth quickly because of its great specific heat, but gives up its accumulated heat slowly because of feeble thermal conductivity. It transforms the heat absorbed from radiation into molecular energy by evaporation, into chemical energy through the living mater which permeates it, and into mechanical energy by waves and ocean currents. The heating and cooling of rivers, air masses and other meteorological phenomena are of analogous force and scale.

The biosphere's essential sources of energy do not lie in the ultraviolet and infrared spectral regions, which have only an indirect action on its chemical processes. It is *living matter*—the Earth's sum total of living organisms—that transforms the radiant energy of the sun into the active chemical energy of the biosphere.

Living matter creates innumerable new chemical compounds by photosynthesis, and extends the biosphere at incredible speed as a thick layer of new molecular systems. These compounds are rich in free energy in the thermodynamic field of the biosphere. Many of the compounds, however, are unstable, and are continuously converted to more stable forms.

These kinds of transformers contrast sharply with terrestrial matter, which is within the field of transformation of short and long solar rays through a fundamentally different mechanism. The transformation of ultraviolet *and infrared* radiation takes place by action on atomic and molecular substances that were created entirely independently of the radiation itself.

Photosynthesis, on the other hand, proceeds by means of complicated, specific mechanisms *created by photosynthesis itself*. Note, however, that photosynthesis can proceed only if ultraviolet and infrared processes are occurring simultaneously, transforming the energy in these wavelengths into active terrestrial energy.

Living organisms are distinct from all other atomic, ionic, or molecular systems in the Earth's crust, both within and outside the biosphere. The structures of living organisms are analogous to those of inert matter, only more complex. Due to the changes that living organisms affect on the chemical processes of the biosphere, however, living structures must not be considered simply as agglomerations of inert stuff. Their energetic character, as manifested in multiplication, cannot be compared geochemically with the static chemistry of the molecular structures of which inert (and *once*-living) matter are composed.

While the chemical mechanisms of living matter are still unknown, it is now clear that photosynthesis, regarded as an energetic phenomenon in living matter, takes place in a particular chemical environment, and also within a thermodynamic field that differs from that of the biosphere's. Compounds that are stable within the thermodynamic field of living matter become unstable when, following death of the organism, they enter the thermodynamic field of the biosphere and become a source of free energy.*

Living Matter in the Biosphere

Life exists only in the biosphere; organisms are found only in the thin outer layer of the Earth's crust, and are always separated from the surrounding inert matter by a clear and firm boundary. Living organisms have never been produced by inert matter. In its life, its death, and its decomposition an organism circulates its atoms through the biosphere over and over again, but living matter is always generated from life itself.

A considerable portion of the atoms in the Earth's surface are united in life, and these are in perpetual motion. Millions of diverse compounds are constantly being created, in a process that has been continuing, essentially unchanged, since the early Archean, four billion years ago. Because no chemical force on Earth is more constant than living organisms taken in aggregate, none is more powerful in the long run. The more we learn, the more convinced we become that biospheric chemical phenomena never occur independent of life. . . .

It is evident that if life were to cease the great chemical processes connected with it would disappear, both from the biosphere and probably also from the crust. All minerals in the upper crust—the free alumino-silicious acids (clays), the carbonates (limestones and dolomites), the hydrated oxides of iron and aluminum (limonites and bauxites), as well as hundreds of others, are continuously created by the influence of life. In the absence of life, the elements in these

---

* The domain of phenomena within an organism ("the field of living matter") is different, thermodynamically and chemically, from "the field of the biosphere." (V.V.)

minerals would immediately form new chemical groups corresponding to the new conditions. Their previous mineral forms would disappear permanently, and there would be no energy in the Earth's crust capable of continuous generation of new chemical compounds.

A stable equilibrium, a chemical calm, would be permanently established, troubled from time to time only by the appearance of matter from the depths of the Earth at certain points (e.g., emanations of gas, thermal springs, and volcanic eruptions). But this freshly-appearing matter would, relatively quickly, adopt and maintain the stable molecular forms consistent with the lifeless conditions of the Earth's crust. . . .

Water is a powerful chemical agent under the thermodynamic conditions of the biosphere, because life processes cause this "natural" *vadose* water to be rich in chemically active foci, especially microscopic organisms. Such water is altered by the oxygen and carbonic acid dissolved within it. Without these constituents, it is chemically inert at the prevailing temperatures and pressures of the biosphere. In an inert, gaseous environment, the face of the Earth would become as immobile and chemically passive as that of the moon, or the metallic meteorites and cosmic dust particles that fall upon us.

Life is, thus, potently and continuously disturbing the chemical inertia on the surface of our planet. It creates the colors and forms of nature, the associations of animals and plants, and the creative labor of civilized humanity, and also becomes a part of the diverse chemical processes of the Earth's crust. There is no substantial chemical equilibrium on the crust in which the influence of life is not evident, and in which chemistry does not display life's work.

Life is, therefore, not an external or accidental phenomenon of the Earth's crust. It is closely bound to the structure of the crust, forms part of its mechanism, and fulfills functions of prime importance to the existence of this mechanism. Without life, the crustal mechanism of the Earth would not exist.

All living matter can be regarded as a single entity in the mechanism of the biosphere, but only one part of life, *green vegetation*, the carrier of chlorophyll, makes direct use of solar radiation. Through photosynthesis, chlorophyll produces chemical compounds that, following the death of the organism of which they are part, are unstable in the biosphere's thermodynamic field.

The whole living world is connected to this green part of life by a direct and unbreakable link. The matter of animals and plants that do not contain chlorophyll has developed from the chemical compounds produced by green life. One possible exception might be autotrophic bacteria, but even these bacteria are in some way connected to green plants by a genetic link in their past. We can therefore consider this part of living nature as a development that came after the transformation of solar energy into active planetary forces. Animals and fungi accumulate nitrogen-rich substances which, as centers of chemical free energy, become even more powerful agents of change. Their energy is also released through decomposition when, after death, they leave the thermodynamic field in which they were stable, and enter the thermodynamic field of the biosphere.

Living matter as a whole—the totality of living organisms—is therefore a unique system, which accumulates chemical free energy in the biosphere by the transformation of solar radiation. . . .

The firm connection between solar radiation and the world of verdant creatures is demonstrated by empirical observation that conditions ensure that this radiation will always encounter a green plant to transform the energy it carries. Normally, the energy of all the sun's rays will be transformed. This transformation of energy can be considered as *a property* of living matter, its *function* in the biosphere. If a green plant is unable to fulfill its proper function, one must find an explanation for this abnormal case.

The Multiplication of Organisms and Geochemical Energy in Living Matter

The diffusion of living matter *by multiplication*, a characteristic of all living matter, is the most important manifestation of life in the biosphere and is the essential feature by which we distinguish life from death. It is a means by which the energy of life unifies the biosphere. It becomes apparent through the *ubiquity of life*, which occupies all free space if no insurmountable obstacles are met. The whole surface of the planet is the domain of life, and if any part should become barren, it would soon be reoccupied by living things. In each geological period (representing only a brief interval in the planet's history), organisms have developed and adapted to conditions which were initially fatal to them. Thus, the limits of life seem to expand with geological time. In any event, during the entirety of geological history life has tended to take possession of, and utilize, all possible space.

This tendency of life is clearly inherent; it is not an indication of an external force, such as is seen, for example, in the dispersal of a heap of sand or a glacier by the force of gravity.

The diffusion of life is a sign of internal energy—of the chemical work life performs—and is analogous to the diffusion of a gas. It is caused, not by gravity, but by the separate energetic movements of its component particles. The diffusion of living matter on the planet's surface is an inevitable movement caused by new organisms, which derive from multiplication and occupy new places in the biosphere; this diffusion is the autonomous energy of life in the biosphere, and becomes known through the transformation of chemical elements and the creation of new matter from them. We shall call this energy *the geochemical energy of life in the biosphere*.

The uninterrupted movement resulting from the multiplication of living organisms is executed with an inexorable and astonishing mathematical regularity, and is the most characteristic and essential trait of the biosphere. It occurs on the land surfaces, penetrates all of the hydrosphere, and can be observed in each level of the troposphere. It even penetrates the interior of living matter itself, in the form of parasites. Throughout myriads of years, it accomplishes a colossal geochemical labor, and provides a means for both the penetration and distribution of solar energy on our planet.

It thus not only transports matter, but also transmits energy. The transport of matter by multiplication thus becomes a process *sui generis*. It is not an ordinary, mechanical displacement of the Earth's surface matter, independent of the environment in which the movement occurs. The environment resists this movement, causing a friction analogous to that which arises in the motion of matter caused by forces of electrostatic attraction. But life is connected with the environment in a deeper sense, since it can occur only through a gaseous exchange between the moving matter and the medium in which it moves. The more intense the exchange of gases, the more rapid the movement, and when the exchange of gases stops, the movement also stops. This exchange is the *breathing* of organisms; and, as we shall see, it exerts a strong, controlling influence on multiplication. Movement due to multiplication is therefore of great geochemical importance in the mechanisms of the biosphere and, like respiration, is a manifestation of solar radiation.

Although this movement is continually taking place around us, we hardly notice it, grasping only the general result that nature offers us—the beauty and diversity of form, color, and movement. We view the fields and forests with their flora and fauna, and the lakes, seas, and soil with their abundance of life, as though the movement did not exist. We see the static result of the dynamic equilibrium of these movements, but only rarely can we observe them directly.

Let us dwell for a moment on some examples of this movement, the creator of living nature, which plays such an essential yet invisible role. From time to time, we observe the disappearance of higher plant life from locally restricted areas. Forest fires, burning steppes, plowed or abandoned fields, newly-formed islands, solidified lava flows, land covered by volcanic dust or created by glaciers and fluvial basins, and new soil formed by lichens and mosses on rocks are all examples of phenomena that, for a time, create an absence of grass and trees in particular places. But this vacancy does not last; life quickly regains its rights, as green grasses, and then arboreal vegetation, reinhabit the area. The new vegetation enters partially from the outside, through seeds carried by the wind or by mobile organisms; but it also comes from the store of seeds lying latent in the soil, sometimes for centuries.

The development of vegetation in a disturbed environment clearly requires seeds, but even more critical is the geochemical energy of multiplication. The speed at which equilibrium is reestablished is a function of the transmission of geochemical energy of higher green plants. The careful observer can witness this movement of life, and even sense its pressure, when defending his fields and open spaces against it. In the impact of a forest on the steppe, or in a mass of lichens moving up from the tundra to stifle a forest, we see the actual movement of solar energy being transformed into the chemical energy of our planet.

The Limits of Life in the Biosphere

The terrestrial envelope occupied by living matter, which can be regarded as the entire field of existence of life, is a continuous envelope, and should be differentiated from discontinuous envelopes such as the hydrosphere.

The field of vital stability is, of course, far from completely occupied by living matter; we can see that a slow penetration of life into new regions has occurred during geological time.

Two regions of the field of vital stability must be distinguished: 1. the region of temporary penetration, where organisms are not subject to sudden annihilation; and 2. the region of stable existence of life, where multiplication can occur.

The extreme limits of life in the biosphere probably represent absolute conditions for all organisms. These limits are reached when any one of these conditions, which can be expressed as independent variables of equilibrium, becomes insurmountable for living matter; it might be temperature, chemical composition, ionization of the medium, or the wavelength of radiations.

Definitions of this kind are not absolute, since adaptation gives organisms immense ability to protect themselves against harmful environmental conditions. The limits of adaptation are unknown, but are increasing with time on a planetary scale.

Establishing such limits on the basis of known adaptations of life requires guesswork, always a hazardous and uncertain undertaking. Man, in particular, being endowed with understanding and the ability to direct his will, can reach places that are inaccessible to any other living organisms.

Given the indissoluble unity of all living beings, an insight flashes upon us. When we view life as a planetary phenomenon, this capacity of *Homo sapiens* cannot be regarded as accidental. It follows that the question of unchanging limits of life in the biosphere must be treated with caution.

The boundaries of life, based upon the range of existence of contemporary organisms and their powers of adaptation, clearly show that the biosphere is a terrestrial *envelope*. For the conditions that make life impossible occur simultaneously over the whole planet. It is therefore sufficient to determine only the upper and lower limits of the vital field.

The upper limit is determined by the radiant energy which eliminates life. The lower limit is formed by temperatures so high that life becomes impossible. Between these limits, life embraces (though not completely) a single thermodynamic envelope, three chemical envelopes, and three envelopes of states of matter. The above limits clearly reveal the importance of the last three, the troposphere, hydrosphere, and upper lithosphere.

The Relationship Between the Living Films and Concentrations of the Hydrosphere and Those of Land

It follows from the preceding that life presents an indivisible and indissoluble whole, in which all parts are interconnected both among themselves and with the inert medium of the biosphere. In the future, this picture will no doubt rest upon a precise and quantitative basis. At the moment, we are only able to follow certain general outlines, but the foundations of this approach seem solid.

The principal fact is that the biosphere *has existed throughout all geological periods*, from the most ancient indications of the Archean.

In its essential traits, the biosphere has always been constituted in the same way. One and the same chemical apparatus, created and kept active by living matter, has been functioning continuously in the biosphere throughout geologic times, driven by the uninterrupted current of radiant solar energy. This apparatus is composed of definite vital concentrations which occupy the same places in the terrestrial envelopes of the biosphere, while constantly being transformed.

These vital films and concentrations form definite secondary subdivisions of the terrestrial envelopes. They maintain a generally concentric character, through never covering the whole planet in an uninterrupted layer. They are the planet's active chemical regions and contain the diverse, stable, dynamic equilibrium systems of the terrestrial chemical elements.

These are the regions where the radiant energy of the sun is transformed into free, terrestrial chemical energy. These regions depend, on the one hand, upon the energy they receive from the sun; and on the other, upon the properties of living matter, the accumulator and transformer of energy. The transformation occurs in different degrees for different elements, and the properties and the distribution of the elements themselves play an important role.

All the living concentrations are closely related to one another, and cannot exist independently. The link between the living films and concentrations, and their unchanging character throughout time, is an eternal characteristic of the mechanism of the Earth's crust. . . .

The land and the ocean have coexisted since the most remote geological times. This coexistence is basically linked with the geochemical history of the biosphere, and is a fundamental characteristic of its mechanism. From this point of view, discussions on the marine origin of continental life seem vain and fantastic. Subaerial life must be just as ancient as marine life, within the limits of geological times; its forms evolve and change, but the change always takes place on the Earth's surface and not in the ocean. If it were otherwise, a sudden revolutionary change would have had to occur in the mechanism of the biosphere, and the study of geochemical processes would have revealed this. But from Archean times until the present day, the mechanism of the planet and its biosphere has remained unchanged in its essential characteristics.

Recent discoveries in paleobotany seem to be changing current opinions in the way indicated above. The earliest plants, of basal Paleozoic age, have an unexpected complexity which indicates a drawn-out history of subaerial evolution.

Life remains unalterable in its essential traits throughout all geological times, and changes only in form. All the vital films (plankton, bottom, and soil) and all the vital concentrations (littoral, sargassic, and fresh water) have always existed. Their mutual relationships, and the quantities of matter connected with them, have changed from time to time; but these modifications could not have been large, because the energy input from the sun has been constant, or nearly so, throughout geological time, and because the distribution of this energy in the vital films and concentrations can only have been determined by living matter — the fundamental part, and the only variable part, of the thermodynamic field of the biosphere.

Commentary

Vladimir Vernadsky, *The Biosphere* (1926)

PEY-YI CHU

Before there was "the environment," there was the biosphere.

For twenty-first century writers, scientists, and educators, "the environment" is a term that embraces all of Earth, encompassing natural resources as well as organic life. Conceptually, it can be subdivided into domains that map superficially onto the various states of matter. These include the lithosphere, or the planet's solid shell; the hydrosphere, its liquid blanket of water; and the atmosphere, its gaseous outer covering. Within this scheme, the biosphere represents the realm of life taken as an aggregate. It consists of the plants, animals, and microbes that inhabit the lithosphere, hydrosphere, and atmosphere. But once upon a time in the history of scientific ideas, the biosphere was more than simply one sphere among the many that make up "the environment." Rather, it denoted the living world together with its surroundings, or organisms and those parts of the lithosphere, hydrosphere, and atmosphere relevant to their existence. Before "the environment," the biosphere was a way of imagining the system of life on earth in its totality.

In order to recover this earlier conception of the biosphere, historians turn to Vladimir Ivanovich Vernadsky (1863–1945). Born in St. Petersburg, then the capital of the Russian empire, Vernadsky was a naturalist and mineralogist who became known as one of the founders of the field of geochemistry. In 1926, his book *The Biosphere* was published for the first time, while he occupied a prominent position in the Academy of Sciences of the newly formed Soviet Union. *The Biosphere* put forth Vernadsky's view of the earth as a system in which life was the primary moving force. He adopted the term from Austrian geologist Eduard Suess, who in 1875 used the term "biosphere" in passing to refer to the planet's organic life. Crediting Suess with the word, Vernadsky went further and elevated it to a major concept by broadening its scope and describing its characteristics and behavior in dynamic terms. *The Biosphere* was therefore both a work of synthesis, in which Vernadsky encapsulated the latest research of his time about the natural world, and a work of creation, in which he advanced a framework for understanding the unity of all matter in the known universe.

Vernadsky was one of many individuals in the early twentieth century preoccupied with conceptualizing the relationship between the planet and its living organisms. Scientists from fields such as physiology, biology, and geology sought to identify systems that consisted of entities linked by exchanges of energy and matter. The "ecosystem," a concept that emerged in the 1930s (see the Tansley essay in Part 5), was one such supposedly naturally occurring arrangement. Vernadsky's idea of the biosphere was another. Unlike ecosystems, however, which were local or regional manifestations, Vernadsky's biosphere operated on a cosmic scale. In his view, it covered the entire surface of the earth and extended into the crust as well as the atmosphere. It was everywhere in which radiation that reached the planet from space became transformed into heat, light, movement, and new molecules. According to Vernadsky, the bodies that performed this crucial work were both inert, like water, and alive, like insects. A special role belonged to photosynthesizing plants, which were capable of directly utilizing the sun's rays in order to create the nutrients that sustained other forms of life.

Given the similarities between the biosphere and concepts in the emerging field of ecology, it seems fitting that ecologists were among the most enthusiastic popularizers of Vernadsky's ideas. Historical contingencies were also important in shaping this uneven process. For a brief period after the Bolshevik Revolution, Vernadsky, who was politically unsympathetic to both the tsarist and socialist regimes in Russia, lived in Europe with his family. Although he returned to the Soviet Union and assumed a leadership role in organizing new institutes for scientific research, his son George remained abroad, eventually joining the faculty at Yale University as a historian. There, the younger Vernadsky met ecologist George Evelyn Hutchinson, with whom he collaborated to publish selections from his father's writings about the biosphere in the 1940s. Until then, very little of Vernadsky's work had been available to English-speaking audiences, although a wider selection had been translated into French and German. When scientists James Lovelock and Lynn Margulis formulated their conception of the earth as a natural system, which they called Gaia, in 1972, they were unaware of the degree to which it resonated with Vernadsky's biosphere. They acknowledged the connection more fully in the 1980s, when an American press issued an abridged version of *The Biosphere*. Although its ideas had been percolating internationally for over half a century, a complete English translation of the first edition was not published until 1997.

Belatedly and posthumously, Vernadsky became associated with the concept of the biosphere, which some historians and scientists now identify as a precursor to contemporary understandings of "the environment." But *The Biosphere* was not simply a step along the way in the progressive evolution of an idea. It was also a distinct text, aspects of which were subsequently ignored, underemphasized, or rejected. The richness and complexity of *The Biosphere* endowed it with the potential of becoming as much an alternative as a predecessor to later intellectual developments. One of its distinguishing characteristics was the emphasis on a holistic view of the biosphere, a perspective shaped by Vasily Dokuchaev, one of Vernadsky's most influential teachers at St. Petersburg University and a founder of the field of soil science. Vernadsky's biosphere, like Dokuchaev's notion of the soil, consisted of both living and nonliving entities that were clearly separate yet mutually dependent. Although life was the crucial animating feature of the biosphere, it was nevertheless constantly involved in chemical exchanges with its surroundings. Vernadsky's commitment to holism entailed an affirmation that both live and inert bodies were necessary components of the one biosphere.

Vernadsky's holism complemented his uniformitarianism, and both met with skepticism from other scientists. Not only did he insist that living matter was physically distinct from nonliving matter, but he also maintained that its total mass had remained constant throughout geological history. Although it changed forms and moved into new spaces, life had always existed on earth, without a beginning, and its origins were a mystery. Today, many scientists would reject these claims. They make room for the possibility that organic matter can emerge from inorganic matter, that the planet's biomass has increased over time, and that major and relatively sudden changes, such as the appearance of free oxygen in the atmosphere, characterized the distant past.

On the other hand, in a time when people are ever more conscious of humankind's degrading of "the environment," Vernadsky's writings might serve as a reminder of the biosphere's ultimate longevity and unknowability. Vernadsky's ideas about the place of humans in the cosmos evolved over time. In writing about life in *The Biosphere*, however, Vernadsky had in mind not so much humans as the sum total of living matter, including the smallest bacteria. People may be capable of influencing the biosphere, but they formed only one small part of a much larger system. Although their behavior may render the biosphere unsuitable for certain species, the whole would undoubtedly endure. Animated by this spirit of humility, we may contemplate the place of humans in "the environment" and what is truly at stake in our choices.

Further Reading

Bailes, Kendall E. 1990. *Science and Russian culture in an age of revolutions: V. I. Vernadsky and his scientific school, 1863–1945*. Bloomington: Indiana University Press.

Grinevald, Jacques. 1997. "Introduction: The invisibility of the Vernadskian revolution." In Mark McMenamin (ed.), *The biosphere*. New York: Copernicus.

Lapo, Andrey V. 1987. *Traces of bygone biospheres*. Moscow: Mir Publishers.

Samson, Paul R., and David Pitt (eds.). 1999. *The biosphere and noösphere reader: Global environment, society and change*. London: Routledge.

Vernadsky, V. 1945. "The biosphere and the noösphere." *American Science* 33:1–12.

Deserts on the March

PAUL SEARS

[165] Twelve years have passed since the preceding chapters were first printed. Since then, a score of excellent books have been written about the resources which have made this nation great and about our obligation to conserve them. At least sixty millions of acres have, with the technical supervision of the Soil Conservation Service, been put under proper land use and management, halting erosion and restoring lost fertility. The care of millions of additional acres has been influenced for the better by our growing consciousness of danger. These measures have not only arrested damage—they have contributed vastly to farm prosperity and to the feeding of the world during its present distress. The estimated increase in production on land so safeguarded is at least 20 per cent.

The land restored has included a generous portion of the high plains which in 1935 had been devastated by wind erosion. The financial cost of this particular restoration was about one dollar an acre, but the decisive factor was not financial. It was the determination of communities of men in Texas, Oklahoma, and other western states to collaborate on the solution of a common problem.

Since the beginning of the century there have been increasing numbers of societies devoted to the conservation of particular resources. These have included groups of sportsmen and lovers of trees, birds, wild flowers, water, and more recently, of the soil. At first each group pressed for a solution [166] of its individual problem, mostly by the passage of laws that were merely temporary expedients.

Typical was the breeding of countless fish to be released in waters which civilization had rendered unfit to sustain them—as though, by some sort of biological mass action, the fish might come to overwhelm the metallic poisons and biological filth from cities and the silt from eroded farms which continued meanwhile to pour unabated into the arteries of our landscape.

Paul Sears. (1935) 1949. "Deserts in Retreat." Chapter 18 in *Deserts on the march*, 2nd ed., 165–177. London: Routledge and Kegan Paul [original page numbers in square brackets].

Here too the past decade has witnessed a quiet but profound change. Each group, following its problem to the source, has bumped into the others—as explorers working up the fingers of a delta must ultimately meet at the parent stream. You cannot have fish without an abundance of clean water. You cannot have water, either for fisheries or industrial use, without forests and well-managed farms which will regulate the flow of water after it falls. You cannot have an adequate supply of timber without an intelligent program of land use. Wildlife must have suitable conditions in which to live and breed. Game animals, songbirds and wild flowers require areas of native vegetation—whether forest, grass, or desert—not required for other use. To feed and clothe itself, and certainly to engage in world commerce, this nation must protect its soil from destruction by wind and water. (We can thank a forthright Southerner, Bob Montgomery, of Texas, for telling the southern planters that every bale of cotton traded abroad for gold has cost them one hundred and thirty tons of soil carried out to the Gulf of Mexico.)

Today, while the secretariat of each of the several conservation groups still retains its pride of accomplishment, all of them are working together as never before. They, and the members who stand back of [behind] them, sense that the problem is manifold only in its symptoms. In essence it is one. That essence can be stated simply: All renewable natural resources are linked into a common pattern of relationship. We can save any one of them only by measures which save [167] them all. And we are a part of the whole which must be conserved.

This is sound science and an intensely practical matter. We have been deceived by the glib statement that science has given man control over nature. What does it mean, for example, to say that a rider controls his horse? Bit, whip, and spur, however valuable, are accessories—useful in proportion to the self-discipline, the judgment and understanding of the rider. The latter must know horses in general and his own mount as an individual. Then horse and rider become a system, and the fabled Centaur symbolized that fact.

We do not and cannot manipulate nature from the outside. We must work our will by knowing laws and conforming to them, never forgetting that we are a part of that upon which we work, as horse and rider are interrelated. Our bodies are composed of the elements of earth and air, and every breath we draw is an interchange of those materials. The work we do is an expression of the energy of the sun, fixed into foods by plant life, perhaps elaborated into new forms by animals. We are not independent of the forces of nature—at best interdependent, at worst, pathetically dependent. The marvelous perfection of the internal combustion engines which draw a giant bomber through the skies is possible only because their designers have known and scrupulously respected the orderly budget of energy and material transformation.

These laws which serve us so well in the mechanical realm apply quite as rigorously to landscape. They are simple in principle. They suggest the rules of accounting, upon which business is built. Whether you are dealing with money, or water, or minerals, or energy in the form of the products of plants and animals, you can take what is there, no more. An ordered landscape, farm-wide or nation-wide, is not a tap to be left open and drained as fast and completely as we please. It must be rather a system, sturdy yet in delicate balance, so [168] managed as to have reserves against the great cyclic swing of need and less favorable conditions.

It is quite as possible for the scientist to assay the budget of a landscape as it is for an examiner to determine the condition of a bank, or for a physician to determine whether a human body is gaining or losing in its struggle to survive. It is a matter of cold science, which no amount of political ranting or shallow enterprise can settle one way or another, any more than respectability can save a bank or love alone can stay the hand of Death.

Even so, the scientist has to guard himself, for he is human and has his pre-dilections. Engineers and biologists, for example, speak a very different language. Excepting agricultural engineers, very few of these highly trained specialists have any understanding of the laws which govern living things and systems, such as the soil, of which they are a part. Even if courses in biology were included in their schools, we could not be certain that they would remedy the deficiency, for many biologists are too concerned with evolution and what has been called "the deadly grammar of the dead cat" to consider the interrelations of living nature here and now. On the other hand, working biologists in the field—foresters, public health specialists, soil conservators—see so much of the damage wrought by industrial-ism gone wild that they are impatient with the needs of our technological society.

The result is that when the two groups view a problem such as the growing water shortage in the industrial eastern states, they do not see eye to eye. The engineer is inclined to blame fluctuating rainfall, to rely upon man-made de-vices, and to absolve industry and urban development. The biologist, conscious of the fact that the control of water begins where it falls, sees his answer in our interference with the balance of nature. This divergence of view is evident, too, in the problem of controlling the Mississippi and other areas of flood hazard. It is plain that these two groups of public servants must hammer away at each other until they reach a [169] mutual understanding before they can enlist the efforts of laymen.

Can the essential qualities of balance, order, and reserve which exist in nature be maintained under human dominance? For western civilization as a whole, the case remains to be proved. It is technically possible; the real problem is moral. One of the most profitable experiences of the past decade has been the chance to see, here and there, communities which have done reasonably well in establishing a permanent relationship with the land. These include most of the Amish and Mennonite communities and numerous other settlements of continental European origin. They generally combine the ancient peasant traditions of stewardship with common religious bonds. This certainly makes sense; obligation to the land is fundamentally a matter of faith, and cooperation has the quality of spiritual fellowship.

Many of these groups are feeling the strain of an alien industrial culture about them. Children rebel against traditional ways and their elders show increasing signs of emotional stress. Thus it is that, for our secular world, the great proving ground of the Tennessee Valley may have more to offer. It is a unit, but not an island. It seems to be absorbing the industrial age, using it without being overwhelmed by it. Perhaps the most encouraging side light on the TVA is the fact that it is warmly defended by Southerners of conservative temperament and that the attacks on it seem to come chiefly from those who would obtain political advantage by wrecking it. If it lives up to its beginnings, it will demonstrate collaboration between the local community and government at its best, with the individual keeping his self-respect and initiative.

I have said that the moral problem of conservation is far more serious than the technical. I am using the word moral in its broad and ancient sense, as including anything that involves human choice. There are signs of increasing will toward protection of our resources. Granting that this exists, [170] the problem of political action remains to be solved, and it seems just now to center about the relationship between central government, the local community, and the individual—a problem so well on its way toward solution in the Tennessee Valley.

Elsewhere the situation is less happy. The average farmer, upon whom falls the custody of all of our arable land and no small share of our other resources, is confused and resentful. He is overworked at present, but that is offset by his increased income. His complaint is against government, which seems curious, since he has never before had one which has tried to do so much for him. His dissatisfaction is not a product of war, for it has been growing in volume for a quarter of a century. He wants his relations with government simplified and wants to do as much for himself as he can.

Farm prosperity is good conservation. Where a man's treasure is, there his heart will be, and generally speaking, the landscape begins to wear away at the fringes of prosperity. Government solicitude for farm distress is much more than expedient politics, it is sound statesmanship. In the dark days of 1933 the urgent problem seemed to be to get more money into the farmer's hands. Destruction of surplus was not a happy solution. Seafarers do not enjoy the jettisoning of valuable cargo from a distressed vessel. Such measures are emergency actions. The government did get money to the farmer, while reducing the apparent surplus;

but destruction went against his convictions and those of the Supreme Court, too. The result was a wiser measure, combining farm subsidy with better care of the soil. This is the present Agricultural Adjustment Administration, the most gigantic effort at agricultural reform in our history and with one exception, in world history. The worst thing about it, in broad principle, is the thing which makes it tick—paying the farmer to do what he generally ought to do for selfish reasons. There is a better way, but it takes courage, patience, faith, and skill. It is the method of those leaders who strive to make people desire [171] their own betterment until they themselves begin to work for it, and who, when the people then request counsel, are ready with the wisest counsel that can be given. This is the essence of enlightened democracy, as against the doctrine, however benevolent, of regulation.

The surpluses of farm products which caused depressed farm income had been obtained not by farming, but by mining the soil. Especially hurtful were the three great cash crops, corn, tobacco, and cotton, each a major source of farmland erosion along its clean-tilled rows. A farm, to be balanced somewhat as a natural landscape is balanced, ought to have a fair proportion of legumes and meadow, and often native vegetation such as woodlot or prairie. Moreover, its layout should be planned.

Again, in a great industrial country like our own, where the urban population exceeds the rural, the production of essential food and fiber cannot be left to happy chance and the farmer's best guess. The fabric of our economy is too tightly interwoven for that. A growing season is very different from an assembly line. The economy of a factory can be shifted in a few weeks or months; that of a farm requires a cycle of years, usually four or five. Some way has to be found to apportion each farmer his task, within reason.

During the Hoover administration a brief attempt had been made to issue up-to-the-minute information regarding supply and demand to the farm public in the hope that, under free enterprise, this would guide the farmer in apportioning his fields to various uses. It was the sort of inside information which lies behind the daily market quotations. Perhaps it went too far inside—at any rate there was pressure by selfish interests on the government and the service was promptly throttled down for slow and sedate release after it had ceased to be news and had become history.

Later, farm relief was started. It began with a single barrel and ended as a battery. Committees elected in each county studied the farms in that county, on the ground and from [172] aerial photographs, trying to work out the scientific use of each field in the light of the nation's needs and the permanent welfare of the soil. This activity was guided by state land-grant experts working with Washington, but the actual work was local. By a combination of rewards and penalties, the county committees urged farmers to fall in line. Compulsion, however benign, does not please the rural American.

Meanwhile, other factors have added to the farmer's confusion. In earlier days his friendly government, with its magnificent technical services, was personified to him by the county agent, working out from the land-grant college of his state. The quiet labors of these agents were slowly moving in the direction of better land use and management, through education and demonstration. The county agent still remains a separate agency, overworked as ever, yet vital contacts between government and farmer have been routed past and around him.

In places where the soil erosion problem was critical, farmers have organized soil conservation districts which make available to them directly the help of the Soil Conservation Service. These districts have worked well, hampered only by the shortage of technically trained supervisors. Here again is a contact between government and farmer, wise in its intent and beneficial in its effects on the land; yet paralleling, if not duplicating the others which have been mentioned. No wonder the farmer is puzzled, at times morose, or even rebellious. Nor did it add to his peace of mind when the War Food Administration was taken from the hands of the Secretary of Agriculture, who is in fact the man responsible for food production in our country.

The government cannot stay out of farming, any more than the farmer can stay out of government. The bond is primordial. Civilized government is as truly a farm product as corn meal. It was not until farming made possible cheap and abundant food that men had energy to spare to become civilized. The technical knowledge needed for farming is biological [173] knowledge, based upon intricate, costly, and prolonged research which only a government can sustain. The needs of a civilized country for food and raw materials are too serious a matter to be left to chance. The farmer needs the government and the government needs the farmer.

Yet it would be folly to suppose that the farmer and the government can work out salvation by themselves. The men and women who live in cities and towns and who now compose the majority of our population are involved quite as deeply, if less directly. Their good will and understanding is necessary to the farmer's stewardship. Their food supplies and industrial raw materials are at stake, likewise a respectable proportion of their investments. Urban owners have title to much of our cropland, in amounts varying from a quarter to more than half of it in different states of the Union. In Oklahoma there are more than 200,000 farms. About 60 per cent of them are owned by individuals who do not live upon them and whose responsibility should be inescapable.

Tenancy of itself need not be an evil thing, providing it represents a partnership and not a system for exploiting land and tenant. Doubtless more landowners are guilty of inattention to business than of greed, for if gain be the object, the sure road to it is the prosperity of the tenant and the care of the land. No matter how hard the tenant works, he must have back of him the capital, the collaboration, and frequently the inducements to treat the land as though it were his own. The basis of relations between owner and tenant is the lease. Leases ought to be so framed that the tenant will have assurance of tenure if he is to do his part. They ought to provide for a fair sharing of both hazard and advantage, and they ought to insure the tenant a living from the farm, regardless of financial ups and downs. Leases should provide, too, for reasonable facilities and improvements so that the tenant may do his work without lost motion and may be housed adequately. They should embody a plan for operation that will conserve the soil and minimize risk—framed not in terms of single [174] years but of the cycles of years which are involved in any intelligent system of land use and management. The reform which would strike most directly at abuse of the soil in the states where it is worst, would certainly attack the present haphazard relationship of tenant and urban landlord. There remains a threat more subtle than the palpable disorder of neglected tenant farm operations. It is not confined to marginal land where erosion is most obvious and spectacular; nor to absentee-owned farms. On the contrary, it involves the best of our farm land and many of our most prosperous farmers. It is sheet erosion, the slow removal of thin layers of surface soil—a loss whose effects can, for a long time, be palliated by energetic tillage, rotation, and use of fertilizer. It is a danger augmented by the fact that the good, thrifty, hardworking farmers on whose land it is occurring can scarcely be convinced that anything is wrong.

This situation exists particularly in the north central and northeastern states, where the bedrock is covered by a thick mantle of glacial drift. When the topsoil is gone, there still remains weathered mineral substance that can be plowed and worked. In the prairie states, which have between three and five feet of rich black topsoil, conditions are still more deceptive, or there may be no color change to show that many inches have been lost.

Much of our fine farmland is gently rolling or so slightly sloping that it appears level and seemingly immune to erosion, thus further adding to the farmer's false sense of security. This is true in the rich Black Swamp Area of northwestern Ohio. Yet the rains of late winter carry out from here sufficient farm soil to muddy Lake Erie half way to Put-in-Bay and the cumulative effects of this silt have all but ruined the spawning grounds of the more desirable species of fish.

The operators of our richest general farming areas must realize that soil loss is not something for hill folk alone to worry about. Frequently the remedy is very simple, and profitable, too. On my own Ohio farm we have merely [175] changed the old rectangular fields into forms that fit the topography. We now plow parallel to the river which runs through the farm and are saving, not only our soil and fertilizer, but the water which we need to keep on the ridge tops. The benefit to our crops was obvious at the end of the first growing season.

Under our system there probably will always be questions as to authority and responsibility. How are these to be divided between central and local government, between government and private agencies? The answers, in most instances will come from an actual showing of results, rather than from political disputation. Whatever local government and individual enterprise will do superbly well is not likely to be taken out of their hands by federal agencies. Whatever they neglect is likely to be lost to them.

This is just what is happening with respect to local banks and farm finance. The purchase, rehabilitation, and equipment of farms requires long-time loans at a lower rate of interest than banks have cared to offer. As a result the government entered the field, borrowing money from the banks at a still lower interest rate! Again many farm loans require careful supervision of and advice to the farmers who need to be set on their feet. Many banks have been unwilling to go to this trouble and expense, even though their accumulated deposits go begging for borrowers.

It is idle to talk of private enterprise unless men are enterprising. Banking organizations and individual banks here and there realize the possibilities of mutual benefit to farm and bank wherever the latter measures up to its opportunity. The finest traditions of American banking are built upon "character" loans and the friendly, responsible interest of the banker in his community and in his client's business. Now that chain stores are widespread and finance themselves, the rural banker finds much less opportunity to employ his resources in making loans to local merchants. His logical outlet today is outward from his village or town into the modern [176] type of scientific, scrupulously managed family farm. If prepares himself for it (and it is no task for an amateur), the possibilities are enormous. Through the power of supervised, inexpensive credit he may do much to the growing confusion, aid in bringing his clients the pensable information from government agencies, and raise the standards of land use and management in each community. If our economic system is good enough to defend, it is good enough to be practised realistically.

The principles of good land use and management are simple and general, but their application is a local problem with infinite variations. No one can do it so well as the man on the spot and it ought to be the first responsibility of local agencies which control the flow of credit.

All that we have said thus far leads us to one of the most bedevilled words in our language—education. Clearly this means more than the learning acquired in eight or twelve or sixteen years of schooling. It is a lifetime enterprise. Judged by the standards of a less strenuous age, we are doing well enough. The American people are rapidly becoming more conscious of their interdependence with the physical and biological worlds and even of their obligation to conserve its resources. But in this mechanized age we are no longer geared to the leisurely pendulum of grandfather's clock. Events have a way of outstripping knowledge, unless we are prepared for them. One of the most effective educational services in agriculture—agricultural extension as personified by the county agent—has thus been overtaken by the momentum of change. We have already described the resulting improvisation, duplication, and confusion of agencies.

This is not the place to add to the tonnage of discussion on education. Two comments must suffice, both of them reiterations. First, conservation of our resources is not a subject. It is a moral attitude in the employment of technical resources and in our way of living. The old adage, "What you do speaks so loudly I cannot hear what you say," applies [177] here. It follows that those who are to educate us—as children and adults—must be imbued with this moral attitude. If they possess it, it makes little difference what *subject* they expound—the conviction will radiate from them to their audience. I can think of no discipline in the ordinary curriculum—certainly not honest mathematics—which does not involve an opportunity to spread the principles of conservation.

Second, there is a body of knowledge, a point of view, which peculiarly implies all that is meant by conservation, and much more. It is at present neglected in most of our schools. It has been called by H. G. Wells, who cannot be accused of a failure to anticipate events, the science of prophecy. Certainly it is the science of perspective. It is the basis of the philosophy of Jan Smuts, one of the greatest and most humane figures of our day. It is the approach to biological knowledge which is called ecology.

Commentary

Paul Sears, *Deserts on the March* (1935)

LIBBY ROBIN

This is very much a book of the Dust Bowl, the disaster that struck America's midwestern agricultural heartlands in the 1930s, resulting in dust storms that darkened the skies in New York. The Dust Bowl was more than an environmental disaster; it was also a social disaster, as John Steinbeck's novel *The Grapes of Wrath* (1939) explored. The "Okies," the small block farmers who watched their dreams blow away with the top soil, packed up their belongings and headed west in their thousands in search of work. The political leaders, who in the 1920s had seen the expanding Midwest wheat fields as "lands of opportunity," were confronted with the challenge of soil erosion: this was a disaster both for the land and for a civilization that depended on that soil.

The extract chosen here, written in 1947, had the benefit of twelve years' reflection on the disaster that gave rise in the United States to the Soil Conservation Service, and to similar authorities in Australia and in the Union of Soviet Socialist Republics, where 1930s dust storms had also raged. Soil science (pedology) flourished as a significant adjunct to agricultural management. The idea that the health of the land was essential to civilization became widespread. "The dust-clouds are carrying with them the material that should be taking the shapes and forms of life," wrote Elyne Mitchell, Australia author of *Soil and Civilization* (1946, 53). She wrote to "retrieve the land" for the sake of "civilization as we know it," fearing that civilization itself would be "gone under the tide of man-made deserts" (1).

The problem with soil conservation was that land is more than just soil, as Aldo Leopold identified in "the land ethic." An emphasis on economic motivation ignores so much of the biotic community: "Of the 22,000 higher plants and animals native to Wisconsin, it is doubtful whether more than 5 per cent can be sold, fed, eaten, or otherwise put to economic use. Yet these creatures are members of the biotic community, and if . . . its stability depends on its integrity, they are entitled to continuance," Leopold argued ([1949] 1987, 210).

Deserts on the March opens with an historical analysis of "desertification," the creation of anthropogenic deserts through poor agricultural practices. It was not just evident in dust-bowl America, but also the Roman Empire, China's Hwang (Yellow) River, the Nile Valley in Egypt, India, and Sudan. Sears describes the bond between agriculture and civilization as "primordial": it is a feature of human civilizations everywhere and across the centuries. He recounted these histories in a bid to suggest a global approach to the marching deserts of the postwar years. Other countries were also concerned. India and Israel urged the United Nations Educational, Scientific and Cultural Organization (UNESCO) to establish an Advisory Council on Arid Zone Research in 1951. The focus of the international and intergovernmental scientific program was first the deserts created by human activities, particularly the Sahara in northern Africa, and in the Middle East and India, and Peru and Australia, places where western agriculture had a shorter history.

The western idea of civilization is closely allied with a certain sort of agricultural settlement. It was a shock to find the land itself limiting the aspirations of the civilized. Francis Ratcliffe, writing in Australia's dust bowl years, commented, "The essential features of white pastoral settlement—a stable home, a circumscribed area of land, and a flock or herd maintained on the land year-in and year-out—are a heritage of life in the reliable kindly climate of Europe. In the drought-risky semi-desert Australian inland they tend to make settlement self-destructive" (Ratcliffe 1938, 332). Such places challenge the civilized, permanent relationship with the land practiced by Amish and Mennonite communities and praised by Sears, yet only possible in "reliable climates." Farming made food sufficiently abundant in order for humanity to have "energy to spare to become civilized," as Sears commented. If "civilized government is as truly a farm product as corn meal" (Sears [1935] 1949, 172), then government becomes an essential part of farming, and this book promotes that relationship for an American audience, but very much in a global context.

In Australia, government-funded science for agriculture was urged from the 1920s on, not just to feed its own citizens, but to "feed the world." Sears also adopts this mantra in his opening paragraph, writing of "feeding of the world during its present distress." Growing food is a (civilized) national obligation contributing to global humanity, an aspiration continued in the rhetoric of "food security" in the present. The global project of feeding the world through cooperative agriculture, research, and new technologies inspired the original International Institute of Agriculture, established in Rome in 1905, which in 1945 became the Food and Agriculture Organization (FAO). The question of conserving soil was just one part of this greater mission.

How might the U.S. federal government help the Midwest in its time of need? Not through taking away the first responsibility from the "man on the spot," but rather supplementing his work with the professional skills offered by a soil conservation service. Sears argued that only the federal government can create the conditions for the long-term loans that are needed to rebuild the land and its soil. Banks cannot support their loans with "supervision and advice," but they can work with government science to ensure that their "enterprise" is not lost in poorly managed farms. The future of rural banking, Sears argues, is a "scientific, scrupulously managed family farm."

For Sears, conservation is not a topic — it is a moral way of living. You cannot teach it, but you can offer leadership through a science of perspective and holism that he called "ecology." The symptoms appear unrelated, but the problem needs holistic treatment. Sears invoked the holism of the South African statesman and author of the preamble to the United Nations Charter, Jan Smuts "one of the greatest and most humane figures of our day," remembering that people are intertwined with both the problem and the solution. "All renewable natural resources are linked into a common pattern of relationship. We can save any one of them only by measures which save them all. And *we* are a part of the whole which must be conserved" (p. 175, my emphasis).

In the 1940s, South Africa, the United States, and Australia all adopted holistic understanding of natural resource management. In 1946 Royal Commissioner Judge Leonard Stretton wrote eloquently in the context of Australian fire management: "Amongst the many subjects which fill the field of this inquiry, three stand pre-eminent, in an inseparable trinity — Forest, Soil and Water. . . . No one of them without the others, can prosper. . . . Destroy any one of them, and by the inexorable cycle which works for health or disease within this fundamental syllogism of the productive physical world, you destroy the well-being of your people. . . . Civilizations have perished, leaving only the monuments of man's pretentiousness to mock their memory, because in ignorance or wantonness man's impious hand has disturbed the delicate balance which nature would maintain between forest, soil and water" (Stretton in Robin 1998, 31–32).

The idea of conserving particular resources: birds, wild flowers, water, or whatever, has been the mission of various societies, but as Sears notes, they work best together: "You cannot have an adequate supply of timber without an intelligent program of land use" (p. 175). Conservation is about partnerships. By the end of this chapter, beyond the extract, Sears recommends a "trained ecologist in each community," and argues that the Soil Conservation Service exemplifies the benefits of the "ecologist at work." The *work* is important: "working biologists in the field" understand damage wrought by industrialism gone wild, where ivory-tower specialists focused on "the deadly grammar of the dead cat" do not (p. 176). Soil is part of the applied ecologist's brief, as soil science is just one of the subdisciplines of ecology, a meta-discipline of holistic principles in Sears's definition. His book seeks solutions to "deserts on the march." Its manifesto is for better anticipation of events, through a "science of prophecy" (p. 182), the integrated meta-discipline of applied ecology.

Further Reading

Leopold, Aldo. (1949) 1987. "A land ethic." In *A Sand County almanac*, 201–228, ed. Robert Finch. New York: Oxford University Press.

Mitchell, Elyne. 1946. *Soil and civilization*. Sydney: Angus and Robertson.

Moon, David. 2005. "The environmental history of the Russian steppes." *Transactions of the Royal Historical Society* 15:149–174.

Muir, Cameron. 2010. "Feeding the world: Our great myth." *Griffith Review* 27:59–73.

Ratcliffe, Francis. 1938. *Flying fox and drifting sand*. London: Chatto and Windus.

Robin, Libby. 1998. *Defending the Little Desert*. Melbourne: Melbourne University Press.

Road to Survival

WILLIAM VOGT

Chapter 12: History of Our Future

By excessive breeding and abuse of the land mankind has backed itself into an ecological trap. By a lopsided use of applied science it has been living on promissory notes. Now, all over the world, the notes are falling due.

Payment cannot be postponed much longer. Fortunately, we still may choose between payment and utterly disastrous bankruptcy on a world scale. It will certainly be more intelligent to pull in our belts and accept a long period of austerity and rebuilding than to wait for a catastrophic crash of our civilization. In hard fact, we have no other choice.

When I write "we" I do not mean the other fellow. I mean every person who reads a newspaper printed on pulp from vanishing forests. I mean every man and woman who eats a meal drawn from steadily shrinking lands. Everyone who flushes a toilet, and thereby pollutes a river, wastes fertile organic matter and helps lower a water table. Everyone who puts on a wool garment derived from overgrazed ranges that have been cut by little hoofs and gullied by the rains, sending runoff and topsoil hundreds of miles away. Especially do I mean men and women in overpopulated countries who produce excessive numbers of children who, unhappily, cannot escape their fate as hostages to the forces of misery and disaster that lower upon the horizon of our future.

If we ourselves do not govern our destiny, firmly and courageously, no one is going to do it for us. To regain ecological freedom for our civilization will be a heavy task. It will frequently require arduous and uncomfortable measures. It will cost considerable sums of money. Democratic governments are not likely to set forth on such a steep and rocky path unless the people lead the way. Nations with lower educational standards than ours, nations that are technologically retarded,

William Vogt. 1948. "History of our future." Chapter 12 in *Road to survival*. New York: Sloane Associates.

are still less likely to move. In our own interest we must accept the responsibility for this leadership, as we have in the spheres of economics and politics.

Drastic measures are inescapable. Above everything else, we must reorganize our thinking. If we are to escape the crash we must abandon all thought of living unto ourselves. We form an earth-company, and the lot of the Indiana farmer can no longer be isolated from that of the Bantu. This is true, not only in John Donne's mystical sense, in the meaning of brotherhood that makes starving babies in Hindustan the concern of Americans; but in a direct, physical sense. An eroding hillside in Mexico or Yugoslavia affects the living standard and probability of survival of the American people. Irresponsible breeding makes amelioration of the condition of the Greeks—or the Italians or Indians or Chinese—difficult, if not impossible; it imposes a drain on the world's wealth, especially that of the United States, when this wealth might be used to improve living standards and survival chances for less people. We cannot escape our responsibility, since it is a responsibility to ourselves.

We must equally abandon any philosophy of "Sufficient unto the day—." We are paying for the foolishness of yesterday while we shape our own tomorrow. Today's white bread may force a break in the levees, and flood New Orleans next spring. This year's wheat from Australia's eroding slopes may flare into a Japanese war three decades hence. Comic books from the flanks of the Nevado de Toluca in 1948 may close Mexico City's factories in 1955. The freebooting, rugged individualist, whose vigor, imagination, and courage contributed so much of good to the building of our country (along with the bad), we must now recognize, where his activities destroy resources, as the Enemy of the People he has become. The exploiting lumberman of Madagascar was beheaded; we should impose at least as effective, if kinder, controls. We must develop our sense of time, and think of the availability of beefsteaks not only for this Saturday but for the Saturdays of our old age, and of our children and grandchildren's youth. The day has long since passed when a senator may callously demand, "What has posterity ever done for me?" Posterity is of our making, as is the world in which it will have to live.

Above all, we must learn to know—to feel to the core of our beings—our dependence upon the earth and the riches with which it sustains us. We can no longer believe valid our assumption that we live in independence. No longer can we rest secure in the certainty that somehow, from somewhere, our wants will be supplied. We, even we fortunate Americans, are pressing hard on our means of subsistence. Our neighbours on five continents know what it means to find their cupboards bare. There is no phase of our civilization that is not touched by wasting dearth. There is hardly an aspect of human activity, through all the complex span of our lives, that does not in some open or occult manner, feel the chill of scarcity's damp breath.

We must—all of us, man, women and children—reorient ourselves with relation to the world in which we live. We must learn to weigh the daily news in terms of man's subsistence. We must come to understand our past, our history, in terms of the soil and water and forests and grasses that have made it what it is. We must see the years to come in the frame that makes space and time one, that will keep us strong only as, like Antaeus, we draw our strength from the earth. Our education must be reshaped, as the story of our existence in an environment as completely subjected to physical laws as is a ball we let drop from our hands. Our philosophies must be rewritten to remove them from the domain of words and "ideas," and to plant their roots firmly in the earth. Above all, we must weigh our place in the society of nations and our future through the decades to come in the scale of our total environment.

The history of our future is already written, at least for some decades. As we are crowded together, two and a quarter billions of us, on the shrinking surface of the globe, we have set in motion historical forces that are directed by our total environment.

We might symbolize these forces by graphs. One of them is the curve of human populations that, after centuries of relative equilibrium, suddenly began to mount, and in the past fifty years has been climbing at a vertiginous rate.

The other graph is that of our resources. It represents the area and thickness of our topsoil, the abundance of our forests, available waters, life-giving grasslands, and the biophysical web that holds them together. This curve, except for local depressions, also maintained a high degree of regularity through the centuries. But it, too, has had its direction sharply diverted, especially during the past hundred and fifty years, and it is plunging downward like a rapid [waterfall].

These two curves—of population and the means of survival—have long since crossed. Ever more rapidly they are drawing apart. The farther they are separated the more difficult will it be to draw them together again.

Everywhere, or nearly everywhere, about the earth we see the results of their divergence. The crumbling ruins of two wars mark their passing. The crumbling ruins of two wars make their passing. The swollen bellies of hungry babies, from San Salvador to Bengal, dot the space between them. Parching fevers and racking coughs, from Osorno to Seoul, cry aloud the cleavage between these curves. The angry muttering of mobs, like the champing of jungle peccaries, is a swelling echo of their passing.

The direction of these curves and the misery they write across the earth are not likely to be changed in the proximate future. Their direction is fixed for some decades. Great masses of people have a preponderantly young population; as they come into the breeding age we must, despite all possible efforts short of generalized slaughter, expect human numbers to increase for a time. The drag imposed by ignorance, selfishness, nationalism, custom, etc., is certain to retard, by some decades, any effective or substantial improvement of resource management.

So that the people shall not delude themselves, find further frustration through quack nostrums, fight their way into blind alleys, it is imperative that this world-wide dilemma be made known to all mankind. The human race is caught in a situation as concrete as a pair of shoes two sizes too small. We must understand that, and stop blaming economic systems, the weather, bad luck, or callous saints. This is the beginning of wisdom, and the first step on the long road back.

The second step is dual—the control of populations and the restoration of resources.

Unless we take these steps and begin to swing into them soon—unless, in short, man readjusts his way of living, in its fullest sense, to the imperatives imposed by the *limited* resources of his environment—we may as well give up all hope of continuing civilized life. Like Gadarene swine, we shall rush down a war-torn slope to a barbarian existence in the blackened rubble.

Commentary

William Vogt, *Road to Survival* (1948)

SVERKER SÖRLIN

"The world is sick," wrote William Vogt in the foreword that he penned in August 1948 for the British edition of his best-selling *Road to Survival*. This was starker pitch than in the original United States edition, perhaps because he could contrast it with "British decency," what he argued the world needed to involve both the brain and the heart in facing its "biophysical dilemma."

William Vogt brought a scientific background to his cause, but this was no scientific book. Trained in ecology at St. Stephens (now Bard) College, he was a dedicated ornithologist and a lifelong nature protectionist. But this book is fundamentally a book of concern, of passion, and of outcry. The world of *Road to Survival* is ill with the Malthusian disease of overpopulation. Its symptoms are soil destruction, resource depletion, food scarcity, starvation, famine, disease, and, ultimately, war. Vogt's diagnosis is based on his own ecological expertise. He sought an "ecological balance" to restore health and enable humanity to flourish. "Man has moved into an untenable position," he wrote, "by protracted and wholesale violation of certain natural laws; to re-establish himself he needs only to bring his behavior into conformity with natural limitations" (264). The world is sick with modern humanity, which has failed to see that there are limits to what the human enterprise could comprise. In 1948 it is not yet a sickness of the physical planet itself.

Road to Survival quickly became an international best seller, translated into nine languages and popularized by the *Reader's Digest*. It has been estimated that the book reached nearly thirty million readers. It raised concern about population in the 1950s and 1960s: Vogt became a voice for population control, and was appointed national director of the Planned Parenthood Federation of America from 1951 to 1962. He also served as a scientific representative to the United Nations.

The book's success can only be partly explained by its scientific content, which was impressionistic rather than rigorous. *Road to Survival* purportedly portrayed a true and accurate state of the world. It had the trappings of science: facts and figures, tables and graphs, a visual language of seeming precision, projections and predictions, and it talked of trends and tendencies, almost invariably gloomy and ominous. But it was not probing, skeptical, or analytic, nor did it test hypotheses. It had a different purpose. Its uncompromising message about the road to survival was buttressed with persuasively arranged information. The book was a rallying cry. If ever there was a neo-Malthusian manifesto, this is it, but without ideological affinities. Vogt, as much as any American on the threshold of the Cold War, is strongly anti-Soviet. He is also deeply distrustful of capitalism.

Vogt was not the first conservationist "gospel singer." He drew harmonies from soil conservationist Hugh Bennett and forester Gifford Pinchot, who had created major government conservation initiatives. His was still the climax ecology of Frederic Clements; ideas of disturbances and feedbacks were yet to come. He was not the first Malthusian thinker, either, coming 150 years after the English priest whose fallen reputation Vogt made it his project to restore. The importance of *Road to Survival* lies in the fact that it, really for the first time, integrates the full range of "survival problems" together as one comprehensive mega-disease and for which human societies should be held responsible. None of the issues that he raised was new. The novelty lay in a comprehensive notion of a common, ever-spreading sickness across the entire world. This disease did not come from the heavens as a metaphysical punishment for human sins or misdemeanors. It was caused by a humanity that suffered from a lack of knowledge, or, more precisely, willfully ignored, the kind of expertise that Vogt himself offered. Vogt devoted much of the book to arguments for more and freer research and better education. He identified long-term environmental history as a key to understanding the future: "the history of Babylon, Assyria, Carthage, China, Spain, Britain—and . . . the United States" would be "as distorted as a Picasso drawing" (19) unless the treatment of plants and habitats was integrated with the story of humans.

The book privileges the views of the expert ecologist, combining Vogt's own theoretical insights with his wide travels and rich experience from the field. Vogt had spent several years in Latin America as chief of the Conservation Section of the Pan American Union, surveying population and resources providing him with examples from El Salvador, Costa Rica, Mexico, Venezuela, and Peru. Vogt echoed his contemporary Paul B. Sears in arguing that ecology was the new chief candidate for the role as provider of the overarching knowledge for world affairs. But, Vogt added, it must be assisted by the social sciences, "the radar that can avert disastrous crashes" (271).

What is the book's main legacy? It has, for more than half a century, been cast as the neo-Malthusian blueprint. But today, after the rise of global change science and the acknowledgment of the Anthropocene, the time is ripe for a reinterpretation. The most important contribution of *Road to Survival* was its style of thought within the newly secular realm on the fate of the world. *Road to Survival* cited many strands of ecology, including animal ecology, mathematical population models, and studies of soil degradation. Otherwise Vogt did not betray much of his intellectual ancestry; the book is more a geographical corollary of global problems, continent by continent, country by country, than an attempt to delineate the roots of the understanding that informed Vogt's voice as a sovereign prophet of the future fate of the world.

Road to Survival is special because it was written just in anticipation of the soon-to-emerge genre of "futures studies." Virtually nothing of what it tried to say had happened at the time of writing, which is also why the title of its last chapter (excerpted here) is doubly telling: "History of Our Future." Vogt wrote the history of a demise of the world to come. The prophetic qualities of his book resound despite the lack of the now well-known metaphors of our own "environmental age." For Vogt there was of course no "Spaceship Earth," no self-organizing "Gaia," nor any "ecological footprint" or "ecosystem services" to measure. There was certainly no "sustainability" to soothe mankind in a utopian future when it had finally cured itself of short-sighted hubris. Vogt wrote, unknowingly, in the moment before the storm.

The word Vogt did use, in the sense that it was to fully acquire in the years to come, was "the environment." He linked it not just to the local and its influence on the individual organism, which the environmental determinists had done, but to the global. In a broad one-page canvas he paints the world as an increasingly unified place, starting with the age of exploration and ending in the modern understanding of a world of interdependencies (another word he could have used, had he known it): "that we live in one world in an ecological—an environmental—sense" (14–15). What made Vogt's book possible was the unifying experience of the war. Conflict resolution—saving the world—was a fully possible enterprise given an overview and the necessary tools. If totalitarianism had been the old enemy, it was now people, calories, hunger, and greed. Things would just be likely to get worse, unless the world heeded the call of ecology.

What had not yet entered Vogt's intellectual universe were the systems ecology of the Odum brothers and the cybernetics of Norbert Wiener, nor had he much appreciated the environmental understanding that was emerging in the geophysical realm, so much favored by the American military in the Cold War years. Vogt married together, while it was still possible, an old conservationist gospel with geographic and demographic tropes that had emerged into center stage in the interwar period. He offered a highly original environmentalist world view, with sincere scientism, an obsession with the future, a compelling rhetoric of gloom, a tone of self-prophesying righteousness, and a "view from nowhere" that just happened to be the West, his own perspective on humanity.

Further Reading

Burch, G. I., and E. Pendell. 1945. *Population roads to peace or war*. Washington, DC: Population Reference Bureau.

Harroy, J.-P. 1944. *Afrique: Terre qui meurt: La degradation des sols Africains sous l'influence de la colonisation.* Bruxelles: Hayez.

Pearson, F. A., and F. A. Harper. 1945. *The world's hunger.* Ithaca, NY: Cornell University Press.

Sears, Paul B. 1947. "Importance of ecology in the training of engineers." *Science* 106(2740):1–3.

Silent Spring

RACHEL CARSON

Chapter 1: A Fable for Tomorrow

[21] There was once a town in the heart of America where all life seemed to live in harmony with its surroundings. The town lay in the midst of a checkerboard of prosperous farms, with fields of grain and hillsides of orchards where, in spring, white clouds of bloom drifted above the green fields. In autumn, oak and maple and birch set up a blaze of colour that flamed and flickered across a backdrop of pines. Then foxes barked in the hills and deer silently crossed the fields, half hidden in the mists of the autumn mornings.

Along the roads, laurel, viburnum and alder, great ferns and wildflowers delighted the traveller's eye through much of the year. Even in winter the roadsides were places of beauty, where countless birds came to feed on the berries and on the seed heads of the dried weeds rising above the snow. The countryside was, in fact, famous for the abundance and variety of its bird life, and when the flood of migrants was pouring through in spring and autumn people travelled from great distances to observe them. Others came to fish the streams, which flowed clear and cold out of the hills and contained shady pools where trout lay. So it had been from the days many years ago when the first settlers raised their houses, sank their wells, and built their barns.

[22] Then a strange blight crept over the area and everything began to change. Some evil spell had settled on the community: mysterious maladies swept the flocks of chickens; the cattle and sheep sickened and died. Everywhere was a shadow of death. The farmers spoke of much illness among their families. In the town the doctors had become more and more puzzled by new kinds of sickness appearing among their patients. There had been several sudden and unexplained deaths, not only among adults but even among children, who would be stricken suddenly while at play and die within a few hours.

Rachel Carson. 1962. Excerpts from *Silent spring*, 21–30, 256–257. New York: Ballantine.

There was a strange stillness. The birds, for example — where had they gone? Many people spoke of them, puzzled and disturbed. The feeding stations in the backyards were deserted. The few birds seen anywhere were moribund; they trembled violently and could not fly. It was a spring without voices. On the mornings that had once throbbed with the dawn chorus of robins, catbirds, doves, jays, wrens, and scores of other bird voices there was now no sound; only silence lay over the fields and woods and marsh.

On the farms the hens brooded, but no chicks hatched. The farmers complained that they were unable to raise any pigs — the litters were small and the young survived only a few days. The apple trees were coming into bloom but no bees droned among the blossoms, so there was no pollination and there would be no fruit.

The roadsides, once so attractive, were now lined with browned and withered vegetation as though swept by fire. These, too, were silent, deserted by all living things. Even the streams were now lifeless. Anglers no longer visited them, for all the fish had died.

In the gutters under the eaves and between the shingles of the roofs, a white granular powder still showed a few patches; some weeks before it had fallen like snow upon the roofs and the lawns, the fields and streams.

No witchcraft, no enemy action had silenced the rebirth of new life in this stricken world. The people had done it themselves. . . .

This town does not actually exist, but it might easily have a thousand counterparts in America or elsewhere in the world. I know of no community that has experienced all the misfortunes I describe. Yet every one of these disasters has actually happened somewhere, and many real communities have already suffered a substantial number of them. A grim spectre has crept upon us almost unnoticed, and this imagined tragedy may easily become a stark reality we all shall know. . . .

What has already silenced the voices of spring in countless towns in America? This book is an attempt to explain.

Chapter 2: The Obligation to Endure

[23] The history of life on earth has been a history of interaction between living things and their surroundings. To a large extent, the physical form and the habits of the earth's vegetation and its animal life have been moulded by the environment. Considering the whole span of earthly time, the opposite effect, in which life actually modifies its surroundings, has been relatively slight. Only within the moment of time represented by the present century has one species — man — acquired significant power to alter the nature of his world.

During the past quarter-century this power has not only increased to one of disturbing magnitude but it has changed in character. The most alarming of all man's assaults upon the environment is the contamination of air, earth, rivers, and sea with dangerous and even lethal materials. This pollution is for the most part irrecoverable; the chain of evil it initiates not only in the world that must support life but in living tissues is for the most part irreversible. In this now universal contamination of the environment, chemicals are the sinister and little-recognized partners of radiation changing the very nature of the world—the very nature of its life. Strontium 90, released through nuclear explosions into the air, comes to earth in rain or drifts down as fallout, lodges in soil, enters into the grass or corn or wheat grown there, and in time takes up its abode in the bones of a human being, there to remain until his death. Similarly, chemicals sprayed on croplands or forests or gardens lie long in soil, entering into living organisms, passing from one to another in a chain of poisoning and death. Or they pass mysteriously by underground streams until they emerge and, through the alchemy of air and sunlight, combine into new forms that kill vegetation, sicken cattle, and work unknown harm on those who drink from once-pure wells. As Albert Schweitzer has said, "Man can hardly even recognize the devils of his own creation."

[24] It took hundreds of millions of years to produce the life that now inhabits the earth—aeons of time in which that developing and evolving and diversifying life reached a state of adjustment and balance with its surroundings. The environment, rigorously shaping and directing the life it supported, contained elements that were hostile as well as supporting. Certain rocks gave out dangerous radiation; even within the light of the sun, from which all life draws its energy, there were short-wave radiations with power to injure. Given time—time not in years but in millennia—life adjusts, and a balance has been reached. For time is the essential ingredient; but in the modern world there is no time.

The rapidity of change and the speed with which new situations are created follow the impetuous and heedless pace of man rather than the deliberate pace of nature. Radiation is no longer merely the background radiation of rocks, the bombardment of cosmic rays, the ultra-violet of the sun that have existed before there was any life on earth; radiation is now the unnatural creation of man's tampering with the atom. The chemicals to which life is asked to make its adjustment are no longer merely the calcium and silica and copper and all the rest of the minerals washed out of the rocks and carried in rivers to the sea; they are the synthetic creations of man's inventive mind, brewed in his laboratories, and having no counterparts in nature.

To adjust to these chemicals would require time on the scale that is nature's; it would require not merely the years of a man's life but the life of generations. And even this, were it by some miracle possible, would be futile, for the new chemicals come from our laboratories in an endless stream; almost five hundred annually find their way into actual use in the United States alone. The figure is staggering and its implications are not easily grasped—five hundred new chemicals to which the bodies of men and animals are required somehow to adapt each year, chemicals totally outside the limits of biologic experience.

Among them are many that are used in man's war against nature. Since the mid 1940s over two hundred basic chemicals have been created for use in killing insects, weeds, rodents, and other organisms described in the modern vernacular as "pests"; and they are sold under several thousand different brand names.

[25] These sprays, dusts and aerosols are now applied almost universally to farms, gardens forests, and homes—non-selective chemicals that have the power to kill every insect, the "good" and the "bad," to still the song of birds and the leaping of fish in the streams, to coat the leaves with a deadly film, and to linger on in soil—all this though the intended target may be only a few weeds or insects. Can anyone believe it is possible to lay down such a barrage of poisons on the surface of the earth without making it unfit for all life? They should not be called "insecticides," but "biocides."

The whole process of spraying seems caught up in an endless spiral. Since DDT was released for civilian use, a process of escalation has been going on in which ever more toxic materials must be found. This has happened because insects, in a triumphant vindication of Darwin's principle of the survival of the fittest, have evolved super races immune to the particular insecticide used, hence a deadlier one has always to be developed—and then a deadlier one than that. It has happened also because, for reasons to be described later, destructive insects often undergo a "flareback," or resurgence, after spraying, in numbers greater than before. Thus the chemical war is never won, and all life is caught in its violent crossfire.

Along with the possibility of the extinction of mankind by nuclear war, the central problem of our age has therefore become the contamination of man's total environment with such substances of incredible potential for harm—substances that accumulate in the tissues of plants and animals and even penetrate the germ cells to shatter or alter the very material of heredity upon which the shape of the future depends.

Some would-be architects of our future look towards a time when it will be possible to alter the human germ plasm by design. But we may easily be doing so now by inadvertence, for many chemicals, like radiation, bring about gene mutations. It is ironic to think that man might determine his own future by something so seemingly trivial as the choice of an insect spray . . .

[26] All this is not to say there is no insect problem and no need of control. I am saying, rather, that control must be geared to realities, not to mythical situ-

ations, and that the methods employed must be such that they do not destroy us along with the insects. . . .

[28] The importation of plants is the primary agent in the modern spread of species, for animals have almost invariably gone along with the plants, quarantine being a comparatively recent and not completely effective innovation. The United States Office of Plant Introduction alone has introduced almost 200,000 species and varieties of plants from all over the world. Nearly half of the 180 or so major insect enemies of plants in the United States are accidental imports from abroad, and most of them have come as hitch-hikers on plants. . . .

These invasions, both the naturally occurring and those dependent on human assistance, are likely to continue indefinitely. Quarantine and massive chemical campaigns are only extremely expensive ways of buying time. We are faced, according to Dr Elton, "with a life-and-death need not just to find new technological means of suppressing this plant or that animal"; instead we need the basic knowledge of animal populations and their relations to their surroundings that will "promote an even balance and damp down the explosive power of outbreaks and new invasions."

Much of the necessary knowledge is now available but we do not use it. We train ecologists in our universities and even employ them in our governmental agencies but we seldom take their advice. We allow the chemical death rain to fall as though there were no alternative, whereas in fact there are many, and our ingenuity could soon discover many more if given opportunity.

Have we fallen into a mesmerized state that makes us accept as inevitable that which is inferior or detrimental, as though having lost the will or the vision to demand that which is good? . . .

[29] The crusade to create a chemically sterile, insect-free world seems to have engendered a fanatic zeal on the part of many specialists and most of the so-called control agencies. On every hand there is evidence that those engaged in spraying operations exercise a ruthless power.

[30] It is not my contention that chemical insecticides must never be used. I do contend that we have put poisonous and biologically potent chemicals indiscriminately into the hands of persons largely or wholly ignorant of their potentials for harm. We have subjected enormous numbers of people to contact with these poisons, without their consent and often without their knowledge. If the Bill of Rights contains no guarantee that a citizen shall be secure against lethal poisons distributed either by private individuals or by public officials, it is surely only because our forefathers, despite their considerable wisdom and foresight, could conceive of no such problem.

I contend, furthermore, that we have allowed these chemicals to be used with little or no advance investigation of their effect on soil, water, wildlife, and man himself. Future generations are unlikely to condone our lack of prudent concern for the integrity of the natural world that supports all life.

There is still very limited awareness of the nature of the threat. This is an era of specialists, each of whom sees his own problem and is unaware of or intolerant of the larger frame into which it fits. It is also an era dominated by industry, in which the right to make a dollar at whatever cost is seldom challenged. When the public protests, confronted with some obvious evidence of damaging results of pesticide applications, it is fed little tranquillizing pills of half truth. We urgently need an end to these false assurances, to the sugar coating of unpalatable facts. It is the public that is being asked to assume the risks that the insect controllers calculate. The public must decide whether it wishes to continue on the present road, and it can do so only when in full possession of the facts. In the words of Jean Rostand, "The obligation to endure gives us the right to know."

Chapter 17: The Other Road

[256] Through all [the] new, imaginative, and creative approaches to the problem of sharing our earth with other creatures there runs a constant theme, the awareness that we are dealing with life—with living populations and all their pressures and counterpressures, their surges and recessions. Only by taking account of such life forces and by cautiously seeking to guide them into channels favourable to ourselves can we hope to achieve a reasonable accommodation between the insect hordes and ourselves.

[257] The current vogue for poisons has failed utterly to take into account these most fundamental considerations. As crude a weapon as the cave man's club, the chemical barrage has been hurled against the fabric of life—a fabric on the one hand delicate and destructible, on the other miraculously tough and resilient, and capable of striking back in unexpected ways. These extraordinary capacities of life have been ignored by the practitioners of chemical control who have brought to their task no "high-minded orientation," no humility before the vast forces with which they tamper.

The "control of nature" is a phrase conceived in arrogance, born of the Neanderthal age of biology and philosophy, when it was supposed that nature exists for the convenience of man. The concepts and practices of applied entomology for the most part date from that Stone Age of science. It is our alarming misfortune that so primitive a science has armed itself with the most modern and terrible weapons, and that in turning them against the insects it has also turned them against the earth.

Commentary
Rachel Carson, *Silent Spring* (1962)

CHRISTOF MAUCH

Perhaps no other book from the United States has caused as strong a stir as Rachel Carson's *Silent Spring*. Like a tsunami, it shattered established worldviews not just in the United States, but around the globe. The book's message about the threat of pesticide abuse reached a wide audience; there is evidence that the so-called ecological revolution was caused in no small part by the 1962 publication of Carson's book. *Silent Spring* became an immediate best seller and remained on the *New York Times* list for thirty-one weeks. *Silent Spring* led to new environmental awareness and a vision that translated into tangible political action. Where did its explosive power come from?

One reason why Carson's arguments were so compelling was her language—a language that manages to be, in equal measure, both gentle and insistent. The first chapter, which describes a town "in the heart of America" struck by "mysterious maladies," sets the tone for the whole book. Her writing style is as clear as it is hauntingly poetic. Carson captivated her readers not just by reciting facts, but by subtly alerting them to the implications of her analysis. Not many scientists are able to present their research in such a gripping way.

A second reason for the vast success of the book is the fierce attacks that Carson withstood at the hands of the powerful pesticide industry and "big agriculture." The balance of power between Carson and her opponents seemed so uneven that it invokes the biblical story of David and Goliath—or rather (because gender matters here) the lesser known story of Jael, the Canaanite heroine who hunted General Sisera down. Taking on DDT required a huge amount of courage. After all, it was just over ten years since Swiss scientist Paul Hermann Müller had been awarded the Nobel Prize in Medicine for his discovery of the insecticidal properties of DDT; the response of the science community had been overwhelmingly positive. Moreover, the world of science in 1960s America was almost exclusively male, and the fact that a female biologist challenged established views lent the dispute a very peculiar quality. No wonder that gender was regularly used by Carson's male critics to denounce her findings. "Why was a spinster so worried about genetics?" is what Secretary of Agriculture Ezra Taft Benson asked President Eisenhower in a private conversation. Others identified women with irrationality (and men with rationality), using derogatory phrases such as "hysterical" and "emotional" or "bird and bunny lover" to disparage Rachel Carson. The showdown between Carson and her opponents reached a high point in a CBS prime-time special in April 1963. Robert White-Stevens, a chemist representing the American Cyanamid Company, attacked Carson, claiming that her worldview would lead humans back to the "dark ages." However, by the end of the broadcast, television audiences all across America took sides with Carson, who, calm and collected, seemed to personify the voice of reason. A couple of weeks later President Kennedy's Science Advisory Committee called for an end to pesticide use in the United States.

A third reason that helps us to understand the impact of *Silent Spring* is the cultural and political climate of the early 1960s. Carson's book appeared at the height of the Cold War, shortly after the construction of the Berlin Wall and only a month before the Cuban missile crisis. The threat of nuclear fallout created anxiety in the United States and beyond. Futurists like Hermann Kahn imagined "the unthinkable"—the brutal realities of a contaminated world. Apocalyptic novels and films such as Nevil Shute's *On the Beach* contributed to a deeper engagement with poisons that could accumulate in body tissues and impact the health of animals and humans. Moreover, the works of social critic Vance Packard about the manipulative forces of excessive advertisement (Carson mentions Packard's work) had undermined trust in authority. Carson was able to draw on the anxiety and skepticism of the American public to direct her readers to her own concept of human vulnerability and risk.

Silent Spring is a very American book, using almost exclusively American examples. And yet the book was translated into two dozen languages and had an impact in many countries around the globe. In Great Britain, the House of Lords held an unprecedented five-hour session on *Silent Spring*. In Sweden, the word "biocid" entered the language as a direct result of the publication of *Silent Spring*. And in Cuba, Fidel Castro launched a book series, "Ediciones Revolucionarias," comprising the most important books for university education; *Silent Spring* became the first volume in the collection. Carson's book was just as controversial abroad as at home. Scientists and politicians instrumentalized the message for their own ends. Eastern Europeans denounced pesticide problems as a consequence and symptom of capitalism, and some Western Europeans, particularly in Spain and Ireland, claimed that excessive use of insecticides was limited to U.S. agribusiness and did not apply to their own country. Clearly, the immediate reaction to *Silent Spring* in the 1960s was on a large scale. But how can we explain the fact that the book still resonates with readers—even in countries such as Turkey and China, far from the United States—fifty years after it was first published?

Perhaps it is because Rachel Carson is a prophet as much as a writer or a scientist; a *poeta vates*, as the Romans might have called her, a creative writer with the energy and the inspiration to imagine the future. Tellingly, she dedicated her book to Albert Schweitzer and his ominous lament that "man has lost the capacity to foresee and to forestall."

Indeed, the way Carson structures her arguments is reminiscent of biblical prophecy: she presents a problem, identifies the causes, provides illustrative examples, and ends with an exhortation to avert future calamities. Her warnings are both vivid and universal. Because the idyllic town in the opening pages of her book is fictional, it could be anywhere and any time. When we read Carson's book today it is not only about DDT: it is the story about everything that has ever come out of spray guns; more broadly, it is about the human "assaults on the natural world."

Unlike George Perkins Marsh, whose magisterial *Man and Nature* was published a century earlier, Carson did not believe that nature and man are separate from one another. Humanity and its environment are both part of a delicate system of living organisms. Some of the more famous photos of Rachel Carson depict her bent over a microscope; others show her at the beach or in the forest with binoculars. This ability to move between the microscopic and the macroscopic—from manipulated atoms to Texan farmers, from egg to reptile, from fish to bird to flower to food, from land to water to air—is what characterizes many of Carson's observations. Her view is one of the world as a complex organic system—a comprehensive and dynamic view that one could call ecoscopic, in which everything is connected to everything else. Other scientists may have had similar views. But nobody else revealed the role of humans in manipulating nature in such a powerful way. Nobody else showed so clearly that the composition of the chemicals "in the tissues of the unborn child," the fate of future generations, and of humanity as a whole, lies in the hands of those who have the authority to define risk today.

Further Reading

Carson, R. 1951. *The sea around us.* New York: Oxford University Press.

Lear, L. J. 1997. *Rachel Carson: Witness for nature.* New York: H. Holt.

Lytle, M. H. 2007. *The gentle subversive: Rachel Carson,* Silent Spring, *and the rise of the environmental movement.* New York: Oxford University Press.

Sideris, L. H., and K. D. Moore (eds.). 2008. *Rachel Carson: Legacy and challenge.* Albany: State University of New York Press.

PART 5

Ecology

How Do We Understand Natural Systems?

In this section we focus on the "economy of nature," the science of ecology. In 1866, Ernst Haeckel defined ecology (*Ökologie* in German) as a science for understanding how organisms relate to their "external world." We are not so much concerned here with "the" ecology (a term sometimes used to refer to the natural world), as the changing worldview created by the methods of this particular science. As it developed in the twentieth century, its concepts influenced thinking about policy making for global change. Later, in Part 8, "Diversity," we focus on ideas about biological diversity and its management, but here we are concerned not so much about nature as how science chose to study it.

Ecologists are still important experts in shaping global change science. Historically, their systems thinking, their understanding of environment to include abiotic elements, and their capacity to apply their science to society all made their conceptual frameworks attractive to policy makers and people seeking practical applied ways to think globally about the world. Ecology was conceptually a "systems science," concerned with feedback loops as early as the 1920s.

In *Animal Ecology*, Charles Elton had argued that "ecology" is a new name for a very old subject, defining it as "scientific natural history" (Elton 1927, 1). Elton was explicitly interested in the relations within an animal "community," including relations with other animals. In particular, Elton focused on *food chains* and the *food cycle* as part of the preconditions for life itself. Elton also defined *niche*, or the place of the animal in its community.

Ecologist Paul Sears linked the idea of soil with civilization in the 1930s. Ecology was important in developing global agricultural ideas. In particular, concerns about food security (the push to "feed the world") arising from issues of population and health, drove questions about what was possible to "improve" in the natural world. Soil science (pedology) was an important branch of ecology, as both Sears and Tansley argued. But perhaps equally important, the rationale for global development in agriculture in the interwar years created a niche for ecological concepts in global thinking.

It is important, however, not just to focus on postwar ecological ideas as we trace the long-term origins of ecological concepts in global change science. This part considers Alexander von Humboldt, a nineteenth-century theorist; Arthur Tansley, an interwar theorist; and two postwar theorists, Eugene Odum and C. S. Holling, all of whom have shaped the conceptual direction of ecology, and in turn have become influential in global change science.

German biogeographer and explorer of the Americas Alexander von Humboldt (1769–1859) theorized the relationships between climate and life systems around the world. His 1840s publications on *The Cosmos* were arguably among the earliest works that explicitly explore the idea of "global" in the modern sense. Here we look at his earlier *Essay on the Geography of Plants* (1807), where we can recognize early ecological thinking in his analysis of global patterns. He writes of how life systems are shaped by climate and elevation, and related to atmosphere, oceans, and soils—biotic and abiotic factors together. Humboldt began theorizing the distribution of plants and their physical and spatial contexts, which much later developed into the science of biogeography. In a sense, this document could equally have been considered as part of the ideas of "geographies" (Part 3 in this book), but its focus on plants and how they survive under various conditions is truly ecological. Indeed it is clear that Humboldt's ideas contributed to the concept of "Oekologie," as coined by his countryman Ernst Haeckel. Humboldt's theorizing opened up the new question of how varieties of life were related to one another and how they had become distributed around the world: questions of evolutionary theory and regions of flora and fauna interested other, later-traveling naturalists, including Charles Darwin and Alfred Russel Wallace. By the early twentieth century, theories like Alfred Wegener's "continental drift" were put forward to account historically for the evolving spatial distribution of life around the world—*where* organisms live and why.

The physiological question of *how* organisms live also became a focus of twentieth-century thinking. Systems had been important to thinking about physiology since William Harvey's seventeenth-century concept of the systemic circulation of the blood in the human body. From the turn of the twentieth century, relationships between organisms and their environments began to be described physiologically. Some treated the whole plant community as an "organism." The community was a particularly important concept in Danish and German plant science, and ideas from this were further developed in the North American prairies by Frederic Clements and others.

The systems approach also came to the fore as a way of conceptualizing the relationships between plants and environments, as a device to get beyond the assumption of a plant community as an "organism" as popularized by Clements. British botanist Arthur Tansley used the term "ecosystem" in the 1935 paper reproduced here, as a response to the American idea of community. Tansley wanted to move botany beyond treating vegetation theory in isolation (like a "body" or organism), to theorize the relationships between the plants and "the environment," which he saw as "biotic and abiotic elements together." He was particularly interested in soil ("edaphic factors" in ecosystem functioning). Tansley writes of "the whole complex of physical factors forming what we call the environment of the biome—the habitat factors in the widest sense" (p. 225). He was also explicitly interested in human-modified landscapes and in anthropogenic forces for change. Tansley's new concept, the *ecosystem*, moved ecology away from community-focused theories, like those of Frederic Clements (for vegetation) and Karl Möbius in Germany (for animals). The concept of the ecosystem transformed botany into a new, increasingly quantitative "ecology" that included not just plants, but also soils, human behavior, and climate.

Ecological systems thinking and the meta-discipline of ecology developed further in the postwar years through Eugene Odum's *Fundamentals of Ecology*, which he first published (with contributions from H. T. [Tom] Odum, his brother) in 1953. In the excerpt here, which outlines principles and concepts pertaining to the ecosystem and biogeochemical cycles, the relationship between the biotic and abiotic worlds is operationalized. Through the idea of "energy," Charles Elton's earlier ideas about food chains became expressed quantitatively. *Fundamentals of Ecology* was an introductory textbook for ecologists for the rest of the century. Indeed, a new edition was published in the opening decade of the twenty-first century. The use of energy budgets is a way of quantifying the physical in the ecosystem, and this notion theorized by Odum has become pervasive within ecology, and made ecology comparable with cybernetics and physical system sciences, something that was important to later global systems science.

Animal ecologists were also concerned with whole ecosystems and often defined them in terms of predator-prey relations. What made the symbiotic relations between the hunter and the hunted successful? And what happened when they were stressed by external factors—climate, anthropogenic change, and other elements beyond the "ecosystem" as defined by the hunting relationship? Questions of managing recovery after shocks to ecosystems led to C. S. Holling's definition of "resilience," a key concept in global change thinking that has particular applications to systems that include human societies. Holling's 1973 paper, reproduced here, is still regularly referenced by scientists today. Resilience, once just an ecological concept, has become a key concept of global change science. It has become particularly important in theorizing how ecological-social systems collapse—or on the contrary, survive stress and become resilient.

Odum's work has been very influential in the consciously global "Big Ecology" of the later twentieth century. Alfred J. Lotka's *Elements of Mathematical Biology* (originally 1925, but republished 1956), which included energetics and energy transformation, was influential at the same time as Odum, and was important to C. S. Holling's work on resilience. Both Lotka and Holling focused on animal ecology and sought mathematical representations for predator-prey relationships (Golley 1973). Odum's textbook was particularly influential for teaching the "holistic ecosystem-oriented approach" (Coleman 2010, 4). Big Ecology was a framework that fostered global scientific projects such as the International Biological Program (IBP) of the 1960s and 1970s, and the U.S. National Science Foundation's Long-Term Ecological Research (LTER) in the 1980s, a project that from the 1990s included urban ecologies (with sites in Baltimore and Phoenix) (Kingsland 2005). The international version of LTER (ILTER) is something that has twenty-first-century manifestations in other places—including, for example, China (Chinese Ecosystem Research Network). The turn to the "mathematical" in ecology was important to international collaborations. Coleman, in his review of his own science, *Big Ecology* (2010), explored global scientific collaborations fostered by mathematical ecosystem science, showing how expertise in ecology was fostered by national and international policy makers and scientific diplomats. "Ecosystem science" was, in his terms, the most important framework for setting up large-scale and global research programs.

Essay on the Geography of Plants

ALEXANDER VON HUMBOLDT AND

AIMÉ BONPLAND

Physical Tableau of Equatorial Regions

Based on measurements and observations performed on location, from the tenth degree of boreal latitude to the tenth degree of austral latitude in the years 1799, 1800, 1801, 1802, and 1803.

When one ascends from sea level to the peaks of high mountains, one can see a gradual change in the appearance of the land and in the various physical phenomena in the atmosphere. The plants in the plains are gradually replaced by very different ones: woody plants decrease little by little and are replaced by herbaceous and alpine plants; higher still, one finds only grasses and cryptogams. Rocks are covered with a few lichens, even in the regions of perpetual snow. As the appearance of the vegetation changes, so does the form of the animals: the mammals living in the woods, the birds flying in the air, even the insects gnawing at the roots in the soil are all different according to the elevation of the land.

By looking carefully at the nature of rocks of the earth's crust, the observer can also see changes in them as he climbs above sea level. Sometimes, the more recent formations covering the granite in the plains reach only a certain altitude; and near the mountain peaks this same primitive rock reappears that is the basis for all the others, and which constitutes the interior of our planet, so far as our feeble endeavors have allowed us to penetrate it.[1] Sometimes, this granitic rock remains hidden under other more recent formations. Peaks over 4,000 meters (2,053 toises) above today's sea level contain strata of shells and petrified corals. Sometimes, small scattered cones made of basalt, greenstone (*Grünstein*), and

Alexander von Humboldt and Aimé Bonpland. (1807) 2009. Excerpt from *Essay on the geography of plants*. Trans. S. Romanowski, ed. S. T. Jackson, with accompanying essays and supplementary material by S. T. Jackson and S. Romanowski. Chicago: University of Chicago Press.

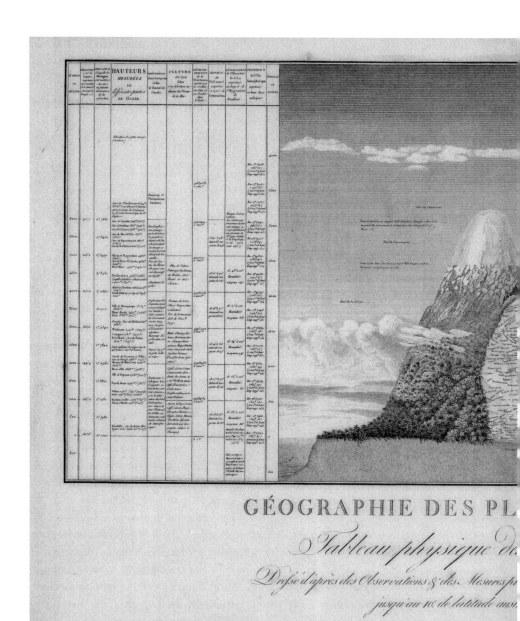

Figure 1. The complete *Tableau physique* from Humboldt's 1807 *Essay on the Geography of Plants*.

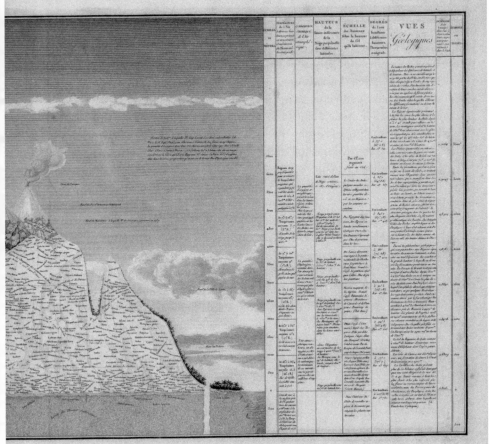

ANTES ÉQUINOXIALES.

s Andes et Pays voisins

ises Sur les lieux depuis le 10.e degré de latitude boréale

rale en 1799, 1800, 1801, 1802 et 1803.

PAR

OLDT et AIMÉ BONPLAND.

Tiré à Paris model, gravé par Bouquet, la Lettre par Beublé, imprimé par Langlois

porphyric schist crown the tops of high mountains, thus posing difficult enigmas for geologists. The mineralogist can see variations according to the elevation of the ground, just as the naturalist can see variations in the plants and the animals: furthermore, the air, this mixture of gaseous fluids of unknown size enveloping our planet, is no less subject to striking variations. As we go further away from sea level, the temperature and pressure of the air diminish; at the same time, its dryness and its electrical tension increase: the blue of the sky seems deeper according to one's altitude. This same altitude also influences the decrease of gravity, the temperature of boiling water, the intensity of the sun's rays traversing the atmosphere and the refraction of the rays as they travel through it. Thus the observer who leaves the center of the earth by an infinitely small amount compared to the radius can reach a new world, so to speak, and he can observe more variations in the aspect of the soil and more modifications in the atmosphere than he would if he were to pass from one latitude to another.

These variations are found in every region where nature made mountain ranges and high plateaus above sea level; but they are less prominent in temperate zones than at the equator where the Cordilleras have an altitude of 5,000 to 6,000 meters (2,565 to 3,078 toises), and where there is a uniform and constant temperature at each elevation. Near the north pole there are mountains almost as colossal as those found in the Quito kingdom, and whose grouping has all too often been attributed to the earth's rotation. Mount Saint Elias, situated on the American coast opposite the Asian coast, at 60°21 of boreal latitude, is 5,512 meters high (2,829 toises); Mount Fairweather, situated at the 59th degree of boreal latitude, is 4,547 meters high (2,334 toises).[2] In our average latitude of 45 degrees, the Mont-Blanc has a height of 4,754 meters (2,440 toises), and one can consider it to be the highest mountain in the Old Continent, until brave explorers can measure the range of mountains in northwest China and which has been affirmed to be higher than Chimborazo. But in the northern regions, in the temperate zones at 45 degrees, the limit of permanent snow, which is also the limit for all organized life, is only at 2,533 meters above sea level (1,300 toises). The result is that on mountains in temperate zones, nature can develop the variety of organized beings and meteorological phenomena on only half the surface offered by tropical regions, where vegetation ceases to exist only at 4,793 meters (2,460 toises). In our northern latitudes, the slant of the sun's rays and the unequal length of the days raise the temperature in the mountain air so much that the difference between the temperature in the plains and the temperature at 1,500 meters is often imperceptible: for this reason, many plants that grow at the foot of our Alps are also found at great heights. The rigors of the cold temperature during autumn nights do not destroy their organization; they would undergo the same decrease in temperature a few months later in the plains. A few alpine plants in the Pyrenees grow at very low elevations in the valleys; there they find a temperature to which they might be exposed sometimes also at a higher elevation.

In the tropics, on the contrary, on the vast surface of up to 4,800 meters, on this steep surface climbing from the ocean level to the perpetual snows, various climates follow one another and are superimposed, so to speak. At each elevation the air temperature varies only slightly; the pressure of the atmosphere, the hygroscopic state of the air, its electrical charge, all these follow unalterable laws that are all the more easy to recognize that the phenomena are less complicated there. As a result, each elevation has its own specific conditions, and therefore produces differently according to these circumstances, so that in the Andes of Quito in a region with a breadth of 2,000 meters (1,000 toises) one can discover a greater variety of life forms than in an equal zone on the slopes of the Pyrenees.

I have attempted to gather in one single tableau the sum of the physical phenomena present in equinoctial regions, from the sea level of the South Sea to the very highest peak of the Andes. This tableau contains:

The vegetation;

The animals;

Geological phenomena;

Cultivation;

The air temperature;

The limit of perpetual snow;

The chemical composition of the atmosphere;

Its electrical tension;

Its barometric pressure;

The decrease in gravity;

The intensity of the azure color of the sky;

The weakening of light as it passes through the strata of the atmosphere;

The horizontal refractions, and the temperature of boiling water at various altitudes.

In order to facilitate the comparison of these phenomena with those of temperate zones, we have added a great number of elevations measured in various parts of the planet and the distance at which these elevations can be seen from the sea, not taking into account the earth's refraction.

This tableau contains almost the entirety of the research I carried out during my expedition in the tropics. It is the result of a large number of works that I am preparing for the public and in which I will develop what I can only outline here. I dared to think that this essay would be interesting not only for physicists, but even more for all those interested in general physics, to whom it may suggest further comparisons and analogies. This science, no doubt one of the highest achievements of human knowledge, can progress only by individual studies and by connecting together all the phenomena and productions on the surface of the earth. In this great chain of causes and effects, no single fact can be considered in isolation. The general equilibrium obtaining in the midst of these disturbances and apparent disorder is the result of an infinite number of mechanical forces and chemical attractions which balance each other; and while each series of facts must be examined separately in order to recognize a specific law, the study of nature, which is the main problem of general physics, demands the gathering together of all the knowledge dealing with modifications of matter.

I thought that if my tableau were capable of suggesting unexpected analogies to those who will study its details, it would also be capable of speaking to the imagination and providing the pleasure that comes from contemplating a beneficial as well as majestic nature. So many objects are capable of seizing our imagination and lifting us to the most sublime considerations: the multitude of forms developed on the slopes of any one of the Cordilleras; the variety of living structures adapted to the climate of each elevation and to its barometric pressure; the layer of perpetual snow that poses insurmountable obstacles to the spreading of vegetation, but whose limit at the very equator is at 2,300 meters (1,200 toises) higher than under our climates; the volcanic fire that sometimes makes its way to the surface in low hills like Mount Vesuvius, sometimes at elevations five times higher, like the cone of Cotopaxi; the petrified shells found on the peaks of the highest mountains recalling the great catastrophes of our planet; finally, those elevated regions of the atmosphere where the aeronautical physicist [Gay-Lussac] went, guided by his bold courage and his noble zeal. By speaking both to our imagination and our spirit at the same time, a physical tableau of the equatorial regions could not only be of interest to those in the field of physical sciences, but could also stimulate people to study it who do not yet know all the pleasures associated with developing our intelligence.

In stating these ideas, I have been concerned not so much with the tableau that I am presenting here, whose imperfections I am well aware, but more with the breadth of which this kind of work is capable. The public, who is so well disposed towards me, will be indulgent toward this work which has been written in the midst of many very heterogeneous occupations. If the new projects I am preparing leave me enough time, I hope to be able eventually to bring this tableau to a higher degree of perfection: for botanical maps are like the ones we call exclusively geographical; we can make them more exact only inasmuch as we accumulate a greater number of good observations. . . .

Notes

1. The greatest vertical depth of mines in Europe is 408 meters (209 toises): the large mine in Valenciana in Mexico has a depth of 516 meters (266 toises).
2. *Viaje al estrecho de Fuca* [Voyage to the Strait of Fuca], by Don Dionisio Galeano y Don Cajetano Valdes; p. LXV.

Commentary

Alexander von Humboldt and Aimé Bonpland, *Essay on the Geography of Plants* (1807)

STEPHEN T. JACKSON

In June 1804, Alexander von Humboldt and Aimé Bonpland stepped off the frigate *La Favorite* in Bordeaux, completing a five-year voyage that took them through what is now Venezuela, Cuba, Colombia, Peru, Ecuador, Mexico, and the United States. They brought with them forty crates containing thousands of plant and animal specimens, along with records of tens of thousands of measurements and observations of physical, chemical, geological, biological, and cultural features of the lands, airs, and waters they traveled through. Humboldt spent most of the next twenty-two years in Paris, preparing a forty-five-volume report of his travels and observations. The first volume, comprising 155 pages and a single plate, appeared in 1807. Modestly titled *Essay on the Geography of Plants*, it was the most comprehensive work of the entire series, setting forth Humboldt's vision of a unified *general physics of the earth*, in which the various constituents of earth's surface—its atmosphere, oceans, soils, bedrock, vegetation, rivers and lakes, animals, and human societies—were linked at local, regional, and global scales. Although this vision guided nearly everything Humboldt wrote until his death in 1859, including his celebrated *Cosmos, Personal Narrative,* and *Views of Nature,* the 1807 *Essay* remains its clearest expression.

The central element of the *Essay* is the *Tableau physique,* a large plate (90 × 66 cm) depicting a profile of the Andean volcano Chimborazo, together with vegetation zones, plant species, snowline, and cloud patterns (Figure 1). The profile is flanked by tables describing elevational patterns in properties of the atmosphere (temperature, pressure, humidity, light refraction, intensity of sky color, light intensity, chemical composition) and the land surface (bedrock and surface geology, snowline at different latitudes, fauna, agriculture). Most of the *Essay* is devoted to description and discussion of the plate's many elements.

Humboldt's synthesis had multiple roots, three of them involving close friendships formed in the 1790s. Karl Ludwig Willdenow introduced Humboldt to botany, and to his pioneering thinking concerning climate and plant geography. Georg Forster passed on ideas developed with his father (Johann Reinhold Forster) concerning the effect of climate on vegetation and human society, and the interrelations of diverse natural phenomena in space. Johann Wolfgang von Goethe discussed the relationship between plant form and function and between morphology and environment, as well as his notion of archetypal forms.

Humboldt fused all these ideas with Immanuel Kant's conception of a science of physical geography and enriched them with his own observations from travels in Europe. By the time of his 1799 departure from Spain to explore the Americas with Bonpland, Humboldt was formulating a unique and general synthesis of climate, vegetation, and geography.

Humboldt developed an important insight during his 1802 exploration of the Andes and partial ascent of Chimborazo. Humboldt recognized that Chimborazo, the highest summit in the equatorial Andes and within two degrees of the equator, could serve as an archetype for the relationships among the various elements of the earth surface and its atmosphere. Here was the broadest elevational and climatic gradient on earth, ranging from rainforests of the tropical lowlands all the way to glaciers near the summit. Humboldt grasped the analogy between the elevational transect on Chimborazo and the latitudinal transect between the equator and the poles, but also recognized two critical contrasts: that higher latitudes experienced more seasonal variation in climate regardless of elevation, and that the position of continents, oceans, mountain ranges, and other surface features, in addition to latitude, influenced climate realizations.

The *Tableau physique* and accompanying text represents a watershed in the development of the environmental sciences. Humboldt applied novel graphical techniques to provide a vivid depiction of vegetation structure (from the palm and banana-type trees visible near sea level to the barren boulder fields just below snowline) and composition (the species names and vegetation zones printed on the right-hand side). He discussed the relationship between elevation, climate, vegetation, and biogeography on Chimborazo and in Europe, and compared and contrasted elevation and latitude from both climatological and vegetational standpoints. These relationships are fundamental to ecology and biogeography, receiving particular attention now with concerns about climate change and its ecological consequences.

Humboldt achieved two particularly important insights in the *Essay* and *Tableau physique*, reaching far beyond ecology. First, he delineated a synthetic view of the earth system as a whole, with its many elements—the physics and chemistry of the atmosphere, the circulation and salinity of the oceans, the structure and composition of vegetation, the chemistry and discharge of rivers, the placement of mountain ranges and geology of bedrock, the economic and agricultural activities of peoples—interacting and influencing each other. This view, encapsulated in his *general physics of the earth*, lay dormant or foreshortened during the disciplinary specializations of the late nineteenth and twentieth centuries, but has reemerged in the past two decades as earth system science. Although earth system science has its origins in Humboldt's work, his role in laying its foundation is largely unacknowledged.

Humboldt receives more credit for his second insight: that spatial and temporal patterns of earth-surface properties can reveal underlying processes and relationships. Humboldt's plant and animal specimens, his diverse physical and chemical measurements, and his observations of geology, biota, and culture were all accompanied by precise delineation of location—latitude, longitude, and elevation—as well as date and often time. Humboldt saw that distributed point-measurements could behave like individual tiles in a mosaic, revealing larger patterns. He also recognized that temporally and spatially referenced observations could, in ensemble, reveal synchrony and sequence, providing understanding of such dynamic phenomena as ocean currents, atmospheric circulation, animal migration, volcanic activity, and commercial trade. Although many of these insights were elaborated more fully in his later writings, they first appeared in the *Essay*.

Strong aesthetic and moral sensibilities are apparent in the *Essay*, the *Tableau physique*, and indeed in Humboldt's other writings and illustrations. Humboldt believed that the dry measurements and observations of science should be coupled to the imagination, that systematic contemplation of nature should bring inspiration, pleasure, and joy, and that the arts and sciences could be fused to provide a full appreciation and understanding of the cosmos. He also expressed a deep commitment to using scientific knowledge to advance human welfare and freedom, and was a pioneer in organizing international cooperation in scientific and other pursuits.

The modern environmental sciences are a Humboldtian enterprise. Humboldt influenced and inspired the great naturalists of the nineteenth century, including Charles Darwin, Alfred Russel Wallace, Henry David Thoreau, George Perkins Marsh, C. H. Merriam, Louis Agassiz, Peter Kropotkin, Ernst Haeckel, and Eugenius Warming, whose work led to the development of the science of ecology at the turn of the twentieth century, and to the emergence of the environmental sciences in the mid-twentieth century. His intellectual legacy continues in the recognition of climate as a critical control of ecological and societal phenomena, in the deployment and utilization of environmental sensor networks across the globe, and in the continuing effort to understand the dynamic interactions of the various components of the earth system. His legacy continues too in widespread international cooperation in environmental science and (sometimes) policy, and in the aesthetic and moral commitments of environmental scientists. Most practicing ecologists and environmental scientists derive both sensory and intellectual pleasure in studying nature, and their work is frequently motivated by a sense of moral obligation to nature and to humankind.

Further Reading

Dettelbach, Michael. 2005. "The stimulations of travel: Humboldt's physiological construction of the tropics." In Felix Driver and Luciana Martins (eds.). *Tropical visions in an age of empire*, 43–85. Chicago: University of Chicago Press.

Humboldt, A. von. 1848–1858. *Cosmos, personal narrative, and views of nature*. Trans. E. C. Otté. London: H. C. Bohn.

Nicolson, Malcolm. 1996. Humboldtian plant geography after Humboldt: The link to ecology. *British Journal for the History of Science* 29:289–310.

Sachs, Aaron. 2006. *The Humboldt current: Nineteenth-century exploration and the roots of American environmentalism*. New York: Viking.

Walls, Laura Dassow. 2009. *The passage to cosmos: Alexander von Humboldt and the shaping of America*. Chicago: University of Chicago Press.

The Use and Abuse of Vegetational Concepts and Terms

A. G. TANSLEY

It is now generally admitted by plant ecologists, not only that vegetation is constantly undergoing various kinds of change, but that the increasing habit of concentrating attention on these changes instead of studying plant communities as if they were static entities is leading to a far deeper insight into the nature of vegetation and the parts it plays in the world. A great part of vegetational change is generally known as *succession*, which has become a recognised technical term in ecology, though there still seems to be some difference of opinion as to the proper limits of its connotation. . . .

Succession

The concept of succession can be given useful scientific significance only if we can trace in the sequences of vegetation "certain uniformities which we can make the subject of investigation, comparison, and the formulation of laws" (Tansley 1929). . . . The concept of succession involves not *merely* change, but the recognition of a *sequence of phases* . . . subject to ascertainable laws. . . . Catastrophic changes due to external factors do not form parts of succession. . . . I should say they were clearly *interruptions*, each initiating a new succession (sere). . . . Succession . . . is continuous, but it may be interrupted by catastrophes unrelated to successional processes, which last are subject to ascertainable laws.

I distinguish between *autogenic succession*, in which the successive changes are brought about by the action of the plants themselves on the habitat, and *allogenic succession* in which the changes are brought about by external factors. . . . Retrogressive succession include[s] the continuous effect of grazing animals which may gradually reduce forest to grassland, the gradual leaching

A. G. Tansley. 1935. Excerpt from "The use and abuse of vegetational concepts and terms." *Ecology* 16(3):284–307.

and concomitant raw humus formation which may ultimately reduce forest to heath, gradual increase of drainage leading to the replacement of a more luxuriant and mesophytic by a poorer and more xerophytic vegetation, or a gradual waterlogging which also leads to a change of type and usually the replacement of a "higher" by a "lower" one.

[Clements (1916) insists] on the prime importance of the water-relations in succession [but denies] the possibility of a gradual change in the soil factors as a result of progressive leaching without change of climate. We may agree with Clements that strict proof of the reality of a retrogression caused in this way must be lacking unless and until we have the results of long-continued observation and properly controlled experiment with the appropriate quantitative data. . . . The invasion and destruction of forest (or heath) by Sphagnum bog is not properly considered as retrogression. I should call it the conquest and suppression of a "higher" type of community by a "lower" one, owing to the peculiar nature of the latter. . . . But such events cannot quite be *equated*, as Clements would equate them, with the formation of new "bare" (water) areas. Sphagnum is after all a plant, and the dominant of very extensive and important communities.

Catastrophic destruction, whether by "natural" agencies or by man, does, I think, remove the phenomena from the field of the proper connotation of succession, because catastrophes are unrelated to the causes of the vegetational changes involved in the actual process of succession. They are only initiating causes . . . : they clear the field . . . for a new succession. Gradualness is a character of succession; [it] . . . is the mark of the action of *continuative* causes.

Development and the Quasi-Organism

The word development may be used in a very wide sense: thus we speak of the development of a theme or of the development of a situation, though always, I think, with the implication of becoming more complex or more explicit. Always, too, it is some kind of *entity* which develops, and in biology it is particularly to the growth and differentiation of that peculiarly well-defined entity the individual organism that we apply the term. . . . As part of the theory of vegetation, a body of well-established and generally acceptable concepts and laws, [it applies] only if we can recognise sufficiently well-defined [vegetation] entities whose development we can trace, and the laws of whose development we can formulate.

In 1920 I enquired whether we could recognise such entities in vegetation. . . . Briefly my conclusion was that mature well-integrated plant communities (which I identified with plant associations) had enough of the characters of organisms to be considered as *quasi-organisms*, in the same way that human societies are habitually so considered. Though plant communities are not and cannot be so highly integrated as human societies and still less than certain animal communities (. . . termites, ants and social bees), the comparison with an organism is firmly based on the close interrelations of the parts of their structure, on their behaviour as wholes. . . . Another important similarity . . . that greatly strengthens the comparison between plant community and organism [is] the remarkable correspondence between the species of a plant community and the genes of an organism, both aggregates owing their "phenotypic" expression to development in the presence of all the other members of the aggregate and within a certain range of environmental conditions.

But for Clements and Phillips . . . the plant community (the "biotic community") *is* an organism. . . . Phillips writes as if he believed words to have perfectly precise and invariable meanings, and that a given verbal proposition *must* either be true or not true, whereas in fact a proposition obviously has different meanings according to the exact connotation of the words employed. The word organism can be applied very widely indeed. . . .

There is no need to weary the reader with a list of the points in which the biotic community does not resemble the single animal or plant. . . . Clements and Phillips argue that no one asserts that the plant community is an *individual* organism. [Rather], it is a "complex organism"—a thoroughly bad term, for it is firmly associated in the minds of biologists with the higher animals and plants. . . . Whether it is true or untrue depends entirely on the connotation of "organism." . . . The present generation of biologists [use the] word for the peculiarly definite, sharply limited and unique type of organisation embodied in the individual animal or plant.

Some argue "quasi-organism" is unnecessary if we keep the concept of "climax," [but] . . . climax does not suggest *organisation*, and the organisation of a mature complex plant association is a very real thing. The relatively stable climax community is a complex whole with more or less definite structure, *i.e.*, interrelation of parts adjusted to exist in the given habitat and to co-exist with one another. It has come into being through a series of stages which have approximated more and more to dynamic equilibrium in these relations. This surely is . . . organisation of the same type as . . . that of the single animal or plant. The organizing factors are on the one hand the total net action of the effective environmental factors, on the other the combined actions of the individual organisms themselves.

. . . *Development* is something very different from the ontogeny of a plant or animal (though even here there are also striking similarities). The adult quasi-organism can develop from beginnings which are totally opposed—a phenomenon completely alien from the ontogeny of a plant or animal—it can be hydrarch or xerarch [i.e., originate in a wet or dry habitat]; and the constituents of the *developmental stages* are quite different from the constituents of the adult. Starting from the type of the individual organism we have here something so different that it is no wonder there is refusal to call it by the same name, but at the same time something like enough to justify a related name. I can only conclude that the term "quasi-organism" is justified in its application to vegetation, but that the terms "organism" or "complex organism" are not.

Climaxes

Just as [Phillips] will only have one kind of succession, which is always progressive, and entirely caused by the "biotic reactions" of the community, so he will have only one kind of climax, the climatic climax, of which there is only one in each climatic region [the "monoclimax" doctrine]. . . . There are [however] some ecologists who believe there may be more than one climax in a climatic region, each with distinct dominants. This so-called "polyclimax theory" . . . takes what appear to be permanent types of vegetation under given conditions and calls them climaxes, because they are culminations of successions.

Under the typical climatic conditions of the region and on the most favourable soils the climatic climax is reached by the succession; but on less favourable soils of special character, different kinds of stable vegetation are developed and remain in possession of the ground, to all appearance as permanently as the climatic climax. These are called *edaphic climaxes*, because the differentiating factor is a special soil type. Similarly special local climates determined by topography (*i.e.* land relief) determine *physiographic climaxes*.

But we may go farther than this. . . . A decisive "biotic factor" such as the continuous grazing of animals may determine a *biotic climax*. A *fire climax* [occurs in] a region swept by constantly recurrent fires [leaving] vegetation consisting only of species able to survive under these trying conditions of life. . . . A *mowing climax* [may] result from regular periodic cutting of grasses or sedges. In each case the vegetation appears to be in equilibrium with *all* the effective factors present, including of course the climatic factors. . . .

Here we encounter a complication . . . [*viz.*] the influence of the modern theory of soil development on the theory and classification of vegetation. It is a simple and attractive idea that development of the soil profile runs *pari passu* with development of the vegetation it bears, and that consequently the mature climatic soil type corresponds and co-exists with the climatic climax community. . . . Even when profile development under the influence of climate is perfectly normal and regular, the climatic climax community may establish itself long before the soil is mature, and may not be substantially altered by the later stages of profile maturation. Again a climatic climax may establish itself on a soil which is *kept immature* by geological and physiographic causes, as on a steep slope. And finally it is now generally agreed by pedologists that some rocks, owing to the simplicity of their composition, produce soils which can *never* form the normal climatic mature profile, and these may or may not bear the typical climatic climax vegetation. . . . I plead for empirical method and terminology. . . .

"The Complex Organism"

On linguistic grounds I dislike the term biotic *community*. A "community" . . . implies *members*, and it seems to me that to lump animals and plants together as *members* of a community is to put on an equal footing things which in their whole nature and behaviour are too different. Animals and plants are not common members of anything except the organic world. . . . One would not speak of the potato plants and ornamental trees and flowers in the gardens of a human community as *members* of that community, although they certainly enter into its constitution—it would be different without them. There must be some sort of *similarity*, though not of course *identity*, of nature and status between the members of a community. . . .

Animal ecologists in their field work constantly find it necessary to speak of *different* animal communities living in or on a given plant community, and this is a much more natural conception, formed in the proper empirical manner as a direct description of experience, than the "biotic community." Some of the animals belonging to these various animal communities have very restricted habitats, others much wider ones, while others again such as the larger and more active predaceous birds and mammals range freely not only through an entire plant community but far outside its limits. For these reasons also, the practical necessity in field work of separating and independently studying the animals communities of a "biome," and for some purposes the necessity of regarding them as external factors acting on the plant community—I cannot accept the concept of the *biotic* community. This refusal is however far from meaning that I do not realise that various "biomes," the whole webs of life adjusted to particular complexes of environmental factors, are real "wholes," often highly integrated wholes, which are the living nuclei of *systems* in the sense of the physicist. Only I do not think they are properly described as "organisms." I prefer to regard them, together with the whole of the effective physical factors involved, simply as *systems*.

I have already criticised the term "organism" as applied to communities of plants or animals, or to "communities" of plants and animals, on the ground that while these aggregations have some of the qualities of organisms (in the biological sense) they are too different from these to receive the same unqualified appellation. And I have criticised the term "complex organism" on the ground that it is already commonly applied to the species or individuals of the higher animals and plants. . . . [Phillips is] justified in calling the whole formed by an integrated aggregate of animals and plants (the "biocenosis," to use the continental term) an "organism," provided that he includes the *physical factors of the habitat* in his conception. But then he must also call the universe an organism, and the solar system, and the sugar molecule and the ion or free atom. They are all organised "wholes." The nature of what biologists call living organisms is wholly irrelevant to this concept. They are merely a special kind of "organism." . . .

The Ecosystem

Clements' earlier term "biome" for the whole complex of organisms inhabiting a given region is unobjectionable, and for some purposes convenient. But the more fundamental conception is, as it seems to me, the whole *system* (in the sense of physics), including not only the organism-complex, but also the whole complex of physical factors forming what we call the environment of the biome—the habitat factors in the widest sense. Though the organisms may claim our primary interest, when we are trying to think fundamentally we cannot separate them from their special environment, with which they form one physical system.

It is the systems so formed which, from the point of view of the ecologist, are the basic units of nature on the face of the earth. Our natural human prejudices force us to consider the organisms (in the sense of the biologist) as the most important parts of these systems, but certainly the inorganic factors are also parts— there could be no systems without them, and there is constant interchange of the most various kinds within each system, not only between the organisms but between the organic and the inorganic. These *ecosystems*, as we may call them, are of the most various kinds and sizes. They form one category of the multitudinous physical systems of the universe, which range from the universe as a whole down to the atom. The whole method of science . . . is to isolate systems mentally for the purposes of study, so that the series of *isolates* we make become the actual objects of our study, whether the isolate be a solar system, a planet, a climatic region, a plant or animal community, an individual organism, an organic molecule or an atom. Actually the systems we isolate mentally are not only included as parts of larger ones, but they also overlap, interlock and interact with one another. The isolation is partly artificial, but is the only possible way in which we can proceed.

Some of the systems are more isolated in nature, more autonomous, than others, [but] all show organisation, which is the inevitable result of the interactions and consequent mutual adjustment of their components. If organisation of the possible elements of a system does not result, no system forms or an incipient system breaks up. There is in fact a kind of natural selection of incipient systems, and those which can attain the most stable equilibrium survive the longest. It is in this way that the dynamic equilibrium . . . is attained. The universal tendency to the evolution of dynamic equilibria has long been recognized. A corresponding idea was fully worked out by Hume and even stated by Lucretius. The more relatively separate and autonomous the system, the more highly integrated it is, and the greater the stability of its dynamic equilibrium.

Some systems develop gradually, steadily becoming more highly integrated and delicately adjusted in equilibrium. Ecosystems are of this kind. Normal autogenic succession is a progress towards greater integration and stability. The "climax" represents the highest stage of integration and the nearest approach to perfect dynamic equilibrium that can be attained in a system developed under the given conditions and with the available components.

The great regional climatic complexes of the world are important determinants of the primary terrestrial ecosystems, and they contribute *parts* (components) to the systems, just as do the soils and the organisms. In any fundamental consideration of the ecosystem it is arbitrary and misleading to abstract the climatic factors, though for purposes of separation and classification of systems it is a legitimate procedure. In fact the climatic complex has more effect on the organisms and on the soil of an ecosystem than these have on the climatic complex, but the reciprocal action is not wholly absent. Climate acts on the ecosystem rather like an acid or an alkaline "buffer" on a chemical soil complex.

Next comes the soil complex which is created and developed partly by the subjacent rock, partly by climate, and partly by the biome. Relative maturity of the soil complex, conditioned alike by climate, by subsoil, by physiography and by the vegetation, may be reached at a different time from that at which the vegetation attains its climax. Owing to the much greater local variation of subsoil and physiography than of climate, and to the fact that some of the existing variants prevent the climatic factors from playing the full part of which they are capable, the developing soil complex, jointly with climate, may determine variants of the biome. . . .

Finally comes the organism-complex or biome, in which the vegetation is of primary importance, except in certain cases, for example many marine ecosystems. The primary importance of vegetation is what we should expect when we consider the complete dependence, direct or indirect, of animals upon plants. . . . This is not to say that animals may not have important effects on the vegetation and thus on the whole organism-complex. . . . By all means let animal and plant ecologists study the composition, structure, and behaviour of the biome together. . . . But is it really necessary to formulate the unnatural conception of biotic *community* to get such co-operative work carried out? I think not. What we have to deal with is a *system*, of which plants and animals are components, though not the only components. The biome is determined by climate and soil and in its turn reacts, sometimes and to some extent on climate, always on soil.

. . . Systems [are] in dynamic equilibrium. The equilibrium attained is however never quite perfect: its degree of perfection is measured by its stability. The atoms of the chemical elements of low atomic number are examples of exceptionally stable systems—they have existed for many millions of millennia: those of the radio-active elements are decidedly less stable. But the order of stability of all the chemical elements is of course immensely higher than that of an ecosystem, which consists of components that are themselves more or less unstable—climate, soil and organisms. Relatively to the more stable systems the ecosystems are extremely vulnerable, both on account of their own unstable components and because they are very liable to invasion by the components of other systems. Nevertheless some of the fully developed systems—the "climaxes"—have actually maintained themselves for thousands of years. In others there are elements whose slow change will ultimately bring about the disintegration of the system.

This relative instability of the ecosystem, due to the imperfections of its equilibrium, is of all degrees of magnitude, and our means of appreciating and measuring it are still very rudimentary. Many systems (represented by vegetation climaxes) which appear to be stable during the period for which they have been under accurate observation may in reality have been slowly changing all the time, because the changes effected have been too slight to be noted by observers. Many ecologists hold that *all* vegetation is *always* changing. It may be so: we do not know enough either to affirm or to deny so sweeping a statement. But there may clearly be minor changes within a system which do not bring about the destruction of the system as such.

Owing to the position of the climate-complexes as primary determinants of the major ecosystems, a marked change of climate must bring about destruction of the ecosystem of any given geographical region, and its replacement by another. This is the *clisere* of Clements (1916). If a continental icesheet slowly and continuously advances or recedes over a considerable period of time all the zoned climaxes which are subjected to the decreasing or increasing temperature will, according to Clements' conception, move across the continent "as if they were strung on a string," much as the plant communities zoned round a lake will move towards its centre as the lake fills up. If on the other hand a whole continent desiccates or freezes many of the ecosystems which formerly occupied it will be destroyed altogether. Thus whereas the *prisere* is the development of a single ecosystem *in situ*, the *clisere* involves their destruction or bodily shifting. . . .

Biotic Factors

As an ecological factor acting on vegetation, the effect of grazing heavy enough to prevent the development of woody plants is essentially the same effect wherever it occurs. If such grazing exists the grazing animals are an important factor in the biome actually present whether they came by themselves or were introduced by man. The dynamic equilibrium maintained is primarily an equilibrium between the grazing animals and the grasses. . . .

Forest may be converted into grassland by grazing animals. The substitution of the one type of vegetation for the other involves destruction . . . and gradual establishment of new vegetation. It is a successional process culminating in a climax under the influence of the actual combination of factors present. . . . When man introduces sheep and cattle, . . . he artificially maintains [an] ecosystem whose essential feature is the equilibrium between the grassland and the grazing animals. He may also alter the equilibrium by feeding his animals not only on the pasture but also partly away from it, so that their dung represents food for the grassland brought from outside, and the floristic composition of the grassland is thereby altered. In such ways *anthropogenic ecosystems* differ from those developed independently of man, but the essential formative processes of the vegetation are the same.

We must have a system of ecological concepts which will allow of the inclusion of *all* forms of vegetational expression and activity. We cannot confine ourselves to the so-called "natural" entities and ignore the processes and expressions of vegetation now so abundantly provided us by the activities of man. Such a course is not scientifically sound, because scientific analysis must penetrate beneath the forms of the "natural" entities, and it is not practically useful because ecology must be applied to conditions brought about by human activity. . . . Natural and anthropogenic derivates alike must be analysed in terms of the most appropriate concepts. Plant community, succession, development, climax, used in their wider and not in specialised senses, represent such concepts.

Conclusions

Succession is a continuous process of change in vegetation which can be separated into a series of phases. When the dominating factors of change depend directly on the activities of the plants themselves . . . the succession is *autogenic*: when the dominating factors are external to the plants . . . it is *allogenic*. The successions (priseres) which lead from bare substrata to the highest types of vegetation actually present in a climatic region (progressive) are primarily autogenic.

Those which lead away from these higher forms of vegetation (retrogressive) are largely allogenic, though both types of factor enter into all successions.

A *climax* is a relatively stable phase reached by successional change. Change may still be proceeding within a climax, but if it is too slow to appreciate or too small to affect the general nature of the vegetation, the apparently stable phase must still be called a climax. The highest types of vegetation characteristic of a climatic region and limited only by climate form the *climatic climax*. Other climaxes may be determined by other factors such as certain soil types, grazing animals, fire and the like.

The term *development* may be applied, as in ordinary speech, to the appearance of any well-defined vegetational entity; but the term is more strictly applied to the autogenic successions leading to climaxes, which have several features in common with the development of organisms. Such climaxes may be considered as *quasi-organisms*.

The concept of the "biotic community" is unnatural because animals and plants are too different in nature to be considered as members of the same community. The whole complex of organisms present in an ecological unit may be called the *biome*.

The concept of the "complex organism" as applied to the biome is objectionable both because the term is already in common use for an individual higher animal or plant, and because the biome is not an organism except in the sense in which inorganic systems are organisms.

The fundamental concept appropriate to the biome considered together with all the effective inorganic factors of its environment is the *ecosystem*, which is a particular category among the physical systems that make up the universe. In an ecosystem the organisms and the inorganic factors alike are *components* which are in relatively stable dynamic equilibrium. Succession and development are instances of the universal processes tending towards the creation of such equilibrated systems.

From the standpoint of vegetation biotic factors, in the sense of decisive influences of animal action, are a legitimate and useful conception. Of these biotic factors heavy and continuous grazing which changes and stabilises the vegetation is an outstanding example.

The supposed methodological value of the ideas of the biotic community and the complex organism is illusory, unlike the values of plant community, succession, development, climax and ecosystem, the concepts of which form the essential framework into which detailed studies of successional processes must be fitted.

Commentary

Arthur Tansley, "The Use and Abuse of Vegetational
Concepts and Terms" (1935)

LIBBY ROBIN

This paper is part of a long and fruitful transatlantic conversation that resulted
in the definition of the "ecosystem" as a tool for modern understanding of the
function and organization of nature. British and North American ecology had developed in rather different ways, and in 1911, Arthur George Tansley (1871–1955)
organized a new international event, the first International Phytogeographical
Excursion (IPE) to the Norfolk Broads in England, to facilitate dialogue between
perspectives from both sides of the Atlantic. The excursion was also part of the
British Vegetation Survey (which in 1913 became the British Ecological Society).

Henry Chandler Cowles, from Chicago, commented that Tansley's IPE had
"internationalized" plant geography and served to "divest it of . . . provincialism."
In this paper Tansley warmly acknowledged the importance of Cowles in theorizing the concept of *succession* (original, p. 284), and much of the paper is directed
at differences between the prairie-driven ecological theories of North America
(particularly those of Frederic Clements) and the Old World ecology of Britain
and continental Europe. It was written to clarify the vegetational concepts used
in what Tansley liked to call the "New Ecology."

"Use and Abuse" is abridged here to about a third of its original length.
The omissions mostly relate to an extended discussion of the philosophical basis for vegetational concepts put forward by South African philosopher-biologist
John Phillips, in his 1934 paper "Succession, Development, the Climax and the
Complex Organism: An Analysis of Concepts." Phillips's ecology was founded
on the ideas of ecologist Frederic Clements, and its holism drew on the ideas of
South Africa's famous international statesman Jan Smuts. Tansley was a scientist inspired by modernity, who rejected the notion of a plant community as an
"organism," as expressed by Phillips and Clements. He favored a more physical
conceptualization of nature and resisted analogies with sociological and holistic
concepts. His "idea of the new" was evident in his early journal *The New Phytolo-
gist* (1902) and extended to his work in the "new psychology" in the 1920s with
Sigmund Freud in Vienna.

The abridged paper sheds light on the key concept of the "ecosystem," and
the question of how to conceptualize landscapes modified by human action.
These have proved the most long-lasting contributions to global change thinking. Like his compatriot, animal ecologist Charles Elton (pp. 367–377), Tansley
was conscious that humans were just one animal among many, and that animal

communities and human communities changed the environment in ways that concepts like the ecosystem needed to take into account. Tansley considered both a human and a plant community to be *quasi-organisms*, but neither was an *organism*. Even the universe itself became an organism, under the broad definitions suggested by Phillips: thus the term effectively became too broadly conceived and inclusive to be of practical scientific use to the modernist Tansley.

A. R. Clapham, a crop physiologist at Rothamsted Experimental Station, coined the term *ecosystem*, but Tansley defined it functionally for the discipline of ecology, including inorganic as well as organic factors. Soils and climates were important elements in an ecosystem, "the habitat factors in the widest sense" (p. 225). More than a biome, an ecosystem became properly the subject for study by physicists, soil scientists, and chemists, along with biologists, thereby promoting ecology itself as a meta-discipline, rather than a mere subdiscipline of biology. The close attention to soils (echoed in Sears pp. 174–182) is part of the rapid growth of pedology (soil science) in this era. "Edaphic" (soil) factors included non-living elements such as nitrogen, rock types, and slope, which determined rates of erosion. While climate had been recognized as significant as far back as von Humboldt (see the previous essay), the ecosystem was a new tool that integrated the study of plants, soils, and climate into one realm of knowledge. While ecologists since the 1990s, such as Daniel Botkin, now dispute the idea of *climax* ecosystems, or the "perfect dynamic equilibrium" discussed by Tansley, the ecosystem concept itself has remained remarkably robust for seven decades.

Natural and human-modified systems alike need the same conceptual treatment, Tansley argued: "We cannot confine ourselves to the so-called 'natural' entities and ignore the processes and expressions of vegetation now so abundantly provided us by the activities of man. Such a course is not scientifically sound, because scientific analysis must penetrate beneath the forms of the 'natural' entities, and it is not practically useful because ecology must be applied to conditions brought about by human activity" (p. 228).

Ecosystems and their successional processes include anthropogenic change, but not catastrophic events (like elephants and volcanoes). Change must be along a continuum to be properly the subject of ecological laws, but the forces of change can be natural or human. Catastrophes are unrelated to the causes of vegetational changes. Tansley resisted the idea of an idealized scientific concept not relevant to the working situations where scientists found themselves. If ecologists were to become "experts" for working landscapes, they needed theoretical concepts inclusive of applied situations.

Tansley was deeply interested in human society, especially its evolutionary past (the *herd instinct*). In *The New Psychology* (1922), he argued that the mind is a highly evolved organism, but "of nature," not apart from it. Its most fundamental activities are non-rational and unconscious, "inherited from primitive man and from man's non-human forerunners" (Tansley 1922: 23). Three decades later, in his preface to *Mind and Life*, Tansley held the same line: since humans *are* animals, biology must come before psychology (Tansley 1952: vi).

Vegetational science had always had an applied and practical edge, which dated back to the *Survey and Study of British Vegetation* begun in 1904. This continued in Tansley's definition of "wild nature," relevant to the later part of his career, when he became chairman of the new Nature Conservancy in 1949. In Tansley's thinking, wild nature must be understood in terms of its human history, so his nature conservation was cast in a "heritage" mold. Tansley was not drawn to New World wilderness thinking, nor to transcendental understanding of landscapes. He recognized the "unspoilt landscapes" of Britain as the handwork of generations of farmers, rather than nature evolving in isolation from people. In 1950, he was awarded a knighthood that acknowledged the usefulness of his science to the nation in the practical planning and management of the countryside.

Further Reading

Cameron, Laura. 2008. "Sir Arthur George Tansley." In *New dictionary of scientific biography*, ed. Noretta Koertge, 7:3–9. Farmington Hills, MI: Charles Scribner's Sons.

Clements, F. E. 1916. *Plant succession*. Publication 242. Washington, DC: Carnegie Institution.

Phillips, John. 1934. "Succession, development, the climax and the complex organism: An analysis of concepts." *International Journal of Ecology* 22:554–71.

Tansley, A. G. 1920. "The classification of vegetation and the concept of development." *Journal of Ecology* 8:118–44.

Tansley, A. G. 1922. *The new psychology and its relation to life*. London: Allen & Unwin.

Tansley, A. G. 1929. "Succession: The concept and its values." *Proceedings of the International Congress of Plant Sciences, 1926*, 677–86.

Tansley, A. G. 1952. *Mind and life: An essay in simplification*. London: Allen & Unwin.

Principles and Concepts Pertaining to the Ecosystem and Biogeochemical Cycles

EUGENE P. ODUM

1. Concept of the Ecosystem

STATEMENT

Living organisms and their nonliving (abiotic) environment are inseparably interrelated and interact upon each other. Any entity or natural unit that includes living and nonliving parts interacting to produce a stable system in which the exchange of materials between the living and nonliving parts follows circular paths is an ecological system or ecosystem.[1] The ecosystem is the largest functional unit in ecology, since it includes both organisms (biotic communities) and abiotic environment, each influencing the properties of the other and both necessary for maintenance of life as we have it on the earth. A lake is an example of an ecosystem.

EXPLANATION

Since no organism can exist by itself or without an environment, our first principle may well deal with the "interrelation" part of our basic definition of ecology. . . . The portion of the earth which contains living organisms and, hence, in which ecosystems operate, is known as the *biosphere*. Since life extends for only a relatively few feet below the earth's surface, the biosphere is the thin outer shell of the earth, including the oceans and the atmosphere. The biosphere is important not only as a place where living organisms can exist but also as a region where the incoming radiation energy of the sun brings about fundamental chemical and physical changes in the inert material of the earth. These changes result chiefly from the functioning of various ecosystems.

E. P. Odum (in collaboration with H. T. Odum). 1953. "Principles and concepts pertaining to the ecosystem and biogeochemical cycles." Chapter 2 in *Fundamentals of ecology*, 9–21. Philadelphia: W. S. Saunders.

The concept of the ecosystem is and should be a broad one, its main function in ecological thought being to emphasize obligatory relationships, interdependence, and causal relationships. Ecosystems may be conceived and studied in various sizes. Thus, the entire biosphere may be one vast ecosystem with numerous more or less circular systems within it. Obviously, smaller systems must be studied before the entire biosphere can be understood. A pond, a lake, a tract of forest, or a chemical cycle is a convenient unit of study.

It is axiomatic that animals and most non-green plants are dependent upon green plants for foods, and that the latter depend on basic raw materials and the energy of the sun. Just how chlorophyll-bearing plants manufacture carbohydrates, proteins, fats, and other complex materials is not yet fully understood, and the process will probably not be duplicated in the laboratory for some time. However, the simplified photosynthesis formula is one of the first things learned in elementary biology. Written in word form, it goes something like this:

| **Carbon dioxide** | plus | **water** | plus | **light energy** in the presence of enzyme systems associated with chlorophyll | results in | **glucose** plus oxygen |

Chemically the photosynthetic process involves the storage of a part of the sunlight energy as potential or "bound" energy of food. The process is much more complicated than indicated in the above word formula, but the basic idea is easily understood. The photosynthetic process of food manufacture is often called the "business of green plants." It is now believed that the synthesis of amino acids, proteins, and other vital materials occurs simultaneously with the synthesis of carbohydrates (glucose), some of the basic steps involved being the same. The reverse of photosynthesis, or respiration, results in the oxidation of foods with the release of energy (making possible growth, movement, heat production, etc.) and is, of course, the "business of all organisms." While it is conceivable that we might have a sizable ecosystem containing only plants, such seems rarely, if ever, to occur on our planet. Plants live with, are influenced and controlled by, and are often even dependent on animals. Both plants and animals are dependent on bacteria and other organisms not readily classified as plants or animals (this group of organisms is conveniently designated as "Protista"). The vital role of bacteria in the functioning of the ecosystem will be discussed later.

Thus, although everyone realizes that the abiotic environment ("physical factors") controls the activities of organisms, it is not always realized that organisms influence and control the abiotic environment in many ways. Changes in the physical and chemical nature of inert materials are constantly being effected by organisms which return new compounds and isotopes to the nonliving environment. Plants growing on a sand dune build up a soil radically different from the original substrate. A South Pacific coral island is a striking example of how organisms influence their abiotic environment. From simple raw materials of the sea, whole islands are built as the result of the activities of animals (corals, etc.) and plants. The very composition of our atmosphere is controlled by organisms. Indeed, without living organisms our world probably would be a relatively unchanging mass composed of fewer kinds of materials. Like our moon, it would be a very dull world indeed.

. . .

As a result of the evolution of the central nervous system and brain, man has gradually become the most powerful organism, as far as the ability to modify the operation of ecosystems is concerned.[2] Man's power to change and control seems to be increasing faster than man's realization and understanding of the results of the profound changes of which he is now capable. As many writers have pointed out, this is a dangerous situation, because tinkering with basic ecosystems of the world can result either in a glorious future for mankind, or in his complete destruction if too many large-scale mistakes are made. The idea of the ecosystem and the realization that mankind is a part of complex "biogeochemical" cycles with increasing power to modify the cycles are concepts basic to modern ecology and are also points of view of extreme importance in human affairs generally. Conservation of natural resources, the most important practical application of ecology, must be built around these viewpoints. Thus, if understanding of ecological systems and moral responsibility among mankind can keep pace with man's power to effect changes, the present day concept of "unlimited exploitation of resources" will give way to "unlimited ingenuity in perpetuating a cyclic abundance of resources." Hutchinson (1948) has aptly expressed this viewpoint something as follows: the ecologist should be able to show that it is just as much fun and just as important to repair the biosphere as to mend the radio or the family car. For additional discussions of the ecosystem concept see Forbes' (1887) classic essay, Tansley (1935), and especially the important paper by Lindeman (1942).

EXAMPLE

One of the best ways to begin the study of ecology is to go out and study a small pond. Ecologists have spent many years studying individual ponds and lakes without investigating all the details or understanding all the process, but the basic principles can be grasped by the student in a short time without resorting to a detailed study. Let us consider the pond as a whole as an ecosystem, leaving the study of smaller systems within the pond for the second section of this book. The inseparability of living organisms and the nonliving environment is at once apparent with the first sample collected. Not only is the pond a place where animals and plants live, but animals and plants make the pond what it is. Thus, a bottle full of the pond water or a scoop full of bottom mud is a mixture of living organisms, both plant and animal, and inorganic and organic compounds. Some of the larger animals and plants can be separated from the sample for study or counting, but it would be difficult to completely separate the myriad of small living things from the nonliving matrix without changing the character of the fluid. True, one could autoclave the sample of water or bottom mud so that only nonliving material remained, but this residue would then no longer be pond water or pond soil but would have entirely different appearances and characteristics.

Despite the complexities, the pond ecosystem may be reduced to four basic units, as follows:

1. *Abiotic substances*—basic inorganic and organic compounds, such as water, carbon dioxide, oxygen, calcium and phosphorus salts, amino and humic acids, etc.
2. *Producer organisms*—green plants which are able to manufacture food from simple inorganic substances. In a pond the producers may be of two main types: (1) rooted or large floating plants generally growing in shallow water only: and (2) minute floating plants, usually algae, called *phytoplankton*, distributed throughout the pond as deep as light penetrates. In abundance, the phytoplankton gives the water a greenish color: otherwise, these producers are not visible to the casual observer, and their presence is not suspected by the layman. Yet phytoplankton is generally far more important than is rooted vegetation in the production of basic food for the pond ecosystem.
3. *Consumer organisms*—animals, such as insect larvae, crustacea, and fish, and some non-green plants. The primary consumers (herbivores) feed directly on plants, and the secondary consumers (carnivores) feed on the primary consumers, etc.
4. *Reducer (decomposer) organisms*—chiefly bacteria and fungi which break down the complex compounds of dead protoplasm into simple substances usable by green plants. Essentially, these organisms bring about the "mineralization" of organic matter.

The first unit constitutes the abiotic environment, and the other three units comprise the *biotic community*. These two parts of the ecosystem are quite distinct, in so far as characteristics are concerned, but are difficult to separate in practice, as already indicated.

2. Concepts of Habitat and Ecological Niche
STATEMENT

The habitat of an organism is the place where it lives, or the place where one would go to find it. The ecological niche, on the other hand, is the position or status of an organism within its community and ecosystem resulting from the organism's structural adaptations, physiological responses, and specific behavior (inherited and/or learned). The ecological niche of an organism depends not only on where it lives but also on what it does. By analogy, it may be said that the habitat is the organism's "address," and the niche is its "profession," biologically speaking.

EXPLANATION

The term habitat is widely used, not only in ecology but elsewhere. It is generally understood to mean simply the place where an organism lives. Thus, the habitat of the water "backswimmer," *Notonecta*, is the shallow, vegetation-choked areas (littoral zone) of ponds and lakes; that is where one would go to collect this particular organism. The habitat of a *Trillium* plant is a moist, shaded situation in a mature deciduous forest; that is where one would go to find *Trillium* plants. Different species in the genus *Notonecta* or *Trillium* may occur in the same general habitat but exhibit small differences in location, in which event we would say that the *microhabitat* is different. Other species in these genera exhibit large habitat, or *macrohabitat*, differences.

Habitat may also refer to the place occupied by an entire community. For example, the habitat of the "sand sage grassland community" is the series of ridges of sandy soil occurring along the north sides of rivers in the southern Great Plains region of the United States. Habitat in this case consists mostly of physical or abiotic complexes, whereas habitat as used with reference to *Notonecta* and *Trillium*, mentioned above, includes living as well as nonliving objects. Thus, the habitat of an organism or group of organisms (population) includes other organsims as well as the abiotic environment. A description of the habitat of the community would include only the latter. It is important to recognize these two possible uses of the term habitat in order to avoid confusion.

Ecological niche is a more recent concept and is not so generally understood outside the field of ecology. It is, however, a very important concept. Charles

Elton in England and Joseph Grinnell in California began using the term "niche" for the "status of an organism in its community" at about the same time; it has gradually become more generally accepted as it became clear that niche is by no means a synonym for habitat. Let us return, for the moment, to the simple analogy of the "address" and the "profession" mentioned above. If we wished to become acquainted with some person in our human community we would need to know, first of all, his address, that is, where he could be found. To really get to know him, however, we would have to learn more than the neighborhood where he lives or works. We would want to know something about his occupation, his interests, his associates, and the part he plays in general community life. So it is with the study of organisms; learning the habitat is just the beginning. To determine the organism's status in the natural community we would need to know something of its activities, especially its food relations—what it eats and what eats it; also its range of movement, its effect on other organisms with which it comes into contact, and the extent to which it modifies or is capable of modifying important operations in the ecosystem.

. . .

3. Biogeochemical Cycles
STATEMENT

The chemical elements, including all the essential elements of protoplasm, tend to circulate in the biosphere in characteristic paths from environment to organisms and back to the environment. These more or less circular paths are known as "inorganic-organic cycles," or *biogeochemical cycles*.

EXPLANATION

The study of ecosystems may be approached not only from the consideration of the organism, but also from that of the abiotic environment. In other words, it is profitable in ecology to study not only organisms and their environmental relations but also the basic nonliving environment in relation to organisms. Of the 90-odd elements known to occur in nature, between 30 and 40 are known to be required by living organisms. Some elements such as carbon, hydrogen, oxygen, and nitrogen are needed in large quantities; others are needed in small, or even minute, quantities. Whatever the need may be, essential elements (as well as nonessential elements) appear to exhibit definite biogeochemical cycles. Some cycles, such as the one involving carbon, are more perfect than others; that is, the material is returned to the environment as fast as it is removed, and although critical "shortages" may occur in certain environments there is little or

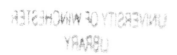

no permanent change in the distribution of the element in the various ecosystems of the biospheres. Other cycles may be less perfect; that is, some portion of the supply may get "lost" for long periods of time in places or in chemical forms inaccessible to organisms.

Hutchinson (1948) points out that man is unique in that not only does he require the 40 essential elements but also in his complex culture he uses nearly all the other elements and the newer synthetic ones as well. He has so speeded up the movement of many materials that the cycles tend to become imperfect, or the process becomes "acyclic," with the result that man is continually running into shortages. The aim of conservation of natural resources in the broadest sense is to make acyclic processes more cyclic. Fortunately, the more perfect cycles have so many compensating mechanisms that man has not yet done too much to modify them. Disturbance or manipulation of a cycle involving a vital element, however, could conceivably be much more dangerous than the disturbance of the less perfect cycles, because if such a cycle were disturbed beyond its compensatory powers, the whole thing might go completely to pieces.

EXAMPLES

Two examples will suffice to illustrate the principle of inorganic-organic cycles. The nitrogen cycle (Fig. 5) is an example of a very complex but more or less perfect cycle; the phosphorus cycle is an example of a simpler, possibly less perfect, one. . . . Both these elements are often very important factors, limiting or controlling the abundance of organisms, and hence they have received much attention and study.

As shown in Figure 5 the nitrogen in protoplasm is broken down from organic to inorganic form by a series of reducing bacteria, each specialized for a particular part of the job. Some of this nitrogen ends up as nitrate, the form most readily used by green plants (although some organisms can use nitrogen in other forms as illustrated), thus completing the cycle. The air, which contains 80 per cent nitrogen, is the greatest reservoir and safety valve of the system. Nitrogen is continually entering the air by the action of denitrifying bacteria and continually returning to the cycle through the action of nitrogen-fixing bacteria or blue-green algae and through the action of lightning (i.e., electrification). The importance of the nitrogen-fixing bacteria associated with legumes is well known, of course, and in modern agriculture continuous fertility of a field is maintained as much by crop rotation involving legumes as by the application of nitrogen fertilizers. Other nitrogen-fixing bacteria live free in the soil.

The self-regulating, feedback mechanisms, shown in a very simplified way by the arrows in the diagram (Fig. 5), make the nitrogen cycle a relatively perfect

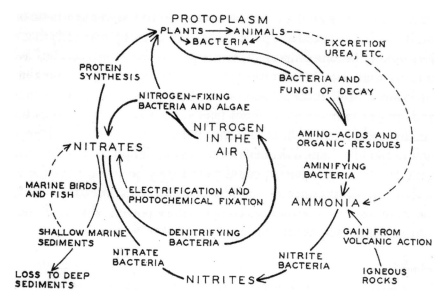

Figure 5. The nitrogen biochemical cycle, an example of a relatively perfect, self-regulating cycle in which there is little over-all change in available nitrogen in *large* ecosystems or in the biosphere as a whole, despite rapid circulation of materials. Some quantitative estimates of interest are as follows:

1. The loss of nitrogen from atmosphere to sediments is apparently balanced by the gain from volcanic action; in fact, the nitrogen content of the air may possibly have increased throughout geological time.
2. According to Hutchinson (1944a), the amount of nitrogen fixed from the air (non-cyclic nitrogen) is estimated to lie between 140 and 700 mg. per square meter, or between 1 and 6 pounds per acre per year. Most of this is biological; only a small portion (not more than 35 mg. per square meter per year in temperate regions) is non-biological (electrification or photochemical fixation).
3. The capacity to fix atmospheric nitrogen was thought, until recently, to be limited to a few, but abundant, organisms, as follows:

   Free-living bacteria—*Azotobacter* and *Clostridium*;
   Symbiotic nodule bacteria on legume plants—*Rhizobium*;
   Blue-green algae—*Anabaena*, *Nostoc*, and probably others.

In 1949 it was discovered that the purple bacterium *Rhodospirillum* and many other representatives of the photosynthetic bacteria are nitrogen fixers (see Kamen and Gest, 1949, and Kamen, 1953). It seems likely that ability to fix nitrogen will prove to be more widespread than was previously believed. Biological fixation on land with legumes may reach 1231 mg. per square meter per year, or about 11 pounds per acre, of which a little less than half is due to nodule bacteria, the rest to free-living forms. Increased use of legumes helps balance increased loss by erosion resulting from the activities of man. Means of increasing the activity of free-living nitrogen fixers will perhaps prove feasible in the future.

one, when large areas of the biosphere as a whole is considered. Thus, increased movement of materials along one path is quickly compensated for by adjustments along other paths.

Some nitrogen from heavily populated regions of land, fresh water, and shallow seas is lost to the deep ocean sediments and thus gets out of circulation, at least for a while (a few million years perhaps). This loss is compensated for by nitrogen entering the air from volcanic gases. Thus, volcanic action is not to be entirely deplored but has some use after all! If nothing else, ecology teaches us not to make snap judgments as to whether a thing or an organism is "useful" or "harmful." One must consider all the aspects of a problem before arriving at a judgment.

Notes

1. The term ecosystem was first proposed by Tansley in 1935, but the concept is by no means so recent. Microcosm (Forbes, 1887), holocoen, biosystem and bioinert body are terms which have been used to express similar ideas.

2. So important is man's role becoming as "a mighty geological agent" that Vernadsky has suggested that we think of the "noösphere" (from Greek *noös*, mind), or the world dominated by the mind of man, as gradually replacing the biosphere, the naturally evolving world which has existed for billions of years. This is dangerous philosophy because it is based on the assumption that mankind is now wise enough to understand the results of all his actions. When the reader has finished with this book I am sure he will agree that we have yet much to learn before we can safely take over the management of everything!

Commentary

Eugene P. Odum, *Fundamentals of Ecology* (1953)

STEPHEN BOCKING

In the early 1950s Eugene Odum surveyed ecological science and found it wanting. Ecologists pursued incompatible perspectives on nature, enjoyed only a marginal status among scientists, and were still largely unable to persuade society of the value of their advice. Odum's response was *Fundamentals of Ecology*: a manifesto for a reconstituted discipline.

Designed to serve as both textbook and handbook, *Fundamentals* opened with chapters presenting the general principles of ecology: the factors limiting organisms, the dynamics of animal populations, competition and succession in communities. These chapters combined insights from diverse specialties: plant ecology, animal ecology, limnology, marine ecology—exemplifying his ambition to synthesize half a century of ecological research. He then turned to the ecology of specific environments: freshwater, the oceans, the land. Ecology, he insisted, must remain grounded in the challenges and opportunities of field work in these distinctive habitats. And finally, Odum presented practical applications. Humans, he asserted, are also part of ecology. Ecologists could contribute to diverse fields: from wildlife management to land use to pollution control and public health, guided by due regard for the ethical and cultural dimensions of our relationship with nature.

Above all, Odum aimed to present ecology "as a whole": unifying ecologists' disparate approaches and study of diverse taxonomic groups. This required a new view of nature—one that ecologists could claim as their own. This would be the role of the ecosystem concept. Several features defined the ecosystem as the basic unit of ecology: it encompassed plants and animals, as well as bacteria, fungi, and other essential but often neglected organisms; its functional components—producers, consumers, decomposers—transcended taxonomy; its cycles of energy and nutrients blurred the distinction between living and nonliving; finally, the ecosystem exhibited distinctive behavior, including a tendency toward stability. And only ecologists had the necessary expertise with which to interpret this behavior.

The ecosystem concept was also, Odum explained, essential to applying ecology to practical issues. If, he argued, ecosystems could be understood in terms of cycles, so could their management: maintaining cycles would ensure a balance between resource harvesting and renewal. More generally, the ecosystem concept underpinned a rational, integrated approach to managing the environment and human activities.

Odum's ambitious synthesis drew widely on the ecological literature. In recent years several ecologists had written textbooks surveying plant or animal ecology: Odum was evidently not alone in perceiving a need to order the unkempt strands of their discipline. His arguments regarding ecosystems invoked diverse elements of ecological thought: holistic approaches to habitats and communities, such as Stephen Forbes's classic 1887 work on the lake as a microcosm, Arthur Tansley's definition in 1935 of the term "ecosystem," Victor Shelford's organismic view of plant and animal ecology, and Charles Elton's perspective on food webs. Odum also drew on essential work by Raymond Lindeman and G. Evelyn Hutchinson on the cycling of elements and nutrients in ecosystems.

Odum and his brother and colleague H. T. (Tom) Odum were themselves developing research along these lines. In the late 1940s H. T. Odum had studied the cycling of strontium with Hutchinson. In subsequent work the brothers stressed an integrative perspective on the "metabolism" of ecosystems, with H. T. Odum going furthest in reducing this to flows of energy.

For a quarter century, and through revised editions published in 1959 and 1971, *Fundamentals* was the most influential textbook of ecology, especially in North America (it was also translated into several other languages). Presentation was a factor. Odum's writing was clear, accessible, even friendly. It was also well organized: each chapter presented a principle, explained its significance, and provided relevant data and examples. Even his diagrams were teaching tools: schematic representations of ecosystems, functional components, and flows of energy and elements, they became ecological icons.

Odum also had excellent timing. *Fundamentals* appeared just as ecology was poised for takeoff. Government support for science was expanding, particularly in America. Ecologists were ready for a new identity: distinctive, yet acceptable to institutions that defined credibility in terms of the physical sciences. Now that ecologists were talking about systems, energy, and elements, even physicists could see that they were not simply butterfly collectors. Odum also emphasized ecology's practical relevance just as demand was growing for systemic perspectives on environmental management and protection. No longer a marginal field, many now perceived ecology as the intellectual foundation of the environmental movement.

Ecosystem ecology also gained a powerful ally. In the 1950s Odum and other ecologists persuaded the Atomic Energy Commission that they could advance the application of nuclear technology to scientific research, while managing the environmental and health impacts of radiation. Thanks in part to the Commission, ecosystem ecology developed especially vigorously in the United States — exemplifying the formation of an ecology compatible with technocracy and the national security state. *Fundamentals* tracked this formation: the 1959 edition included a chapter on radiation ecology.

But ecosystem ecology was also only one of several schools of ecological research. Many ecologists pursued research on populations, evolutionary change, and multispecies communities, focusing on competition and other phenomena that ecosystem ecology tended to neglect. Some ecologists also resisted identifying their discipline with environmental protection. Applied environmental practice itself shifted away from integrated management—the particular province of ecosystem ecology. By the 1980s ecosystem ecology had been to some extent eclipsed, and *Fundamentals* was displaced by textbooks urging other approaches.

Nevertheless, the approach represented in *Fundamentals* remains essential, with many ecologists integrating it with study of individuals, species, and other levels of organization. Its emphasis on systems and biogeochemical cycles has flowed far beyond ecology, evident in efforts to predict the future of global systems and the movement and transformation of elements and other substances, including contaminants. Odum expressed an integrative vision that remains essential to our understanding of the natural world. His synthesis served more than any other single text to codify just what postwar ecologists were about, and why they deserved attention.

Further Reading

Bocking, S. 1997. *Ecologists and environmental politics: A history of contemporary ecology.* New Haven: Yale University Press.

Bocking, S. 2004. *Nature's experts: Science, politics, and the environment.* New Brunswick, NJ: Rutgers University Press.

Forbes, S. A. 1887. "The lake as a microcosm." *Bulletin of the Scientific Association,* 77–87. Reprinted 1925 in *Illinois Natural History Survey Bulletin* 15(9):537–550.

Hutchinson, E. 1948. "Circular causal systems in ecology." *Annals of the New York Academy of Sciences* 50:221–246.

Kingsland, S. E. 2005. *The evolution of American ecology, 1890–2000.* Baltimore: Johns Hopkins University Press.

Resilience and Stability of Ecological Systems

C. S. HOLLING

Introduction

Individuals die, populations disappear, and species become extinct. That is one view of the world. But another view of the world concentrates not so much on presence or absence as upon the numbers of organisms and the degree of constancy of their numbers. These are two very different ways of viewing the behavior of systems and the usefulness of the view depends very much on the properties of the system concerned. If we are examining a particular device designed by the engineer to perform specific tasks under a rather narrow range of predictable external conditions, we are likely to be more concerned with consistent non variable performance in which slight departures from the performance goal are immediately counteracted. A quantitative view of the behavior of the system is, therefore, essential. With attention focused upon achieving constancy, the critical events seem to be the amplitude and frequency of oscillations. But if we are dealing with a system profoundly affected by changes external to it, and continually confronted by the unexpected, the constancy of its behavior becomes less important than the persistence of the relationships. Attention shifts, therefore, to the qualitative and to questions of existence or not. . . .

The purpose of this review is to explore both ecological theory and the behaviour of natural systems to see if different perspectives of their behavior can yield different insights useful for both theory and practice. . . .

Theoretical models generally have not done well in simultaneously incorporating realistic behavior of the processes involved, randomness, spatial heterogeneity, and an adequate number of dimensions or state variables. This situation is changing very rapidly as theory and empirical studies develop a closer technical

C. S. Holling. 1973. Excerpt from "Resilience and stability of ecological systems." *Annual Review of Ecology and Systematics* 4:1–23.

partnership. In what follows I refer to real world examples to determine how the four elements that tend to be left out might further affect the behavior of ecological systems.

Some Real World Examples
SELF-CONTAINED ECOSYSTEMS

In the broadest sense, the closest approximation we could make of a real world example that did not grossly depart from the assumptions of the theoretical models would be a self-contained system that was fairly homogenous and in which climatic fluctuations were reasonably small. If such systems could be discovered they would reveal how the more realistic interaction of real world processes could modify the patterns of systems behavior described above. Very close approximations to any of these conditions are not likely to be found, but if any exist, they are apt to be fresh water aquatic ones. Fresh water lakes are reasonably contained systems, at least within their watersheds; the fish show considerable mobility throughout, and the properties of the water buffer the more extreme effects of climate. Moreover, there have been enough documented man-made disturbances to liken them to perturbed systems in which either the parameter values or the levels of the constituent populations are changed. . . . Two major classes of disturbances have occurred: first, the impact of nutrient enrichment from man's domestic and industrial wastes, and second, changes in fish populations by harvesting.

The paleolimnologists have been remarkably successful in tracing the impact of man's activities on lake systems over surprisingly long periods. For example, Hutchinson (1970) has reconstructed the series of events occurring in a small crater lake in Italy from the last glacial period in the Alps (2000 to 1800 BC) to the present. Between the beginning of the record and Roman times the lake had established a trophic equilibrium with a low level of productivity which persisted in spite of dramatic changes in surroundings from Artemesia steppe, through grassland, to fir and mixed oak forest. Then suddenly the whole aquatic system altered. This alteration towards eutrophication seems to have been initiated by the construction of the Via Cassia about 171 BC, which caused a subtle change in the hydrographic regime. The whole sequence of environmental changes can be viewed as changes in parameters or driving variables, and the long persistence in the face of these major changes suggests that natural systems have a high capacity to absorb change without dramatically altering. But this resilient character has its limits, and when the limits are passed, as by the construction of the Roman highway, the system rapidly changes to another condition.

More recently the activities of man have accelerated and limnologists have recorded some of the responses to these changes. The most dramatic change consists of blooms of algae in surface waters, an extraordinary growth triggered, in most instances, by nutrient additions from agricultural and domestic sources. . . .

The history of the Great Lakes provides not only some particularly good information on responses to man-made enrichment, but also on responses of fish populations to fishing pressure. . . . Prior to 1930, before eutrophication complicated the story, the lake sturgeon in all the Great Lakes, the lake herring in Lake Erie, and the lake whitefish in Lake Huron were intensively fished. In each case the pattern was similar: a period of intense exploitation during which there was a prolonged high level harvest, followed by a sudden and precipitous drop in populations. Most significantly, even though fishing pressure was then relaxed, none of these populations showed any sign of returning to their previous levels of abundance. This is not unexpected for sturgeon because of their slow growth and late maturity, but it is unexpected for herring and whitefish. The maintenance of these low populations in recent times might be attributed to the increasingly unfavorable chemical or biological environment, but in the case of the herring, at least, the declines took place in the early 1920s before the major deterioration in environment occurred. It is as if the population had been shifted by fishing pressure from a domain with a high equilibrium to one with a lower one. This is clearly not a condition of neutral stability . . . since once the populations were lowered to a certain point the decline continued even though fishing pressure was relaxed. It can be better interpreted as [a situation] where populations have been moved from one domain of attraction to another.

Since 1940 there has been a series of similar catastrophic changes in the Great Lakes that has led to major changes in the fish stocks [of lake trout, whitefish, herring, walleye, sauger, and blue pike]. After sustained but fluctuating levels of harvest, the catch dropped dramatically in a span of a very few years, covering a range of from one to four orders of magnitude. In a number of examples particularly high catches were obtained just before the drop. . . .

The explanations for these changes have been explored in part, and involve various combinations of intense fishing pressure, changes in the physical and chemical environment, and the appearance of a foreign predator (the sea lamprey) and foreign competitors (the alewife and carp). For our purpose the specific cause is of less interest than the inferences that can be drawn concerning the resilience of these systems and their stability behavior. . . . Whatever the specific causes, it is clear that the precondition for the collapse was set by the harvesting of fish, even though during a long period there were no obvious signs of problems. The fishing activity, however, progressively reduced the resilience of the system so that when the inevitable unexpected event occurred, the populations collapsed. If it had not been the lamprey, it would have been something else: a change in climate as part of the normal pattern of fluctuation, a change in the chemical or physical environment, or a change in competitors or predators. These examples again suggest distinct domains of attraction in which the populations forced close to the boundary of the domain can then flip over it. . . .

The same patterns have even been suggested for terrestrial systems. Many of the arid cattle grazing lands of the western United States have gradually become invaded and dominated by shrubs and trees like mesquite and cholla. In some instances grazing and the reduced incidence of fire through fire prevention programs allowed invasion and establishment of shrubs and trees at the expense of grass. Nevertheless, Glendening (1952) has demonstrated, from data collected in a 17-year experiment in which intensity of grazing was manipulated, that once the trees have gained sufficient size and density to completely utilize or materially reduce the moisture supply, elimination of grazing will not result in the grassland re-establishing itself. In short, there is a level of the state variable "trees" that, once achieved, moves the system from one domain of attraction to another. Return to the original domain can only be made by an explicit reduction of the trees and shrubs.

These examples point to one or more distinct domains of attraction in which the important point is not so much how stable they are within the domain, but how likely it is for the system to move from one domain into another and so persist in a changed configuration. . . .

PROCESS ANALYSIS

One way to represent the combined effects of processes like fecundity, predation, and competition is by using Ricker's (1954) reproduction curves. These simply represent the population in one generation as a function of the population in the previous generation. . . . In the simplest form, the one most used in practical fisheries management . . . , the reproduction curve is dome-shaped. . . . It is extremely difficult to detect the precise form of such curves in nature, however; variability is high, typically data are only available for parts of any one curve, and the treatment really only applies to situations where there are no lags. It is possible to deduce various forms of reproduction curves, however, by disaggregating the contributions of fecundity and mortality. . . .

Mortality from predation, for example, has been shown to take a number of classic forms (Holling 1961, 1965). The individual attack by predators as a function of prey density (the functional response to prey density) can increase with a linear rise to a plateau (type I), a concave or negatively accelerated rise to a plateau (type 2), or an S-shaped rise to a plateau (type 3). The resulting contribution to mortality from these responses can therefore show ranges of prey density in which there is direct density dependence . . . , density independence . . . , and inverse dependence. . . .

The fecundity curves are likely to be more complex, however, since . . . at some very low densities fecundity will decline because of difficulties in finding mates and the reduced effect of a variety of social facilitation behaviors. We might even logically conclude that for many species there is a minimum density below which fecundity is zero. A fecundity curve of this Allee-type has been empirically demonstrated for a number of insects (Watt 1968). . . . If populations slip below [a lower density] they will proceed inexorably to extinction. The extinction threshold is particularly likely since it has been shown mathematically that each of the three functional response curves will intersect with the ordinate of percent predation at a value above zero. . . .

The behavior of systems in phase space cannot be completely understood by the[se] graphical representations presented above, [which are] appropriate only when effects are immediate. The graphical treatment of Rosenzweig & MacArthur (1963) to a degree can accommodate these lags and cyclic behaviour, [by dividing] phase planes . . . into various regions of increasing and decreasing x and y populations. . . . The predator isocline, in the simplest condition, is presumed to be vertical, assuming that only one fixed level of prey is necessary to just maintain the predator population at a zero instantaneous rate of change. . . . Considerable importance is attached to whether the predator isocline intersects the rising or falling portion of the prey isocline, [but] these techniques are only appropriate near equilibrium . . . , and the presumed unstable conditions in fact generate stable limit cycles. . . .

The empirical evidence described above shows that realistic fecundity and mortality (particularly predation) processes will generate forms that the theorists might tend to identify as special subsets of more general conditions. But it is just these special subsets that separate the real world from all possible ones, and these more realistic forms will modify the general conclusions of simpler theory. . . . More realistic forms of functional response change this pattern in degree only. . . . Limitations in the predator's numerical response and thresholds for reproduction of predators, similar to those for prey, could further change the form of the domain. . . . The essential point, however, is that these systems are not globally stable but can have distinct domains of attraction. So long as the populations remain within one domain they have a consistent and regular form of behavior. If populations pass a boundary to the domain by chance or through intervention of man, then the behavior suddenly changes. . . .

THE RANDOM WORLD

To this point, I have argued as if the world were completely deterministic. In fact, the behavior of ecological systems is profoundly affected by random events. It is important, therefore, to add another level of realism at this point to determine how the above arguments may be modified. Again, it is applied ecology that tends to supply the best information from field studies since it is only in such situations that data have been collected in a sufficiently intensive and extensive manner. As one example, for 28 years there has been a major and intensive study of the spruce budworm and its interaction with the spruce-fir forests of eastern Canada (Morris 1963). There have been six outbreaks of the spruce budworm since the early 1700s . . . and between these outbreaks the budworm has been an exceedingly rare species. When the outbreaks occur there is major destruction of balsam fir in all the mature forests, leaving only the less susceptible spruce, the non-susceptible white birch, and a dense regeneration of fir and spruce. . . . This process evolves to produce stands of mature and overmature trees with fir a predominant feature. . . . The fir persists because of its regenerative powers and the interplay of forest growth rates and climatic conditions that determine the timing of bud worm outbreaks . . . If we view the budworm only in relation to its associated predators and parasites we might argue that it is highly unstable in the sense that populations fluctuate widely. But these very fluctuations are essential features that maintain persistence of the budworm, together with its natural enemies and its host and associated trees. By so fluctuating, successive generations of forests are replaced, assuring a continued food supply for future generations of budworm and the persistence of the system.

Until now I have avoided formal identification of different kinds of behavior of ecological systems. The more realistic situations like budworm, however, make it necessary to begin to give more formal definition to their behavior. It is useful to distinguish two kinds of behavior. One can be termed stability, which represents the ability of a system to return·to an equilibrium state after a temporary disturbance; the more rapidly it returns and the less it fluctuates, the more stable it would be. But there is another property, termed resilience, that is a measure of the persistence of systems and of their ability to absorb change and disturbance and still maintain the same relationships between populations or state variables. In this sense, the budworm forest community is highly unstable and it is because of this instability that it has an enormous resilience. . . .

Random events, of course, are not exclusively climatic. The impact of fires on terrestrial ecosystems is particularly illuminating . . . and the periodic appearance of fires has played a decisive role in the persistence of grasslands as well as certain forest communities. As an example, the random perturbation caused by fires in Wisconsin forests (Loucks 1970) has resulted in a sequence of transient changes that move forest communities from one domain of attraction to another. The apparent instability of this forest community is best viewed not as an unstable condition alone, but as one that produces a highly resilient system capable of repeating itself and persisting over time until a disturbance restarts the sequence.

In summary, these examples of the influence of random events upon natural systems further confirm the existence of domains of attraction. Most importantly they suggest that instability, in the sense of large fluctuations, may introduce a resilience and a capacity to persist. It points out the very different view of the world that can be obtained if we concentrate on the boundaries to the domain of attraction rather than on equilibrium states. Although the equilibrium-centered view is analytically more tractable, it does not always provide a realistic understanding of the systems' behavior. Moreover. if this perspective is used as the exclusive guide to the management activities of man, exactly the reverse behavior and result can be produced than is expected.

THE SPATIAL MOSAIC

To this point, I have proceeded in a series of steps to gradually add more and more reality. I started with self-contained closed systems, proceeded to a more detailed explanation of how ecological processes operate, and then considered the influence of random events, which introduced heterogeneity over time.

The final step is now to recognize that the natural world is not very homogeneous over space, as well, but consists of a mosaic of spatial elements with distinct biological, physical, and chemical characteristics that are linked by mechanisms of biological and physical transport. The role of spatial heterogeneity has not been well explored in ecology because of the enormous logistic difficulties. Its importance, however, was revealed in a classic experiment that involved the interaction between a predatory mite, its phytophagous mite prey, and the prey's food source (Huffaker et al 1963). Briefly, in the relatively small enclosures used, when there was unimpeded movement throughout the experimental universe, the system was unstable and oscillations increased in amplitude. When barriers were introduced to impede dispersal between parts of the universe, however, the interaction persisted. Thus populations in one small locale that suffer chance extinctions could be re-established by invasion from other populations having high numbers. . . .

Gilbert & Hughes (1971) . . . combined an insightful field study of the interaction between aphids and their parasites with a simulation model, concentrating upon a specific locale and the events within it under different conditions of immigration from other locales. The important focus was upon persistence rather than degree of fluctuation. They found that specific features of the parasite-host interaction allowed the parasite to make full use of its aphid resources just short of driving the host to extinction. It is particularly intriguing that the parasite and its host were introduced into Australia from Europe and in the short period that the parasite has been present in Australia there have been dramatic changes in its developmental rate and fecundity. The other major difference between conditions in Europe and Australia is that the immigration rate of the host in England is considerably higher than in Australia. If the immigration rate in Australia increased to the English level, then, according to the model the parasite should increase its fecundity from the Australian level to the English to make the most of its opportunity short of extinction. This study provides, therefore, a remarkable example of a parasite and its host evolving together to permit persistence, and further confirms the importance of systems resilience as distinct from systems stability.

Synthesis
SOME DEFINITIONS

Traditionally, discussion and analyses of stability have essentially equated stability to systems behavior. In ecology, at least, this has caused confusion since, in mathematical analyses, stability has tended to assume definitions that relate to conditions very near equilibrium points. This is a simple convenience dictated by the enormous analytical difficulties of treating the behavior of nonlinear systems at some distance from equilibrium. On the other hand, more general treatments have touched on questions of persistence and the probability of extinction, defining these measures as aspects of stability as well. To avoid this confusion I propose that the behaviour of ecological systems could well be defined by two distinct properties: resilience and stability.

Resilience determines the persistence of relationships within a system and is a measure of the ability of these systems to absorb changes of state variables, driving variables, and parameters, and still persist. In this definition resilience is the property of the system and persistence or probability of extinction is the result. Stability, on the other hand, is the ability of a system to return to an equilibrium state after a temporary disturbance. The more rapidly it returns, and with the least fluctuation, the more stable it is. In this definition stability is the property of the system and the degree of fluctuation around specific states the result.

RESILIENCE VERSUS STABILITY

The balance between resilience and stability is clearly a product of the evolutionary history of . . . systems in the face of the range of random fluctuations they have experienced.

In Slobodkin's terms evolution is like a game, but a distinctive one in which the only payoff is to stay in the game (Slobodkin 1964). Therefore, a major strategy selected is not one maximizing either efficiency or a particular reward, but one which allows persistence by maintaining flexibility above all else. A population responds to any environmental change by the initiation of a series of physiological, behavioral, ecological, and genetic changes that restore its ability to respond to subsequent unpredictable environmental changes. Variability over space and time results in variability in numbers, and with this variability the population can simultaneously retain genetic and behavioral types that can maintain their existence in low populations together with others that can capitalize on chance opportunities for dramatic increase. The more homogeneous the environment in space and time, the more likely is the system to have low fluctuations and low resilience. It is not surprising, therefore, that the commercial fishery systems of the Great Lakes have provided a vivid example of the sensitivity of ecological systems to disruption by man, for they represent climatically buffered, fairly homogeneous and self-contained systems with relatively low variability and hence high stability and low resilience. Moreover, the goal of producing a maximum sustained yield may result in a more stable system of reduced resilience.

Nor is it surprising that however readily fish stocks in lakes can be driven to extinction, it has been extremely difficult to do the same to insect pests of man's crops. Pest systems are highly variable in space and time; as open systems they are much affected by dispersal and therefore have a high resilience. Similarly, some Arctic ecosystems thought of as fragile may be highly resilient, although unstable. . . .

The notion of an interplay between resilience and stability might also resolve the conflicting views of the role of diversity and stability of ecological communities. Elton (1958 q.v.) and MacArthur (1955) have argued cogently from empirical and theoretical points of view that stability is roughly proportional to the number of links between species in a trophic web. In essence, if there are a variety of trophic links the same flow of energy and nutrients will be maintained through alternate links when a species becomes rare. . . .

MEASUREMENT

If there is a worthwhile distinction between resilience and stability it is important that both be measurable. . . . There are two components that are important: one that concerns the cyclic behavior and its frequency and amplitude, and one that concerns the configuration of forces caused by the positive and negative feedback relations.

. . . At least for more realistic models parameter values can be discovered that do generate neutrally stable orbits. In the complex predator-prey model of Holling (1971), if a range of parameters is chosen to explore the effects of different degrees of contagion of attack, the interaction is unstable when attack is random and stable when it is contagious. We have recently shown that there is a critical level of contagion between these extremes that generates neutrally stable orbits. These orbits, then, have a certain frequency and amplitude and the departure of more realistic trajectories from these referent ones should allow the computation of the vector of forces. If these were integrated a potential field would be represented with peaks and valleys. If the whole potential field were a shallow bowl the system would be globally stable and all trajectories would spiral to the bottom of the bowl, the equilibrium point. But if, at a minimum, there were a lower extinction threshold for prey then, in effect, the bowl would have a slice taken out of one side, as suggested in Figure 4.

Trajectories that initiated far up on the side of the bowl would have amplitude that would carry the trajectory over the slice cut out of it. Only those trajectories that just avoided the lowest point of the gap formed by the slice would spiral in to the bowl's bottom. If we termed the bowl the basin of attraction . . . then the domain of attraction would be determined by both the cyclic behavior

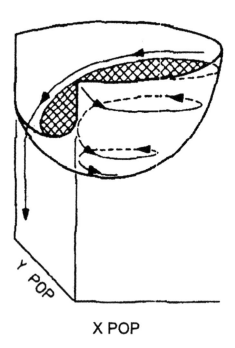

X POP

Fig. 4

and the configuration of forces. It would be confined to a smaller portion of the bottom of the bowl, and one edge would touch the bottom portion of the slice taken out of the basin.

This approach, then, suggests ways to measure relative amounts of resilience and stability. There are two resilience measures: Since resilience is concerned with probabilities of extinction, firstly, the overall area of the domain of attraction will in part determine whether chance shifts in state variables will move trajectories outside the domain. Secondly, the height of the lowest point of the basin of attraction (e.g. the bottom of the slice described above) above equilibrium will be a measure of how much the forces have to be changed before all trajectories move to extinction of one or more of the state variables.

The measures of stability would be designed in just the opposite way from those that measure resilience. They would be centered on the equilibrium rather than on the boundary of the domain, and could be represented by a frequency distribution of the slopes of the potential field and by the velocity of the neutral orbits around the equilib-rium. . . .

Application

The resilience and stability viewpoints of the behavior of ecological systems can yield very different approaches to the management of resources. The stability view emphasizes the equilibrium, the maintenance of a predictable world, and the harvesting of nature's excess production with as little fluctuation as possible. The resilience view emphasizes domains of attraction and the need for persistence. But extinction is not purely a random event; it results from the interaction of random events with those deterministic forces that define the shape, size, and characteristics of the domain of attraction. The very approach, therefore, that assures a stable maximum sustained yield of a renewable resource might so change these deterministic conditions that the resilience is lost or reduced so that a chance and rare event that previously could be absorbed can trigger a sudden dramatic change and loss of structural integrity of the system.

A management approach based on resilience, on the other hand, would emphasize the need to keep options open, the need to view events in a regional rather than a local context, and the need to emphasize heterogeneity. Flowing from this would be not the presumption of sufficient knowledge, but the recognition of our ignorance; not the assumption that future events are expected, but that they will be unexpected. The resilience framework can accommodate this shift of perspective, for it does not require a precise capacity to predict the future, but only a qualitative capacity to devise systems that can absorb and accommodate future events in whatever unexpected form they may take.

Literature Cited

[*The list includes only references cited in the excerpt.*]

Baskerville, GL 1971. The Fir-Spruce-Birch Forest and the Budworm. Forestry Service, Canada Dept. Environ., Fredericton, NB. Unpublished

Elton, CS 1958. *The Ecology of Invasions by Animals and Plants.* London: Methuen

Gilbert, N, Hughes, RD 1971 A model of an aphid population-three adventures. *J. Anim. Ecol.* 40: 525–34

Glendening, G 1952. Some quantitative data on the increase of mesquite and cactus on a desert grassland range in southern Arizona. *Ecology* 33: 319–28

Holling, CS 1961. Principles of insect predation. *Ann. Rev. Entomol.* 6: 163–82

Holling, CS 1965. The functional response of predators to prey density and its role in mimicry and population regulations. *Mem. Entomol. Soc. Can.* 45: 1–60

Holling, CS, Ewing, S 1971. Blind man's buff: exploring the response space generated by realistic ecological simulation models. *Proc. Int. Symp. Statist. Ecol.* New Haven, Conn.: Yale Univ. Press 2: 207–29

Huffaker, CD, Shea, KP, Herman, SS 1963. Experimental studies on predation. Complex dispersion and levels of food in an acarine predator-prey interaction. *Hilgardia* 34: 305–30

Hughes, R.D., Gilbert, N. 1968. A model of an aphid population—a general statement. *J. Anim. Ecol.* 40: 525–34

Hutchinson, GE 1970. Ianula: an account of the history and development of the Lago di Monterosi, Latium, Italy. Trans. Am. Phil. Soc. 60: 1–178

Loucks, OL 1970. Evolution of diversity, efficiency and community stability. *Am. Zool.* 10: 17–25

MacArthur, R 1955. Fluctuations of animal populations and a measure of community stability. *Ecology* 36: 533–6

Morris, RF 1963. The dynamics of epidemic spruce bud worm populations. *Mem. Entomol. Soc. Can.* 31: 1–332

Ricker, WE 1954. Stock and recruitment. *J. Fish. Res. Bd. Can.* 11: 559–623

Rosenzweig, ML, MacArthur, RH 1963. Graphical representation and stability condition of predator-prey interactions. *Am. Natur.* 97: 209–23

Slobodkin, LB 1964. The strategy of evolution. *Am. Sci.* 52:342–57

Watt, KEF 1968. A computer approach to analysis of data on weather, population fluctuations, and disease. *Biometeorology*, ed. W. P. Lowry. Corvallis, Oregon: Oregon State UP.

Commentary

C. S. Holling, *Resilience and Stability of Ecological Systems* (1973)

LIBBY ROBIN

Crawford Stanley (Buzz) Holling (1930–) defined the notion of resilience as a "measure of the persistence of systems and their ability to absorb change and disturbance" while maintaining their systemic structure (14). Holling's 1973 survey paper, "Resilience and Stability of Ecological Systems," is still frequently cited today, because it defines the ecological concept of resilience. It has become the touchstone paper for the "resilience community," a community of interdisciplinary global change scholars working on policy and science for society. The first international Resilience conference was held in 2008 in Stockholm, Sweden, and the second in 2011 in the United States in Tempe, Arizona. Resilience is a property not just of ecological systems but also of "social-ecological systems" (SES). In discussions of the human dimensions of climate change and correlated social changes on a planetary scale, the ecological concept of resilience is helpful to policy making and for the management of natural and social resources under conditions of global change.

For historians of ecology, the paper is also interesting as a reflection on the state of the science and limits of ecological modeling at this time, before large-scale computer technologies became widely available: Holling is concerned that theoretical models generally failed to engage with real-world behavior and particularly *randomness* and *spatial heterogeneity* in natural systems. This is one of the important papers arguing against the assumption of equilibrium in nature, and in theorizing non-equilibrium systems.

Resilience: An Ecological Concept for Managers and Policy Makers

Resilience was a key concept in shifting ecosystem thinking from science into management in the 1970s. The generation of ecologists who had grown up with Eugene Odum's *Fundamentals of Ecology* were familiar with working with whole ecosystems, rather than focusing just on single animals or plants. Odum's "energy system" of nutrient cycling linked biological and physical environments. The physiology of the animal was nested in a cycle that depended on plants, soils, climate, and other abiotic factors, and converted these physical elements into energy. In the same period, Alfred J. Lotka's predator-prey curves provided mathematical ways to represent relations between an hierarchy of biological elements in the system. Systems had a history, and this history was a force in the evolutionary pathways of biota.

It was difficult to testing theories of "whole ecosystems." They were complex and dynamic, and their borders were often porous. Islands sometimes offered the microcosm that made it possible to consider a range of elements interacting in complex ways. Robert MacArthur and E. O. Wilson's *Theory of Island Biogeography* (1967) was important in creating a theoretical framework for island field studies beloved by practical ecologists. Perhaps the unexpected emergence of a brand new island, Surtsey, near the coast of Iceland on the transatlantic ridge in 1963 was one of the factors that gave island research new impetus around this time. Surtsey was a case study for primary colonization. What elements appeared first? What made life possible?

Another "whole ecosystem" was the lacustrine (lake) environment, the summer haunt of prairie and forest dwellers above and below the forty-ninth parallel in North America. The lake provided a classic opportunity to observe a system of biotic and abiotic elements working together. Stephen Forbes, as early as 1887, defined the lake as a microcosm, and studied its "complete and independent equilibrium of organic life and activity" (Forbes 1887, 77). Evelyn Hutchinson's work on freshwater ecosystems was important here too, increasingly introducing more precise measures of physical elements such as water quality, acidity, and temperature, and considering how these triggered breeding events. Rather than an emphasis on a "steady-state" ecology (the "equilibrium" of Forbes), Holling was interested in theorizing the dynamics of change, using real-life examples, including anthropogenic changes. Holling commented, "Freshwater lakes are reasonably contained systems, at least within their watersheds; the fish show considerable mobility throughout, and the properties of the water buffer the more extreme effects of climate. Moreover, there have been enough documented man-made disturbances to liken them to perturbed systems" (p. 246).

Holling's mission in this paper is to define a concept that works in real-world ecosystems, an idea that "simultaneously incorporates realistic behavior of the processes involved, randomness, spatial heterogeneity, and an adequate number of dimensions or state variables." His holy grail was a "not too simple but just sufficient" concept that captured the notion of adaptation and learning in response to environmental change. In this sense this paper very much emerges from the limits to growth debate and the strong sense among ecologists that the planet could not just keep bouncing back to equilibrium whatever happened to it. Even the simplest real-world systems were rarely self-contained, homogenous, and subject to minimum climatic fluctuations.

Sustainability in Society

The most important implications of the paper are that it offers a model for a society that is based not on efficiency and optimization, the ruthless ideals of the economic world, but rather on an ecology of "persistence"—or survival of the species in a changing context. In Holling's words, "A major strategy selected is not one maximizing either efficiency or a particular reward, but one which allows persistence by maintaining flexibility above all else" (p. 253). In particular, the idea of an optimized "sustained yield" is not the same thing as the "sustainability" of the system as a whole: "The very approach, therefore, that assures a stable maximum sustained yield of a renewable resource might so change these deterministic conditions that the resilience is lost or reduced so that a chance and rare event that previously could be absorbed can trigger a sudden dramatic change and loss of structural integrity of the system," Holling explains (p. 255). Thus, a resilience approach to management urges strategies that keep options open for the whole system at a regional, not just a local, scale. Instead of a focus on *predicting* the future, managers of natural resources should rather focus on maximizing the range of possible futures, given the unexpected nature of the system under unknown future conditions.

Further Reading

Forbes, S. A. 1887. "The lake as a microcosm." *Bulletin of the Scientific Association*, 77–87.
Kingsland, S. E. 2005. *The evolution of American ecology, 1890–2000*. Baltimore: Johns Hopkins University Press,
MacArthur, R. J., and E. O. Wilson. 1967. *The theory of island biogeography*. Princeton, NJ: Princeton University Press.
Salt, D., and B. Walker. 2006. *Resilience thinking: Sustaining ecosystems and people in a changing world*. Washington, DC: Island Press.

PART 6

Technology

Does Technology Create More Problems Than It Solves?

Technology has long been a double-edged sword when it comes to human relations with nature. It is in large part through technology that we have come to our modern understanding of the environment, giving us the capacity to measure, analyze, and view it from a microscopic to a global level. In this sense, we can even say that technology has been a crucial element in constructing the very idea of the environment.

At the same time, technology has been a major agent in the human transformation of the global environment—often in a way seen as damaging. This has been ongoing since the first deliberate use of fire and edged tools, but has changed vastly in scale and scope with the advent of modern technology powered by fossil fuels—chainsaws, trucks, aircraft, electricity networks, pesticides, and so on. Technology has also, at times, been identified as a distinguishing feature of humanity that sets us apart from the rest of nature—both a virtue and problem, depending on one's viewpoint. In modern debates, it remains both plague and panacea—driver of economic growth (variously viewed as good or bad for the environment), population expansion (through medical care, but also braking it with contraception), offering possibilities of geo-engineering and efficiency improvement to "save the planet." In contrast, pessimists argue that a culture of continuing technological fixes will always, eventually, bring us back to the same impasse. They see faith in technology as the problem.

Technological change may be almost as old as humanity, but an expectation of continuous change—that technology will define the future, and constantly redefine society—is relatively recent. Early modern thinkers who expressed optimism about the human capacity to improve the world (the most famous of whom is English statesman Francis Bacon, 1561–1626) were largely preoccupied with changes in *knowledge* rather than changes in the *instruments* by which the environment can be transformed. The pace of change accelerated rapidly with the Industrial Revolution, a transformation in the energy regime toward fossil fuels, bringing urbanization, pollution, social dislocation, and an expectation of constant, relentless innovation—and in the long run, greater wealth for all.

But even at this time, not everyone thought that science was progress. Often the reaction against industrialization was more a case of nostalgia, or discomfort amid the new, populous, and polluted urban society. But in 1855, we meet Eugène Huzar, who was deeply concerned that scientific and technological advance would lead to catastrophe and the end of the world as he knew it. What is interesting about this paper is its imaginative predictiveness: he invents scenarios—and reflects on options. Huzar's idea of preventing scientific "progress" from running out of control was in fact a form of globalization, a "world council" that could watch over the "globe's harmony." For those that think "techno-pessimism" was invented in the wake of Hiroshima and the Atomic Age, this document from a century earlier gives pause.

From the nineteenth century the idea was established that the world was being profoundly shaped, more than ever before, by technology—and as Huzar wondered, people pondered if they had unleashed a force that was beyond their own capacity to regulate. Technological change wrought an unalloyed future-orientation, a sense of "modernity," that however the future was, technology would be a major determining element, expressed as much in literature and art as engineering and planning. The history of society—and by implication that of nature—would, henceforth, be in part, if not largely, a function of the history of technology.

The next paper takes us into the postwar period, and with it a very different set of preoccupations about determinism, and expertise for the future. Alva Myrdal was a Swedish politician whose internationally famous career straddled the middle of the twentieth century. She was commissioned to produce a report on "futures studies" in 1971. This was at the same time a reaction to, and part of, a postwar wave of futurology, the most influential of which was closely related to the interests of United States military and commerce, often employing pioneering computer modeling. Swedish critics were suspicious of what they saw as an expansionist and controlling agenda; they sensed a new form of determinism that came not from environmental factors, but from military ways of modeling change. At the same time, Myrdal's report belonged to an era of postwar optimism in development: it was a time when it was reasonable to study how the future could and should be better. It belonged to a broad consensus that destiny (and what had once been called "civilization") would be shaped by technology. Myrdal asserted, "Futures must not be left to a new breed of specialists," but rather be subject to "democratic control." The report expressed ambivalence about aspects of human progress, and the desire for narratives about the future and nature to be tied to political ideals. Early and mid-twentieth-century debates about resources had expressed the fear that technological advance was outstripping the capacity of the planet to supply human demands; there had been waves of anxiety and new evidence about pollutants produced from modern technology, coal smoke, and hydrochloric acid in the nineteenth century, chemicals and radiation in the twentieth. Myrdal's report was part of a wave of thinking that did not reject technology, but feared that a technocratic world was eroding what was most valuable in human society.

More generally, however, futurologists were optimistic about the prospects of technological transformation. Theories about the future have frequently been theories about the dynamics of technological change. This is the world of systems theorists Cesare Marchetti and Nebojsa Nakicenovic, who applied systems analysis to technological and social systems. Rather than the cause of environmental catastrophe, environmental and resource problems could be conceived as the system outcomes of previous technological choices that simply generated new waves of innovation and transition. Limits should not be applied to growth; scarcities and challenges generated and perpetuated the dynamic of growth, and in this sense environmental solutions were the inevitable outcome of environmental problems.

Few have doubted, over the past century, that whatever the future of nature, technology would have a major role to play (not least in enabling us to find out what was happening). It continues to be a lightning conductor for ambivalent and opposing views about change in human society, and our impact in the world. Does it distance us from fellow nature, and our better virtues? Does it restrict to open up choice? Who controls its development and use? Is it the best expression of human ingenuity and distinctiveness? What these documents reveal is how modern technology has been felt as a force apart—of humans, but not human, natural and unnatural—increasingly part of the "anthropocene" global system.

The Tree of Science

EUGÈNE HUZAR

Progress and the Perils of Purely Experimental Science

Modern civilization, unlike Greek or Roman civilizations, is not limited to a town or an empire, it can be found in Europe, Asia, Africa and the South Sea Islands, as such it is a very specific phenomenon to which nothing can be compared. . . . All these historical civilizations have fallen through invasions, wars, political or religious upheavals; modern civilization however will fall in harmony, not through conquest or invasions but under the weight of its own force and power, in an excited wave of immense scientific discovery, and for the present we cannot possibly conceive the repercussions of this.

. . .

Can we predict the dire influence that deforestation will have on the organic kingdom and on our health? Not at all, but instead of destroying them we should be conserving them religiously. Have we forgotten that vegetation is not only essential to man for food, for heating and cooking his food, but above all for breathing, which is the first condition of organic life. By absorbing carbon dioxide and releasing oxygen, vegetation removes from the air noxious gases and restores to man's lungs breathable gas, oxygen; this is the only way that hematosis can occur. Have we forgotten that air laden with carbon dioxide is unsafe for human beings?

When will man's foolhardy hand hit the organic kingdom with the most violence? Precisely when man most needs it, as the air becomes gradually more polluted by the day. The large quantities of carbon dioxide and carbon monoxide, two toxic gases already present in the air, are increasing at an alarming rate.

Mr. Péligot[1] tells us somewhere that emissions of carbon dioxide from coal and other fossil fuels, of which we are now extracting alarming quantities, now exceed 550 billion kilograms per year in Europe alone. If we take that these fuels contain on average 80% carbon, their use thrusts into the atmosphere 80 billion

Eugène Huzar. 1857. Excerpts from *L'Arbre de la science*. Paris: Dentu. New translation with scholarly and contextual notes by J.-B. Fressoz and M.-R. Cheadle.

cubic meters of carbon dioxide per year. He also adds that it is likely that the production of carbon dioxide outweighs the amount consumed by plants. Now if you consider that the forests of America are experiencing deforestation, and that the forests in Europe and the civilized world are also seeing high levels of deforestation, and that the proportion of carbon dioxide and carbon monoxide in the atmosphere keeps increasing ad infinitum, as man becomes more involved with industry and uses more coal, one can predict that in a hundred or two hundred years, the world crisscrossed by railways and steamboats, and covered by factories and industrial plants, will emit trillions of tons of carbon dioxide and carbon monoxide. And since the forests will have been destroyed, all this eliminated, all this carbon dioxide and carbon monoxide could well disturb the harmony of the world somewhat.

Man disturbing the globe's harmony is like the story of the drop of water that finally penetrates the rock face; or the story of the aneurysm that finally reaches the bone; we should never ask "What is going on?" Could 15 trillion kilograms tilt the axis of rotation and change the center of gravity of the Earth?

We should add to that another cause for concern, the clearing of the isthmuses. Who can predict that the seas will not one day, once the sea walls have been broken by man, rush to one corner of the world, rather than another, thus breaking the balance of the seas and tilt the axis of rotation of the planet and make whole continents disappear under water?[2]

. . .

We know that laws of nature work for the better in the best of all possible worlds. But do we know what will happen when we have taken control?

Man, playing in this way with the complex machine of nature, prompts me to think of a blind man, who, ignorant of the fundamentals of mechanics, dismantles all the parts of a working clock, only to put it together again according to his whims and fancies. Some will say that what man does to nature is like a scratch on the skin of a strong and healthy man. But who does not know that sometimes the smallest scratch can lead to death? Think what happens in the tropics.

. . .

I would understand if a native of South America who had never left his forest, came up to me and said that the earth is infinite and therefore man cannot disturb its harmony. But today, with science, this proposition is inverted; it is man who has infinite capacities and the planet that is very finite. Because of steam and electricity, space and time do not exist any further.

For us men of the 19th century, the earth is not the same as it was to man in ancient times. Man can now go round the world 40 or 50 times in his life; ancient man had never even measured its circumference. For us our planet is limited, very limited, as it takes us less time to go round it than it took a Greek to go around Attica. . . . When one sees something as limited as the Earth and a power as unlimited as that of man armed with the lever of science, one can only wonder what impact this power will have one day on our poor small Earth.

. . .

From a medical and therapeutic standpoint, think how many experiments and schools we have not done! Are we not on the verge of cursing today, what we worshipped yesterday? Is it not the case that a new study has just railed against the vaccine, and then taken upon itself the responsibility of the degeneration and withering of the human race?

A book with the daring title of *The Physical and Moral Decline of Mankind Through Vaccination*[3] accuses Jenner[4] of contributing toward the degeneration of our European family. Should we also attribute to this reckless innovation the increased frequency of gastritis, tuberculosis, pulmonary consumption and above all typhoid fever? There is so much hot air and literature around! Vaccination has been practiced and glorified for so long now and we have only just started to ask ourselves if perhaps we made a big mistake and that we might even have inflicted on mankind diseases a thousand times worse than those we seek to destroy. Will we constantly see science contradicting itself in this way? The Academy of Science will take on the task of resolving this question, as it is serious enough to appear on their meeting agenda.

. . .

As a result of scientific discoveries, death is now becoming collective, and it will continue in this way as long as we continue to play with ever greater forces. Fatalities used to be sporadic and only befell isolated individuals; now with science they are endemic and affect whole peoples: the floods of last year are proof of this.[5] Once isthmuses have been broken and our ocean currents disturbed, death could even strike whole continents. The globe, so unsettled by this ignorant and purely experimental science, will one day be endangered by the very laws which govern its harmony today.

. . .

What are the palliative measures I propose?

1. In the future, man should not attempt major and decisive experiments, without the assurance that they will not trouble the harmony of the laws of nature
2. In the future, we should look to the creation of new science which will study the laws which govern the world's equilibrium
3. In the future, we should also create a world council to regulate humanity's work, in such a way that nothing decisive or crucial, such

as the deforestation of a continent or the cutting of an isthmus etc. can be allowed to take place without the authorization of the world council. This council will sit in one of the world's great cities and will be composed of the very best of scientists from across the whole world. Each member will be appointed by his fellow countrymen.

The councillors will be the top magistrates in the world, and whenever a nation wishes to embark upon one of these daring experiments, which could disturb the world's harmony, it should seek counsel from the councillors, who can grant or refuse the permission, as they will be there to watch over the globe's harmony.

The nation who dares contravene orders of the council will be ostracized by the other nations, and charged with the crime of less-humanity.

So if a nation wants to clear the trees from its forests, it should seek permission from the world council. Equally, if a nation wishes to cut an isthmus, it should also seek their permission; in other words, every time a nation wishes to perform an important experiment which could trouble the natural balance of the planet, it should seek the permission of the whole of humanity, as represented by the councillors.

This will be man's solidarity in the future. To those of you reading my words, the idea of a world council must seem absurd, and yet it is already part of our daily existence. Do we not already have in France, a smaller model of what I propose for the World? Is there not a law which grants homeowners the right to use and enjoy their property but not to exploit it? Can a man legally set fire to his own house? No. Why? Because the whole town could be affected by this single abuse of property.

To watch over the world's harmony and ensure that it is never disturbed, this will be the aim of this foremost institution in the world. Some will object to this, saying that it is an infringement of individual freedom; but no, it only serves to prevent the abuse of the freedom that could compromise the general harmony. But for this to work, I will be told, we have to admit the unity of mankind and that all people are brothers.

In response I say, that nowadays only ignorant people can really believe that hatred will exist forever between nations. Any sensible man faced with what is happening around him will see what the unity of mankind will be in a few centuries time. Barriers between countries will be replaced by railway tracks; electric wires, which are suspended like lyres in space, will be the strings of future world harmony.

I am not saying that the world council can be formed today, as at present nobody can see the need for it. But I tell you that a few centuries from now, after witnessing disasters caused in some parts of the world by the unbridled ravages of industrial science, we will be able to justify such an institution.

It is the whole of mankind who will facilitate the implementation of the world council as I see it. One hundred thousand miles of electric telegraph cable will link up all the corners of the world in a few seconds, and thus be in direct contact with the council. The court, like a spider in its web, will receive requests for authorization by the thousands through the strands of its web and it will be able to respond as soon as all matters have been duly considered.

One day, science will be the queen of the world; everything else will vanish in its wake. Her responsibility will be vast; she will have to take care of the world's souls, it will be the largest pontificate ever.

Henceforth, man's work will no longer be given free reign, and it will not be left to chance to disrupt the world's harmony and to trample with both feet on the eternal laws of nature.

. . .

My intention in this book was not to wage war on either science or progress, but to show that I am the implacable enemy of an ignorant science, of a blind progress that walks with no guide and no compass and risks disturbing all the laws of nature. I fear that man, using experimental science as he does, risks one day falling victim to the incalculable forces of nature.

Notes

1. Eugène Melchior Peligot (1811–1890) was professor of applied chemistry at the Conservatoire National des Arts et Métiers (Institute for Engineering Studies) from 1841, prior to that he was a tutor at the École Polytechnique. He became a member of the French Academy of Sciences in 1852.

2. The project to build a canal in the Suez caused a controversy: the land leveling carried out by Lapère, a member of Bonaparte's scientific expedition to Egypt, measured a difference in level of eight meters between the Red Sea and the Mediterranean. See Anne Montel, "To establish the scientific truth of the 19th century, the controversy surrounding the difference in level of the two seas (1799–1869)," Genèses 32 (September 1998): 86–109.

3. This is about the smallpox vaccination, introduced in 1798 by English doctor Edward Jenner. European governments encouraged and often imposed vaccination. An anti-vaccination movement surfaced in the early days of innovation, particularly in England, where vaccination was compulsory from 1852 (1902 in France). Huzar is referring to the book by Dr. Verdé-Delisle, Physical and moral decline of mankind through vaccination (Paris: Charpentier, 1855). The author, looking at the anti-vaccination theories exposed by Carnot, Duché, Bayard, and Ancelon, uses statistics to prove that vaccination has

increased death through typhoid and pulmonary consumption. He refers to the then fashionable theory of population degeneration to favor the anti-vaccination theories.

4. Eugène Huzar writes Genner.

5. Throughout May and June 1856, exceptional floods devastated the basins of the Loire, the Rhône, and their tributaries, prompting a growth of doomsday literature as well as political and legislative actions to fight against this type of disaster.

Commentary

Eugène Huzar, *The Tree of Science* (1857)

JEAN-BAPTISTE FRESSOZ

In April 1855, in Paris, as people flocked into the universal exhibition to admire all sorts of inventions, Eugène Huzar, an unknown lawyer, published a provocative book: *La Fin du monde par la science* [The End of the World by Science], which is arguably the first critique of progress founded on technological catastrophism. Two years later, encouraged by the success of his short book, Huzar offered a longer sequel with *L'Arbre de la science*.

Huzar is not a romantic writer railing against the ugliness of industrialization. On the contrary, he defends himself from being in any way reactionary: "I wage war on neither science nor progress, but I am the implacable enemy of an ignorant science, of a blind progress that walks with no guide and no compass" (*L'Arbre de la science*, 40). According to Huzar, being "anti-progress" would in any event be useless because it is the driving force of the world. His theory fully integrated the usual discourse of progress: technological advances would accelerate "as the square law" as society itself became increasingly organized to produce innovations. Knowledge democratization and the increasing interaction between science and industry would establish the conditions for an unlimited progress. National rivalries would cease, races would fuse, and one language would integrate all other tongues. In short, humanity would be unified in its quest for technological progress.

Nevertheless, after singing the praises of the future, Huzar warned his readers that however powerful science may become, it will always remain *experimental*—that is, it can, by definition, learn only a posteriori and therefore cannot anticipate the far-reaching consequences of increasingly powerful technologies. He called this the principle of "science impresciente"—that is, science cannot predict the results of its own actions. And the gulf between technological ability and limited foresight would cause the apocalypse.

Huzar was actually a very accomplished prophet of doom: he imagined an impressive list of human-induced global disasters: who knows whether, by extracting ton after ton of coal, man will not change the center of gravity of the Earth and tilt its axis of rotation? Who knows whether, by digging transoceanic canals, man will not cause a perturbation of maritime currents and terrible flooding? Who knows if the carbon dioxide emitted by industries coupled with deforestation will not unbalance the equilibrium of the atmosphere? According to Huzar, the best candidate for the apocalypse is a future substance that could set fire to water, burn the oceans, melt the soil, and destroy organic life on Earth.

Huzar proposed that on such important matters the burden of proof should be shifted. Scientists must show that canals, mines, deforestation, or any other proposed innovations are perfectly harmless. "If we are so demanding toward science it is because nowadays science tends to substitute its blind action for nature's; but before doing so, it is necessary to prove that science will do better than nature" (110).

The prophylactic measure proposed by Huzar was the global governance of nature and innovation by science. First, a new science "will have to be created so as to determine and study the laws that govern the globe's equilibrium" (275). Second, a world council (*édilité planétaire*) should be instituted as the first authority on Earth for "regulating humanity's work" and "watching over the globe's harmony" (275). It will grant authorizations to scientists to perform important experiments, or to nations to deforest, extract coal, or cut an isthmus. However, Huzar thought that the world council, being guided by experimental science, could only delay rather than prevent the apocalypse. Something different from science had to emerge: a kind of "prescience" or "intuition." He remained rather elliptic on this point because it was supposed to be the subject of a third book (*L'Arbre de vie*), which was never published. But he hinted at a new utopia, the construction of a desirable future that would allow mankind to shape the present and resist the destructive logic of progress, with its tendency to replace nature by artifacts.

As the press was saturated with eulogies of Progress and Industry, the book was obviously intended to be provocative. And it succeeded: every major periodical reviewed it, and most of them spoke of it very highly: "This is the book I have always dreamt of." "Mr. Huzar's system does not lack grandeur or truth." "This is a completely new system which, although strange, is based on facts" (2–3). This first book by an obscure lawyer with some dubious scientific background was unquestionably a success, and by 1865 it had been through three editions. Celebrated literary and religious figures took some inspiration from it; indeed Alphonse de Lamartine was accused of having plagiarized Huzar in proposing that Eden was in fact an advanced civilization. His *L'Arbre de la science* also got excellent reviews. To the renowned numismatist and historian Felix de Saulcy, writing in the *Courrier de Paris* on 21 October 1857, it was "one of the most enticing books [he] had ever read." The poet Auguste de Vaucelle, in *L'Artiste* on 9 August the same year, went so far as to praise "one of the most remarkable books of this century . . . a book of capital interest for humanity."

If Huzar is worth reading today, it is because his writings demonstrate with perfect clarity that the techno-scientific revolution of the 1800s was not accomplished in a fog of careless modernism. Huzar was not the ignored Cassandra he pretended to be: his catastrophist theory was actually founded on the controversies of his times: the multiplication of railroad accidents, the insalubrity brought by industrialization, the climatic consequences of deforestation, or the possible degeneration of population caused by vaccination. His originality lies in having grouped all these debates in a counternarrative of progress: mastery over nature leads to a loss of mastery over technology. We need to take on board the disturbing fact that the risks entailed by technological modernity have been run knowingly. Our generation is not the first to question the sense of progress. Ulrich Beck's "risk society" is probably not a new phenomenon.

Further Reading

Fressoz, J.-B. 2007. "Beck back in the 19th century: Towards a genealogy of risk society." *History and Technology* 23(4):333–350.

Fressoz, J.-B. 2012. *L'Apocalypse joyeuse une histoire du risque technologique*. Paris: Le Seuil.

Huzar, E. 1855. *La Fin du monde par la science*. Paris: Dentu.

To Choose a Future

ALVA MYRDAL

Foreword

[7–8] The concept of the welfare state has hitherto been seen as closely harnessed to the rapid development of science and technology. However, this development is now also being recognized as a source of danger for our societies because of over-hasty transformations and unforeseen negative effects. There is broad agreement that scientific and technological capacity will be a crucial factor in our common efforts to bring acceptable material standards of living to all people. But are the present large-scale technologies really furthering this aim? What alternative choices exist? How are cumulative technological innovatory processes to be reconciled with slower changes in social values and institutions? These are the questions to which we ought to devote more attention, if the total development efforts within and between countries are to bring us nearer the all-embracing goal of an improved quality of life for all mankind.

A less naïve approach towards the role of science and technology in the promotion of development is emerging. This holds for rich and poor countries alike. An international debate that has been going on for several years has stressed the importance of determined efforts to analyze and "create" the future. The present tendency to know more and more about less and less creates problems for society as a whole; the era of extreme specialization must come to an end. . . .

One main idea behind the report is the need for public participation in the work on future studies. We must always devote much energy to making complex problems commonly understandable. Future studies and the discussion about different futures must not be left to a new breed of specialists or to an élite who claim to know what is best for everybody. The democratic control of this work

Swedish Institute (Alva Myrdal, chair). (1972) 1974. Extracts from *To choose a future: A basis for discussion and deliberations on future studies in Sweden*, 7–8, 14–18. Swedish version of the original published as *Att välja framtid*. Stockholm: Swedish Institute.

must never weaken, and public participation in vital long-range decisions must be safe-guarded and deepened.

International events during the last year have underscored the urgency of the pair of ideas—future-orientation and internationalism—that are basic in "To Choose a Future." I have, however, learned with great satisfaction that international organizations, such as UNESCO and UNITAR, have decided to establish and promote activity in the field of future studies. It is my sincere hope that the international community will extend its responsibility—also in practical terms— to include future generations.

[signed] Alva Myrdal

2 Future Studies—Main Problems and Basic Concepts
2.1 FUTURE STUDIES AND SOCIETY. SOME FUNDAMENTAL STARTING POINTS

[14] A growing awareness that many of the decisions taken today will have very long-ranging consequences makes it increasingly necessary for both the general public and decision-makers to anticipate and take stands on their future responsibility. At the same time it can be noted that certain decisions assume very large-scale or sweeping technical and social effects, and that the pace of changes which directly impinge on individuals is accelerating.

Every discussion of social conditions proceeds, explicitly or implicitly, from certain premises as to how society works and how it ought to work. So does the present review of the knowledge needed to make future assessments, and in this section we shall formulate some fundamental starting points which we consider relevant to Swedish society when seen in a future perspective.

- The future development of society is assumed to take place under democratic control and to move towards goals that are laid down by democratic means.

For various reasons, however, the concept of democracy is of itself problematic in the futuristic context. For one thing, by virtue of planning decisions now being taken, the conditions under which people live are shaped for those not yet entitled to vote, and perhaps even for those who have yet to be born. In many cases there is a real conflict between the short-term interests of those now living and their responsibility to coming generations. In our opinion such conflicts must be brought out in the open. The responsibility for dealing with them must ultimately rest with democratically elected bodies and take place under political responsibility:

- In our democratic society it is a task for the political bodies to represent the interests of coming generations.

[15] It is therefore desirable to have the decision-making processes and the organs of government shaped so that they can uphold interests longer-ranging than those of individuals in the immediate moment of decision. But this also assumes that the citizens will have "future awareness" of the kind which enables them to carry this responsibility in the political process and to accept long-term decisions.

To this aspect of "the strong society" can be added the observation that the democratic element of long-range social changes will be confronted with several problems in the real world. In the first place, a great many material and controlling decisions, especially of an economic nature, are taken by bodies which come not at all or only very indirectly under democratic control or which are open to public scrutiny. In the second place, various measures are taken in different sections of the community without sufficient insight into their repercussions on other areas. In the third place, owing to the complexity of many issues popular rule tends to become a mere formality—even in areas where decision-making responsibility rests with democratically elected organs of government. Summing up:

- Future-creating activity must contain the following important elements: a deepening of the democratic process, a greater exchange of information between different actors, and guarantees of participation by citizens in decision-making.

Future studies entail, first, the production and collection of knowledge, and second, the modelling of various kinds of scenarios. But knowledge is power and scenarios have great persuasiveness. Hence the studies here envisioned can become extremely important factors in the power play between different groups in the society. In particular, certain types of statements about the future can acquire an unfortunate self-fulfilling character.

- The democratic state has a special responsibility for bringing out source data on behalf of the long-range public interest and to render service to the weaker groups and individuals, as well as to ensure that free, independent groups also have access to relevant information.

Yet another vital social interest is to make realistic the public dialogue about the future:

[16]

- Future studies must identify the real constraints on our prospects for shaping the future, but at the same time present a data base so complete that all favourable prospects can be taken into consideration.

The constraints are of course primarily imposed by the availability of resources, and not least by international circumstances over which we have very little if any control. Further constraints are determined by conflicts between different interests and groups in society as well as by the "legacy" which derives from the inertia and lags in many social mechanisms.

Lastly, it should be said that the most fundamental motive for studying and analyzing remote futures is to enable us to take genuine decisions at the right time, i.e. to shape the future rather than let it be passively shaped for us. But if this is going to make sense there must be more than one alternative from which to choose:

- One of the criteria that all future studies of conditions which can still be influenced must satisfy is to provide for several, *alternative futures*.

This is the only way to guard against the risk that future studies will stand out as descriptions of inevitable and fatalistic events. On the contrary, future studies should give pointers indicating how decisions and measures taken at different points in time can act on the course of events.

2.2 Terminology and Taxonomy

Terms such as *future research, futurology* and *Futurologie* are in fairly widespread international use, though no closely specified meaning attaches to any one of them. The word *framtidsforskning* (future research) has been introduced in Sweden, but several good arguments have also been advanced here in favour of a less pretentious label, e.g. *framtidsstudier* (future studies). One compelling reason is that the domain of future studies must accommodate a great many activities which have to do more with the application and development of known methods than with the search after new knowledge. Another is that there can be no justification to have prestige-laden words such as "research" or "science" cover up uncertainty, methodological obscurity and inevitable subjectivity in the

making of statements. [17] It is also possible to discern a risk that linguistic usage will reinforce a tendency to regard future studies as the private preserve of
specialists (investigators, planners), thereby eroding the democratic and political
element. Lastly, of course, statements about the future pose other epistemological problems than, say, empirical or logical assertions.

Many different principles can be used to structure and classify the domain
of future research or future studies. The growing literature in this field certainly
contains a profusion of taxonomy. For present purposes we shall refer to no more
than a couple of distinctions which seem to be of general interest.

One distinction may be drawn between *passive* and *creative* future studies.
Included in the passive category are studies which assume that no major changes
can be brought about by one's own decisions. This conclusion may be totally
unchallengeable (as in forecasting astronomic events!), plausible for the most
part (as in evaluating the conduct of the great powers to provide background for
Swedish security policy), but also highly dubious (e.g. certain energy forecasts
or population forecasts). A sort of antipole to the passive future studies are the
creative ones, whose hallmark is their attempt to identify qualitatively new alternatives, to "invent the future" as it were.

Closely related to creative future studies is the conceptual dyad called
explorative–normative, a classification that denotes the degree of evaluative direction imparted to the studies: a normative study is governed by a predetermined
ideology or vision, whereas an explorative study tries to present different alternatives for the realization of a later value-determined choice (cf. 2.1).

One determination that has played a certain role in the international discussion is the contrast between settled or *establishment* futurology and *critical*
futurology. Establishment futurology is portrayed in these texts as sharing a community of interests with a monolithic capitalistic-military power structure, politically conservative and purported to be scientifically "objective" in the positivist
tradition. Critical futurology, on the other hand, views the society in conflict
perspective (often in terms of class analysis) and concentrates on the feasibility of
changes in a radical, anti-technocratic direction.

Yet another factor that distinguishes between different future studies/predic-
tions is the *time scale* or *horizon* they employ. We shall analyze the need of future
studies in conjunction with the concept of *planning*, and it may clarify matters
at the present stage if we use the classification by time period that is customarily
used for planning [18] purposes. Alongside the immediate planning, e.g. in the
form of preparing annual budgets, reference is usually made to two other types of
planning: the *medium-term* and the *long-term*. Whereas medium-term planning
can be described as prolonged activity planning, admittedly complicated by uncertainty but still based on the same conceptual structure of activities, planning

that is going to be called long-term should span across a time period that is long enough to eliminate a great many of the short-term commitments, i.e. so that real decisions can be taken as to qualitatively different activity. For a large manufacturing firm this means that existing machinery and technical know-how will be largely renewable, and in similar situations the "long-term" will take in anywhere from 7 to 10 years. For the Swedish armed forces the time is determined by two factors, the turnover of conscripts and the service life of the most expensive materiel, and it has been deemed that 15 years is a reasonable horizon for long-term planning. . . . In respect of physical planning, both as regards the spatial distribution of residents and the location of such things as roads, railways and bridges, considerably longer periods have to be reckoned with, perhaps 40–60 years. The long-term effect of education can, in certain respects, be said to extend across a half-century. Then again, of course, there are social aspects, e.g. as regards values and preferences, where the changes can unfold so rapidly that even 2–3 years may qualify as "long-term" (such that predictions over longer periods can also appear to be quite pointless).

It should be established that the time horizon for *future studies* is equivalent to the definition given here for "long-term": the relative lack of major commitments so that consideration can be given to qualitative changes. It follows from this that the "natural" time horizons for different sectoral and selective future studies will be different.

Commentary

Alva Myrdal, *To Choose a Future* (1972)

ARNE KAIJSER

In 1972 the book called *Limits to Growth* (see Part 2) presented future scenarios for the world generated by a sophisticated computer model. Most of the scenarios were very depressing, predicting resource depletion, massive pollution, and even mass starvation if concerted action on a global level was not soon initiated to change current trends. The research at the Massachusetts Institute of Technology (MIT) had been initiated and financed by an organization called the Club of Rome, which had been formed in 1969 by prominent businessmen and civil servants. The Club of Rome was very skillful in promoting and disseminating *Limits to Growth*, which had enormous media impact and was widely discussed in the following years in many different places.

In 1972 another report on future studies with a very different approach and message was also published. It was called *Att välja framtid* [To Choose a Future] and had been produced by a Working Party set up the previous year by the Swedish Prime Minister Olof Palme to deal with the question of future studies in Sweden. This report—in Swedish—did of course not have a similar impact as *Limits to Growth*, but it was much debated in Sweden, and it was translated into English in 1974. It became the founding document for a new discipline of "future studies" in Sweden.

Why was there such an interest in future studies in the early 1970s? First, a general shift in the public debate had occurred in the 1960s, with a growing awareness of environmental problems and global inequalities. Many came to the conclusion that to come to grips with these challenges, a long-term perspective was needed. Furthermore, a new professional group of experts on the future came to the fore in the 1960s offering their services. The roots of this new profession went back to U.S. "think tanks" established after World War II to help the U.S. armed forces plan for the future. The most prominent of these was the RAND Corporation, which developed new mathematically founded methods for long-range planning like systems analysis. These methods were crucial for the development of U.S. weapon systems during the Cold War, and later they were also adopted by large U.S. multinational corporations.

The new planning methods were also transferred abroad. In Sweden, the Defence Research Institute was the first to introduce them, and in the 1960s the institute developed an elaborate planning and budgeting system for the whole military sector. It was based on fifteen-year "perspective plans" outlining different scenarios that the Swedish Defence would have to be able to cope with. In

Sweden, too, these planning methods were gradually transferred to large corpo-rations, and there was a growing group of domestic "future experts" trained in systems analysis and other mathematical methods in the late 1960s. However, these new "future experts" also met opposition. In a famous article in 1971 with the title "Colonizing the Future," two Swedish left-wing intellectuals launched a forceful verbal attack on all these endeavors by powerful organizations to plan for a progressivist future that served their own ends.

Olof Palme was clearly influenced by this critique. When he publicly pre-sented the working group on future studies in 1971, he outlined its task in the following way in a press release: "Future studies are now going on in many places throughout the world. . . . Many such foreign research projects are directed by military and large-scale industrial interests. . . . What must be emphasized, how-ever, are the clear risks that arise . . . if future research turns into a sort of un-controlled monopoly wielded by a handful of especially powerful interests. One-sided views of the future contrived in various ways can dangerously call forth a moulding public opinion as to a certain fatalistic development for countries and peoples, when actually it is the citizens themselves who can and must determine the future developments of their society" (Myrdal 1972, 10).

The members of the working group on future studies were carefully chosen: five of its seven members were prominent professors of history, psychology, geog-raphy, economics, and mathematics; one was previously general director of the Defence Research Institute; and most important, its chair was a world-renowned politician, diplomat, and social reformer, Alva Myrdal.

By 1971, Alva Myrdal was already a grand old lady of the Swedish Social Democratic Party. At the time she was sixty-nine years old, but she was still very active as minister for disarmament and head of the Swedish delegation at the disarmament negotiations in Geneva. She became famous in 1934, when she and her husband, Gunnar Myrdal, published the book *Kris i befolkningsfrågan* [Crisis in the Population Question], which forcefully argued that the low birthrate in Sweden at the time could be increased only through a number of social reforms, including government support for better housing for families with children, child benefits, and improved schools and nurseries. This book became very influential for much of the reform programs launched by the Social Democratic Party in the following decades.

Alva Myrdal had worked in many different arenas, including in global insti-tutions. She was first a social reformer in Sweden, and later head of UNESCO's division for social sciences. She was Swedish ambassador to India, a member of parliament, and finally minister for disarmament. She also continued writing a number of influential books and articles on social policies and on disarmament. Alva and Gunnar Myrdal are primarily known as a couple, probably Sweden's best-known couple of the twentieth century. Gunnar was a professor of economics at the Stockholm Business School who, like his wife, at times was a minister and high U.N. official. He wrote very influential books like *An American Dilemma* and *Asian Drama*. The peak of their fame was when they became Nobel laureates; Gunnar in 1974 (Economics) and Alva in 1982 (Peace).

In many respects, Alva Myrdal was thus a perfect choice as chair of the working group for future studies. She had a broad experience of long-term reforms in different sectors both nationally and internationally. And with the confidence and fame she enjoyed, she gave credibility to its final product, *Choosing a Future*.

Further Reading

Andersson, J. 2006. "Choosing futures: Alva Myrdal and the construction of Swedish futures studies, 1967–1972." *International Review of Social History* 51:277–295.

Bok, S. 1991. *Alva Myrdal: A daughter's memoir*. Reading, MA: Addison-Wesley.

Carlson, A. 1990. *The Swedish experiment in family politics: The Myrdals and the interwar population crisis*. New Brunswick, NJ: Transaction.

Hirdman, Y. 2008. *Alva Myrdal: The passionate mind*. Translated by Linda Schenck. Bloomington: Indiana University Press.

The Dynamics of Energy Systems and the Logistic Substitution Model

CESARE MARCHETTI AND NEBOJSA NAKICENOVIC

One of the objectives of IIASA's Energy Systems Program is to improve the methodology of medium- and long-range forecasting in the areas of the energy market and energy use, demands, supply opportunities and constraints. This is commonly accomplished with models that capture and put into equations the numerous relationships and feedbacks characterizing the operation of an economic system or parts of it. Such an approach encounters many difficulties, which are linked to the extreme complexity of the system and the fairly short-term variation of the parameters and even of the equations used. Consequently, these models lend themselves to short- and perhaps medium-range predictions, but normally fail to be useful for predictions over a period of about 50 years, the time horizon that the Energy Systems Program has chosen for study.

Following the current scheme of attacking similar problems in the physical sciences, we have left aside all details and interactions, and have attempted a macroscopic description of the system via the discovery of long-term invariants. Heuristically, this approach is certainly not new. In a broad sense, the sciences can be seen as a systematic search for invariants.

This work is dedicated to the empirical testing and theoretical formulation of an invariant, the logistic learning curve, as it applies to the structural evolution of energy systems and systems related to energy, such as coal mining. The great success of the model in organizing past data, and the insensitivity to major political and economic perturbations of the structures obtained seem to lend great predictive power to this invariant.

. . .

C. Marchetti and N. Nakicenovic. 1979. Excerpt from *The dynamics of energy systems and the logistic substitution model*. Working Paper RR-79-13. International Institute for Applied Systems Analysis, Laxenburg, Austria.

1 Introduction

Four years ago, the International Institute for Applied Systems Analysis began a study of energy systems using the techniques of market penetration analysis. The basic hypothesis—which has proved very fruitful and powerful—is that *primary energies, secondary energies, and energy distribution systems are just different technologies competing for a market* and should behave accordingly.

Previous analysis of market competition had always been performed for only two competitors. But it is a peculiarity of energy systems over the last hundred years that most of the time more than two competitors took important shares of the market. Thus, we had to modify the original rules by introducing new constraints that permitted us to deal with more complicated cases. These constraints were defined empirically from a few cases, but proved very successful in dealing with virtually all the cases that we analyzed. A mathematical formulation of the substitution process is given below and the manual for the software package is given in Nakicenovic (1979).

2 The Logistic Function and Substitution Dynamics

Substitution of a new way for the old way of satisfying a given need has been the subject of a large number of studies. One general finding is that almost all binary substitution processes, expressed in fractional terms, follow characteristic S-shaped curves, which have been used for forecasting further competition between the two alternative technologies or products, and also the final takeover by the new competitor.

. . .

The widespread empirical applications of the logistic function as a means of describing growth phenomena also originated in the studies of human population, biology, and chemistry. The first reference to the logistic function can be found in Verhulst (1838, 1845, 1847). Pearl (1924, 1925) rediscovered the function and used it extensively to describe the growth of populations, including human population. From then on, numerous studies have been conducted only to confirm the logistic property of most growth processes. Robertson (1923) was the first to use the function to describe the growth process in a single organism or individual. Later, the function found application in work concerning bioassays (see e.g., Emmens 1941, Wilson and Worcester 1942, and Bergson 1944), and in work on the growth of bacterial cultures in a feeding solution, autocatalyzed chemical reactions, and so on.

. . .

One of the most notable models of binary technological substitution, which extended Mansfield's findings, was formulated by Fisher and Pry (1970). This model uses the two-parameter logistic function to describe the substitution process. The basic assumption postulated by Fisher and Pry is that once a substitution of the new for the old has progressed as far as a few percent, it will proceed to completion along a logistic substitution curve :

$$\frac{f}{1-f} = \exp(\alpha t + \beta)$$

where t is the independent variable usually representing some unit of time, α and β are constants, f is the fractional market share of the new competitor, and $1-f$ that of the old one. *The coefficients α and β are sufficient to describe the whole substitution process.* They cannot be directly observed; they can, however, be estimated from the historical data.

. . .

Every given technology undergoes three distinct substitution phases: growth, saturation, and decline. The growth phase is similar to the Fisher-Pry binary logistic substitution, but it usually terminates before full substitution is reached. It is followed by the saturation phase which is not logistic, but which encompasses the slowing of growth and the beginning of decline. After the saturation phase of a technology, its market share proceeds to decline logistically.

. . .

The most important predictions of our model that differ from those in the current literature are that there will be:

- A relatively rapid phaseout of coal as a primary energy source
- A quite important role for natural gas in the next 50 years
- A negligible role in the next 50 years for new sources such as geothermal energy, solar energy, and fusion because of the very long lead times *intrinsic* to the system

The curious fact about the last point is that the flourish of very expensive research on these sources implies a fairly low discounting factor in decisions on the allocation of funds for energy R&D. This appears to be very wise, if not internally consistent, because the lead times of the systems are so long that nothing could be started rationally if higher discounting rates were used.

These and many other predictions (like the compatibility of resources with demand), although extremely interesting, are not really part of our research task;

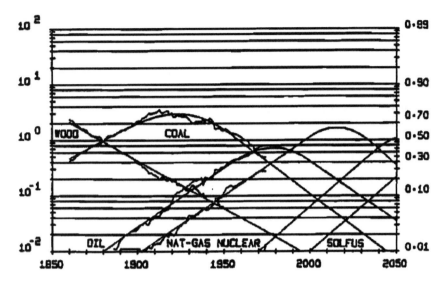

Figure 1

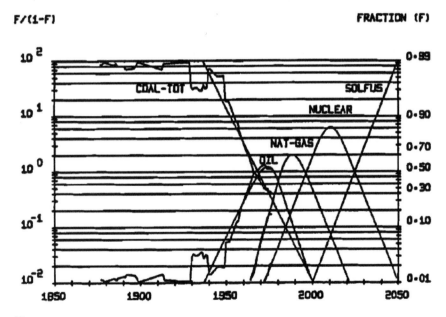

Figure 2

our work is centered in the past, where we try to find order and which we try to understand rationally.

. . .

The curves of the preceding figure [Figure 1] are now plotted as log [f/(1–f)]; the logistic curves appear as straight lines, greatly helping visual inspection and formal considerations. The first fact to be observed is the *extreme regularity and slowness* of the substitution. It takes about 100 years to go from 1 percent to 50 percent of the market. We call this length of time the *time constant* of the system.

The regularity refers not only to the fact that the rate of penetration (defined as constant a in the equation and corresponding to the slope of the curves) remains constant over such very long periods when so many perturbing processes seem to take place, but also to the fact that all perturbations are reabsorbed elastically without influencing the trend. It is as though *the system had a schedule, a will, and a clock*.

. . .

As the statistics on fuel wood are often unreliable, we have eliminated wood and analyzed how the other fuels share the market for commercial energy sources. Oil remains at a level of 1 percent for half a century and shows again that actual logistic market penetration does not start until the market has been penetrated by a few percent. An extraordinary feature of the predictive side of the graph is that oil as a primary source of energy will virtually disappear in the year 2000, a feature common to the UK, the Netherlands, and Belgium. If this happens to be true, what will automobiles run on? Perhaps on LNG, Hz, or methanol.

Commentary

Cesare Marchetti and Nebojsa Nakicenovic, "The Dynamics of Energy Systems and the Logistic Substitution Model" (1979)

PAUL WARDE

The 1970s is commonly remembered, among other things, as the decade of energy crisis, albeit the energy crisis whose bark was worse than its bite—at least seen in the context of the subsequent two decades. The energy crisis's cause is now mostly attributed to a political response to the Yom Kippur war of 1973, when the predominately Arab governments in OPEC (the Organization of Petroleum Exporting Countries) flexed their collective muscle in price hikes and embargoes. Even among those who now predict "peak oil," the 1970s crisis is nowadays less frequently seen as a precursor to current dilemmas. This popular memory is, however, misleading for a number of reasons, not least that the necessity of an energy transition was already being discussed in the years prior to 1973, especially in the United States, where the main issue was security and the prospect of dependence on imported oil. It was in fact a commonplace, although a controversial one, that the price rises of autumn 1973 were part of a longer-term trend toward shortages. Another driver was worry about pollution; in that case, not the excess of oil and its by-products causing global warming as today, but the direct consequences of combusting fossil fuels for human health. But in the form of arguments about resource scarcity and health impacts—neither of course new in themselves—the debates of the 1970s significantly laid the ground for later concerns about energy.

Certainly the events of autumn 1973 provided a mighty stimulus to efforts to predict the longevity of oil supplies. Already in 1956 oil executive M. King Hubbert had (accurately, as it turned out) predicted "peak oil," the point after which production would start to decline, for the United States as coming around 1970. Predictions and their impact on gas prices became highly politicized in mid-1970s America: were the long queues at gas stations and high prices the result of Arab diplomacy, long-term pressures on supply, or manipulation and profiteering by the oil companies? Debates raged on the reliability of data on reserves, eventually leading to an all-encompassing, but relatively soon forgotten energy policy published by the Carter administration in 1978, aiming at a transition to "alternative" fuels and increased energy efficiency. It is in this context that energy predictions and the work of systems and transitions theorists such as Cesare Marchetti and Nebojsa Nakicenovic won a moment of traction.

The authors belonged, however, to a different postwar predictive tradition: rather than simple amelioration of scarcity, they were aiming to realize the promise of abundance, based on brand-new technologies such as nuclear power and hydrogen. Marchetti began his career as a nuclear physicist, and he became interested in both the problem of how to expand the potential market for nuclear energy (which seemed limited at the time if it was only ever going to supply electricity, a minority share of the total energy market) and the possibility of using the vast waste heat of reactors to produce hydrogen fuels. By the 1970s he supplemented work on European-funded projects with work as a predictive analyst for General Electric. His reaction to the Club of Rome report of 1972 (see Part 2) was to produce a hypothetical paper as to how humanity, if it gave the right direction to its technical ambitions, would inhabit an Earth that was not only far from its limits, but could support no less than a trillion people. He was not a pessimist.

Thus the "crisis" of the 1970s, understood in terms of security and scarcity, stimulated interest in energy *transitions* and also led to an examination of the experience of the past in switching to coal, oil, and natural gas. Yet despite the apparently economic drivers of wider interest in the problem, the response of some technicians working in these fields was not to see energy transition as being the outcome of economic questions of price (that derived in the end from the balance between rates of consumption and the discovery of new reserves) but rather as a process contingent on technology and innovation, a response to market saturation and trends in scientific research. Marchetti saw the past and the future not in terms of geopolitical dynamics, or an intertwined series of demand, supply, and price, but as waves of innovation, a largely autonomous process that drove the development and substitution of new products. "Transition" was thus largely a problem of knowledge.

The paper of Marchetti and Nakicenovic is typical of the new world of modeling, mathematical and increasingly computerized, that argued for standard patterns in the uptake and eclipse of technology. Often this was linked to "boosterism" for the new: the cost of novel inventions would decrease "in time by exponential laws," and therefore current cost was an unsound guide to rational choices. In 1970 it could be predicted that breeder reactors would soon become cheaper than fossil fuels whose prices had ceased to fall. The emergent technology of the time was supposed to be on a set trajectory of cheapening and eventual adoption by society. The particular tool adopted to predict in this case was the Fischer-Pry equation, an algorithm predicting that the proportional rate at which a product enters a market is itself in proportion to its current share of that market. New technologies will grow in usage, and thus cheapen, rapidly. The process will be smooth, and to fail to see this is to deny a kind of logic to history. Implicitly, as the future will follow the pattern already set, the past must supply evidence of the same deterministic logic. Energy systems were subject to the general properties of systems where substitution took place (much like competing brands of soap). "The whole destiny of an energy source," wrote Marchetti, "seems to be completely determined in early childhood."

Indeed, some of the modeling of past transitions (albeit at times based on conjectural data) seemed to confirm the intuitions. But these were generally "transitions" between very limited choices (wood or coal) or, indeed, where the new energy carrier actually mostly provided a new kind of service rather than a direct substitute for the old (so not just the straight swap of coal replacing wood in warming a house, but a new good to enjoy, such as oil powering motor vehicles). By the 1970s, predictors were facing a much more diversified and complex energy market, with many options, but only some of which were substitutable with each other. The application of the method to country-by-country predictions that Marchetti and Nakicenovic attempted for the much more complex and diversified energy market of the 1970s market produced very strange results; neighboring countries, for example, would abandon using gasoline for automobiles at very different dates. Oil, they noted with surprise, would be entirely abandoned by Germany during the 1990s. Thus, deterministic systems analysis ran into major issues of what scale they might be plausible on when confronted with the messy reality of the world. But overall, the lesson that Marchetti drew was clear: one had to "go nuclear or go bust"; the nature and pace of transition already in large part decided that nuclear was next, and other alternative energy sources were as yet too undeveloped to have a major influence for many decades.

It is hardly necessary to observe that the "boosterism" for nuclear power was proven wrong (or at least, was somehow spoiled by the contingency of history). But this predictive work nevertheless had its important legacy, establishing the notion of an "energy transition" before wider concerns about anthropogenic global warming appeared, and viewing energy history according to a kind of stage-theory of reiterated transition, conforming to expectations of innovation in both the technical and commercial worlds. Transition might have many proximate causes, but in the end had to rely on ingenuity. The future could be thought into existence by ingenious people, but nevertheless they were still subject to waves of activity and progress that largely related to the dynamics of knowledge creation itself rather than the development of the economy. The transition experts were the inventors and innovators in the laboratories, even if they themselves could not fully perceive the forces that drove them.

Further Reading

Adelman, M. A. 1995. *The genie out of the bottle: World oil since 1970.* Cambridge, MA: MIT Press.

Bergson, J. 1944. "Application of the logistic function to bio-assay." *Journal of the American Statistical Association* 39:357–365.

Emmens, C. W. 1941. "The dose-response relation for certain principles of the pituitary gland, and of the serum and urine of pregnancy." *Journal of Endocrinology* 2:194–225.

Fisher, J. C., and R. H. Pry. 1970. "A simple substitution model of technological change." Report 70-C-215. General Electric Company, Research and Development Center, Schenectady, New York, Technical Information Series. Also in *Technological Forecasting and Social Change* 3(1971):57–88.

Kander, A., P. Malanima, and P. Warde. 2013. *Power to the people: Energy in Europe over the last five centuries.* Princeton, NJ: Princeton University Press.

Marchetti, C. 1979. "Ten to the twelfth: A check on the earth-carrying capacity for man." *Energy* 4:1107–1117.

Marchetti, C., and N. Nakicenovic. 1979. "The dynamics of energy systems and the logistic substitution model." Working paper rr-79–13, International Institute for Applied Systems Analysis, Laxenburg, Austria.

Mitchell, Timothy. 2011. *Carbon democracy: Political power in the age of oil.* New York: Verso.

Pearl, R. 1924. *Studies in human biology.* Baltimore: Williams and Wilkins.

Pearl, R. 1925. *The biology of population growth.* New York: Alfred A. Knopf.

Robertson, T. B. 1923. *The chemical basis of growth and senescence.* Philadelphia: J. B. Lippincott.

Smil, Vaclav. 2008. *Energy in nature and society: General energetics of complex systems.* Cambridge, MA: MIT Press.

Verhulst, P.-F. 1838. "Notice sur la loi que la population poursuit dans son accroissement." *Corresp. Math. Phys.* 10:113–121.

Verhulst, P. F. 1845. "Recherches mathématiques sur la loi d'accroissement de la population." *Nouveaux Mémoires de l'Académie Royale des Sciences et Belles-Lettres de Bruxelles,* 18: 1–38.

Verhulst, P. F. 1847. "Deuxième mémoire sur la loi d'accroissement de la population." *Mémoires de l'Académie Royale des Sciences et Belles-Lettres de Bruxelles* 20:1–32.

Wilson, E. B., and J. Worcester. 1942. "The determination of L.D.50 and its sampling error in bio-assay." *Proceedings of the National Academy of Sciences* 29:79–85.

PART 7

Climate

How Can We Predict Change?

Climate, a word used as long ago as the time of the ancient Greeks, was formerly understood as a local phenomenon, the climate of a place. Only recently has its meaning changed to refer to global climate, mainly defined as global mean temperature, to which local climates contribute only partially. Carbon dioxide has played a major role in this globalization of climate, from early theorizing about atmospheric chemistry in the nineteenth century, to measuring the concentration of this gas throughout the entire atmosphere encircling the earth. Carbon dioxide is now seen as playing a major role in steering the climate of the planet, and each and every locality within it.

Climate change caused by greenhouse gas concentrations caught media attention and stimulated the political imagination not only because its consequences are sinister, but also because it is so mono-causally predictive. Current orthodoxy holds that every increase of greenhouse gases, chiefly carbon dioxide (originally known as carbonic acid), leads to an almost linear growth of global average temperature. Regional effects, however, are unevenly distributed, with amplifications in the polar regions and lesser effects in the tropical and temperate zones. This new orthodoxy dates from the late 1970s, when a report to the National Academy of Sciences in the United States suggested that anthropogenic climate change was a reality (Charney et al. 1979), later corroborated by reports from the Intergovernmental Panel on Climate Change (IPCC), founded in 1988.

By the late 1970s, the science behind the idea was already old. Climate change was an idea that developed in the nineteenth century. Advances in climatic knowledge combined theory with observation in the field and work in the laboratory. After fundamental theoretical advances by Joseph Fourier in the 1820s, further seminal work was done by Irish chemist John Tyndall, who experimentally confirmed the role of water vapor as a greenhouse gas, absorbing infrared radiation from the Earth's surface, and rightly drew the conclusion, in 1859, that he had found ultimate proof of why the surface of the Earth was such a warm place and so well suited for life. A few decades later, the Swede Svante Arrhenius, while adding nothing to the theoretical understanding, provided detailed and, as it turned out, very accurate calculations of the correlation between atmospheric carbon dioxide concentrations and global average surface temperature under certain assumptions.

It is one of the enigmas of climate science history that it took such a long time for the insights in the work by Tyndall and Arrhenius to inform more general understanding of the role of humans' use of fossil fuels in forcing global warming. With few exceptions, Thomas C. Chamberlin (1899, 1907) being one, nobody tried their hand at closer examination of the relationship of carbon dioxide concentrations and global temperature changes in the first decades of the twentieth century. Instead of climate as something affected by human action, the contrary was explored in this period. Climate, following the tradition dating back to the ancient Greeks, was regarded as a determining factor on humans, societies, and "civilizations" (see Part 3), thus privileging the local and regional dimension of climate and downplaying climate as a global concern, common to all of humanity.

The first real attempt at establishing an empirical measure of the relationship between carbon dioxide and global climate change was published in 1938 by British steam engineer G. S. Callendar, whose results paralleled those made by Arrhenius. Callendar also made the point that fossil fuel combustion had already brought the secular (background) level of atmospheric carbon dioxide up considerably, by 10 percent, since 1900. Callendar's work did not receive much attention. At the time he was marginalized by mainstream climate scientists, who regarded him as an amateur. Others questioned the quality of his data, as it happened that the years following his predictions were all cool.

In fact, Callendar's work is among the early examples of a true "global change science." He treated the world as a whole as a system, and used quantitative data and theory to explain a large-scale rate of change in the system. His ideas were poorly received because there was no established way of thinking of humans (nor indeed humanity as a species) as capable of altering natural phenomena on a global scale. In this respect Callendar diverges from the great early traditions of Fourier, Tyndall, and Arrhenius. Callendar pioneered a new tradition, turning climate science into a global change enterprise. He refocused its interest toward the *future* instead of the past, which had been the main preoccupation of his predecessors, who had concentrated on explaining the atmospheric physics of past ice ages.

It took almost two more decades until Callendar's work was recognized as groundbreaking, and even then, action was taken not only, nor even chiefly, because of his work. The Cold War stimulated renewed interest in atmospheric change, as the military in many places felt the need for better weather predictions for possible air warfare and for monitoring nuclear fallout from atom bomb tests. This made numerical weather predictions at a global scale highly desirable, and by the mid-1950s the first computer-generated models for weather prediction were successfully implemented, paving the way for more complex global climate models and, subsequently, comprehensive, multi-parameter global change models. The first computer model–based climate change prediction was published in 1967 by S. Manabe and R. T. Wetherald.

Several of the scientists who developed an understanding of climate change based on the carbon dioxide hypothesis were trained as physicists and worked on meteorology projects. Their thinking was increasingly large scale, moving toward the global. British meteorologists like Gilbert T. Walker had already been working on large-scale ocean-based phenomena such as the Southern Oscillation, advancing the work begun more than a century earlier by their countrymen, and like them using the vast imperial infrastructure of observation stations. Other groups of climate scientists came from the earth sciences, collecting data on local and regional scales. Key to the new theoretical understanding were measurements of carbon dioxide that had been taken in various locations since the nineteenth century but were reinforced from the early to mid-1950s onward and developed methodologically to test the renewed carbon dioxide hypothesis of global climate change (Bohn 2011). Charles David Keeling started carbon dioxide measurements on the Mauna Loa mountain in Hawaii as part of the International Geophysical Year 1957–58 (Keeling 1978). His work generated the most reliable global time series, the now iconic "Keeling Curve" that plots the ongoing increase in concentration of carbon dioxide in Earth's atmosphere since 1958.

There has been a paradigm shift in human thought in the twentieth century, as John McNeill argued in *Something New Under the Sun* (2000). Going from a world where humanity was deemed "powerless to do . . . any titanic damage" to "the Creator's handiwork" (Millikan 1930), to a world where this is not only possible, but where the *rate of this change* can be predicted, reveals a major shift in perspective. Such a shift was conceived by Callendar in the 1930s, but it was not until the information technology revolution made computerized models possible that a vision for a science of global change emerged. Its technical nature made global change science initially inaccessible to the policy world, which was slower to adopt thinking on this scale.

Yet, even so, the understanding of climate went from a deterministic and local mode to a global predictive in a remarkable fifty years. The predictive element has now gained such hegemony that climate has started to appear as a predictor itself. Climate is not just the dependent variable that is affected by human action; it has started to appear in global change science and in public understanding as a "driver" determining changes of other environmental parameters, such as bio-geography and species extinctions. Climate determinism has made a return in a new guise, and it remains to be seen how this will unfold.

The idea that changing climate drives significant global change was boosted by evidence from ice cores, the frozen records of climate change retrieved from the Greenland and Antarctic ice sheets in the 1970s and 1980s onward, papers from which era are included here. Ice cores provided the most reliable and de-tailed record available of climate variations over very long periods in the past, covering multiple ice ages. Even more, they showed the possibility of dramatic and abrupt shifts of temperatures, changes that took place in just a few decades over a global scale. The two polar extremes of the world spoke to each other as distant factors, yet acting in the same integrated global system. The implications for humanity's current relation to Earth's climate were tremendous. If abrupt change could happen in the past, perhaps it could happen again? And if variation in the concentration of greenhouse gases was one of the triggers of drastic climate change, human-induced global atmospheric change was very likely, almost cer-tain, to be a trigger for sudden "unpleasant surprises in the greenhouse," as earth scientist Wallace Broecker put it in 1987.

On the Transmission of Heat

JOHN TYNDALL

Weekly Evening Meeting,
Friday, June 10, 1859.
THE PRINCE CONSORT, Vice-Patron, in the Chair.
JOHN TYNDALL, Esq. F.R.S. Professor of Natural Philosophy, Royal Institution.

On the Transmission of Heat of different qualities through Gases of different kinds.

Some analogies between sound and light were first pointed out: a spectrum from the electric light was thrown upon a screen—the spectrum was to the eye what an orchestra was to the ear—the different colours were analogous to notes of different pitch. But beyond the visible spectrum in both directions there were rays which excited no impression of light. Those at the red end excited heat, and the reason why they failed to excite light probably was that they never reached the retina at all. This followed from the experiments of Brücke and Knoblauch. These obscure rays had been discovered by Sir Wm. Herschel, and the speaker demonstrated their existence by placing a thermo-electric pile near to the red end of the spectrum, but still outside of it. The needle of the large galvanometer connected with the pile was deflected and came to rest in a position about 45 degrees from zero. A glass cell, containing the transparent vitreous humour of the eye of an ox, was now placed in the path of the rays: the *light* of the spectrum was not perceptibly diminished, but the needle of the galvanometer fell to zero, thus proving that the obscure rays of the spectrum, to which the galvanometric deflection was due, were wholly absorbed by the humours of the eye.

John Tyndall. 1859. "On the transmission of heat of different qualities through gases of different kinds." *Proceedings of the Royal Institution (London)* 3:155–158.

Reference was made of the excellent researches of Melloni. In a simple and ingenious manner he had proved the law of inverse squares to be true of radiant heat passing through air, and the eminent Italian inferred from his experiments that for a distance of 18 or 20 feet, the action of air upon radiant heat was totally inappreciable. This is the only experimental result now known regarding the transmission of radiant heat from terrestrial sources through air; with regard to its transmission through other gases it was believed that we were without any information.

It was, however, very desirable to examine the action of such media—desirable on purely scientific grounds, and also on account of certain speculations which had been based upon the supposed deportment of the atmosphere as regards radiant heat. These speculations were originated by Fourier; but it was to M. Pouillet's celebrated Memoir, and the recent excellent paper of Mr. Hopkins, to which we were indebted for their chief development. It was supposed that the rays from the sun and fixed stars could reach the earth through the atmosphere more easily than the rays emanating from the earth could get back into space. This view required experimental verification, and the more so, as the experiment we possessed was the negative one of Melloni, to which reference has been already made.

The energetic action of the solid and liquid compounds into which the element hydrogen enters, suggested the thought that hydrogen gas might act more powerfully than air, and the following means were devised to test this idea. A tube was constructed, having its ends stopped air-tight by polished plates of rock-salt held between suitable washers, which salt is known to be transparent to heat of all kinds; the tube could be attached to an air-pump and exhausted, and any required gas or vapour could be admitted into it. A thermo-electric pile being placed at one end of the tube, and a course of heat at the other, the needle of an extremely sensitive galvanometer connected with the pile was deflected. After it had come to rest, the air was pumped from the tube, and the needle was carefully observed to see whether the removal of the air had any influence on the transmission of the heat. No such influence showed itself—the needle remained perfectly steady. A similar result was obtained when hydrogen gas was used instead of air.

Thus foiled, the speaker put his questions to Nature in the following way: a source of heat, having a temperature of about 300°C, was placed at one end of the tube, and a thermo-electric pile at the other—a large deflection was the consequence. Round the astatic needle, however, a second wire was coiled, thus forming a so-called differential galvanometer; a second pile was connected with this second wire, so that the current from it circulated round the needle in a direction opposed to that of the current from the first pile. The second pile was caused to approach the source of heat until both currents exactly neutralised each other, and the needle stood zero. Here then we had two powerful forces in equilibrium, and the question now was whether the removal of the air from the tube would disturb this balance. A few strokes of the air-pump decided the question, and on the entire removal of the air the current from the pile at the end of the tube predominated over its antagonist from 40° to 50°. On re-admitting the air the needle again fell to zero; thus proving beyond a doubt that the air within the tube intercepted a portion of the radiant heat.

The same method was applied with other gases, and with most remarkable results. Gases differ probably as much among themselves with regard to their action upon radiant heat as liquids and solids do. Some gases bear the same relation to others that alum does to rock salt. The speaker compared the action of perfectly transparent coal-gas with perfectly transparent atmospheric air. To render the effect visible to the audience, a large plano-convex lens was fixed between two upright stands at a certain height above a delicate galvanometer. The dial of the instrument was illuminated by a sheaf of rays from an electric lamp, the sheaf being sent through a solution of alum to sift it of its heat, and thus avoid the formation of air-currents within the glass shade of the instrument. Above the lens was placed a looking-glass, so inclined that the magnified image of the dial was thrown upon a screen, where the movements of the needle could be distinctly observed by the whole audience. Air was first examined, the currents from the two piles being equilibrated in the manner described, the tube was exhausted, and a small but perfectly sensible deflection was the result. It was next arranged that the current from the pile at the end of the tube predominated greatly over its antagonist. Dry coal-gas was now admitted into the tube, and its action upon the radiant heat was so energetic, the quality of heat which it cut off was so great, that the needle of the galvanometer was seen to move from about 80° on one side of zero to 80° on the other. On exhausting the tube the radiant heat passed copiously through it, and the needle returned to its first position.

Similar differences have also been established in the case of vapours. As representatives of this diverse action, the vapour of ether and of bisulphide of carbon may be taken. For equal volumes, the quantity of heat intercepted by the former is enormously greater than that intercepted by the latter.

To test the influence of *quality*, the following experiment was devised. A powerful lime light was placed at one end of the tube, and the rays from it, concentrated by a convex lens, were sent through the tube, having previously been caused to pass through a thin layer of pure water. The heat of the luminous beam excited a thermo-electric current in the pile at the end of the exhausted tube;

and this current being neutralised by the current from the second pile, coal-gas was admitted. This powerful gas, however, has no sensible effect upon the heat selected from the lime light; while the same quantity of heat, from an obscure source,* was strongly affected.

The bearing of this experiment upon the action of planetary atmospheres is obvious. The solar heat possesses, in a far higher degree than that of the lime light, the power of crossing an atmosphere; but, and when the heat is absorbed by the planet, it is so changed in quality that the rays emanating from the planet cannot get with the same freedom back into space. Thus the atmosphere admits of the entrance of the solar heat, but checks its exit; and the result is a tendency to accumulate heat at the surface of the planet.

In the admirable paper of M. Pouillet already referred to, this action is regarded as the cause of the lower atmospheric strata being warmer than the higher ones; and Mr. Hopkins has shown the possible influence of such atmospheres upon the life of a planet situated at a great distance from the sun. We have hitherto confined our attention to solar heat; but were the sun abolished, and did stellar heat alone remain, it is possible that an atmosphere which permits advance, and cuts off retreat, might eventually cause such an accumulation of small savings as to render a planet withdrawn entirely from the influence of the sun a warm dwelling-place. But whatever be the fate of the speculation, the experimental fact abides—that gases absorb radiant heat of different qualities in different degrees; and the action of the atmosphere is merely a particular case of the inquiry in which the speaker was at present engaged.†

---

* The *quality* of the heat is measured by the amount of the galvanometric deflection which it produces; its power of passing through media may be taken as a test of *quality*.
† While correcting the proof of this abstract, I learned that Dr. Franz had arrived at the conclusion that an absorption of 3.54 per cent of the heat passing through a column of air 30 centimeters long takes place; for coloured gases he finds the absorption greater; but all colourless gases he assumes show no marked divergence from the atmosphere.
—Poggendorff's Annalen, xciv. P.337.

Commentary

John Tyndall, "On the Transmission of Heat" (1859)

MIKE HULME

Friday, 10 June 1859, had been a rather cool, cloudy, and showery day in London. The national papers were full of news about the Italian wars of independence and the long-running violent trade dispute with China. The local news story in London was of the accidental deaths of seven construction laborers working on the luxury Westminster Palace Hotel. Nearby in central London, early-evening visitors began arriving at the Royal Institution on Albemarle Street to attend one of the Institute's regular Friday night discourses. Among them was Prince Albert, Queen Victoria's husband, the vice-patron of the Royal Institution and chair of the event. Those arriving that cool midsummer evening—an audience number-ing perhaps about 150 people—were about to hear reported the first scientific evidence for what later became known as "the planetary greenhouse effect."

John Tyndall—one of the rising scientific personalities of mid-Victorian London and an incumbent scientist at the Royal Institution—was delivering the lecture. He not only gave a verbal account of his own very recently completed experiments about the radiative properties of different gases. He also included a public demonstration of those same experiments, which a few days earlier he had first performed in the basement laboratory of the building. The article reprinted here is the Institute's official record of that evening discourse, written by Tyndall himself, and the first public record of his experimental findings.

Why had Tyndall been engaged in such experimental work? There had been speculation by physicists for several decades concerning how heat arriv-ing from the sun passed through the Earth's atmosphere and, in particular, how much of the re-radiated heat was lost back to space. Ferdinand De Saussure, Joseph Fourier, and Claude Pouillet had earlier proposed that the gases in the air must somehow act to control these fluxes of energy, but no one had evidence of how, nor which, gases might be involved. Irish scientist John Tyndall had be-come fascinated by this question, prompted partly through his close observations while on regular Alpine expeditions of vertical atmospheric heat gradients and glacier melt rates. Being a practiced experimenter, Tyndall by the spring of 1859 had established his own apparatus in the Royal Institution that was to enable him to "interrogate Nature" (as he put it) to find out the role of different gases in this theory of radiant heat.

Just three weeks before this June evening, thirty-eight-year-old Tyndall had recorded in his journal his greatest success to date: *"Experimented all day; the subject is completely in my hands!"* Tyndall was fully aware at this time of the potential significance of his experimental findings. Just a few days later—on 1 June—he wrote to one of his correspondents, German physicist and mathematician Rudolf Clausius, outlining the purpose and results of these experiments. And when on 10 June he demonstrated and delivered his lecture at the Royal Institution, he also outlined their significance to his public audience: "*The bearing of this experiment upon the action of planetary atmospheres is obvious . . . the atmosphere admits of the entrance of the solar heat, but checks its exit; and the result is a tendency to accumulate heat at the surface of the planet.*" Fourier and Pouillet's theory of the regulation of planetary temperatures through the differential actions of atmospheric gases now had experimental evidence in its favor. Tyndall's remarks that evening also reveal his high view of laboratory experimentation: "*But whatever be the fate of speculation, the experimental fact abides—that gases absorb radiant heat of different qualities in different degrees.*"

Tyndall continued these experiments for a number of years, and nearly two years later, in February 1861, he was invited to deliver the prestigious Bakerian Lecture at London's Royal Society. By now, he was able to give a much fuller account of the behavior of different gases and vapors in the atmosphere, and his lecture was published in the *Philosophical Magazine* of London. He was also in a position to deduce from these findings that changes in the amount of any of the radiatively active constituents of the atmosphere—especially water vapor, carbon dioxide, and methane, "*may have produced all the mutations of climate which the researches of geologists reveal*" (Tyndall 1861, 277).

This was a bold claim to make at this time of changing scientific ideas about the stability of nature. In November 1859, just six months after Tyndall's initial experimental results, Charles Darwin's book *On the Origin of Species* was published in London. Along with Darwin's challenge to the prevailing orthodoxy of a fixed biological creation, scientists were also grappling with the equally revolutionary implications of Louis Agassiz's 1837 ice age theory. Trying to understand the causes of these great ice age fluctuations in climate was one of the issues of the day. The ideas of Darwin and Agassiz were assaulting traditional conceptions of time and stability in, respectively, biological and climatic history.

John Tyndall was intimately connected with these debates. He became a close friend of Thomas Huxley, Joseph Hooker, and other scientists in Darwin's circle and was one of the members of the X-Club, an exclusive scientific dining club of nine members founded in 1864 and out of whose discussions the journal *Nature* was first published in 1869. Tyndall was also consulted by geologist Charles Lyell about whether his new radiative theory of climatic change could help unravel the causal mystery of Agassiz's ice ages. Lyell, for example, was trying to evaluate James Croll's newly published orbital theory of climate change.

On 1 June 1866, in his reply to Lyell, Tyndall remarked that he thought changes in radiative properties alone were *unlikely* to be the root causes of glacial epochs. Nevertheless, Tyndall's experimental work made more plausible the idea of an eternally varying climate, modified by the changing constituent gases of the Earth's atmosphere itself.

This exchange between Lyell and Tyndall about theories of climatic change presaged much later argument about the interplay between different factors involved in the modification of global climate, arguments that continue today. Although the direct association of coal burning with Tyndall's "greenhouse effect" and global climate change was to await nearly forty years—until Svante Arrhenius—John Tyndall is deservedly credited with establishing the experimental basis for the putative greenhouse effect. For example, Gilbert Plass in his classic 1956 article in *Tellus* that gave widespread visibility and impetus in the postwar era to the carbon dioxide theory of climate change also quoted from Tyndall's 1861 Bakerian Lecture. He recognized Tyndall as the first to attempt a calculation of the infrared flux of the atmosphere. And Tyndall was also correct in identifying the fundamental role of water vapor in atmospheric dynamics, which, he claimed, must form one of the chief foundation stones of the science of meteorology.

The differential radiative absorption properties of the gases and vapors revealed by Tyndall's interrogation of nature—a suite of gases now expanded to include a group of artificial gases unknown to Tyndall, the halocarbons—remain central to the idea of anthropogenic climate change. Subsequent work has established the global warming potentials of each of these gases with some level of precision, calculations that are pivotal in efforts to quantify the extent of human influence on the world's temperature and in efforts to reduce and manage those consequences.

John Tyndall's experimental work in the spring of 1859 in the basement of a renowned London scientific institution may not be remembered in the same way as is Darwin's written masterpiece *On the Origin of Species*. Yet in its own way the legacy of Tyndall's work is just as significant for contemporary cultural and scientific debates. And Tyndall's public "performances" of his science also speak to us today. He put his expertise on display: first, in front of one of the royal celebrities of his day, just days after he had acquired new scientific evidence; and then, within two years, to the country's, and to many of the world's, scientific elite at the Royal Society in London. John Tyndall combined the meticulous determination of the experimental scientist with the bravado of the public celebrity. And he was the first to glimpse—just fleetingly—the potential entanglement of human activities in the fashioning of global climates.

Further Reading

Fleming, J. R. (1998) 2005. *Historical perspectives on climate change*, 2nd ed. Oxford: Oxford University Press.

Hulme, M. 2009. On the origin of "the greenhouse effect": John Tyndall's 1859 interrogation of nature. *Weather* 64(5):122–124.

Lightman, B. 2007. *Victorian popularisers of science: Designing nature for new audiences* Chicago: University of Chicago Press.

Tyndall, J. 1861. On the absorption and radiation of heat by gases and vapours. *Philosophical Magazine* 22:169–194, 273–285.

On the Influence of Carbonic Acid in the Air upon the Temperature of the Ground

SVANTE ARRHENIUS

I. Introduction: Observations of Langley on Atmospherical Absorption

A great deal has been written on the influence of the absorption of the atmosphere upon the climate. Tyndall in particular has pointed out the enormous importance of this question. To him it was chiefly the diurnal and annual variations of the temperature that were lessened by this circumstance. Another side of the question, that has long attracted the attention of physicists, is this: Is the mean temperature of the ground in any way influenced by the presence of heat-absorbing gases in the atmosphere? Fourier maintained that the atmosphere acts like the glass of a hot-house, because it lets through the light rays of the sun but retains the dark rays from the ground. This idea was elaborated by Pouillet; and Langley was by some of his researches led to the view, that "the temperature of the earth under direct sunshine, even though our atmosphere were present as now, would probably fall to −200°C., if that atmosphere did not possess the quality of selective absorption" (Langley 1884: 123). This view, which was founded on too wide a use of Newton's law of cooling, must be abandoned, as Langley himself . . . showed that the full moon, which certainly does not posses any sensible heat-absorbing atmosphere, has a "mean effective temperature" of about 45°C (Langley 1890: 193).

The air retains heat (light or dark) in two different ways. On the one hand, the heat suffers a selective diffusion on its passage through the air; on the other hand, some of the atmospheric gases absorb considerable quantities of heat.

Svante Arrhenius. 1896. "On the influence of carbonic acid in the air upon the temperature of the ground." *Philosophical Magazine and the Journal of Science*, series 5, 41 (April): 237–276. Extract from a paper presented to the Royal Swedish Academy of Sciences, 11 December 1895.

These two actions are very different. The selective diffusion is extraordinarily great for the ultra-violet rays, and diminishes continuously with increasing wavelength of the light, so that it is insensible for the rays that form the chief part of the radiation from a body of the mean temperature of the earth.

The selective absorption of the atmosphere is, according to the researches of Tyndall, Lecher and Pernter, Röntgen, Heine, Langley, Ångström, Paschen, and others, of a wholly different kind. It is not exerted by the chief mass of the air, but in a high degree by aqueous vapour and carbonic acid, which are present in the air in small quantities. Further, this absorption is not continuous over the whole spectrum, but nearly insensible in the light part of it, and chiefly limited to the long-waved part, where it manifests itself in very well-defined absorption-bands, which fall off rapidly on both sides. The influence of this absorption is comparatively small on the heat from the sun, but must be of great importance in the transmission of rays from the earth. Tyndall held the opinion that the water-vapour has the greatest influence, whilst other authors, for instance Lecher and Pernter, are inclined to think that the carbonic acid plays the more important part. The researches of Paschen show that these gases are both very effective, so that probably sometimes the one, sometimes the other, may have the greater effect according to the circumstances.

. . .

5. Geological Consequences.

I should certainly not have undertaken these tedious calculations if an extraordinary interest had not been connected with them. In the Physical Society of Stockholm there have been occasionally very lively discussions on the probable causes of the Ice Age; and these discussions have, in my opinion, led to the conclusion that there exists as yet no satisfactory hypothesis that could explain how the climatic conditions for an ice age could be realized in so short a time as that which has elapsed from the days of the glacial epoch. The common view hitherto has been that the earth has cooled in the lapse of time; and if one did not know that the reverse has been the case, one would certainly assert that this cooling must go on continuously. Conversations with my friend and colleague Professor Högbom, together with the discussions above referred to, led me to make a preliminary estimate of the probable effect of a variation of the atmospheric carbonic acid on the temperature of the earth. As this estimation led to the belief that one might in this way probably find an explanation for temperature variations of 5°–10°C., I worked out the calculation more in detail, and lay it now before the public and the critics.

From geological researches the fact is well established that in Tertiary times there existed a vegetation and an animal life in the temperate and Arctic zones that must have been conditioned by a much higher temperature than the present in the same regions. The temperature in the Arctic zones appears to have exceeded the present temperature by about 8 or 9 degrees. To this genial time the ice age succeeded, and this was one or more times interrupted by interglacial periods with a climate of about the same character as the present, sometimes even milder. When the ice age had its greatest extent, the countries that now enjoy the highest civilization were covered with ice. This was the case with Ireland, Britain (except a small part in the south), Holland, Denmark, Sweden and Norway, Russia (to Kiev, Orel, and Nijni-Novgorod), Germany and Austria (to the Harz, Erz-Gebirge, Dresden, and Cracow). At the same time an ice-cap from the Alps covered Switzerland, parts of France, Bavaria south of the Danube, the Tyrol, Styria, and other Austrian countries, and descended into the northern part of Italy. Simultaneously, too, North America was covered with ice on the west coast to the 47th parallel, on the east coast to the 40th, and in the central part to the 37th (confluence of the Mississippi and Ohio rivers). In the most different parts of the world, too, we have found traces of a great ice age, as in the Caucasus, Asia Minor, Syria, the Himalayas, India, Thian Shan, Altai, Atlas, on Mount Kenya and Kilimanjaro (both very near to the equator), in South Africa, Australia, New Zealand, Kerguelen, Falkland Islands, Patagonia and other parts of South America. The geologists in general are inclined to think that these glaciations were simultaneous on the whole earth; and this most natural view would probably have been generally accepted, if the theory of Croll, which demands a genial age on the Southern Hemisphere at the same time as an ice age on the Northern and vice versa, had not influenced opinion. By measurements of the displacement of the snow-line we arrive at the result,—and this is very concordant for different places—that the temperature at that time must have been 4°–5°C. lower than at present. The last glaciation must have taken place in rather recent times, geologically speaking, so that the human race certainly had appeared at that period. Certain American geologists hold the opinion that since the close of the ice age only some 7000 to 10,000 years have elapsed, but this most probably is greatly underestimated.

One may now ask how much must the carbonic acid vary according to our figures, in order that the temperature should attain the same values as in the Tertiary and Ice ages respectively? A simple calculation shows that the temperature in the Arctic regions would rise about 8° to 9°C., if the carbonic acid increased to 2.5 or 3 times its present value. In order to get the temperature of the ice age between the 40th and 50th parallels, the carbonic acid in the air should sink to 0.62–0.55 of its present value (lowering of temperature 4°–5°C.). The demands of the geologists, that at the genial epochs the climate should be more uniform than now, accords very well with our theory. The geographical annual and diurnal ranges of temperature would be partly smoothed away, if the quantity of carbonic acid was augmented. The reverse would be the case (at least to a latitude of 50° from the equator), if the carbonic acid diminished in amount. But in both these cases I incline to think that the secondary action . . . due to the regress or the progress of the snow-covering would play the most important rôle. The theory demands also that, roughly speaking, the whole earth should have undergone about the same variations of temperature, so that according to it genial or glacial epochs must have occurred simultaneously on the whole earth. Because of the greater nebulosity of the Southern hemisphere, the variations must there have been a little less (about 15 per cent.) than in the Northern hemisphere. The ocean currents, too, must there, as at the present time, have effaced the differences in temperature at different latitudes to a greater extent than in the Northern hemisphere. This effect also results from the greater nebulosity in the Arctic zones than in the neighbourhood of the equator.

There is now an important question which should be answered, namely: — Is it probable that such great variations in the quantity of carbonic acid as our theory requires have occurred in relatively short geological times? The answer to this question is given by Prof. Högbom. As his memoir on this question may not be accessible to most readers of these pages, I have summed up and translated his utterances which are of most importance to our subject:

> Although it is not possible to obtain exact quantitative expressions for the reactions in nature by which carbonic acid is developed or consumed, nevertheless there are some factors, of which one may get an approximately true estimate, and from which certain conclusions that throw light on the question may be drawn. In the first place, it seems to be of importance to compare the quantity of carbonic acid now present in the air with the quantities that are being transformed. If the former is insignificant in comparison with the latter, then the probability for variations is wholly other than in the opposite case.
>
> On the supposition that the mean quantity of carbonic acid in the air reaches 0.03 vol. per cent., this number represents 0.045 per cent. by weight, or 0.342 millim. partial pressure, or 0.466 gramme of carbonic acid for every cm.^{2} of the earth's surface. The quantity of carbon that is fixed in the living organic world can certainly not be estimated with the same degree of exactness; but it is evident that the numbers that might express this quantity ought to be of the same order of magnitude, so that the carbon in the air can neither be conceived of as very great nor as very

little, in comparison with the quantity of carbon occurring in organisms. With regard to the great rapidity with which the transformation in organic nature proceeds, the disposable quantity of carbonic acid is not so excessive that changes caused by climatological or other reasons in the velocity and value of that transformation might be not able to cause displacements of the equilibrium.

The following calculation is also very instructive for the appreciation of the relation between the quantity of carbonic acid in the air and the quantities that are transformed. The world's present production of coal reaches in round numbers 500 millions of tons per annum, or 1 ton per km.[2] of the earth's surface. Transformed into carbonic acid, this quantity would correspond to about a thousandth part of the carbonic acid in the atmosphere. It represents a layer of limestone of 0.003 mm. thickness over the whole globe, or 1.5 km.[3] in cubic measure. This quantity of carbonic acid, which is supplied to the atmosphere chiefly by modern industry, may be regarded as completely compensating the quantity of carbonic acid that is consumed in the formation of limestone (or other mineral carbonates) by the weathering or decomposition of silicates. From the determination of the amounts of dissolved substances, especially carbonates, in a number of rivers in different countries and climates, and of the quantity of water flowing in these rivers and of their drainage-surface compared with the land-surface of the globe, it is estimated that the quantities of dissolved carbonates that are supplied to the ocean in the course of a year reach at most the bulk of 3 km.[3] As it is also proved that the rivers the drainage regions of which consist of silicates convey very unimportant quantities of carbonates compared with those that flow through limestone regions, it is permissible to draw the conclusion, which is also strengthened by other reasons, that only an insignificant part of these 3 km.[3] of carbonates is formed directly by decomposition of silicates. In other words, only an unimportant part of this quantity of carbonate of lime can be derived from the process of weathering in a year. Even though the number given were on account of inexact or uncertain assumptions erroneous to the extent of 50 per cent. or more, the comparison instituted is of very great interest, as it proves that the most important of all the processes by means of which carbonic acid has been removed from the atmosphere in all times, namely the chemical weathering of siliceous minerals, is of the same order of magnitude as a process of contrary effect, which is caused by the industrial development of our time, and which must be conceived of as being of a temporary nature.

In comparison with the quantity of carbonic acid which is fixed in limestone (and other carbonates), the carbonic acid of the air vanishes. With regard to the thickness of sedimentary formations and the great part of them that is formed by limestone and other carbonates, it seems not improbable that the total quantity of carbonates would cover the whole earth's surface to a height of hundreds of metres. If we assume 100 metres,—a number that may be inexact in a high degree, but probably is underestimated,—we find that about 25,000 times as much carbonic acid is fixed to lime in the sedimentary formations as exists free in the air. Every molecule of carbonic acid in this mass of limestone has, however, existed in and passed through the atmosphere in the course of time. Although we neglect all other factors which may have influenced the quantity of carbonic acid in the air, this number lends but very slight probability to the hypothesis, that this quantity should in former geological epochs have changed within limits which do not differ much from the present amount. As the process of weathering has consumed quantities of carbonic acid many thousand times greater than the amount now disposable in the air, and as this process from different geographical, climatological and other causes has in all likelihood proceeded with very different intensity at different epochs, the probability of important variations in the quantity of carbonic acid seems to be very great, even if we take into account the compensating processes which, as we shall see in what follows, are called forth as soon as, for one reason or another, the production or consumption of carbonic acid tends to displace the equilibrium to any considerable degree. One often hears the opinion expressed, that the quantity of carbonic acid in the air ought to have been very much greater formerly than now, and that the diminution should arise from the circumstance that carbonic acid has been taken from the air and stored in the earth's crust in the form of coal and carbonates. In many cases this hypothetical diminution is ascribed only to the formation of coal, whilst the much more important formation of carbonates is wholly overlooked. This whole method of reasoning on a continuous diminution of the carbonic acid in the air loses all foundation in fact, notwithstanding that enormous quantities of carbonic acid in the course of time have been fixed in carbonates, if we consider more closely the processes by means of which carbonic acid has in all times been supplied to the atmosphere. From these we may well conclude that enormous variations have occurred, but not that the variation has always proceeded in the same direction.

Carbonic acid is supplied to the atmosphere by the following processes:—(1) volcanic exhalations and geological phenomena connected
therewith; (2) combustion of carbonaceous meteorites in the higher
regions of the atmosphere; (3) combustion and decay of organic bodies; (4) decomposition of carbonates; (5) liberation of carbonic acid
mechanically inclosed in minerals on their fracture or decomposition. The carbonic acid of the air is consumed chiefly by the following
processes:—(6) formation of carbonates from silicates on weathering;
and (7) the consumption of carbonic acid by vegetative processes. The
ocean, too, plays an important rôle as a regulator of the quantity of carbonic acid in the air by means of the absorptive power of its water, which
gives off carbonic acid as its temperature rises and absorbs it as it cools.
The processes named under (4) and (5) are of little significance, so that
they may be omitted. So too the processes (3) and (7), for the circulation
of matter in the organic world goes on so rapidly that their variations
cannot have any sensible influence. From this we must except periods
in which great quantities of organisms were stored up in sedimentary
formations and thus subtracted from the circulation, or in which such
stored-up products were, as now, introduced anew into the circulation.
The source of carbonic acid named in (2) is wholly incalculable.

Thus the processes (1), (2), and (6) chiefly remain as balancing
each other. As the enormous quantities of carbonic acid (representing
a pressure of many atmospheres) that are now fixed in the limestone of
the earth's crust cannot be conceived to have existed in the air but as
an insignificant fraction of the whole at any one time since organic life
appeared on the globe, and since therefore the consumption through
weathering and formation of carbonates must have been compensated
by means of continuous supply, we must regard volcanic exhalations as
the chief source of carbonic acid for the atmosphere.

But this source has not flowed regularly and uniformly. Just as single volcanoes have their periods of variation with alternating relative rest
and intense activity, in the same manner the globe as a whole seems in
certain geological epochs to have exhibited a more violent and general
volcanic activity, whilst other epochs have been marked by a comparative quiescence of the volcanic forces. It seems therefore probable that
the quantity of carbonic acid in the air has undergone nearly simultaneous variations, or at least that this factor has had an important influence.

If we pass the above-mentioned processes for consuming and producing carbonic acid under review, we find that they evidently do not

stand in such a relation to or dependence on one another that any probability exists for the permanence of an equilibrium of the carbonic acid
in the atmosphere. An increase or decrease of the supply continued during geological periods must, although it may not be important, conduce
to remarkable alterations of the quantity of carbonic acid in the air, and
there is no conceivable hindrance to imagining that this might in a certain geological period have been several times greater, or on the other
hand considerably less, than now. (Högbom 1894: 169ff)

As the question of the probability of quantitative variation of the carbonic
acid in the atmosphere is in the most decided manner answered by Prof. Högbom, there remains only one other point to which I wish to draw attention in a
few words, namely: Has no one hitherto proposed any acceptable explanation for
the occurrence of genial and glacial periods? Fortunately, during the progress of
the foregoing calculations, a memoir was published by the distinguished Italian
meteorologist L. De Marchi which relieves me from answering the last question.
He examined in detail the different theories hitherto proposed — astronomical,
physical, or geographical, and of these I here give a short *résumé*. These theories
assert that the occurrence of genial or glacial epochs should depend on one or
other change in the following circumstances:

(1) The temperature of the earth's place in space.
(2) The sun's radiation to the earth (solar constant).
(3) The obliquity of the earth's axis to the ecliptic.
(4) The position of the poles on the earth's surface.
(5) The form of the earth's orbit, especially its eccentricity (Croll).
(6) The shape and extension of continents and oceans.
(7) The covering of the earth's surface (vegetation).
(8) The direction of the oceanic and aërial currents.
(9) The position of the equinoxes.

De Marchi arrives at the conclusion that all these hypotheses must be rejected (De Marchi 1895: 207). On the other hand, he is of the opinion that a
change in the transparency of the atmosphere would possibly give the desired
effect. According to his calculations,

a lowering of this transparency would effect a lowering of the temperature on the whole earth, slight in the equatorial regions, and increasing with the latitude into the 70th parallel, nearer the poles again a

little less. Further, this lowering would, in non-tropical regions, be less on the continents than on the ocean and would diminish the annual variations of the temperature. This diminution of the air's transparency ought chiefly to be attributed to a greater quantity of aqueous vapour in the air, which would cause not only a direct cooling but also copious precipitation of water and snow on the continents. The origin of this greater quantity of water-vapour is not easy to explain. (De Marchi 1895: 207ff)

De Marchi has arrived at wholly other results than myself, because he has not sufficiently considered the important quality of selective absorption which is possessed by aqueous vapour. And, further, he has forgotten that if aqueous vapour is supplied to the atmosphere, it will be condensed till the former condition is reached, if no other change has taken place. As we have seen, the mean relative humidity between the 40th and 60th parallels on the northern hemisphere is 76 per cent. If, then, the mean temperature sank from its actual value +5.3 by 4°–5°C., *i. e.* to +1.3 or +0.3, and the aqueous vapour remained in the air, the relative humidity would increase to 101 or 105 per cent. This is of course impossible, for the relative humidity cannot exceed 100 per cent. in the free air. A *fortiori* it is impossible to assume that the absolute humidity could have been greater than now in the glacial epoch.

As the hypothesis of Croll still seems to enjoy a certain favour with English geologists, it may not be without interest to cite the utterance of De Marchi on this theory, which he, in accordance with its importance, has examined more in detail than the others. He says, and I entirely agree with him on this point:— "Now I think I may conclude that from the point of view of climatology or meteorology, in the present state of these sciences, the hypothesis of Croll seems to be wholly untenable as well in its principles as in its consequences" (De Marchi 1895: 166ff).

It seems that the great advantage which Croll's hypothesis promised to geologists, viz. of giving them a natural chronology, predisposed them in favour of its acceptance. But this circumstance, which at first appeared advantageous, seems with the advance of investigation rather to militate against the theory, because it becomes more and more impossible to reconcile the chronology demanded by Croll's hypothesis with the facts of observation.

I trust that after what has been said the theory proposed in the foregoing pages will prove useful in explaining some points in geological climatology which have hitherto proved most difficult to interpret.

References

[*The list includes only references cited in the excerpt.*]

De Marchi, Luigi. 1895. *Le cause dell' era glaciale*, premiato dal R. Pavia: Istituto Lombardo.

Högbom, A. G.1894. *Svensk Kemisk Tidskrift*, vol. vi.

Langley, Samuel P. 1884. "Professional Papers of the Signal Service," No. 15. *Researches on Solar Heat*, Washington.

Langley, Samuel P. 1890. "The temperature of the moon," *Memoirs of the National Academy of Sciences*, vol. iv, 9th mem.

Commentary

Svante Arrhenius, "On the Influence of Carbonic Acid in the
Air upon the Temperature of the Ground" (1896)

SVERKER SÖRLIN

Svante Arrhenius's paper on the influence of carbonic acid on global climate,
published in 1896 but researched during 1895 and read to the Royal Swedish
Academy of Sciences in December of that year, has become an almost iconic
reference as climate change has become one of the central tropes of current
Weltanschauung in the late twentieth and early twenty-first centuries. The theory
of the greenhouse effect was far from new. It was proposed by Joseph Fourier in
1827 and much discussed in the nineteenth century—for example, by John Tyndall, who emphasized the role of water vapor.

Arrhenius's own contribution to discussions of the greenhouse effect were
not revived until the second half of the twentieth century. His work on this
problem in 1895–1896 was not driven by any attempt to understand anthropogenic global climatic warming, but rather the opposite—namely, to understand
the mechanisms behind ice ages, a central concern of Scandinavian geophysi-
cists. The immediate inspiration came from conversations with geologist Arvid
Gustaf Högbom at Uppsala University, who was a patriotic man of the north
who emphasized the defining features of Sweden as the northern industrial resources, minerals, and forests, and the soil and agricultural conditions that had
been shaped by the ice age—thus, by cold. Ironically, one of Arrhenius's chief
arguments to undertake his calculations was "the fact," "well established" from
"geological researches," that "in Tertiary times there existed a vegetation and an
animal life in the temperate and Arctic zones that must have been conditioned
by a much higher temperature than the present in the same regions" (267). This
was of course written long before Alfred Wegener's theory of continental drift,
first proposed in 1912, let alone its confirmation by the knowledge of seafloor
spreading and plate tectonics in the 1960s. Arrhenius, despite being correct in his
assumption of large prehistoric temperature variations, especially in the Arctic
(now evidenced also by ice cores; see Broecker essay), had no idea that the carbon layers of the Arctic had been transported there over many millions of years
rather than grown on site. So, not only was Arrhenius interested in explaining
global (he cited data from all continents) cooling rather than global warming, he
was also wrong in a cornerstone of his reasoning.

On another crucial point Arrhenius was also off the mark. His theory needed,
of course, an explanation of prehistoric variations of carbon dioxide in the atmosphere, and that explanation he also found in Högbom, who argued that over

long periods of time there would be a minuscule yet significant release of carbonic acid from the Earth's surface minerals. He quotes Högbom's argument at length in his paper.

Still, the essential nature of the problem he addressed remained the same as if he had chosen twenty-first-century knowledge as his point of departure. His calculations of temperature variations following changes in atmospheric carbon dioxide led him to more or less the same range of changes as the ones we have today. He also weighed in the importance of factors such as the reflection of light and energy from snow, now known as the albedo effect. His paper has thus remained cited, despite the large number of papers over the years addressing either atmospheric carbon dioxide or explanations of climate change, or both: Arrhenius made calculations that could be turned into a *predictive* mode of the kind that became dominant when the carbon dioxide hypothesis returned in the second half of the twentieth century. Arrhenius's greenhouse paper was almost completely overlooked until 1938, when Guy Stewart Callendar published a paper on human climate forcing, using the greenhouse connection as a point of departure.

Since the 1970s, Arrhenius's theory has been less valued for its precision, and more because it raises questions about the size of the contribution of anthropogenic greenhouse gases on measured increases in global temperatures. From the late 1980s Arrhenius became more famous as the founding figure of an understanding of global warming than for his discovery of electrolytic dissociation, for which he was awarded the Nobel Prize in 1903. A common feature of the two theories, however, is that they were controversial: ionic theory during his lifetime, the greenhouse effect long after his death. Indeed, almost all important work that Arrhenius did, and some of his less important work as well, drew attention and controversy. He fell out with his erstwhile friend Walther Nernst and continued to contest Nernst's Nobel Prize, awarded in 1920, until that was no longer possible. He quarrelled incessantly with Frankfurt immunologist Paul Ehrlich (1854–1915), another Nobel Prize winner, who was not able to follow Arrhenius's mathematical method, and was more interested in therapy than in theory. Many came to question and even to dislike Arrhenius because he moved easily between chemistry, biology, and physiology, using the tools of physical chemistry. But his fame and standing as a central figure in the expanding power centers of Stockholm made him a local and national celebrity even before he received the Nobel Prize. He was a man who seemed to thrive in battle; it released his energies, it lent eloquence to his vitriolic polemics, and it provided him stamina for fourteen-hour workdays over months and years.

Arrhenius's research style was as expansive as his personality, and after his protracted controversies around the ionic theory he cast his net even wider. He was easily diverted by new inspirations or apparently random contacts or proposals. Much of this newer work was in cosmic physics, a field where already in 1903 he was able to publish a thousand-page *Lehrbuch der kosmischen Physik*. He was inspired by his Stockholm colleagues, including Otto Pettersson, Vilhelm Bjerknes, and numerous other scientists who contributed to the development of inventories of natural resources in northern Sweden, and also by a series of research expeditions to the Arctic. He was himself a member of an expedition to Spitsbergen in 1896, traveling as a "hydrographer." In this way Arrhenius brought his particular skills to bear on a range of new scientific ideas. Some of these ideas, on volcanoes, physiology, serology, and a multitude of other interests, proved short-lived, and some proved marginal or were even considered whimsical, such as his belief in the transportation of living spores (*panspermy*) from outer space to Earth. However, his theory of the influence of carbonic acid on global temperatures, which was also just a passing interest for a few months and based on loose grounds indeed, would in the end prove far from marginal.

Further Reading

Crawford, Elisabeth. 1996. *Arrhenius: From ionic theory to the greenhouse effect*. Cambridge, MA: Science History.

Sörlin, Sverker. 2002. "Rituals and resources of natural history: The north and the Arctic in Swedish scientific nationalism." In Michael T. Bravo and Sverker Sörlin (eds.), *Narrating the Arctic: A cultural history of Nordic scientific practices*. Cambridge, MA: Science History.

Weart, Spencer. (2003) 2008. *The discovery of global warming*, new ed. Cambridge, MA: Harvard University Press.

Seasonal Foreshadowing

GILBERT T. WALKER

[Read at a meeting of the Society on May 21, 1930.]

[359] 1. Statistical methods have been applied to the discovery of relation-ships between weather in many parts of the world, but although the number of coefficients worked out is of the order of ten thousand, satisfactory formulae for predicting the character of seasons have been worked out in very few countries. An effort was made by the author in 1908 in connection with the monsoon rain-fall of India and Australia and the Nile floods, and methods promising greater reliability have since been developed for the summer and winter rainfall of India. Formulae for Southern Rhodesia have also been produced by C. L. Robertson, for Japan by T. Okada, and for various parts of Europe by W. Wiese and F. Baur. The tables of general relationships of world weather published in India[1] and in our *Memoirs*[2] have as yet only been applied by Bliss to the Nile[3] and the river Parana in Brazil (each with a coefficient of .72, *i.e.*, controlling 72 per cent of the variations), and by myself to the Ceara region in Brazil (with a coefficient of .82). There are many parts of the world in which indications of abnormal seasons would have great value, so that it was an obvious duty to try what assistance the new information could offer; and the present paper contains some of the results recently obtained by Mr. Bliss and the author.[3]

2. In Australia some important rain is brought by winter depressions to its western and southern regions, but the chief source is the summer monsoon in the northern and eastern coastal districts. Accordingly the simpler problem has been tackled first, the total summer rainfall of the Kimberley division of West Australia, Northern Territory and Queensland. The data of 28 stations have been

G. T. Walker. 1930. "Seasonal foreshadowing." *Quarterly Journal of the Royal Meteoro-logical Society* 56(237):359–364.

1. *Memoirs of the Indian Meteorological Department*, 24, Pt. 4 (1923) and Pt. 9 (1924).
2. Vol. 2, No. 17 (1928).
3. For details reference may be made to *Memoir No.* 24.

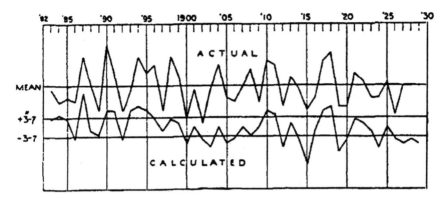

Fig. 1. Departure from normal of Australian rainfall, October to April. R = 79.

used, and the period of rainfall is that of October to April. An abundant monsoon tends to be preceded by high pressure at Honolulu and South America, and by low pressure in Northern Australia;[4] the number of years on which the conclusions rest is 45, and the coefficient of the resulting formula is .79, that of the 1908 formula being .61. In the present result no account is taken of the prejudicial effect of abundant rain in Southern Rhodesia of the previous year, as based on data of 28 years; or of effects due to periodicity which, as H. A. Hunt tells [360] me in a recent letter, are of great importance. It is to be hoped, therefore, that when these matters have been thoroughly sifted a coefficient well over .8 will result.

A comparison with the actual rainfall of the results that would have been given since 1871 by the 79 formula is given in Fig. 1; in the earlier years only eight rainfall stations have been used, but for more recent years the number has been 28. Of 28 years in which the indicated excess or defect exceeded 3.7 the indication would have had the right sign 24 times, two would have been failures, and two years had normal rainfall.

3. In South Africa as in Australia the south-west portion has a winter rainy season, and although the rainfall is hard to predict the winter temperature is controlled to an extraordinary extent by that a year before in Drake's Straits as measured at the island of Ano Nuevo. The coefficient between the two temperatures is .84, and the explanation probably is that during the winter months the ocean temperature controls the air temperature, and the water flowing eastward between Cape Horn and the Antarctic takes about a year to reach the Cape of

4. It may be noted that the three factors used in 1908 have all been confirmed by an additional 20 years of data.

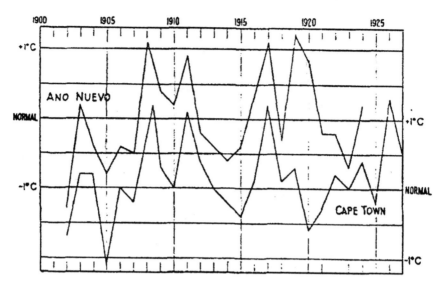

Fig. 2. Departures from normal of Ano Nuevo temperature, June to August, and of Cape Town temperature, June to August, of following year.

Good Hope. The control is the more remarkable in view of the smallness of the seasonal variations, [361] the departure from normal being usually less than 1°C. The past history is shown[3] in Fig. 2.

The summer rainfall of South Africa appears to be provided mainly by moisture from the Indian Ocean; but the rainfall of Natal is determined by causes other than those controlling the surrounding region—the southern Transvaal, the Orange Free State, the Cape Province, except the south-west, and the east of British Bechuanaland. If we adopt 15 of the oldest rainfall stations as representing this area we find that a wet season from October to April tends to be preceded by high pressure at Honolulu, low pressure in Northern Australia, an abundant monsoon in India, and a high Nile. The joint coefficient is ·58, which is less good than for some other countries. If, however, the first eight years, of which the rainfall data are incomplete, are ignored, the coefficient between the actual and the indicated values over the remaining 40 years is ·72. The effect of the formula in the past is given in Fig. 3.

4. The winter temperature of south-west Canada (represented by Winnipeg, Calgary, Edmonton, Prince Albert and Qu'Appelle) proves to be more closely related with known factors than is the mean of the temperatures of Winnipeg, St. Louis and St. Paul, which had previously been taken to represent North America. A warm winter in south-west Canada tends to be preceded by low pressure at

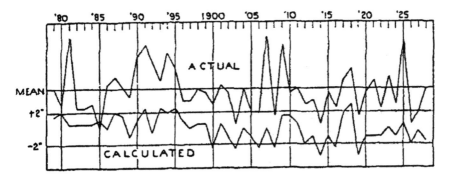

Fig. 3. Departure from normal of South African rainfall, October to April, R = 72.

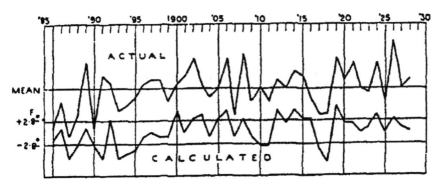

Fig. 4. Departure from normal of winter temperature in south-west Canada, R = 72.

Honolulu and high pressure in Northern Australia, scanty Indian monsoon rains, and a low Nile flood, and high temperature at Madras. The joint coefficient is ·73, and the value for prediction may be gathered from Fig. 4.

[362] For north-west Canada as represented by Dawson the conditions are somewhat different, conditions at Zanzibar now playing a part in addition to those at Honolulu, South America, and Northern Australia. Here also the coefficient is ·72, and the agreement is shown in Fig. 5.

5. It must be admitted that although statistical methods must afford some information as to the likelihood of covering seasonal weather, one or two departments that have tried them have afterwards lost confidence and ceased to issue forecasts. The reason is, I believe, that if the rainfall of a country has a mean departure of, say, 4½in. the probable departure will be about 3in. so that the rainfall is as likely as not to exceed 3in.; and if the forecasting formula has a coefficient

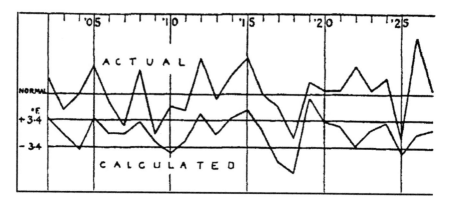

Fig. 5. Departure from normal of Dawson temperature, December to February, $R = 72$.

of ·75 the probable error of the forecast departure will still be 2in. Hence it is unsafe to forecast normal rainfall when a minute departure is indicated, for the actual departure is as likely as not to exceed 2in., and if it is −2in. the lay critic wilt say that the deficiency was not indicated and the forecast was wrong. But if the forecasting formula points to a defect of 2in. a forecast of a deficiency will be right three times out of four; while an indicated deficiency of 2½in. will precede an actual deficiency four times out of five. In general the object of predictions is to assist the layman, and it is the opinion formed by him that decides whether they succeed or fail. Hence I regard it as foolish to issue a prediction except in years when the indications of an excess or defect are so strongly marked as to give a 4:1 chance of success. Even with a coefficient of ·75 this will occur only in about half the years, and as the claim to "forecast" the seasons arouses the expectation of an annual prediction, I advocate the word "foreshadow" as expressing a smaller ambition.

Discussion.

Sir GILBERT WALKER began the discussion by remarking that since the paper was written the arrival of further data had enabled him to put the formulae to a partial test. In order to make the comparison of the predicted departures with the actuals as comprehensive as possible he had added the results of previously published formulæ for the Indian monsoon, the heights of the Parana river and the Nile, and the Ceara rainfall, including only those years in which the indications were sufficient [363] definite to justify an estimate of 4:1 as the probability of correctness in the prediction. Out of 12 of such cases 6 foreshadowings would have been right, 5 wrong and 1 neutral. This disappointing experience would be explained if an extremely low quarterly pressure reading at Port Darwin were due to the passage over it of one or two cyclones; but the clue is probably to be found in the fact that the 3 or 4 factors used in forecasting are selected for their size from the significant factors (all much larger than the probable greatest produced by chance) and hence R instead of being, say ·75, is in reality probably between ·65 and ·7. It seems wiser to adopt the smaller value of R, and so there must be a larger indicated departure to justify the issue of a prediction, namely ·63 times the standard deviation instead of ·56 times it. Accordingly instead of 12 cases there are only 8, out of which 5 are right, 2 wrong and 1 neutral. The effect of the change is of the right order of magnitude, and he proposed to adopt it in future work.

Sir RICHARD GREGORY: There is little I can say upon any specific points in Sir Gilbert Walker's paper, but in common with other Fellows of the Society I am glad he has brought together this digest of recent work done by Mr. Bliss and himself upon some interesting weather relationships. His studies of such relationships as a means of predicting the characters of monsoons in India are well known, and the extension of these to other countries is leading to results of practical as well as scientific value. As I understand him, he suggests that predictions should not be made unless the odds as shown by the forecasting formula are decidedly in their favour. No doubt this is the surest plan of securing public confidence, for the press is more inclined to direct attention to failures than to successes, however sound the principles of arriving at correlation coefficients may be. In some cases, as, for example, that of winter temperatures near Cape Horn and at Cape Town a year later an explanation of the relationship is forthcoming, but in others the physiographic cause of the connexion is not so obvious. Notwithstanding this, Sir Gilbert Walker's patient and persistent inquiries into world weather by precise statistical methods are notable contributions to meteorological science, and he is to be congratulated upon deriving from them some very striking relationships.

Colonel GOLD said he thought it would add to the interest of the paper if Sir Gilbert Walker could supplement his statement of the number of occasions when his calculated figures justified the issue of a forecast by a statement of the number of occasions on which the actual departures were greater than his limiting values, although the calculated differences did not justify the issue of a forecast. He suggested that the name "foreshadowed" was not a very satisfactory term for the purpose for which it was introduced, viz., the issue of a forecast on occasions when the calculated differences were in excess of certain values. The

statements on those occasions would be essentially "forecasts," and he depre-
cated the introduction of a new imperfect word where the existing word could
be reasonably applied.

Sir NAPIER SHAW asked whether any account could be taken of the character
of the variation of the elements within the specified periods.

Professor ALTER: The amount by which the multiple correlations are too
high, due to selection of most highly related stations from among a large number
whose correlations are too low to be of value in prediction, is a function of R and
of the number of stations thus considered, either explicitly or implicitly. It would
seem best to issue each [364] year a prediction of maximum and minimum limits,
between which the rainfall may be expected. A year when the correlation factor
shows a normal expectation is as valuable as any other.

Sir GILBERT WALKER in reply to the remarks of Colonel Gold emphasised
the danger of arousing expectations that would be disappointed. If he claimed to
forecast the character of the seasons the uninformed public would be frequently
disappointed by a statement that a forecast could not be made; and he thought
it better to introduce a less ambitious title to express definitely less ambitious
intention.

In reply to the request for supplementary information there were in all
21 years for comparison of the Indian monsoon 1908 formula, for which R was
only ·58. In these there were 4 calculated departures exceeding the limit of 2·3in.,
and there were 7 years in which the actual departure exceeded 2·3in., when no
prediction would have been made. For the remaining formulae, with R about
·75 in value, there were 19 seasons, with 4 forecasts and 6 in which the actual
departures exceeded the limit without any prediction being made. But probably
Colonel Gold would admit that his demands were high, and that there would be
no cause for serious complaint unless a departure of, say, $5/4 \times$ (standard deviation) was missed as such departures would on the average occur about once in 5
years. With this limit there were 2 cases missed for India out of 21 and 2 cases out
of 19 elsewhere.

With reference to Sir Napier Shaw's question experience seemed to show
that it was more difficult to forecast for a portion of a season than for the whole.

In reply to Prof. Dinsmore Alter the selection is not from a large number of
unrelated factors but from a few significant ones with relationships of ·45 or over.
Sir Gilbert did not know of any simple way of handling the subject mathematically, and had to trust to instinct. With regard to forecasting the limits within
which rainfall might be expected to fall, the kind of year over which there is difficulty is one of normal outlook. If R is ·75 as calculated, and is therefore about
·65 presumably in reality, the mean of the error of the forecast amounts is ·76s,

where s is the s.d. of the original rainfall. The prediction would therefore be that there is a 4:1 chance of the departure not exceeding 1·15 times this, or ·87s. But if we had not any forecast at all we could still say that there was a 4:1 chance of the departure not exceeding 1·15s; and the man in the street would, in my view, be inclined to say that under these conditions a limit of ·87s was of little value to him.

Commentary

Gilbert T. Walker, "Seasonal Foreshadowing" (1930)

NEVILLE NICHOLLS

Sir Gilbert Walker is regarded by practicing meteorologists as the "father" of scientific seasonal-to-interannual climate prediction (or long-range weather forecasting). This short paper provides an overview of his voluminous contributions in this area, and their potential for operational prediction. Of course, such a brief paper cannot do justice to the huge amount of work that Walker completed, but it provides a roadmap to this work and its importance.

The discussion included at the end of Walker's paper by eminent contemporary meteorologists illustrates the high regard in which he was held then and since. What did he do to earn this high regard? Walker was the third person appointed to be head of the India Meteorological Department. The first person to hold the post was Henry Blanford, appointed in 1875, just two years before what we now would call a very strong El Niño event (1877). There was a drought, severe food shortages, and famine across India. Blanford noticed that pressures were high across India and in some other locations of the world. He corresponded with government meteorologists in many parts of the world, asking them about the atmospheric pressures in 1877, and found that pressures were high right across the Indian Ocean and surrounding countries (including Australia). This is one aspect of what Walker came to call the Southern Oscillation, a phenomenon that he would later document more comprehensively.

Blanford looked for various ways of forecasting Indian monsoon rainfall, checking the solar cycle and even snowfall early in the year in the Himalayas as predictors, with little success. His successor, John Eliot, widened the scope, looking for possible precursors to Indian monsoon rainfall in the trade winds over the Indian Ocean, and data from around the world. Walker was appointed to the post of director in 1903, and subsequently expanded on the work of his two predecessors, adopting a more systematic approach. It was this work that revealed the "Southern Oscillation," a seesaw in pressure between the Indian and Pacific Oceans. This phenomenon has a long lifetime: extremes or phases of the phenomenon tend to last about twelve months. Because it is linked to climate variations across lands bordering these two oceans, it provided the foundation of scientific "seasonal-to-interannual" prediction. From 1908, Walker correlated seasonal averages of rainfall, temperature, and pressure measured at many locations across the world, using data provided from a wide range of places. This was the privilege of his position at a hub of the British Empire. He published his results in the *Indian Meteorological Memoirs* and the *Quarterly Journal of the Royal Meteorological Society*.

Walker's interest was in predicting the Indian monsoon rainfall, the rainfall critical to feeding India. He calculated statistical lag relationships of the rainfall with antecedent climate variations within and outside India. He soon found that the better relationships between seasonal conditions occurred inconveniently *after* the peak of the Indian monsoon. Thus the Indian monsoon rainfall was a better predictor of subsequent conditions elsewhere. We now at least partly understand that this is because the extreme phase usually starts around May, near the time of onset of the Indian monsoon. Indices of the Southern Oscillation can be used to predict climate variables that occur later in the year or even at the start of the following calendar year.

The first result Walker notes in this paper is that abundant summer rainfall over northern Australia is generally preceded by high pressures at Honolulu and South America, and low pressure over northern Australia. Such situations are nowadays referred to as La Niña events.

Walker's paper also reported that temperature and rainfall variations in some other parts of the world could be predicted using similar statistical relationships associated with the Southern Oscillation. One example in this paper is that warm winters in southwest Canada tend to follow El Niño events.

Despite the strong lag relationships Walker found between climate variables in many parts of the world, he was nervous about overselling them as a way to forecast the climate. He advocated the term "foreshadow," rather than "forecast," to reduce expectations. Even with the strong relationships Walker uncovered, he knew that these would not result in accurate forecasts in every instance. This issue still plagues seasonal climate prediction today.

Although Walker's work was appreciated at the time, particularly by his successors in the field, it did not lead directly to operational seasonal prediction across the world. This was partly because some of the relationships he found turned out to be less stable than he had hoped, but also because there was no physical understanding of the mechanism for the Southern Oscillation. It was not until the 1980s when the Southern Oscillation was linked empirically with the oceanic phenomenon known as El Niño, and computer models were developed that could reproduce some of its behavior, that seasonal climate prediction became accepted by meteorologists.

Walker was appointed professor of meteorology at the Imperial College of Science and Technology in 1924, was elected president of the Royal Meteorological Society in 1926, and edited the *Quarterly Journal* from 1935 to 1941. He died on 4 November 1958, leaving a lasting legacy of improved understanding of how the world's climates are intertwined. The Walker Circulation, a vertical circulation of wind across the equatorial Pacific, was named after him, because of the complex of interrelationships he documented.

It is rare that a single individual can be considered to have spawned an entire field. Although Walker's work expanded on hints from earlier scientists, it was his careful and comprehensive documentation of the spatial and seasonal relationships between climate variables across the world that revealed to him the pattern of the Southern Oscillation, and other climate variability. His observations also provided the empirical basis for scientific seasonal climate prediction.

Further Reading

Nicholls, N. 2005. "Climatic outlooks: From revolutionary science to orthodoxy." In T. Sherratt, T. Griffiths, and L. Robin (eds.), *A change in the weather: Climate and culture in Australia*. Canberra: National Museum of Australia.

Sheppard, P. A. 1959. "Obituary of Sir Gilbert Walker, CSI, FRS." *Quarterly Journal of the Royal Meteorological Society* 85:186.

Walker, J. M. 1997. "Pen portrait of Sir Gilbert Walker, CSI, MA, ScD, FRS." *Weather* 52:217–220.

The Artificial Production of Carbon Dioxide and Its Influence on Temperature

G. S. CALLENDAR

(Steam technologist to the British Electrical and Allied Industries Research Association.)
(Communicated by Dr. G. M. B. DOBSON, F.R.S.)
[Manuscript received 19 May 1937—read 16 February 1938.]

Summary

By fuel combustion man has added about 150,000 million tons of carbon dioxide to the air during the past half century. The author estimates from the best available data that approximately three quarters of this has remained in the atmosphere.

The radiation absorption coefficients of carbon dioxide and water vapour are used to show the effect of carbon dioxide on "sky radiation." From this the increase in mean temperature, due to the artificial production of carbon dioxide, is estimated to be at the rate of 0.003°C per year at the present time.

The temperature observations at 200 meteorological stations are used to show that world temperatures have actually increased at an average rate of 0.005°C. per year during the past half century.

Few of those familiar with the natural heat exchanges of the atmosphere, which go into the making of our climates and weather, would be prepared to admit that the activities of man could have any influence upon phenomena of so vast a scale.

In the following paper I hope to show that such influence is not only possible, but is actually occurring at the present time.

G. S. Callendar. 1938. "The artificial production of carbon dioxide and its influence on temperature." *Quarterly Journal of the Royal Meteorological Society* 64(275):223–240.

It is well known that the gas carbon dioxide has certain strong absorption bands in the infra-red region of the spectrum, and when this fact was discovered some 70 years ago it soon led to speculation on the effect which changes in the amount of the gas in the air could have on the temperature of the earth's surface. In view of the much larger quantities and absorbing power of atmospheric water vapour it was concluded that the effect of carbon dioxide was probably negligible, although certain experts, notably Svante Arrhenius and T. C. Chamberlin, dissented from this view.

Of recent years much new knowledge has been accumulated which has a direct bearing upon this problem, and it is now possible to make a reasonable estimate of the effect of carbon dioxide on temperatures, and also of the rate at which the gas accumulates in the atmosphere. Amongst important factors in such calculations may be mentioned the temperature-pressure-alkalinity-CO_2 relation for sea water, determined by C. J. Fox (1909), the vapour pressure-atmospheric radiation relation, observed by A. Angstrom (1918) and others, the absorption spectrum of atmospheric water vapour, observed by Fowle (1918), and a full knowledge of the thermal structure of the atmosphere.

This new knowledge has been used in arriving at the conclusions stated in this paper, but for obvious reasons only those parts having a meteorological character will be referred to here.

I. The Rate of Accumulation of Atmospheric Carbon Dioxide

I have examined a very accurate set of observations (Brown and Escombe, 1905), taken about the year 1900, on the amount of carbon dioxide in the free air, in relation to the weather maps of the period. From them I concluded that the amount of carbon dioxide in the free air of the North Atlantic region, at the beginning of this century, was 2.74 ± 0.05 parts in 10,000 by volume of dry air.

A great many factors which influence the carbon cycle in nature have been examined in order to determine the quantitative relation between the natural movements of this gas and the amounts produced by the combustion of fossil fuel. Such factors included the organic deposit of carbon in swamps, etc., the average rate of fixation of the gas by the carbonisation of alkalies from igneous rocks, and so on. The general conclusion from a somewhat lengthy investigation on the natural movements of carbon dioxide was that there is no geological evidence to show that the *net* offtake of the gas is more than a small fraction of the quantity produced from fuel. (The artificial production at present is about 4,500 million tons per year.)

The effect of solution of the gas by the sea water was next considered, because the sea acts as a giant regulator of carbon dioxide and holds some sixty times as much as the atmosphere. The rate at which the sea water could correct an excess of atmospheric carbon dioxide depends mainly upon the fresh volume of water exposed to the air each year, because equilibrium with the atmospheric gases is only established to a depth of about 200 m. during such a period.

. . .

2. Infra-Red Absorption by Carbon Dioxide and Water Vapour

The loss of heat from the earth's surface and atmosphere is nearly all carried upon wave lengths greater than 4μ, the maximum intensity being at about 10μ.

There have been a great many careful and accurate measurements of the absorption and radiation by various gases in this part of the spectrum, but owing to the very great difficulties attending these observations most of the earlier values were highly conflicting. However, considerable accuracy has now been attained, and from a number of considerations, which cannot be detailed, the values observed by Rubens and Aschkinass (1898), of Germany, are used here for the absorption by carbon dioxide on the longer wave lengths.

For water vapour I have made many comparisons between the measurements of F. E. Fowle (1918), who observed the absorption by atmospheric water vapour, and those of Rubens and of Hettner (1918), who used steam at 1 atmos. pressure for their measurements. These comparisons fully support the conclusion arrived at by Fowle, that the absorption by water vapour as it occurs in the air is less than half as great as that found for steam under laboratory conditions. Perhaps the most powerful support for this conclusion comes from a comparison between the observed and calculated atmospheric radiation, which shows that, for dry and cold air conditions, the absorption exponents found for steam lead to much too high a value for atmospheric radiation.

To return to the absorption by carbon dioxide, the three primary bands given by this gas are at 2.4 to 3.0 μ, 4 to 4.6 μ, and 13 to 16 μ, the latter being much the most important for atmospheric conditions because very little low temperature radiation is carried on the small bands. . . .

3. Sky Radiation

The downward radiation from the sky, excluding the direct and scattered short wave radiation from the sun, is usually called the "sky radiation." Valuable papers on this subject have been published by A. Angstrom (1918), W. H. Dines (1927), Simpson (1928), Brunt (1932), and others, and it is not proposed to refer to it at any length here.

For normal conditions near the earth's surface, with a clear sky the downward radiation varies between three and four fifths of that from the surface, the proportion being greatest when the air is warm and carries much water vapour.

The method used to calculate the sky radiation from the absorption coefficients of water vapour and carbon dioxide is simple but laborious. It consists of dividing the air into horizontal layers of known mean temperature, water vapour, and carbon dioxide content, and summing the absorbing power of these layers on the different wave bands, in conjunction with the spectrum distribution of energy at the surface temperature. In this way the perpendicular component of sky radiation is obtained. . . .

4. The Effect of Carbon Dioxide Upon Sky Radiation

For atmospheric conditions the sky radiation on wave band 13 to 16 μ comes from a mixture of water vapour and carbon dioxide. This is true also for band 4 to 4.7 μ, but the energy here is so small that it may be disregarded in relation to the probable error of the large band absorption.

In the case of mixed gases the absorption for the mixture is equal to the difference between the sum and product of their respective absorptions: —

$$Acw = Ac + Aw - AcAw \qquad . \qquad . \qquad . \qquad (3)$$

This relation is true if the respective absorptions are symmetrical in relation to the energy distribution over the wave band to which they refer.

The observed sky radiation results from the infra-red absorption by variable amounts of water vapour and from 3 parts in 10,000 of carbon dioxide, which are normally present in the air. For temperate conditions at vapour pressure 7.5 mm. Hg. I calculate that 95 per cent of the radiation comes from the water vapour; for arctic conditions the carbon dioxide may supply as much as 15 per cent of the total.

For the purpose required here it is necessary to consider the effect of a change in the amount of carbon dioxide, firstly upon sky radiation, and secondly the effect of changes in the latter upon temperatures.

When radiation takes place from a thick layer of gas, the average depth within that layer from which the radiation comes will depend upon the density of the gas. Thus if the density of the atmospheric carbon dioxide is altered it will alter the altitude from which the sky radiation of this gas originates. An increase of carbon dioxide will lower the mean radiation focus, and because the temperature is higher near the surface the radiation is increased, without allowing for any increased absorption by a greater total thickness of the gas.

The change of sky radiation with carbon dioxide depends largely upon this change in the altitude of the radiation focus, because the present quantity in the atmosphere (equal to a layer of 2 m. at N.T.P.) can absorb nearly the maximum of which this gas is capable. The latter assumption depends upon the exponents used, but it is probable that great thicknesses of carbon dioxide would absorb on other wave lengths besides those of the primary bands. . . .

5. The Relation Between Sky Radiation and Temperature

If the whole surface of the earth is considered as a unit upon which a certain amount of heat falls each day, it is obvious that the mean temperature will depend upon the rate at which this heat can escape by radiation, because no other type of heat exchange is possible. For simplicity the reflection loss from clouds and ice surfaces is assumed to be a constant factor.

The radiation loss from the surface and clouds depends upon the fourth power of the absolute temperature and is proportional to the difference between the surface and sky radiation:—

$$H = \sigma \cdot T^4 \cdot (1 - S) \quad . \qquad . \qquad . \qquad (4)$$

where:

H = radiation heat loss from surface.

σ = radiation constant. $10^{-7} \times 1.18$ cal/cm^2/day.

T = temperature of surface, Abs.

S = sky radiation, as proportion of that from the surface.

Suppose that the sky radiation is changed from S_1 to S_2 whilst H remains constant. Then:—

$$T_2 = T_1 \sqrt[4]{[(1 - S_1)/(1 - S_2)]} \quad . \qquad . \qquad . \qquad (5)$$

From this relation it will be seen that the change of temperature for a given change of sky radiation increases rapidly as the latter approaches that from the surface, it being always assumed that the heat supply and loss are constant.

On the earth the supply of water vapour is unlimited over the greater part of the surface, and the actual mean temperature results from a balance reached between the solar "constant" and the properties of water and air. Thus a change of water vapour, sky radiation and temperature is corrected by a change of cloudiness and atmospheric circulation, the former increasing the reflection loss and thus reducing the effective sun heat.

There is also a further loss owing to the scattering of solar energy by the water molecule.

Small changes of atmospheric carbon dioxide do not affect the amount of sun heat which reaches the surface, because the CO_2 absorption bands lie well outside the wave lengths, 0.25 to 1.5 μ, on which nearly all the sun energy is carried. Consequently a change of sky radiation due to this gas can have its full effect upon low level temperatures, provided it does not increase the temperature *differences* on which the atmospheric circulation depends.

An increase of temperature due to sky radiation will be different from that caused by an increase of sun heat; the latter would tend to increase the temperature differences and atmospheric circulation, and the ultimate rise of temperature should not be in proportion to the change of sun heat.

From the change of sky radiation with carbon dioxide, and from expression (5), the resulting change in surface temperature can be obtained. The relation between atmospheric carbon dioxide and surface temperature is shown in Fig. 2 for the temperate air section.

At first sight one would expect that carbon dioxide would have far more effect upon the temperature of arctic regions where the amount of water vapour is very small; this is true as regards its effect on sky radiation, but not on temperatures, because the ratio dT/dS increases rapidly as the sky radiation approaches that from the surface. The result of these opposing changes is to make the quantitative influence of carbon dioxide on temperature remarkably uniform for the different climate zones of the earth.

. . .

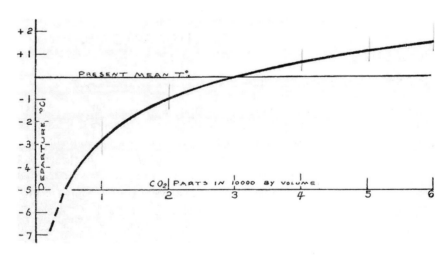

Fig. 2. Change of surface temperature with atmospheric carbon dioxide (H_2O vapour pressure, 75 mm. Hg.)

6. The Observed Temperature Variations on the Earth

Coming now to the actual temperatures which have been observed near the earth's surface during the recent past, these measurements have provided an almost overwhelming mass of statistical detail, including many millions of accurate and standardized readings of temperature. The period to which these standardized observations refer is generally not more than 65 years and often less. It is a matter of opinion whether such a period is sufficiently extended to show a definite trend in world temperatures.

I have relied principally on that valuable Smithsonian publication, "World Weather Records," to obtain the temperature readings summarized here. . . .

In order to represent the temperature anomalies of great regions of the earth's surface I have grouped a number of stations together, and then weighted each group according to the area represented by its stations. In this way the curves for the different zones of the earth, shown in Fig. 2, were obtained.

. . . To return to the world temperature curve, Fig. 2, the dotted line shows the displacement of the mean due to increasing atmospheric carbon dioxide, and it is evident that present temperatures, particularly in the northern hemisphere, are running above the calculated values. The course of world temperatures during the next twenty years should afford valuable evidence as to the accuracy of the calculated effect of atmospheric carbon dioxide.

As regards the long period temperature variations represented by the Ice Ages of the geologically recent past, I have made many calculations to see if the

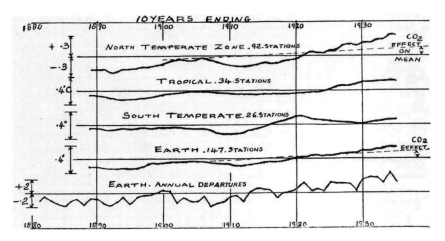

Fig. 4. Temperature variations of the zones and of the earth. Ten-year moving departures from the mean, 1901–1930. °C.

natural movements of carbon dioxide could be rapid enough to account for the great changes of the amount in the atmosphere which would be necessary to give glacial periods with a duration of about 30,000 years. I find it almost impossible to account for movements of the gas of the required order because of the almost inexhaustible supply from the oceans, when its pressure in the air becomes low enough to give a fall of 5 to 8°C. in mean temperatures (see Fig. 2). Of course, if the effect of carbon dioxide on temperatures was considerably greater than supposed, glacial periods might well be accounted for in this way.

In conclusion it may be said that the combustion of fossil fuel, whether it be peat from the surface or oil from 10,000 feet below, is likely to prove beneficial to mankind in several ways, besides the provision of heat and power. For instance the above mentioned small increases of mean temperature would be important at the northern margin of cultivation, and the growth of favourably situated plants is directly proportional to the carbon dioxide pressure (Brown and Escombe, 1905). In any case the return of the deadly glaciers should be delayed indefinitely.

As regards the reserves of fuel these would be sufficient to give at least ten times as much carbon dioxide as there is in the air at present.

References

Angstrom, A.	1918	*Smithson. Misc. Coll.*, **65**, No.3.
Arrhenius, Svante	1903	*Kosmische Physik*, **2**.
Brown, H., and Escombe, F.	1905	*Proc. Roy. Soc.*, B, **76**.
Brunt, D.	1932	*Quart. J.R.Met. Soc.*, **58**.
Carpenter, T. M.	1937	*J. Amer. Chem. Soc.*, **59**.
Dines, W. H.	1927	*Mem. R. Met. Soc.*, **2**, No. 11.
Fox, C. J. B.	1909	*International Council for the Investigation of the Sea. Publications de Circonstance*, No. 44.
Fowle, F. E.	1918	*Smithson. Misc. Coll.*, **68**, No.8.
Hettner, A.	1918	*Ann. Phys., Leipzig*, **55**.
Kincer, J. B.	1933	*Mon. Weath. Rev., Wash.*, **61**.
Mossman, R. C.	1902	*Trans. Roy. Soc., Edin.*, **40**.
Radcliffe Observatory	1930	*Met. Obs.*, **55**.
Rubens, H., and Aschkinass, R.	1898	*Ann. Phys. Chem.*, **64**.
Schmidt, H.	1913	*Ann. Phys., Leipzig*, **42**.
Simpson, G. C.	1928	*Mem. R. Met. Soc.*, **3**, No. 21.
World Weather Records	1927	*Smithson. Misc. Coll.* **79**
World Weather Records	1934	*Smithson. Misc. Coll.* **90**

Commentary

G. S. Callendar, "The Artificial Production of Carbon Dioxide and Its Influence on Temperature" (1938)

JAMES RODGER FLEMING

Beginning in 1938 and continuing throughout his life, British engineer and scientist Guy Stewart Callendar (1898–1964) identified important links between the burning of fossil fuels and global warming. He compiled weather data from stations around the world that clearly indicated a climate warming trend of 0.5 degrees Celsius in the early decades of the twentieth century. He investigated the carbon cycle, including natural and anthropogenic sources and sinks, and the role of glaciers in the Earth's heat budget. His estimate of 290 parts per million (ppm) for the nineteenth-century background concentration of carbon dioxide is still valid, and he documented an increase of 10 percent between 1900 and 1935, which closely matched the amount of fuel burned. Based on new understanding of the infrared spectrum and calculations of the absorption and emission of radiation by trace gases in the atmosphere, Callendar established the carbon dioxide theory of climate change in its recognizably modern form, reviving it from its earlier, physically unrealistic, and moribund status. In 1939 Callendar wrote about the ongoing changes in the atmosphere and their likely effects: "As man is now changing the composition of the atmosphere at a rate which must be very exceptional on the geological time scale, it is natural to seek for the probable effects of such a change. From the best laboratory observations it appears that the principal result of increasing atmospheric carbon dioxide . . . would be a gradual increase in the mean temperature of the colder regions of the earth" (Callendar 1939, 38).

Callendar's landmark studies revived the anthropogenic carbon dioxide theory of climate change. In 1938 he identified links among fuel combustion, rising carbon dioxide levels, increased sky radiation, and the observed rise in world temperatures from two hundred stations: "By fuel combustion man has added about 150,000 million tons of carbon dioxide to the air during the past half century" (Callendar 1938, 223). Callendar estimated from the best available data that approximately three-quarters of this had remained in the atmosphere. He used the radiation absorption coefficients of carbon dioxide and water vapor to show the effect of carbon dioxide on "sky radiation." From this he estimated the increase in mean temperature due to the artificial production of carbon dioxide to be at the rate of 0.003 degrees Celsius per year at the present time. The temperature observations at two hundred meteorological stations were used to show that world temperatures have actually increased at an average rate of 0.005 degrees Celsius per year during the past half-century.

Callendar grouped temperature data in England on a town-by-town basis based on population growth rates. In that way he effectively eliminated the urban heat island effect, which might have skewed his data. Interestingly, Callendar used his home weather station to record an unbroken data set from 1942 to 1964. Callendar's work attracted the attention of leading meteorologists in his day. In 1956 C.-G. Rossby cited it as one of the current problems in meteorology and reproduced Callendar's curve of the increase of carbon dioxide in the atmosphere (Rossby [1956] 1959, 14). Today, the theory that global climate change can be attributed to an enhanced greenhouse effect due to elevated levels of carbon dioxide in the atmosphere from anthropogenic sources, primarily from the combustion of fossil fuels, is called the "Callendar Effect."

Further Reading

Callendar, G. S. 1939. "The composition of the atmosphere through the ages." *Meteorological Magazine* 74:33–39.

Callendar, G. S. 1949. "Can carbon dioxide influence climate?" *Weather* 4:310–314.

Callendar, G. S. 1961. "Temperature fluctuations and trends over the earth." *Quarterly Journal of the Royal Meteorological Society* 87:1–12.

Fleming, James Rodger. 2007. *The Callendar Effect: The life and work of Guy Stewart Callendar (1898–1964), the scientist who established the carbon dioxide theory of climate change*. Boston: American Meteorological Society.

Rossby, C.-G. (1956) 1959. "Aktuella meteorologistka problem." Trans. Svensk Naturvetenskap. "Current problems in meteorology." In Bert Bolin (ed.), *The atmosphere and the sea in motion: Scientific contributions to the Rossby Memorial Volume*. New York: Rockefeller Institute Press and Oxford University Press, 9–50.

Unpleasant Surprises in the Greenhouse?

WALLACE S. BROECKER

[123] The inhabitants of planet Earth are quietly conducting a gigantic environmental experiment. So vast and so sweeping will be the consequences that, were it brought before any responsible council for approval, it would be firmly rejected. Yet it goes on with little interference from any jurisdiction or nation. The experiment in question is the release of CO_2 and other so-called 'greenhouse gases' to the atmosphere. Because these releases are largely by-products of energy and food production, we have little choice but to let the experiment continue. We can perhaps slow its pace by eliminating frivolous production and by making more efficient use of energy from fossil fuels. But beyond this we can only prepare ourselves to cope with its effects.

The task of scientists is to predict the consequences of the build-up of CO_2 and other gases. To be useful these predictions must be reasonably detailed, but we are in no better a position to make them than are medical scientists when asked when and where cancer will strike a specific person. Understanding the operation of the joint hydrosphere–atmosphere–biosphere–cryosphere system is every bit as difficult as understanding the factors that determine whether or not cancerous cells will get the upper hand. Because of our lack of basic knowledge, the range of possibility for the greenhouse effects remains large. It is for this reason that the experiment is a dangerous one. We play Russian roulette with climate, hoping that the future will hold no unpleasant surprises. No one knows what lies in the active chamber of the gun, but I am less optimistic about its contents than many.

My suspicion is that we have been lulled into complacency by model simulations that suggest a gradual warming over a period of about 100 years. If this seemingly logical response to a gradual build-up of greenhouse gases is correct, then one can imagine that man may be able to cope with the coming changes. While I do not have any complaints about how these modelling experiments

W. S. Broecker. 1987. "Unpleasant surprises in the Greenhouse?" *Nature* 328:123–126.

were conducted—indeed they were done by brilliant scientists using the best computers available—the basic architecture of the models denies the possibility of key interactions that occur in the real system. The reason is that we do not yet know how to incorporate such interactions into models.

My impressions are more than educated hunches. They come from viewing the results of experiments nature has conducted on her own. The results of the most recent of them are well portrayed in polar ice, in ocean sediment and in bog mucks. What these records indicate is that Earth's climate does not respond to forcing in a smooth and gradual way. Rather, it responds in sharp jumps which involve large-scale reorganization of Earth's system. If this reading of the natural record is correct, then we must consider the possibility that the main responses of the system to our provocation of the atmosphere will come in jumps whose timing and magnitude are unpredictable. Coping with this type of change is clearly a far more serious matter than coping with a gradual warming.

For more than 100 years scientists have been aware that the Earth is in the midst of a series of cyclic glaciations. Oxygen isotope measurements made on microscopic shells from deep-sea sediments provide a continuous record of these events. They show that cycles averaging 100,000 years in length carried us from warm climates, comparable to today's, to the cold climates of full glacial time. Power spectra for these records point a finger at cycles in the Earth's orbital characteristics as the driving force.[1,2] Seasonality was at times stronger and at times weaker than it is today. Although most scientists now accept the orbital hypothesis, none of the mechanisms that have been proposed to explain the linkage between seasonality and climate is universally accepted. This matter is the subject of much research.

The $^{18}O/^{16}O$ record in deep-sea cores gives the impression that the response of the climatic system to orbital forcing is smooth and gradual (Fig. 1—see over). Only recently have we begun to realize that this impression is a false one. One of the early clues came from studies of the ecology of the remains of planktonic organisms contained in deep-sea cores of rapidly accumulated sediment from the North Atlantic.[3] The ecological changes recorded in these cores probably reflect changes in surface-water temperature. This record does not show the gradual change scientists had become accustomed to. Instead it shows an abrupt end to glacial time and, even more interesting, a brief period of intense cold interrupting the warm period that followed (Fig. 1). Although the two records shown in Fig. 1 are quite different, they are not incompatible. Changes in $^{18}O/^{16}O$ in the shells of marine sediments are largely the result of the waxing and waning of the ^{18}O-deficient continental ice caps. As the response time of global ice caps is thousands of years, the ^{18}O record smooths out the rapid changes in climate.

It took more than this, however, to make us take these abrupt changes seriously. The evidence that turned our heads came from holes drilled through the Greenland ice cap. As a foot or so of ice forms from each year's snowfall, the record captures changes in the ice-cap environment no matter how rapid they have been. These changes are recorded in the ratio of isotopically "heavy" to isotopically "light" water in the ice (a measure of air temperature)[4,5] in the content of particulate matter in the ice (a measure of the dustiness of the air over the ice cap)[6] and in the content of CO_2 and other greenhouse gases contained in the air trapped as bubbles in the ice (a measure of the atmosphere's greenhouse capacity)[7]. Like the ecological record in deep-sea muds from the North Atlantic, the ice core records from Greenland give a dramatically different impression of the manner in which climate changes than do the oxygen isotope records for marine shells. They show that during glacial time climate changed frequently and in great leaps. The typical leap involves a 6°C change in air temperature, a fivefold change in atmospheric dust content and a 20% change in the CO_2 content of air. In cold times the air was dustier and contained less CO_2.

While the record in Greenland's ice shows that climate can change in big leaps, we have to look further for a clue as to why these leaps occur. The last of the events seen in the Greenland record is well documented not only in Greenland ice and in North Atlantic sediment, but also in lake and bog sediments from throughout Western Europe and maritime Canada[8]. On land it is recorded by large shifts in the ecology of plants (as recorded by their pollen grains). During warm periods trees grew in these areas; during cold periods [124] the trees were replaced by tundra shrubs. Although this last of the Greenland events is found in bog and lake sediments throughout northern Europe, it is not seen in similar records from the United States. This geographical distribution suggests the North Atlantic Ocean as the culprit. Based on this clue, we are beginning to see how devilish are the links between components of the climatic system.

Rather than responding smoothly to gradual forcing, the ocean–atmosphere system appears to respond by changing the way it works. India's climate provides an apt analogy. The winters in India are very dry because of the descent of cold air from the Tibetan plateau. Summer heating causes this atmospheric circulation pattern to reverse abruptly. Air rises from the Tibetan plateau, drawing oceanic air across the Indian subcontinent, and the dry conditions give way to monsoonal rains.

The Earth's climatic system currently works in a way beneficial to northern Europe[9]. This region is warmed by heat released from the surface waters of the North Atlantic. The amount is a staggering 30% of that received by the North Atlantic from the Sun! This heat is steadily carried northward, as on a conveyor belt, by the ocean circulation system. In the vicinity of Iceland the warm water meets cold air. The air warms and ameliorates climate on the adjacent land. The water cools, sinks to the abyss and flows as a great river, down the full length of the Atlantic, around Africa, through the southern Indian Ocean and finally up the Pacific Ocean (Fig. 2). This current carries 20 times more water than the world's rivers combined.

In the North Pacific the ocean conveyor belt runs just the other way round. Deep waters move towards the north and rise to the surface and then flow towards the Equator in the upper ocean. So in today's world, the Atlantic conveyor belt carries tropical heat for delivery to the atmosphere at high northern latitudes, while the Pacific conveyor belt forces cold surface waters to move southward, pushing the invading warm waters back towards the Equator.

Why does our ocean operate in this fashion? Although we don't have the complete answer, we do have the first principle. The circulation system is governed by salt. Because of the difference in the circulation patterns, surface waters in the North Atlantic are on average warmer than those in the North Pacific. This allows more water to evaporate from the North Atlantic than from the North Pacific, and in turn gives rise to a net transport of water vapour through the atmosphere from the Atlantic to the Pacific. The North Atlantic is enriched in salt by this process (and the North Pacific waters correspondingly diluted). The enrichment of salt in the North Atlantic must somehow be compensated by a flow of salt through the sea from Atlantic to Pacific. The compensation in today's ocean is achieved by the flow of a deep current of salty water from the Atlantic to the Pacific and a matching flow of correspondingly less salty water around the other way in the upper ocean.

The phenomenon that maintains this situation is, I suspect, a potentially dangerous one. The circulation pattern is self-reinforcing and hence self-stabilizing. The deep current is driven by the extra density supplied to the waters of the North Atlantic by the enrichment of salt. The enrichment of salt is driven by the heat carried by the warm water which flows northward in the upper Atlantic to supply the deep current. A classic chicken and egg situation! Excess evaporation causes the deep current; the deep current causes excess evaporation.

What are the consequences of perturbation of this system? Palaeoclimatic evidence points to a shutdown of the North Atlantic conveyor belt during glacial periods. Such a shutdown would cool the North Atlantic and its adjacent lands by 6–8°C. This in turn would cause the boreal [125] forests in these areas to give way to tundra shrubs. The sparsity of vegetation would permit far more dust to be lifted into the atmosphere. Finally, the only feasible mechanisms scientists have come up with to explain rapid changes in the atmosphere's CO_2 content involve modifications in the ocean's circulation pattern and intensity. Thus we surmise

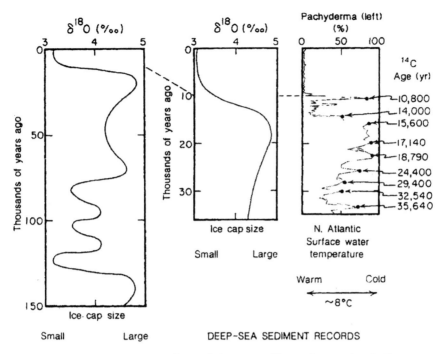

Fig. 1. The ratio of heavy oxygen (^{18}O) to light oxygen (^{16}O) in foraminifera shells preserved in deep-sea sediments provides our best record as to how the great continental ice sheets of the Northern Hemisphere waxed and waned. The reason is that the snow that builds ice caps is depleted in the heavy oxygen isotope. Therefore as ice caps grow, the ocean is correspondingly enriched in heavy oxygen. So also is the oxygen in the calcium carbonate of shells living in these waters.

The oxygen isotope record on the left covers just over one major climate cycle. There have been eight such cycles in the past million years. The diagram in the middle is an enlargement of that on the left, designed to emphasize that these oxygen isotope records indicate a gradual and smooth transition from full glacial to full interglacial conditions. Recent findings demonstrate that this view of natural climate change is misleading. The sluggish retreat of the ice and blurring of the record by burrowing worms combine to give what proves to be the wrong impression.

On the right is the record from a deep-sea sediment obtained west of the British Isles. This record is based on the abundance of the shells of the planktonic herbivore *Pachyderma* (left coiling) which lives in waters adjacent to Greenland. It is the only species of planktonic foraminifera found in these icy waters. During cold periods this species dominated the entire North Atlantic, providing up to 90% of the shells found in the sediments off Britain.

As can be seen, this record tells a different story. The North Atlantic warmed abruptly about 15,000 years ago, and has so remained (except for two brief cold relapses) until the present. It is the abruptness of the warming and of the two brief and intense epochs of renewed cold that are of great interest. The second of these cold epochs was first recognized from study of the pollen records in bogs created in northern Europe during the retreat of the last great ice sheet. These records show that the forests which repopulated this area after ice departed were destroyed by a brief and intensely cold interval of several hundred years' duration. They were replaced by the shrubs of glacial time.

that the reorganization of ocean circulation that accompanied the shutdown of deep-water production in the North Atlantic produced all of the environmental changes recorded in ice cores. The many leaps in climate seen in the glacial part of the Greenland ice core records probably represent flips of the system back and forth between two self-stabilizing modes of operation.

It must be emphasized that the leaps in climate seen in the ice and sediment records for the North Atlantic region were confined to the cold parts of the climatic cycle. Following the transition from cold to warm conditions 10,000 years ago, the climate in this region has remained remarkably constant. Apparently Earth's climatic system has remained firmly locked in its present mode of operation. If so, what is the likelihood that increases in CO_2 and other greenhouse gases will jolt the ocean–atmosphere system out of its current mode into one more suitable to the coming conditions?

Unfortunately, we have little basis for answering this critical question. Nature provides us with no recent analogy to the super-interglacial conditions we are about to generate. The climate of the past 10,000 years is representative of the warmest part of the last several glacial cycles. Hence nature has not explored the super-interglacial climatic regime (at least not in sufficiently recent times that our geological records give an adequate picture of Earth's environment to be useful for future prediction). Further, as we have only recently become aware of the complexity of the linkages which tie together ocean, atmosphere, ice and terrestrial vegetation, we have not even begun to formulate means by which these linkages might be modelled. Indeed, reliable modelling may never be possible.

If we are to get into a position to cope with the effects not only of the greenhouse gases, but also of the various poisons being released to the environment, we must greatly expand our efforts to understand the individual parts of the Earth's great surface system and the interactions among them. The task, however, is every bit as complex as that of preventing cancer.

I see no hope of getting much done in the United States without a fundamental reorganization of our approach to environmental research. I believe that most scientists would agree that the handling of research on greenhouse effects by the Department of Energy and on acid rain by the Environmental Protection Agency has been ineffective. The problem lies partly in the perception by the managers of these programmes that their mandate is to obtain quick fixes, and partly in their inability to grasp the subtleties of the problems placed under their jurisdiction. The result is that these managers avoid long-term strategies designed to build up our base of knowledge. Rather, they are strongly attracted by proposals to re-work existing information even though most scientists agree that it has been squeezed nearly dry. Thus, while the existing programmes serve to keep management informed, they do little to further our ability to predict the conse-

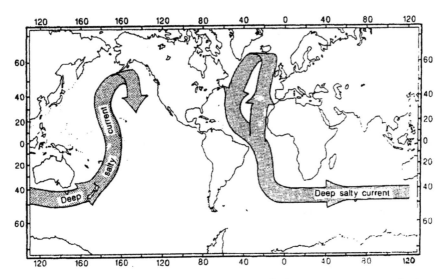

Fig. 2. A large-scale salt transport system operates in today's ocean to compensate for the transport of water (as vapour) through the atmosphere from the Atlantic to the Pacific Ocean. Salty, deep water formed in the North Atlantic flows down the length of the Atlantic, around Africa, through the southern Indian Ocean and finally northward in the deep Pacific Ocean. Some of this water upwells in the North Pacific, bringing with it the salt left behind in the Atlantic due to vapour transport. This atmospheric water vapour and ocean salt transport system is self-stabilizing. Records from ice and sediment tell us that this great conveyor system was somehow disrupted during glacial time and replaced by an alternative mode of operation.

quences of our actions. It is as if the efforts to prevent cancer were based largely on epidemiological information with no biochemical or cellular research. If we are to get out of this rut, the responsibility for basic research on environmental systems must be wrested from these mission-orientated agencies and given instead to an organization which is dedicated to basic research and is isolated from immediate political pressures.

As it now stands, a healthy component of basic research is being carried out only in those areas where there is a strong tradition of federal sponsorship (that is, the atmospheric and ocean sciences). Research on the continental parts of the environmental system (vegetation, soils and waters) remains in the Dark Ages. It's a Catch-22 situation—managers at the National Science Foundation (NSF) feel that basic research in these areas should be sponsored by the appropriate mission-orientated agency, but such agencies shun basic research.

Unique opportunities to expand our knowledge of environmental systems slip by without being fully exploited. A prime example is the invasion of man-made tracers (that is, nuclear-testing tritium and radiocarbon, reactor radiokrypton and industrial freons) into the sea. A detailed knowledge of the patterns and rates of movement of these substances through the sea will prove a tremendous boon to the validation of general circulation models for the atmosphere–ocean system. Although the atmospheric side of such models is fully developed, the oceanic side is still in a primitive state. Further development is hampered by our lack of knowledge of the physics of key processes occurring in the sea. Because of this, ocean models of adequate sophistication are decades away. Despite the obvious value of tracer data, the efforts to map the passage of these substances through the sea's interior are sparse. NSF is concerned that surveying brings too little immediate scientific yield, while the mission-orientated agencies fail to appreciate the fact that if we are to understand the Earth's system we must adopt long-term strategies. With the exception of the French, who have developed a remarkably good tracer programme in the Indian Ocean, other industrial countries have shown no greater interest in ocean tracers than had the United States. Hence future ocean modellers will rue the deficiencies in the documentation of man's great ocean-tracer experiment.

I am not suggesting that things could [126] be altered simply by increasing NSF's budget. NSF is not set up to put strings on the use of its funds. Nor should it be. But to do the kind of research that will be required if we are to deal effectively with long-term environmental problems, there must be some strings.

In this time of budget deficits it may be unrealistic to call for new government entities, but I am convinced that the only way to straighten out the mess in environmental research is to create a new national institute. This institute would have two objectives. First, it would sponsor environmental research; second, it would advise the government on a broad range of environmental questions.

The International Committee of Scientific Unions has recently given its blessing to a new initiative in environmental science, the International Geosphere-Biosphere Program (IGBP). The aim is to further our understanding of the operation of the individual elements of the environmental system and of the linkages between them. The mission-orientated agencies in the United States are already planning how to capture their share of this new pie. Judging by past performance, it would be a mistake to let them have any of it. Instead, the US monies to be spent on IGBP should be funnelled through a new entity, The National Institute for the Environment. The primary responsibility of this institute would be to develop integrated programmes dedicated to furthering our knowledge of environmental systems. In particular, the Institute would focus on those areas that most need attention, namely, the continental components of the climatic system. The programmes would be planned by the most capable scientists in the country and conducted in the most appropriate laboratories (university, industry and government). The institute would be overseen by a board of trustees, consisting of eminent scientists. Its director, chosen by this board, would be a highly capable scientist, skilled in management. While the institute would have its own in-house scientific staff, most of its resources would go towards the funding of external activities.

There are isolated success stories. For example, two NASA managers, Shelby Tilford and Bob Watson, are running a programme designed to determine the sources of atmospheric methane and how the strengths of these sources have changed with time. This programme is a beautiful example of how environmental research should be conducted. It was designed by the nation's most knowledgeable scientists and is being carried out by highly capable groups throughout the country. The cost is modest, about two million dollars a year, and the payoff will be large. It works because Watson and Tilford are themselves outstanding scientists who know what the tractable problems are and where the intellectual resources to tackle them reside.

Most of the research of the new institute should be housed in universities. It is in such settings that excellent research is generated for a minimum cost. By placing graduate education and research under one roof, an optimum environment for both is created. With a few exceptions, environmental research in large laboratories operated by government agencies has, in my opinion, proved to be unproductive and wasteful.

There are, however, two tasks that clearly require facilities not available to universities. One involves satellite observations, which will be the cornerstone of the effort to monitor the climatic system. Such platforms are best operated by NASA. The other is large-scale climate modelling, which is now the basis for all climate prediction. These facilities require large research staffs and immense computer power. Fortunately, a successful format has already been found which involves cooperation between universities and government departments. The National Oceanic and Atmospheric Administration has its primary effort on Princeton's campus. NASA bases such an effort on Columbia's campus. NSF operates such an effort at its National Center for Atmospheric Research in Boulder, Colorado. Because of the great importance of the so-called 'general circulation models' to climate prediction, these existing laboratories should be strengthened. Also, similar efforts must be launched in other university settings.

Five specific areas in which research must be greatly expanded can be singled out for special attention: (1) the large-scale circulation of the ocean; (2) the processes regulating soil moisture; (3) the processes responsible for cloud formation; (4) the role of biogeochemical processes on the atmosphere's composition; (5) the processes regulating sea ice.

Of course, there are already research programmes in these areas, but in my estimation they will not come up with answers quickly enough. In each area we need major new observational programmes to supply the key data that are necessary to develop a better physical understanding. In each area we need a cadre of young scientists with the appropriate training.

We also have to intensify our study of the climatic changes that have taken place over the past 100,000 years. As discussed above, nature herself has conducted large-scale climatic experiments. The response of the system to these natural experiments is recorded in sediments and in ice. By study of these records, we will be able to learn valuable lessons about the interactions that link the various elements of Earth's system together. I should stress once again that it is changes in these linkages which are likely to carry the greatest threats.

Although we don't know nearly enough about the operation of the Earth's climate to make reliable predictions of the consequences of the build-up of greenhouse gases, we do know enough to say that the effects are potentially quite serious. Whatever happens, it seems to me that the Earth's remaining wildlife will be dealt a serious blow. If, as the climatic record in ice and sediment suggests, changes in climate come in leaps rather than gradually, then the greenhouse build-up may threaten our food supply. To date, we have dealt with this problem as if its effects would come in the distant future and so gradually that we could easily cope with them. This is certainly a possibility, but I believe that there is an equal possibility that they will arrive suddenly and dramatically.

To prepare ourselves, we must take the problem of climatic change as seriously as we take those of cancer and nuclear defence. There are no easy solutions, and we must gear up for the long, hard job of working out how Earth's climate operates. To do this will require not only more financial and human resources, but also the administration appropriate to the task. Not only do our current managers lack a proper intellectual grasp of the problem, but they are obsessed with legislatively imposed 'five-year reports', and give little attention to developing a long-term strategy to build the needed base of knowledge. Even with a great intensification of effort, I fear that the effects of the rise in concentration of the greenhouse gases will come largely as surprises. But the greater our knowledge, the greater the wisdom that will be brought to bear if surprises do come.

Literature Cited

1. Hays, J.D. *et al. Science* **194**, 1121–1132 (1976).

2. Imbrie, J. *et al. Milankovitch and Climate* Part I (eds Berger, A.L. *et al.*) 269–305 (Reidel, Dordrecht, 1984).

3. Ruddiman, W.F. & McIntyre, A. *Palaeogeogr. Palaeoclimatol. Palaeoecol.* **35**, 145–214 (1981).

4. Dansgaard, W. *et al. Science* **218**, 1273–1277 (1982).

5. Dansgaard, W. *et al.* in *Green Ice Core: Geophysics. Geochemistry and the Environment* (eds Langway, C.C. *et al.) Am. Geophys. Un. Monogr.* No. 33. 71–76 (1985).

6. Hammer, C.U. *et al.* in *Green Ice Core: Geophysics. Geochemistry and the Environment* (eds Langway, C.C. *et al.) Am. Geophys. Un. Monogr.* No. 33. 90–94 (1985).

7. Stauffer, B. *et al. Ann. Glaciol.* **5**, 160–164 (1984).

8. Rind, D. *et al. Climate Dynamics* **1**, 3–33 (1986).

9. Broecker, W.S. *et al. Nature* **315**, 21–25 (1985).

Climate and Atmospheric History of the Past 420,000 Years from the Vostok Ice Core, Antarctica

J. R. PETIT, J. JOUZEL, D. RAYNAUD, ET AL.

[429] The recent completion of drilling at Vostok station in East Antarctica has allowed the extension of the ice record of atmospheric composition and climate to the past four glacial–interglacial cycles. The succession of changes through each climate cycle and termination was similar, and atmospheric and climate properties oscillated between stable bounds. Interglacial periods differed in temporal evolution and duration. Atmospheric concentrations of carbon dioxide and methane correlate well with Antarctic air-temperature throughout the record. Present-day atmospheric burdens of these two important greenhouse gases seem to have been unprecedented during the past 420,000 years. The late Quaternary period (the past one million years) is punctuated by a series of large glacial–interglacial changes with cycles that last about 100,000 years (ref 1). Glacial–interglacial climate changes are documented by complementary climate records (1,2) largely derived from deep sea sediments, continental deposits of flora, fauna and loess, and ice cores. These studies have documented the wide range of climate variability on Earth. They have shown that much of the variability occurs with periodicities corresponding to that of the precession, obliquity and eccentricity of the Earth's Orbit (1,3). But understanding how the climate system responds to this initial orbital forcing is still an important issue in palaeoclimatology, in particular for the generally strong ~100,000-year (100-kyr) cycle.

J. R. Petit, J. Jouzel, D. Raynaud, N. I. Barkov, J. M. Barnola, I. Basile, M. Bender, J. Chappellaz, M. Davis, G. Delaygue, M. Delmotte, V. M. Kotlyakov, M. Legrand, V. Y. Lipenkov, C. Lorius, L. Pepin, C. Ritz, E. Saltzman, and M. Stievenard. 1999. Excerpts from "Climate and atmospheric history of the past 420,000 years from the Vostok Ice Core, Antarctica." *Nature* 399:429–436.

Ice cores give access to palaeoclimate series that include local temperature and precipitation rate, moisture source conditions, wind strength and aerosol fluxes of marine, volcanic, terrestrial, cosmogenic and anthropogenic origin. They are also unique with their entrapped air inclusions in providing direct records of past changes in atmospheric trace-gas composition. The ice-drilling project undertaken in the framework of a long-term collaboration between Russia, the United States and France at the Russian Vostok station in East Antarctica (788 S, 1068 E, elevation 3,488 m, mean temperature $-55°C$) has already provided a wealth of such information for the past two glacial—interglacial cycles (refs 4–13). Glacial periods in Antarctica are characterized by much colder temperatures, reduced precipitation and more vigorous large-scale atmospheric circulation. There is a close correlation between Antarctic temperature and atmospheric concentrations of CO_2 and CH_4 (refs 5, 9). This discovery suggests that greenhouse gases are important as amplifiers of the initial orbital forcing and may have significantly contributed to the glacial–interglacial changes (14–16). The Vostok ice cores were also used to infer an empirical estimate of the sensitivity of global climate to future anthropogenic increases of greenhouse gas concentrations (15).

The recent completion of the ice-core drilling at Vostok allows us to considerably extend the ice-core record of climate properties at this site. In January 1998, the Vostok project yielded the deepest ice core ever recovered, reaching a depth of 3,623m (ref 17). Drilling then stopped ~120m above the surface of the Vostok lake, a deep subglacial lake which extends below the ice sheet over a large area (18), in order to avoid any risk that drilling fluid would contaminate the lake water. Preliminary data (17) indicated that the Vostok ice-core record extended through four climate cycles, with ice slightly older than 400 kyr at a depth of 3,310 m, thus spanning a period comparable to that covered by numerous oceanic and continental records.

Here we present a series of detailed Vostok records covering this ~400-kyr period. We show that the main features of the more recent Vostok climate cycle resemble those observed in earlier cycles. In particular, we confirm the strong correlation between atmospheric greenhouse-gas concentrations and Antarctic temperature, as well as the strong imprint of obliquity and precession in most of the climate time series. Our records reveal both similarities and differences between the successive interglacial periods. They suggest the lead of Antarctic air temperature, and of atmospheric greenhouse gas concentrations, with respect to global ice volume and Greenland air-temperature changes during glacial terminations.

The Ice Record

The data are shown in Figs. 1, 2, and 3. . . . They include the deuterium content of the ice (δD_{ice}, a proxy of local temperature change), the dust content (desert aerosols), the concentration of sodium (marine aerosol), and from the entrapped air the greenhouse gases CO_2 and CH_4, and the $\delta^{18}O$ of O_2 (hereafter $\delta^{18}O_{atm}$) which reflects changes in global ice volume and in the hydrological cycle (19). (δD and $\delta^{18}O$ are defined in the legends to Figs. 1 and 2, respectively.) All these measurements have been performed using methods previously described except for slight modifications (see figure legends).

The detailed record of δD_{ice} (Fig. 1) confirms the main features of the third and fourth climate cycles previously illustrated by the coarse-resolution record. However, a sudden decrease from interglacial-like to glacial-like values, rapidly followed by an abrupt return to interglacial-like values, occurs between 3,320 and 3,330 m. [430] In addition, a transition from low to high CO_2 and CH_4 values (not shown) occurs at exactly the same depth. In undisturbed ice, the transition in atmospheric composition would be found a few metres lower (due to the difference between the age of the ice and the age of the gas (20)). Also, three volcanic ash layers, just a few centimetres apart but inclined in opposite directions, have been observed—10m above this δD excursion (3,311 m). Similar inclined layers were observed in the deepest part of the GRIP and GISP2 ice cores from central Greenland, where they are believed to be associated with ice flow disturbances.

Vostok climate records are thus probably disturbed below these ash layers, whereas none of the six records show any indication of disturbances above this level. We therefore limit [431] the discussion of our new data sets to the upper

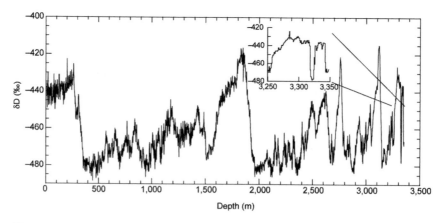

Figure 1

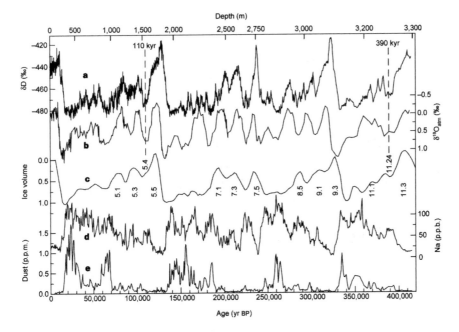

Figure 2

3,310m of the ice core, that is, down to the interglacial corresponding to marine stage 11.3. . . .

Climate and Atmospheric Trends

TEMPERATURE

The overall amplitude of the glacial—interglacial temperature change is about 8 degrees C for ΔT_I (inversion level) and about 12 degrees C for ΔT_S, the temperature at the surface (Fig. 3).

Broad features of this record are thought to be of large geographical significance (Antarctica and part of the Southern Hemisphere), at least qualitatively. When examined in detail, however, the Vostok record may differ from coastal sites in East Antarctica (28) and perhaps from West Antarctica as well.

Jouzel et al. (13) noted that temperature variations estimated from deuterium were similar for the last two glacial periods. The third and fourth climate cycles are of shorter duration than the first two cycles in the Vostok record. The same is true in the deep-sea record, where the third and fourth cycles span four precessional cycles rather than five as for the last two cycles (Fig. 3). Despite this difference, one observes, for all four climate cycles, the same "sawtooth" sequence of

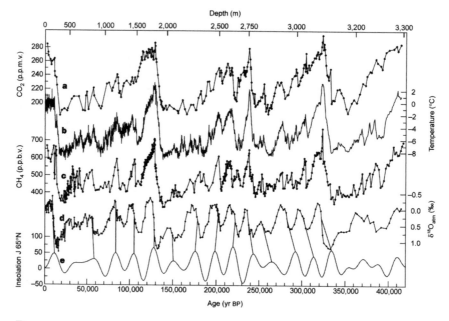

Figure 3

a warm interglacial (stages 11.3, 9.3, 7.5 and 5.5), followed by increasingly colder interstadial events, and ending with a rapid return towards the following intergla-cial. The [432] coolest part of each glacial period occurs just before the glacial ter-mination, except for the third cycle. This may reflect the fact that the June 65°N insolation minimum preceding this transition (255 kyr ago) has higher insolation than the previous one (280 kyr ago), unlike the three other glacial periods. None-theless, minimum temperatures are remarkably similar, within 1 degree C, for the four climate cycles. The new data confirm that the warmest temperature at stage 7.5 was slightly warmer than the Holocene, and show that stage 9.3 (where the highest deuterium value, −414.8‰, is found) was at least as warm as stage 5.5. That part of stage 11.3, which is present in Vostok, does not correspond to a par-ticularly warm climate as suggested for this period by deep-sea sediment records (ref 29). As noted above, however, the Vostok records are probably disturbed be-low 3,310 m, and we may not have sampled the warmest ice of this interglacial. In general, climate cycles are more uniform at Vostok than in deep-sea core records. The climate record makes it unlikely that the West Antarctic ice sheet collapsed during the past 420 kyr (or at least shows a marked insensitivity of the central part of East Antarctica and its climate to such a disintegration).

INSOLATION

[Measure of solar radiation energy received on a given surface area and recorded during a given time; section and Figure 4 omitted.] [433]

GREENHOUSE GASES

The extension of the greenhouse-gas record shows that the main trends of CO_2 and CH_4 concentration changes are similar for each glacial cycle (Fig. 3). Major transitions from the lowest to the highest values are associated with glacial—interglacial transitions. At these times, the atmospheric concentrations of CO_2 rises from 180 to 280–300 p.p.m.v. and that of CH_4 rises from 320–350 to 650–770 p.p.b.v. There are significant differences between the CH_4 concentration change associated with deglaciations. Termination III shows the smallest CH_4 increase, whereas termination IV shows the largest (Fig. 4).

Differences in the changes over deglaciations are less significant for CO_2. The decrease of CO_2 to the minimum values of glacial times is slower than its increase towards interglacial levels, confirming the sawtooth record of this property.

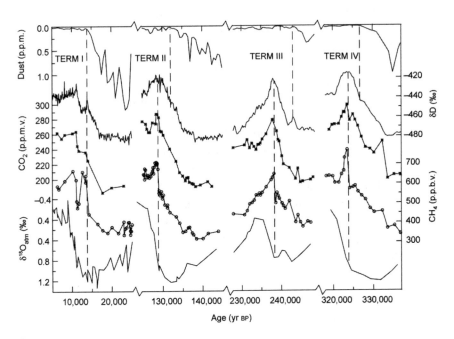

Figure 4

CH_4 also decreases slowly to its background level, but with a series of superimposed peaks whose amplitude decreases during the course of each glaciation. Each CH_4 peak is itself characterized by rapid increases and slower decreases, but our resolution is currently inadequate to capture the detail of millennial-scale CH_4 variations. During glacial inception, Antarctic temperature and CH_4 concentrations decrease in phase. The CO_2 decrease lags the temperature decrease by several kyr and may be either steep (as at the end of interglacials 5.5 and 7.5) or more regular (at the end of interglacials 9.3 and 11.3). . . . The 100-kyr component dominates both CO_2 and CH_4 records. However, the obliquity and precession components are much stronger for CH_4 than for CO_2.

The extension of the greenhouse-gas record shows that present day levels of CO_2 and CH_4 . . . are unprecedented during the past 420 kyr. Pre-industrial Holocene levels . . . are found during all interglacials, while values higher than these are found in stages 5.5, 9.3 and 11.3 (this last stage is probably incomplete), with the highest values during stage 9.

The overall correlation between our CO_2 and CH_4 records and the Antarctic isotopic temperature (refs 5,9,16) is remarkable ($r^2 = 0.71$ and 0.73 for CO_2 and CH_4, respectively). This high correlation indicates that CO_2 and CH_4 may have contributed to the glacial–interglacial changes over this entire period by amplifying the orbital forcing along with albedo, and possibly other changes (refs 15,16). We have calculated the direct radiative forcing corresponding to the CO_2, CH_4 and N_2O changes. The largest CO_2 change, which occurs between stages 10 and 9, implies a direct radiative warming of [434] $\Delta T_{rad} = 0.75$ degrees C. Adding the effects of CH_4 and N_2O at this termination increases the forcing to 0.95 degrees C (here we assume that N_2O varies with climate as during termination I—ref 37). This initial forcing is amplified by positive feedbacks associated with water vapour, sea ice, and possibly clouds (although in a different way for a "doubled CO_2" situation than for a glacial climate—ref 38). The total glacial—interglacial forcing is important ($\sim3Wm^{-2}$), representing 80% of that corresponding to the difference between a "doubled CO_2" world and modern CO_2 climate. Results from various climate simulations (refs 39,40) make it reasonable to assume that greenhouse gases have, at a global scale, contributed significantly (possibly about half, that is, 2–3 degrees C) to the globally averaged glacial–interglacial temperature change.

Glacial Terminations and Interglacials

Our complete Vostok data set allows us to examine all glacial commencements and terminations of the past 420 kyr. . . .

A striking feature of the Vostok deuterium record is that the Holocene, which has already lasted 11 kyr, is, by far, the longest stable warm period recorded in Antarctica during the past 420 kyr (Fig. 4). Interglacials 5.5 and 9.3 are different from the Holocene, but similar to each other in duration, shape and amplitude. During each of these two events, there is a warm period of 4 kyr followed by a relatively rapid cooling and then a slower temperature decrease, rather like some North Atlantic deep-sea core records (ref 42). Stage 7.5 is different in all respects, with a slightly colder maximum, a more spiky shape, and a much shorter duration (7 kyr at mid-transition compared with 17 and 20 kyr for stages 5.5 and 9.3, respectively). This difference between stage 7.5 and stages 5.5 or 9.3 may result from the different configuration of the Earth's orbit (in particular concerning the phase of precession with respect to obliquity). Termination III is also peculiar as far as terrestrial aerosol fallout is concerned. Terminations I, II and IV are marked by a large decrease in dust; high glacial values drop to low interglacial values by the mid-point of the δD_{ice} increase. But for termination III, the dust concentration decreases much earlier, with low interglacial values obtained just before a slight cooling event, as for termination I (for which interglacial values are reached just before the "Antarctic Cold Reversal"). . . .

[435] . . . Our results suggest that the same sequence of climate forcings occurred during each termination: orbital forcing (possibly through local insolation changes, but this is speculative as we have poor absolute dating) followed by two strong amplifiers, with greenhouse gases acting first, and then deglaciation enhancement via ice-albedo feedback. The end of the deglaciation is then characterized by a clear CO_2 maximum for terminations II, III and IV, while this feature is less marked for the Holocene. Comparison of CO2 atmospheric concentration changes with variations of other properties illuminates oceanic processes influencing glacial—interglacial CO_2 changes. The sequence of climate events described above rules out the possibility that rising sea level induces the CO_2 increase at the beginning of terminations. On the other hand, the small CO_2 variations associated with Heinrich events (ref 45) suggest that the formation of North Atlantic Deep Water does not have a large effect on CO_2 concentrations. Our record shows similar relative amplitudes of atmospheric CO_2 and Vostok temperature changes for the four terminations. Also, values of both CO_2 and temperature are significantly higher during stage 7.5 than during stages 7.1 and 7.3, whereas the deep-sea core ice volume record exhibits similar levels for these three stages. These similarities between changes in atmospheric CO_2 and Antarctic temperature suggest that the oceanic area around Antarctica plays a role in the long-term CO_2 change. An influence of high southern latitudes is also suggested by the comparison with the dust profile, which exhibits a maximum

during the periods of lowest CO_2. The link between dust and CO_2 variations could be through the atmospheric input of iron (ref 46). Alternatively, we suggest a link through deep ocean circulation and sea ice extent in the Southern Ocean, both of which play a role in ocean CO_2 ventilation and, as suggested above, in the dust input over East Antarctica.

New Constraints on Past Climate Change

As judged from Vostok records, climate has almost always been in a state of change during the past 420 kyr but within stable bounds (that is, there are maximum and minimum values of climate properties between which climate oscillates). Significant features of the most recent glacial—interglacial cycle are observed in earlier cycles. Spectral analysis emphasises the dominance of the 100-kyr cycle for all six data series except $\delta^{18}O_{atm}$ and a strong imprint of 40-and/or 20-kyr periodicities despite the fact that the glaciological dating is tuned by fitting only two control points in the 100-kyr band.

Properties change in the following sequence during each of the last four glacial terminations, as recorded in Vostok. First, the temperature and atmospheric concentrations of CO_2 and CH_4 rise steadily, whereas the dust input decreases. During the last half of the temperature rise, there is a rapid increase in CH_4. This event coincides with the start of the $\delta^{18}O_{atm}$ decrease. We believe that the rapid CH_4 rise also signifies warming in Greenland, and that the deglacial $\delta^{18}O_{atm}$ decrease records rapid melting of the Northern Hemisphere ice sheets. These results suggest that the same sequence of climate forcing operated during each termination: orbital forcing (with a possible contribution of local insolation changes) followed by two strong amplifiers, greenhouse gases acting first, then deglaciation and ice-albedo feedback. Our data suggest a significant role of the Southern Ocean in regulating the long-term changes of atmospheric CO_2.

The Antarctic temperature was warmer, and atmospheric CO_2 and CH_4 concentrations were higher, during interglacials 5.5 and 9.3 than during the Holocene and interglacial 7.5. The temporal evolution and duration of stages 5.5 and 9.3 are indeed remarkably similar for all properties recorded in Vostok ice and entrapped gases. As judged from the Vostok record, the long, stable Holocene is a unique feature of climate during the past 420 kyr, with possibly profound implications for evolution and the development of civilizations. Finally, CO_2 and CH_4 concentrations are strongly correlated with Antarctic temperatures; this is because, overall, our results support the idea that greenhouse gases have contributed significantly to the glacial–interglacial change. This correlation, together with the uniquely elevated concentrations of these gases today, is of relevance with respect to the continuing debate on the future of Earth's climate.

References

1. Imbrie, J. *et al*. On the structure and origin of major glaciation cycles. 1. Linear responses to Milankovich forcing. *Paleoceanography* 7, 701–738 (1992).

2. Tzedakis, P. C. *et al*. Comparison of terrestrial and marine records of changing climate of the last 500,000 years. *Earth Planet. Sci. Lett.* 150, 171–176 (1997).

3. Berger, A. L. Long-term variations of daily insolation and Quaternary climatic change. *J. Atmos. Sci.* 35, 2362–2367 (1978).

4. Lorius, C. *et al*. A 150,000-year climatic record from Antarctic ice. *Nature* 316, 591–596 (1985).

5. Barnola, J. M., Raynaud, D., Korotkevich, Y. S. & Lorius, C. Vostok ice cores provides 160,000-year record of atmospheric CO_2. *Nature* 329, 408–414 (1987).

6. Jouzel, J. *et al*. Vostok ice core: a continuous isotope temperature record over the last climatic cycle (160,000 years). *Nature* 329, 402–408 (1987).

7. Raisbeck, G. M. *et al*. Evidence for two intervals of enhanced ^{10}Be deposition in Antarctic ice during the last glacial period. *Nature* 326, 273–277 (1987).

8. Legrand, M., Lorius, C., Barkov, N. I. & Petrov, V. N. Vostok (Antarctic ice core): atmospheric chemistry changes over the last climatic cycle (160,000 years). *Atmos. Environ.* 22, 317–331 (1988).

9. Chappellaz, J., Barnola, J.-M., Raynaud, D., Korotkevich, Y. S. & Lorius, C. Ice-core record of atmospheric methane over the past 160,000 years. *Nature* 127–131 (1990).

10. Petit, J. R. *et al*. Paleoclimatological implications of the Vostok core dust record. *Nature* 343, 56–58 (1990).

11. Sowers, T. *et al*. 135 000 year Vostok—SPECMAP common temporal framework. *Paleoceanography* 8, 737–766 (1993).

12. Jouzel, J. *et al*. Extending the Vostok ice-core record of palaeoclimate to the penultimate glacial period. *Nature* 364, 407–412 (1993).

13. Jouzel, J. *et al*. Climatic interpretation of the recently extended Vostok ice records. *Clim. Dyn.* 12, 513–521 (1996).

14. Genthon, C. *et al*. Vostok ice core: climatic response to CO_2 and orbital forcing changes over the last climatic cycle. *Nature* 329, 414–418 (1987).

15. Lorius, C., Jouzel, J., Raynaud, D., Hansen, J. & Le Treut, H. Greenhouse warming, climate sensitivity and ice core data. *Nature* 347, 139–145 (1990).

16. Raynaud, D. *et al*. The ice record of greenhouse gases. *Science* 259, 926–934 (1993).

17. Petit, J. R. *et al*. Four climatic cycles in Vostok ice core. *Nature* 387, 359–360 (1997).

18. Kapitza, A. P., Ridley, J. K., Robin, G. d.Q., Siegert, M. J. & Zotikov, I. A. A large deep freshwater lake beneath the ice of central East Antarctica. *Nature* 381, 684–686 (1996).

19. Bender, M., Sowers, T. & Labeyrie, L. D. The Dole effect and its variation during the last 130,000 years as measured in the Vostok core. *Glob. Biogeochem. Cycles* 8, 363–376 (1994).

20. Barnola, J. M., Pimienta, P., Raynaud, D. & Korotkevich, Y. S. CO_2 climate relationship as deduced from the Vostok ice core: a re-examination based on new measurements and on a re-evaluation of the air dating. *Tellus B* 43, 83–91 (1991).

. . .

28. Steig, E. *et al.* Synchronous climate changes in Antarctica and the North Atlantic. *Science* 282, 92–95 (1998).
29. Howard, W. A warm future in the past. *Nature* 388, 418–419 (1997).

. . .

37. Leuenberger, M. & Siegenthaler, U. Ice-age atmospheric concentration of nitrous oxide from an Antarctic ice core. *Nature* 360, 449–451 (1992).
38. Ramstein, G., Serafini-Le Treut, Y., Le Treut, H., Forichon, M. & Joussaume, S. Cloud processes associated with past and future climate changes. *Clim. Dyn.* 14, 233–247 (1998).
39. Berger, A., Loutre, M. F. & Gallée, H. Sensitivity of the LLN climate model to the astronomical and CO_2 forcings over the last 200 ky. *Clim. Dyn.* 14, 615–629 (1998).
40. Weaver, A. J., Eby, M., Fanning, A. F. & Wilbe, E. C. Simulated influence of carbon dioxide, orbital forcing and ice sheets on the climate of the Last Glacial Maximum. *Nature* 394, 847–853 (1998).

. . .

42. Cortijo, E. *et al.* Eemian cooling in the Norwegian Sea and North Atlantic ocean preceding ice-sheet growth. *Nature* 372, 446–449 (1994).

. . .

45. Stauffer, B. *et al.* Atmospheric CO_2 concentration and millennial-scale climate change during the last glacial period. *Nature* 392, 59–61 (1998).
46. Martin, J. H. Glacial-interglacial CO_2 change: The iron hypothesis. *Paleoceanography* 5, 1–13 (1990).

Commentary

Wallace S. Broecker, "Unpleasant Surprises in the Greenhouse?" (1987)
J. R. Petit, J. Jouzel, D. Raynaud, et al., "Climate and Atmospheric
History of the Last 420,000 Years from the Vostok Ice Core, Antarctica"
(1999)

TOM GRIFFITHS

Ice cores are the holy scripts, the sacred scrolls of our age. When you drill into an ice cap that is kilometers thick, you can extract a core that is layered year by year, a precious archive of deep time. Whereas the oldest Greenland cores go back to the last interglacial period, about 120,000 years ago, the deepest Antarctic cores currently retrieve 800,000 years of climate history. Generally, in Antarctica, there is less precipitation and seasonality and more compression of the layers of ice; resolution is thus traded for time (although the Law Dome in Antarctica has delivered some unusually detailed atmospheric narratives of the recent past). In Greenland the layers tend to be clearer, because of the greater annual accumulation of ice. And so the more discriminating Greenland cores are essential to calibrating the longer, more condensed Antarctic archive. The polar ice caps therefore combine beautifully to give us detailed long-term climate data.

In the final two decades of the twentieth century, ice cores from both Antarctica and Greenland delivered a sense of urgency and crisis about global warming. A brief history of greenhouse ideas reveals that most foundational climate science was curiosity-driven and conducted with little anxiety about its human implications. In the nineteenth century, scientists such as John Tyndall and Svante Arrhenius were chiefly researching the ancient past—the causes of ice ages—rather than the immediate future. And where they did speculate on future global warming, they saw it as generally positive. The prospect of a rise in temperature of several degrees was greeted equably, especially in Britain and northern Europe. If the world were warmer, it might make winters more comfortable and agriculture more productive, or even help stave off the next ice age. And there was also a reassuring assumption that if global warming was happening, it would occur slowly, at a stately pace, as had presumably been the case in the past. For the first two-thirds of the twentieth century, therefore, the global warming trend was often called the "embetterment" of climate or the "recent amelioration." From the 1980s, this complacency changed swiftly, and it was ice cores from Greenland that would deliver the first shock.

Ice core research had been pioneered in Greenland in the postwar years by Danish scientist Willi Dansgaard. When Dansgaard began his research in the

1950s, he speculated that there might be an opportunity, through studying the messages about temperature embedded in ice, to research climate over "the past several hundred years." But the timeline was soon to deepen dramatically. The Cold War, a noted stimulant to science, generated the world's first paleo-climate data from an ice core—one that was 1,390 meters deep, 100,000 years long, and drilled in 1966 at Camp Century, a secret American military base embedded in the northern Greenland ice cap. In the 1970s, clearer ice cores extracted from the summit of the ice cap revealed that climate change over the past 110,000 years had often occurred very quickly, with temperature variations in the North Atlantic of 6 degrees Celsius and more in a few decades. At the end of the last ice age, temperatures in Greenland fluctuated suddenly and rapidly, especially around an event known as the Younger Dryas (so named because *Dryas octopetala* is a plant of the northern tundra that flourishes in the cold). The Younger Dryas ended around 11,500 years ago when temperatures in Greenland warmed about 8 degrees Celsius in a decade.

Wallace Broecker's commentary in *Nature* in 1987 gave clear expression to this new view of the climatic past. There were likely, he declared, to be "unpleasant surprises in the greenhouse." Broecker was inclined to refer to the climate system as an angry beast that we were foolish enough to be poking with sticks. In his article, although he drew also on climate information embedded in marine sediments from the North Atlantic, he explained that "the evidence that turned our heads came from holes drilled through the Greenland ice cap." Broecker's theory was that dramatic swings in temperature were triggered by the sudden shutdown or restarting of what he called the "conveyor belt" of ocean circulation. It was this insight that global warming might halt the Gulf Stream and thereby induce sudden regional cooling that was exaggerated and fictionalized in the North American film *The Day After Tomorrow* (2004).

In his article, Broecker included a version of his simple and famous diagram of the "conveyor belt," which has been described as galvanizing climate research in the late 1980s. The diagram symbolized the timely convergence of his twin passions for climate science and oceanography. It was this very interdisciplinarity—this need to understand the complex connectedness of earth systems, "the linkages which tie together ocean, atmosphere, ice and terrestrial vegetation"—that his article went on to champion. He called for basic research on environmental systems to be wrested from mission-oriented government agencies and pursued instead by institutions free of political pressures that could invest in long-term, interdisciplinary analysis. Although he emphasized the difficulties of prediction in complex systems, he felt that climatic change must be taken as seriously as cancer and nuclear defense.

Antarctica would deliver the next breakthrough in understanding the scale of the climate crisis. As recently as the beginning of the twentieth century, explorers thought they might meet polar bears in Antarctica, and just fifty years ago they were still finding out how much ice there is clamped to the South Pole. Antarctica, it was slowly discovered, was quite unlike the Arctic. Northern hemisphere assumptions foundered on a very different, southern reality: a continent at the pole, an ice age in action, the cold core of Earth's atmosphere, the engine of global climate. Antarctica was Louis Agassiz's dream—or nightmare—discovered on our planet in our own time. It is where nine-tenths of the world's land ice resides.

If a hundred years ago the defining Antarctic journey was the sledging expedition across the surface of the ice, and fifty years ago it was the tractor traverse that, with seismic soundings, measured the volume of the ice sheet, then the defining Antarctic journey of our own era goes straight down, with the help of a drill, from the top of the ice dome to the continental bedrock, a vertical journey back through time. And the ice core extracted enables us to see our civilization in the context of hundreds of thousands of years of climate history. Ice may seem abiotic and inhuman, but it is bound up with our fate.

In the 1990s, a long, 400,000-year Antarctic ice core was extracted from the inland ice sheet near the Russian station, Vostok. The Vostok core, which charted four full cycles of glacial and interglacial periods, established that the carbon dioxide and methane concentration in the atmosphere had moved in lockstep with the ice sheets and the temperature. When the ancient temperature variations were graphed, they seemed to be, in the words of Australian archaeologist Mike Smith, "the oscillations of an unstable system." It also revealed that present-day levels of these greenhouse gases were unprecedented during the past 420,000 years.

Like the 1987 Broecker paper with its depiction of the "conveyor belt" of ocean circulation, the 1999 Vostok paper had a powerful visual signature—the rhythmic, sawtooth graph of ice ages. Will Steffen, earth system scientist (and contributor to this volume), identifies the 1999 publication of the findings of the Vostok core as the moment when climate issues crystallized for him: "To me the Vostok core was the most beautiful piece of evidence of the earth as a single system. . . . For the first time we saw this beautiful rhythmic pattern, how the earth as a whole operated. You saw temperature, you saw gases, you saw dust all dancing to the same tune, all triggered by the earth's orbit around the sun . . . but still a mystery, it couldn't explain the magnitude of those swings." The Vostok core, according to Steffen, demanded recognition of the strong role of biology in a system that had previously been analyzed largely in geophysical terms.

From the late 1980s, therefore, ice core data from both poles helped to transform climate science into the complex, interdisciplinary study of earth systems and also established abrupt climate change as a disturbing past reality. A sense of urgency thus entered scientific and public debates about the climate crisis and coincided with the fast warming of average temperatures since the 1980s. This was the same period in which ecological science abandoned the idea of the "balance of nature" and accepted "disturbance" as normal in ecosystems. It was the same period that "punctuated equilibrium"—the idea of sudden change—reentered debates in evolutionary science. Catastrophism was back.

Right now, in Antarctica, the international race is on again—not for the South Pole, not for the first trans-Antarctic traverse, but for the first million-year ice core. It will take us back to the time when the rhythm of ice ages shifted from a periodicity of 40,000 years to that of 100,000 years depicted in the Vostok core, a pattern that shaped the evolution of humanity.

Further Reading

Broecker, W. S., and R. Kunzig. 2008. *Fixing climate*. New York: Hill and Wang.

Chandler, J. 2011. *Feeling the heat*. Carlton: Melbourne University Press.

Dansgaard, W. 2004. *Frozen annals: Greenland ice cap research*. Copenhagen: University of Copenhagen.

Smith, M. A. 2005. "Palaeoclimates: An archaeology of climate change." In T. Sherratt, T. Griffiths, and L. Robin (eds.), *A change in the weather: Climate and culture in Australia*, 176–186. Canberra: National Museum of Australia Press.

Diversity

*Why Do We Need It, and
Can We Conserve It?*

In Part 8 we explore ideas about global life systems and the biological sciences that inform their management. Cultural diversity also has been an important principle of global change thinking, and has recently been intertwined in current management discussions. In this part, the focus is mostly on *biological* diversity (and after 1986, "biodiversity") as diversity has emerged as a key concept in reorganizing nature conservation historically on a global scale in the postwar years.

Approaches to nature in the eighteenth century were often highly local: one study of the nature of a single parish in southern England was Gilbert White's *The Natural History of Selborne* (1789), one of the best-selling books ever written in the English language. White, a parson-naturalist, was interested in the connections between the changing seasons and the patterns of nature. He recorded the arrival of migratory birds and the flowering of plants on an annual calendar over many years. He understood the natural history of his parish as deeply responsive to climatic events and seasonal shifts, and the natural elements such as plants and animals as interdependent. White was celebratory rather than "systematic," and deeply committed to understanding fully the place where he lived, rather than the whole world.

As we have seen in Part 5, in the nineteenth century Alexander von Humboldt theorized global patterns for the way diversity may be shaped by climate, elevation, and other abiotic factors. *The Origin of Species* (1859) by British natural historian Charles Darwin (1809–1882), is a fundamental treatise in understanding the diversity of life. Darwin's theory of natural selection has been a fundamental building block on which ideas of "species" have developed. His theory built on the idea of "the preservation of favoured races in the struggle for life," the subtitle of his book. The theory of natural selection was enhanced by understanding the mechanism for genetic inheritance, the gene. In the first decade of the twentieth century, the experimental work of German-Austrian Gregor Mendel (1822–1884) was rediscovered by a number of scholars, and the science of "genetics" was formally named by William Bateson (1861–1926).

This part explores some of the key documents that moved thinking from theoretical ideas, such as "natural selection," to global change science, with its emphasis on managing global change.

In the twenty-first century, the International Union for the Conservation of Nature (IUCN) defines "biodiversity" as species diversity, genetic diversity, and ecosystem diversity. The IUCN, founded in 1948, claims to be the world's first "global environmental organization." Sponsored by the United Nations, IUCN now describes itself as a democratic network that connects over one thousand governments and nongovernmental organizations (NGOs) (IUCN 2011). "Species richness (the number of species in a given area) represents a single but important metric that is valuable as the common currency of the diversity of life," its website states.

The spread of plants and animals around the world—biological invasions—has been identified as very much linked with the expansion of Europe: historian Alfred Crosby (1986) argued that without the accompanying biota, European expansion could never have happened. A similar argument was put forward in the 1950s by animal ecologist Charles Elton, in a series of lectures published as *The Ecology of Invasions* (1958). Elton's expertise started with "pest control," but he developed an ecological view of "invaders" that came to define the science of "invasion biology." His global and historical perspectives of invading species have been highly influential in global change thinking. He writes: "We are living in a period of the world's history when the mingling of thousands of kinds of organisms from different parts of the world is setting up terrific dislocations in nature" (18).

Understanding the relationship between extinction events and invasions is fundamental to ideas about biodiversity today. There is no doubt that human introductions of biota, which have accompanied people since the first humans spread out of Africa, are a very important factor. Increases in diversity are created by both colonization and speciation. In *The Theory of Island Biogeography*, MacArthur and Wilson (1967) suggested that isolated habitats, because they are particularly susceptible to invasion by human-introduced species, will tend to lose endemics (local species with limited range) in their total species number. Thus they argue that what we are seeing is increasing diversity in individual places, but an increasing homogenization of the diversity at a global scale. This is referred to as the "cosmopolitanization" of the biota. Local extinctions may not be important, if there are other populations elsewhere, but where invasive species replace endemics, there is a loss to overall global biodiversity. Thus, many individual extinctions together, aggravated by the increasingly important factor of human introductions, constitute a massive global experiment that is running completely out of control, something also observed by Charles Elton.

By the time of the *Global 2000 Report to the President* (1980), the notion of a "crisis" of extinctions at a global level had become widely understood (Robin 2011). Thomas Lovejoy's predictive models for extinction, the section of the document excerpted here, and the idea of "diversity loss" were gathering attention. "Biodiversity" was a term coined in 1986 by Lovejoy and E. O. Wilson and a number of other conservation biologists as a tool to respond to a perceived crisis of extinctions, and a concern for managing human behaviors that contribute to this crisis.

Earlier, the idea of biological diversity was framed as purely scientific. "Biodiversity," by contrast, was an idea of crisis. It is much more than a "new name for nature" (Farnham 2007, 2). It provides a way to *measure* change in nature, human induced and otherwise. Biologists quantify species, measure genetic variation, and consider the pressures on the health of whole ecosystems.

Biodiversity is political as well as scientific and may inform management practice. Thus, biodiversity is both a scientific and a social tool, and a key concept for science, management, and governance on a global scale. Michael Soulé advocated a new science of crisis, conservation biology, at the same time, and his paper is the next one in this part. Ecosystem health and biodiversity are mutually supportive, and more broadly, good for human health and economic opportunity, Soulé argues in the document reproduced here. His focus on "science for crisis" has been very influential beyond biology.

Life systems are now "managed" through global conventions and institutions such as the International Union for Conservation of Nature, which maintains a Red List, "the most comprehensive information source on the status of wild species and their links to livelihoods" (IUCN website). The IUCN published *The Red List of Threatened Species* in 1986, just at Michael Soulé's "science of crisis" moment in conservation biology. This was not the first list of endangered species to be published by IUCN (earlier ones dated to the 1940s), but it was marketed in red, the color of alarm. It aimed to represent a crisis that could shift policy and politics. *The Red List* provided a baseline for global assessments of the conservation status of species, and it has been regularly updated since.

The western way of counting extinctions and measuring nature using biodiversity indicators is sometimes seen as overwhelming the nature conservation goals of traditional societies, particularly in the way people are "outside" natural resource management. Environmental justice demands that conservation includes more than what can be conveniently measured, various social theorists argue. Sheila Jasanoff, for example, recommends "attention to substance and process," and stresses "deliberation as well as analysis" (Jasanoff 2003, 243). Jasanoff calls these "social technologies of humility" that are an important complement to traditional western scientific expertise in managing nature ethically.

A range of important writers from emerging economies have raised concerns about some of the assumptions underpinning western conservation. In this collection we include Indian historian Ramachandra Guha, who offers a critique of the idea of "wilderness," a cornerstone of American conservation in the 1960s and 1970s. When nature conservation concentrates only on places without people, the protection of biodiversity is severely compromised, something increasingly acknowledged by western biologists as well (Robin 2009).

Conservation is an ethical and moral process, much more than just "management." Vandana Shiva critiques the way western agricultural practices threaten both people and biodiversity—for example, through patenting seeds. She reminds us that "in nature's economy, the currency is not money, it is life" (Shiva 2005, 33).

In the past two decades, organizations like Conservation International (CI, established in 1987, at the height of the "biodiversity crisis") have modified their approach to be more inclusive of environmental justice and cultural diversity in managing nature. International NGOs like CI still focus on biodiversity, but it is number six in the initiatives on their website, behind climate, fresh water, food security, health, and cultural services. Global change embraces all of these, and diversity concerns are increasingly nested in broader initiatives. By 2009, when Rockström et al. (pp. 491–501) were defining "planetary boundaries," they considered nine different areas of concern, just one of which was "biodiversity loss," though some of their other concerns, such as ocean acidification, global fresh water use, and changes in land use, also affect biological diversity and its conservation.

The Invaders

CHARLES S. ELTON

[15] . . . Nowadays we live in a very explosive world, and while we may not know where or when the next outburst will be, we might hope to find ways of stopping it or at any rate damping down its force. It is not just nuclear bombs and wars that threaten us, though these rank very high on the list at the moment: there are other sorts of explosions, and this book is about ecological explosions. An ecological explosion means the enormous increase in numbers of some kind of living organism—it may be an infectious virus like influenza, or a bacterium like bubonic plague, or a fungus like that of the potato disease, a green plant like the prickly pear, or an animal like the grey squirrel. I use the word "explosion" deliberately, because it means the bursting out from control of forces that were previously held in restraint by other forces. Indeed the word was originally used to describe the barracking of actors by an audience whom they were no longer able to restrain by the quality of their performance.

Ecological explosions differ from some of the rest by not making such a loud noise and in taking longer to happen. That is to say, they may develop slowly and they may die down slowly; but they can be very impressive in their effects, and many people have been ruined by them, or died or forced to emigrate. At the end of the First World War, pandemic influenza broke out on the Western Front, and thence rolled right round the world, eventually, not sparing even the Eskimos of Labrador and Greenland, and it is reputed to have killed 100 million human beings. Bubonic plague is still pursuing its great modern pandemic that started at the back of China in the end of last century, was carried by ship rats to India, South Africa, and other continents, and now smoulders among hundreds of species of wild rodents there, as well as in its chief original [16] home in Eastern Asia. In China it occasionally flares up on a very large scale in the pneumonic form, resembling the Black Death of medieval Europe. In 1911 about 60,000 people

C. S. Elton. 1958. Excerpts from "The invaders." Chapter 1 in *The ecology of invasions by animals and plants*, 15–32. London: Methuen and Co. [original page numbers in square brackets]

of Manchuria died this way. This form of the disease, which spreads directly from one person to another without the intermediate link of a flea, has mercifully been scarce in the newly invaded continents. Wherever plague has got into natural ecological communities, it is liable to explode on a smaller or larger scale, though by a stroke of fortune for the human race, the train of contacts that starts this up is not very easily fired. In South Africa the gerbilles living on the veld carry the bacteria permanently in many of their populations. Natural epidemics flare up among them frequently. From them the bacteria can pass through a flea to the multimammate mouse; this species, unlike the gerbilles, lives in contact with man's domestic rat; the latter may become infected occasionally and from it isolated human cases of bubonic plague arise. These in turn may spread into a small local [17] epidemic, but often do not. In the United States and Canada a similar underworld of plague (with different species in it) is established over an immense extent of the Western regions, though few outbreaks have happened in man. Here, then, the chain of connexions is weaker even than in South Africa, though the potentiality is present. Although plague-stricken people and plague-infected rats certainly landed from ships in California early this century, it is still possible that the plague organism was already present in North America. Professor Karl Meyer, who started the chief ecological research on sylvatic plague there, says: "The only conclusion one can draw is that the original source and date of the creation of the endemic sylvatic plague area on the North American Continent, inclusive [of] Canada, must remain a matter of further investigation and critical analysis."

Another kind of explosion was that of the potato fungus from Europe that partly emptied Ireland through famine a hundred years ago. Most [18] people have had experience of some kind of invasion by a foreign species, if only on a moderate scale. Though these are silent explosions in themselves, they often make quite a loud noise in the Press, and one may come across banner headlines like "Malaria Epidemic Hits Brazil," "Forest Damage on Cannock Chase," or "Rabbit Disease in Kent." This arrival of rabbit disease—myxomatosis—and its subsequent spread have made one of the biggest ecological explosions Great Britain has had this century, and its ramifying effects will be felt for many years.

But it is not just headlines or a more efficient news service that make such events commoner in our lives than they were last century. They are really happening much more commonly; indeed they are so frequent nowadays in every continent and island, and even in the oceans, that we need to understand what is causing them and try to arrive at some general viewpoint about the whole business. Why should a comfortably placed virus living in Brazilian cotton-tail rabbits suddenly wipe out a great part of the rabbit populations of Western Europe? Why do we have to worry about the Colorado potato beetle now, more than 300 years after the introduction of the potato itself? Why should the pine looper moth break out in Staffordshire and Morayshire pine plantations two years ago? It has been doing this on the Continent for over 150 years; it is not a new introduction to this country.

The examples given above point to two rather different kinds of outbreaks in populations: those that occur because a foreign species successfully invades another country, and those that happen in native or long-established populations. This book is chiefly about the first kind—the invaders. But the interaction of fresh arrivals with the native fauna and flora leads to some consideration of ecological ideas and research about the balance within and between communities as a whole. In other words, the whole matter goes far wider than any technological discussion of pest control, though many of the examples are taken from applied ecology. The real thing is that we are living in a period of the world's history when the mingling of thousands of kinds of organisms from different parts of the world is setting up terrific dislocations in nature. We are seeing huge changes in the natural population balance of the world. Of course, pest control is very important, because we have to preserve our living [19] resources and protect ourselves from diseases and the consequences of economic dislocation. But one should try to see the whole matter on a much broader canvas than that. I like the words of Dr Johnson: "Whatever makes the past, the distant, or the future, predominate over the present advances us in the dignity of thinking beings." The larger ecological explosions have helped to alter the course of world history, and, as will be shown, can often be traced to a breakdown in the isolation of continents and islands built up during the early and middle parts of the Tertiary Period. . . . In order to focus the subject, here are seven case histories of species [20] which were brought from one country and exploded into another. About 1929, a few African mosquitoes accidentally reached the north-east corner of Brazil, having probably been carried from Dakar on a fast French destroyer. They managed to get ashore and founded a small colony in a marsh near the coast—the Mosquito Fathers as it were. At first not much attention was paid to them, though there was a pretty sharp outbreak of malaria in the local town, during which practically every person was infected. For the next few years the insects spread rather quietly along the coastal region, until at a spot about 200 miles farther on explosive malaria blazed up and continued in 1938 and 1939, by which time the mosquitoes were found to have moved a further 200 miles inland up the Jaguaribe River valley. It was one of the worst epidemics that Brazil had ever known, hundreds of thousands of people were ill, some twenty thousand are believed to have died, and the life of the countryside was partially paralysed. The biological reasons for this disaster were horribly simple: there had always been malaria-carrying mosquitoes in the country, but none that regularly flew into houses like the African species, and could also breed so successfully in open sunny pools outside the shade of the forest. Fortunately both these habits made control possible, and the Rockefeller Foundation combined with the Brazil government to wage a really astounding campaign, so thorough and drastic was it, using a staff of over three thousand people who dealt with all the breeding sites and sprayed the inside of houses. This prodigious enterprise succeeded, at a cost of over two million dollars, in completely exterminating *Anopheles gambiae* on the South American continent within three years.

Here we can see three chief elements that recur in this sort of situation. First there is the historical one: this species of mosquito was confined to tropical Africa but got carried to South America by man. Secondly, the ecological features—its method of breeding, and its choice of place to rest and to feed on man. It is quite certain that the campaign could never have succeeded without the intense ecological surveys and study that lay behind the inspection and control methods. The third thing is the disastrous consequences of the introduction. One further consequence was that quarantine inspection of aircraft was started, and in one of these they discovered a tsetse fly, *Glossina palpalis*, the African [21] carrier of sleeping sickness in man, and at the present day not found outside Africa.

The second example is a plant disease. At the beginning of this century sweet chestnut trees in the eastern United States began to be infected by a killing disease caused by a fungus, *Endothia parasitica*, that came to known as the chestnut blight. It was brought from Asia on nursery plants. In 1913 the parasitic fungus was found on its natural host in Asia, where it does no harm to the chestnuts. But the eastern American [22] species, *Castanea dentata*, is so susceptible that it has almost died out over most of its range. This species carries two native species of *Endothia* that do not harm it, occurring also harmlessly on some other trees like oak; one of these two species also comes on the chestnut, C. *sativa*, in Europe. . . . Even by 1911, the outbreak, being [23] through wind-borne spores, had spread to at least ten states, and the losses were calculated to be at least twenty-five million dollars up to that date. In 1926 it was still spreading southwards, and by 1950 most of the chestnuts were dead except in the extreme south; and it is now on the Pacific coast too. So far, the only answer to the invasion has been to introduce the Chinese chestnut, C. *mollissima*, which is highly though not completely immune through having evolved into the same sort of balance with its parasite, as had the American trees with theirs; much as the big game animals of Africa can support trypanosomes in their blood that kill the introduced domestic animals like cattle and horses. The biological dislocation that occurs in this trypanosomiasis is the kind of thing that presumably would have happened also if the American chestnut had been introduced into Asia. The Chinese chestnut is immune both in Asia and America. Already by 1911, the European chestnuts grown in America had been found susceptible. In 1938 the blight appeared in

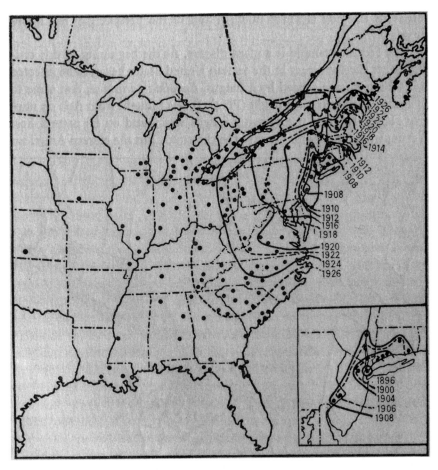

Spread of the breeding range of the European starling, *Sturnus vulgaris*, in the United States and Canada from 1891 to 1926. Dots outside the 1926 line are chiefly winter records of pioneer spread. (From MT Cooke, 1928.)

Italy where it has exploded fast and threatens the chestnut groves that there are grown in pure stands for harvesting the nuts; it has also reached Spain and will very likely reach Britain in the long or short run. Unfortunately the Chinese chestnut will not flourish in Italy, and [24] hopes are placed solely on the eventual breeding of a resistant variety of hybrid.

The third example is the European starling, *Sturnus vulgaris*, which has spread over the United States and Canada within a period of sixty years. (It has also become established in two other continents—South Africa and Australia, as well as in New Zealand.) This subspecies of starling has a natural range extending into Siberia, and from the north of Norway and Russia down to the Mediterranean. We should therefore expect it to be adaptable to a wide variety of continental habitats and climate. Nevertheless, the first few attempts to establish it in the United States were unsuccessful. Then from a stock of about eighty birds put into Central Park, New York, several pairs began to breed in 1891. After this the increase and spread went on steadily, apart from a severe mortality in the very cold winter of 1917–18. But up to 1916 the populations had not established beyond the Allegheny Mountains. Cooke's map (Figure 1) of the position up to the year 1926 shows how the breeding range had extended concentrically, with outlying records of non-breeding birds far beyond the outer breeding limits, which had moved beyond the Alleghenies but nowhere westward of a line running about southwards from Lake Michigan. By 1954 the process was nearly reaching its end, and the starling was to be found, at any rate on migration outside its breeding season, almost all over the United States, though it was not fully entrenched yet in parts of the West coast states. It was penetrating northern Mexico during migration, and in 1953 one starling was seen in Alaska. This was an ecological explosion indeed, starting from a few pairs breeding in a city park; just as the spread of the North American muskrat, *Ondatra zibethica*, over Europe was started from only five individuals kept by a landowner in Czechoslovakia in 1905. The muskrat now inhabits Europe in many millions, and its range has been augmented by subsidiary introductions for fur-breeding, with subsequent establishment of new centres of escaped animals and their progeny (Figure 2). Since 1922, over 200 transplantations of muskrats have been started in Finland, some originally from Czechoslovakia in 1922, and the annual catch is now between 100,000 and 240,000. Independent Soviet introductions have also made the muskrat an important fur animal in [25 (no text); 26] most of the great river systems of Siberia and northern Russia as well as in Kazakstan. In zoogeographical terminology, a purely Palaearctic species (the starling) and a purely Nearctic species (the muskrat) have both become Holarctic within half a century.

The fifth example is a plant that has changed part of our landscape—the tall strong-growing cord-grass or rice-grass, *Spartina townsendii*, that has colonized many stretches of our tidal mud-flats. It is a natural hybrid between a native English species, S. *maritima*, and an American species, S. *alterniflora*, the latter brought over and established on our South coast in the early years of the nineteenth century. The strong hybrid, which breeds true, was first seen in Southampton Water in 1870, and for thirty years was not particularly fast-spreading. But during the present century it has occupied great areas on the Channel coast not only in England but also on the North of France. It has also been planted in some other places in England, and has been introduced into North and South America, Australia and New Zealand. The original American parent has largely

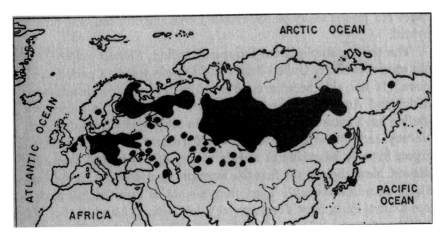

Distribution of the North American muskrat, *Ondatra zibethica*, in Europe and Asia.
(From A De Vos, RH Manville and RG Van Gelder, 1956.)

been suppressed or driven out by the hybrid form. Here is a peculiar result of the
spread of a species by man: the creation of a new polyploid hybrid species, from
parents of Nearctic and Palaearctic range, which then becomes almost cosmopolitan by further human introduction. And it is on the whole a rather useful
plant, because it stabilizes previously bare and mobile mud between tide-marks
on which often no other vascular plant could grow, helps to form new land and
often in the first instance provides salt-marsh grazing. Its effects upon the coastal
pattern are, however, not yet fully understood by physiographers and plant ecologists; but Tansley remarks that "no other species of salt-marsh plant, in northwestern Europe at least, has anything like so rapid and so great an influence in
gaining land from the sea."

Changes of similar magnitude have been taking place in fresh-water lakes
and rivers, as a result of the spread of foreign species. The sixth example given
here concerns the sea lamprey, *Petromyzon marinus*, in the Great Lakes region
of North America. This creature is a North Atlantic river-running species, mainly
living in the sea, and spawning in streams. But in the past it established itself
naturally in Lake Ontario, as well as in some small lakes in New York State. But
Niagara Falls formed an [27] insurmountable barrier to further penetration into
the inner Great Lakes. In 1829 the Welland Ship Canal was completed, providing a by-pass into Lake Erie. But it was a further hundred years or so before any
sea lampreys were observed in that lake. Then the invasion went with explosive
violence. By 1930 lampreys had reached the St Clair River, and by 1937 through
it to Lake Huron and Lake Michigan, where they began to establish spawning

runs in the streams flowing to these lakes. In 1946 they were in Lake Superior. Meanwhile the lampreys were attacking fish, especially the lake trout, *Salvelinus namaycush*, a species of great commercial importance. The sea lamprey is a combination of hunting predator and ectoparasite: it hangs on to a fish, secretes an anticoagulant and lytic fluid into the wound, and rasps and sucks the flesh and juices until the fish is dead, which may be after a few hours or as long as a week. The numbers of lake trout caught had always fluctuated to some extent, and the statistics of the fishery since 1889 have been thoroughly analysed. But never before the recent catastrophe had the catch collapsed so rapidly: in ten years after the lamprey invasion began to take effect, the numbers of lake trout taken in the American waters of Lake Huron and Lake Michigan fell from 8,600,000 lb. to only 26,000 lb. On the Canadian side things were little better. This was not caused by change in fishing pressure. Other species besides the lake trout have also been hard hit. Among these are the lake whitefish, burbot, and suckers, all of which declined in numbers. So, the making of a ship canal to give an outlet for produce from the Middle West has brought about a disaster to the Great Lakes fisheries over a century later. But in Lake Erie lampreys did not multiply, partly because there are not many lake trout there, but probably also because the streams are not right for spawning in.

The seventh example is the Chinese mitten crab, *Eriocheir sinensis*, a two-ounce crab that gets its name from the extraordinary bristly claws that make it look as if it was wearing dark fur mittens. At home it inhabits the rivers of North China, and it has been found over 800 miles up the Yang Tse Kiang. However, it breeds only in the brackish estuaries, performing considerable migrations downstream for the purpose. The females don't move so far away from the sea as the males, and they can lay up to a million eggs in a season, which hatch into a planktonic [28 (no text); 29] larva whose later Megalopa stage migrates up-river again. It is not really known how they got from East to West; they were first seen in the River Weser in 1912. The most likely explanation is that the young stages got into the tanks of a steamer and managed to get out again on arrival. Two large specimens were actually found in the sea-water ballast tanks of a German steamer in 1932, having, it is thought, got in locally from Hamburg Harbour. But these tanks are normally well screened. In the last forty-five years, mitten crabs have colonized other European rivers from the Baltic to the Seine. Those that invaded the Elbe have arrived as far as Prague, like Karel Capek's newts. This crab has not yet taken hold in Britain, though it may very likely do so some day, as one was caught alive in a water-screen of the Metropolitan Water Board at Chelsea in 1935.

These seven examples alone illustrate what man has done in deliberate and accidental introductions, especially across the oceans. Between them all they cover the waters of sea, estuary, river, and lake; the shores of sea and estuary; tropical and temperate forest country, farm land, and towns. In the eighteenth century there were few ocean-going vessels of more than 300 tons. Today there are thousands. A Government map made for one day, 7 March 1936, shows the position of every British Empire ocean-going vessel all over the world. There are 1,462 at sea and 852 in port; and this map does not include purely coasting vessels. Some idea of what this can mean for the spread of animals can be got from the results of an ecological survey done by Myers, a noted tropical entomologist, while travelling on a Rangoon rice ship from Trinidad to Manila in 1929. He amused himself by making a list of every kind of animal on board, from cockroaches and rice beetles to fleas and pet animals. Altogether he found forty-one species of these travellers, mostly insects. And when he unpacked his clothes in the hotel in Manila, he saw some beetles walk out of them. They were *Tribolium castaneum*, a well-known pest of stored flour and grains, which was one of the species living among the rice on the ship.

A hundred years of faster and bigger transport has kept up and intensified this bombardment of every country by foreign species, brought accidentally or on purpose, by vessel and by air, and also overland from [30] places that used to be isolated. Of course, not all the plants and animals carried around the world manage to establish themselves in the places they get to; and not all that do are harmful to man, though they must change the balance among native species in some way. But this worldwide process, gathering momentum every year, is gradually breaking down the sort of distribution that species had even a hundred years ago.

To see the full significance of what is happening, one needs to look back much further still, in fact many millions of years by the geological time-record. It was Alfred Russel Wallace who drew general public attention to the existence of great faunal realms in different parts of the world, corresponding in the main to the continents. These came to be known as Wallace's Realms, though their general distribution had already been pointed out by an ornithologist, PL Sclater. Wallace, however, did the enormous encyclopaedic work of assembling and classifying information about them. He supposed these realms to have been left isolated for such long periods that they had kept or evolved many special groups of animals. When one was a child, this circumstance was very simply summed up in books about animals. The tiger lives in India. The wallaby lives in Australia. The hippopotamus lives in Africa. One might have learned that the coypu or nutria lives in South America. A very advanced book might have speculated that this big water rodent was evolved inside South America, which we now know to be so. But nowadays, it would have to add a footnote to later editions, saying that the coypu is also doing quite well in the States of Washington, Oregon, California, and New Mexico; also in Louisiana (where 374,000 were trapped in one year recently); in south-east USSR; in France; and in the Norfolk Broads of East Anglia. In the Broads it carries a special kind of fur parasite, *Pitrufquenia coypus*, belonging to a family (Gyropidae) that also evolved in South America. These fur lice have antennae shaped like monkey-wrenches, which perhaps explains how they managed to hang on so well all the way from South America. But in very early times, say 100 million years ago in the Cretaceous Period, the world's fauna was much more truly cosmopolitan, not so much separated off by oceans, deserts, and mountains. If there had been a Cretaceous child living at the time the chalk was deposited in the warm [31] shallow seas at Marlborough or Dover, he would have read in his book or slate perhaps: "Very large dinosaurs occur all over the world except in New Zealand; keep out of their way." Or that water monsters occurred in more than one loch in the world. In fact, zoogeographically, it would have been rather a dull book, though the illustrations and accounts of the habits of animals would have been terrifically interesting. There would have been much less use for zoos: you just went out, with suitable precautions, and did dinosaur-watching wherever you were, and made punch-card records of their egg clutch-sizes. But the significance of these dinosaurs for the serious historical evidence is that you couldn't then get an animal the size of a lorry from one continent to another except by land; therefore the continents must have been joined together, at any rate fairly frequently, as geological time is counted.

This early period of more or less cosmopolitan land and fresh-water life was about three times longer than that between the Cretaceous Period and the present day. It was in the later period that Wallace's Realms were formed, because the sea, and later on great obstructions like the Himalaya and the Central Asian deserts, made impassable barriers to so many species. In fact the world had not one, but five or six great faunas, besides innumerable smaller ones evolved on isolated islands like Hawaii or New Zealand or New Caledonia, and in enormous remote lakes like Lake Baikal or Tanganyika. Man was not the first influence to start breaking up this world pattern. A considerable amount of re-mixing has taken place in the few million years before the Ice Age and since then: two big factors in this were the emergence of the Panama Isthmus from the sea, and the passage at various times across what is now Bering Strait. But we are artificially stepping up the whole business, and feeling the manifold consequences.

For thirty years I have read publications about this spate of invasions; and many of them preserve the atmosphere of first-hand reporting by people who have actually seen them happening, and give a feeling of urgency and scale that is absent from the drier summaries of text-books. We must make no mistake: we are seeing one of the great historical convulsions in the world's fauna and flora. We might say, with Professor Challenger, standing on Conan Doyle's "Lost World," with his black [32] beard jutting out: "We have been privileged to be present at one of the typical decisive battles of history—the battles which have determined the fate of the world." But how will it be decisive? Will it be a Lost World? These are questions that ecologists ought to try to answer.

Commentary

Charles S. Elton, "The Invaders" (1958)

LIBBY ROBIN

"The invaders" began its life as a BBC radio broadcast in 1957. "Balance and Barrier" was the title of the program, where Charles Elton, an elder statesman of science, reflected on what it meant to "conserve" the animal kingdom in a global world. In 1949, Elton was (with Arthur Tansley, Max Nicholson, and Dudley Stamp) a founding father of the Nature Conservancy in Britain, which aimed to ensure that good science underpinned national conservation practice (Crowcroft 1991).

The Ecology of Invasions goes beyond the national, taking a truly global view of conservation practice. Like many documents chosen for this book, Elton's "Invaders" is written by a scientist for a general public, grounded in material well beyond his own professional discipline. In another broadcast, Elton examined animal distributions through the work of nineteenth-century explorer Alfred Russel Wallace, arguing, "If we are to understand what is likely to happen to the ecological balance in the world, we need to examine the past as well as the future" (Elton 1958: 33). International conservation policy draws on history and biogeography, as well as ecology, Elton argues (by contrast with Paul Sears [pp. 174–182] who sees ecology itself embracing the other disciplines as a meta-discipline).

Elton's most important work was *Animal Ecology* (1927), a major textbook for most of the century, revised most recently in 2001. This book and his 1920s fieldwork in the Arctic caught the attention of North American businessman Copley Amory, who was concerned about changes in animal numbers, particularly salmon, in the Gulf of St. Lawrence. Amory sponsored a major international biological conference at Matamek, Quebec, Canada, in 1931 and asked Elton to be its secretary. As a result of this event the Bureau of Animal Population (BAP) was established at Oxford, its American-sounding name chosen by analogy with the Bureau of Biological Survey in Washington, D.C., something that appealed to its major sponsors, which included the New York Zoological Society. Elton was director of BAP, a hub for international collaborative research in animal ecology from 1932 until his retirement in 1967. Many of Elton's students found themselves "up against practical problems in the field," as in this era the Colonial Office was the biggest employer of British biologists, and BAP graduates were highly sought after throughout the British colonies and dominions. Francis Ratcliffe was one who went to Australia in 1929 to work on the ecology of flying foxes (fruit bats) and wrote *Flying Fox and Drifting Sand* (1938), a best-selling book about their ecology. Many, like Ratcliffe, returned as long-term visitors to BAP to undertake further research later in their careers, reinforcing the global reach of BAP.

At a time when zoology generally focused more on physiology and evolution, and ecology most often mean "plant ecology," *Animal Ecology* set an international benchmark in the new subdiscipline of zoological ecology. "Ecology is a branch of zoology which is perhaps more able to offer immediate practical help to mankind than any of the others," Elton wrote in its first preface to the book. "In the present rather parlous state of civilisation, it would seem particularly important to include it in the training of young zoologists." Civilization here was to be achieved by agriculture. Overcoming pests and problem animals was central to the mission of BAP. The "sociology and economics" of animals (as Elton termed it) was science in "practical service." Elton's own theories built on observations "made by people working on economic problems, many of whom were not trained as professional zoologists." Thus, when it came to preparing public lectures, Elton was not shy about informing his ecology with other disciplines, and stepping beyond science to serve the public interest that had underpinned *Animal Ecology*.

The Ecology of Invasions speaks of the "explosive" world. Elton is writing in the 1950s, when atomic testing was widespread, and Cold War anxieties were high. But here the explosions are ecological, not nuclear. They "take longer to happen," developing slowly, but "can be very impressive in their effects" (p. 367). Elton's examples draw on human history to show how animal population explosions can include human ones. The biota that live within humans drove the 1919 influenza pandemic, and an explosion in the rat population caused the medieval bubonic plague in Europe and brought on Black Death in Manchuria, China, as late as 1911. Ecological explosions have a global span, but appear in different eras. Time frames matter for practical management concerns. Predicting when a minor change will become an "explosion" is very difficult. Although Elton mentioned ecological explosions within native populations, his focus is primarily on outbreaks "that occur because a foreign species successfully invades another country" (p. 369). Framing this problem in terms of "invasion" invited a human-led military-style solution or management regime for dealing with the animals out of control.

Elton's work anticipated historian Alfred Crosby's *Ecological Imperialism* (1986) by three decades. Crosby's thesis was that European expansion was possible only because of the "avalanche" of the biota that came with the immigrants. Europeans arrived in the New World accompanied by "a grunting, lowing, neighing, crowing, chirping, snarling, buzzing, self-replicating and world-altering avalanche," as Crosby wrote in *Ecological Imperialism* (94). While Elton read history, Crosby apparently did not read Elton at all, although his book appeared at a time when international science was very much focused on invasions as threats to biological diversity. Crosby's work created a distinctive rhetoric of biological exchange, but it has remained better known among environmental historians than among invasion biologists, who revere Elton as the "father" of their discipline (Simberloff 2012).

Global conservation management practice has included managing "ecological explosions" since the Scientific Committee on Problems of the Environment (SCOPE) program launched a worldwide assessment of the ecology of biological invasions under the auspices of the International Council of Scientific Unions (ICSU) in the 1980s (Huenneke et al. 1988). The scientific questions that they identified are still driving research in invasion biology today: (1) What biological characteristics make an invader? (2) What makes a natural ecosystem susceptible to invasion? (3) How do we predict (quantitatively) the outcome of a particular introduction? (4) What is best practice for managing and conserving natural and semi-natural ecosystems? These are the questions that scientifically frame problems on many scales, from the local to the international.

Elton's imaginative leap was to conceptualize biota as *invaders*, actively expanding forces for ecosystem change. The "fight against invasion" was taken up by the scientific generations that followed him. "The Invaders," the document and the radio broadcast, created a framework that shapes our understanding of global change threats to what we now call biodiversity.

Further Reading

Crosby, Alfred W. (1986) 2004. *Ecological imperialism: The biological expansion of Europe, 900–1900.* Cambridge: Cambridge University Press.

Crowcroft, Peter. 1991. *Elton's ecologists: A history of the Bureau of Animal Population.* Chicago: University of Chicago Press.

Elton, Charles S. 1927. *Animal ecology.* London: Sidgwick & Jackson.

Huenneke, Laura, Dennis Glick, F. W. Waweru, Robert L. Brownell, Jr., and R. Goodland. 1988. "SCOPE Program on Biological Invasions: A status report." *Conservation Biology* 2(1):8–10.

Simberloff, Daniel. 2012. "Charles Elton: Pioneer conservation biologist." *Environment and History* 18(2):183–202.

The Forestry Projections and the Environment

Global-Scale Environmental Impacts

Council on Environmental Quality

The second global change implied by the forestry projections is a significant reduction in biotic diversity. The extent to which the diversity of the flora and fauna is maintained provides a basic index to the ecological health of the planet. Presently the world's biota contains an estimated 3–10 million species. Until the present century, the number of species extinguished as a result of human activities was small, and the species so affected were regarded as curiosities. Between now and 2000, however, the number of extinctions caused by human activities will increase rapidly. Loss of wild habitat may be the single most important factor. The projected growth in human population and economic activity can be expected to create enormous economic and political pressure to convert the planet's remaining wild lands to other uses. As a consequence, the extinction rate will accelerate considerably.

The death of an individual is very different from the death of a species. A species is a natural biotic unit—a population or a series of populations of sufficient genetic similarity that successful reproduction between individuals can take place. The death of an individual of a particular species represents the loss of one of a series of similar individuals all capable of reproducing the basic form, while the death of a species represents both the loss of the basic form and its reproductive potential.

Extinction, then, is an irreversible process through which the potential contributions of biological resources are lost forever. In fact, plant and animal species are the only truly nonrenewable resources. Most resources traditionally

Council on Environmental Quality [Gerald O. Barney (ed.)]. 1980. Excerpt from *Global 2000 report to the President of the United States: Entering the 21st century*, 3 vols., 1:149–155. New York: Pergamon Press.

termed "nonrenewable"—minerals and fossil fuels—received that label because they lack the reproductive capability. Yet most nonbiological compounds and elements are at least in theory, fully renewable. Given sufficient energy, non-biological resources can be separated, transformed, and restored to any desired form. By contrast, biotic resources—species (not individuals) and ecosystems— are completely nonrenewable. Once extinguished, species cannot be recreated. When extinct, biotic resources and their contributions are lost forever.

How many extinctions are implied by the Global 2000 Study's forestry projections? An estimate was prepared for the Global 2000 Study by Thomas E. Lovejoy of the World Wildlife Fund. Dr. Lovejoy's analysis, together with a tabular summary of the results, is presented on the next four pages. His figures, while admittedly rough, are frightening in magnitude. *If present trends continue—as they certainly will in many areas—hundreds of thousands of species can be expected to be lost by the year 2000.*

Extinction, of course, is the normal fate of virtually all species. The gradual processes of natural extinction will continue in the years ahead, but the *extinctions projected for the coming decades will be largely human-generated and on a scale that renders natural extinction trivial by comparison. Efforts to meet basic human needs and rising expectations are likely to lead to the extinction of between one-fifth and one-seventh of all species over the next two decades.* A substantial fraction of the extinctions are expected to occur in the tropics.

The lost potential of the earth's biological resources is often neglected in considering the consequences of deforestation in the tropics. Tropical forests contain both the richest variety and the least well known flora and fauna of the world. It would be difficult to overstate the potential value of this huge stock of biological capital, which, if carefully managed, could be a rich, sustainable source of building materials and fuel, as well as medicinal plants, specialty woods, nuts, and fruits. However, if present trends continue, sustained benefits from this capital will never be realized. Unique local plants and animals will be unknowingly and carelessly destroyed. Particularly well-adapted or fast-growing local trees will be cut before their fruits or seeds are collected. Predatory insects and plants with herbicidal or insecticidal properties will be lost for lack of observation and study. Diverse assemblies of gigantic trees, their understories, and their resident communities of mammals, birds, and insects—natural wonders every bit as unique and beautiful as the Grand Canyon—will be irreparably lost. In short, the projected loss of tropical forests represents a massive expenditure of biological capital, an expenditure so sudden and so large that it will surely limit future benefits that even careful management and husbanding can sustain from the remaining biotic resources of the earth.

. . .

A Projection of Species Extinctions

This projection was developed for the Global 2000 Study by
Dr. Thomas E. Lovejoy of the World Wildlife Fund.

Virtually all of the Global 2000 Study's projections—especially the forestry, fisheries, population, and GNP projections—have implications for the extinction of species. Accepting these projections as correct, how many extinctions can be anticipated by 2000?

Probably the largest contribution to extinctions over the next two decades will come as a result of deforestation and forest disruption (e.g., cutting "high-grade" species), especially in the tropics. The forestry projections . . . provide an estimate of the amount of tropical deforestation to be expected. The question then is: What fraction of the species now present will be extinguished as a result of that deforestation?

Possible answers are provided by the curves in Figure 13–8. The curves in this figure do not represent alternative scenarios but rather reflect the uncertainty in the percent of species lost as a result of a given amount of deforestation. The endpoints are known with more accuracy than other points on the curves. Clearly, at zero deforestation the resulting loss of species is zero—and for 100 percent deforestation, the loss approaches 100 percent. The reasons for the high losses at 100 percent deforestation are as follows.

The lush appearance of tropical rain forests masks the fact that these ecosystems are among the most diverse and fragile in the entire world. The diversity[1] of tropical forests stems in part from the tremendous variety of life zones created by altitude, temperature, and rainfall variations. The fragility of tropical forests stems from the fact that, in general, tropical soils contain only a very limited stock of nutrients. Typically, the nutrients in tropical soils are only a small part of the total inventory of nutrients in the tropical ecosystem. Most of the nutrients are in the diverse flora and fauna of the forests themselves. Tropical forests sustain themselves through a rapid and highly efficient recycling of nutrients. Little nutrient is lost when an organism dies, but when extensive areas of forest are cleared, the nutrients are quickly leached out and lost.

Studies have shown that there are a wide variety of tropical soils. Some (such as those in lowland swamps) are rich, but most are either thin, infertile, and highly acidic, or thick and highly leached of nutrients. Recent aerial surveys of the Amazon basin, for example, indicate that only 2 percent of the soils are suitable for sustained agriculture. Once cleared, the recycling of nutrients is interrupted, often permanently. In the absence of the forest cover, the remaining vegetation and exposed soil cannot hold the rainfall and release the water slowly.

The critical nutrients are quickly leached from the soils, and erosion sets in—first, sheet erosion, then gully erosion. In some areas only a few years are required for once dense forest lands to turn to virtual pavements of laterite, exposed rock, base soil, or coarse "weed" grasses. The Maryland-sized area of Bragantina in the Amazon basin is probably the largest and best-known area to have already undergone this process, becoming what has been called a "ghost landscape."

With formerly recycled nutrients lost through deforestation and its after-effects, the capacity of a tropical rain forest to regenerate itself is highly limited and much less than that of a temperate forest. The possibilities for regeneration are limited further by the fact that the reproductive biology of many of the tree species found in mature tropical forests is adapted to recolonizing small patches of disturbed forest rather than the large areas now being cleared. As a result and

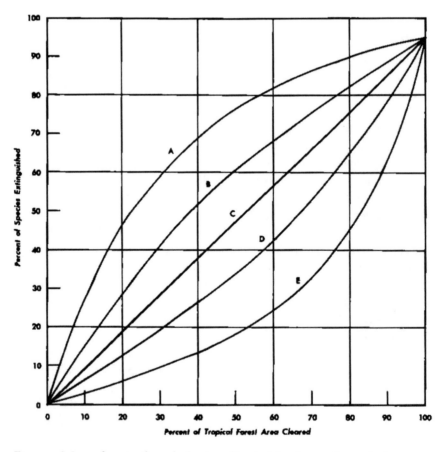

Figure 13-8. Loss of species through clearing of tropical forest areas—five projections.

in spite of rapid succession rates, the disruption and simplification from deforestation of tropical rain forests is, for the most part, irreversible, given the time scale necessary to preserve the present mature and diverse biota. While a few attempts at reforestation to natural tropical forest are being made, only limited success has yet been achieved (e.g., in Puerto Rico).

More typically, tropical forests are either cleared and abandoned, or (if the soil and economics permit) converted to plantation forests of high-growth species such as *Eucalyptus*, *Pinus*, and *Gmelina*, which are not suitable for the diverse local fauna found in mature tropical forests.

As a consequence (and for purposes of the rough calculations here) the rain forest areas modified by deforestation can be expected, with few exceptions, to include a negligible number of the species that were present in the virgin forests. The estimate used for the endpoint of the curves in the figure for 100 percent deforestation is therefore a 95 percent loss of species. So much for the endpoints of the curves.

The general shape of the curves (convex, linear, or concave) depends in large part on the size of the areas needed to preserve the ecosystems on which the species depend. Unfortunately, little is known about the size of these areas. Most tropical species occur at exceedingly low densities. Dispersal (to compensate for local extinctions) is probably an important part of their biology, and this survival strategy is impeded or precluded when forest areas are reduced to isolated reserves.

It is also known that only a limited amount of a rain forest region can be converted to non-forest before local (or regional) changes in climate will occur, endangering the remaining forest areas. For example, current estimates put the percent of precipitation in the Amazon basin generated by the forest (as opposed to the ocean) at slightly over 50 percent. Although this contribution to the precipitation will not be eliminated entirely by deforestation, it will be reduced, and reductions in rainfall beyond a certain point could initiate an irreversible drying trend.

In the face of our limited knowledge, the concept of "refugia" provides one approach to estimating the areas needed to preserve tropical ecosystems and their species. During the Pleistocene glaciations, the climate in the equatorial regions was generally drier and unable to support tropical rain forests. Rain forests did survive, however, in relatively small patches, now termed refugia. It has been found that within the Amazon basin there are areas of concentration of species not occurring elsewhere (centers of endemism), and these areas of concentration are thought by many ecologists to represent vestiges of Pleistocene refugia. Refugia have now been identified in the basin for a few families of organisms (primarily small animals, including some insects and a few plants). There is wide variation in size among the areas of concentration that have been measured, the smallest areas being for the class with the most species (i.e., insects). And, of course, the smaller the area of concentration, the more vulnerable the species are to extinction by reason of deforestation and loss of habitat.

How then should the curves in Figure 13-8 be drawn? Assuming that the refugia concept applies generally to all tropical forests, a convex curve (curve A or B) would result if the refugia turn out to be relatively small and if the refugia should happen to be cleared first. A linear curve (curve C) would result from random cutting if the refugia were not adjacent but overlapping, or were relatively small. (The limited data now available suggest that the refugia are in fact relatively small for classes with the most numerous species.) A concave curve (curve D or E) would result if the refugia should turn out to be small and highly overlapping, and (a) if efforts are made to identify and preserve the refugia, or (b) if the refugia are widely separated from one another.

While curve D is used in the following calculations, it may underestimate the impacts of the projected deforestation. Were it possible—which it is not—to create instantly a minimum system of biological reserves of adequate size and ideal location, the impacts of the projected deforestation would still be on the order of the effects of the Pleistocene glaciations. Even so, such a system would not be secure because of local climatic effects and ever rising political and economic pressures. It is difficult to estimate how much of the world's tropical forest has been cleared and how such pressures will influence future cutting, but considering the amount of forest known to have already been destroyed, it is too late to achieve even the minimum system of reserves. Furthermore, present global conservation plans for rain forests are limited. For example, the *most ambitious proposal* for conservation in the Amazon basin would be site parks and reserves (Table 13.30) in the areas where refugia are thought to have occurred, but the total area of the parks and reserves would comprise only 5 percent of the total land area of the Amazon. Such a system would mimic the distribution of forests at the height of the Pleistocene glaciations.

What then is a reasonable estimate of global extinctions by 2000? Given the amount of tropical forest already lost (which is important but often ignored), the extinctions can be estimated as shown in Table 13-30. In the low deforestation case, approximately 15 percent of the planet's species can be expected to be lost. In the high deforestation case, perhaps as much as 20 percent will be lost. This means that of the 3-10 million species now present on the earth, at least 500,000-600,000 will be extinguished during the next two decades. The largest number of extinctions can be expected in the insect order—many of them beneficial species—simply because there are so very many species of insects. The next highest number of extinctions will be among plants. While the projected extinctions refer to all biota, they are *much* larger than the 1,000 bird and mammal

Table 13–30

Extinctions of Species Implied by the Global 2000 Study's Projections

	Present Species (in thousands)	Projected Deforestation	Loss of Species	Extinctions (in thousands)
	Low Deforestation Case			
Tropical forests				
Latin America	300–1,000	50	33	100–333
Africa	150–500	20	13	20–65
S. and SE. Asia	300–1,000	60	43	129–430
Subtotal	759–2,500			249–828
All other habitats				
Oceans, fresh water, nontropical forests, islands, etc.	2,250–7,500	—	8	188–625
Total	3,000–10,000			437–1,453
	High Deforestation Case			
Tropical Forests				
Latin America	300–1,000	67	50	150–500
Africa	150–500	67	50	75–250
S. and SE. Asia	300–1,000	67	50	150–500
Subtotal	750–2,500			375–1,250
All other habitats				
Oceans, fresh water, nontropical forests, islands, etc.	2,250–7,500	—	8	188–625
Total	3,000–10,000			563–1,875

species now recognized as endangered. Clearly the extinctions caused by human activities will rise to unprecedented rates by 2000.

Note

1. Diversity here is used simply in the sense of number of species. Ecologists sometimes use more complex indices of ecological diversity.

Commentary

Council on Environmental Quality, *The Forestry Projections and the Environment* (1980)

MARK V. BARROW JR.

On 1 January 1970, President Richard Nixon signed the National Environmental Policy Act (NEPA) into law with the declaration that history would remember the 1970s as the "Decade of the Environment." That prediction proved remarkably prescient as the ensuing years saw an outpouring of public support for, scientific research on, and legislative response to a wide range of environmental issues. Soon after the passage of NEPA—which mandated Environmental Impact Statements as part of the project planning process for federal agencies and established the President's Council on Environmental Quality to review those statements— Americans celebrated the first Earth Day, Nixon established the Environmental Protection Agency, and Congress passed the Clean Air Act. A slew of federal environmental initiatives followed, establishing a sweeping legislative and regulatory framework that remains largely in place to this day.

One of many areas of environmental concern during this remarkable decade was declining wildlife. In 1973 Congress passed the Endangered Species Act, one of the most stringent and comprehensive environmental laws ever to be enacted. Representing the culmination of over a century of growing anxiety about the fate of vanishing wildlife and coming on the heels of the increasingly publicized declines of the whooping crane, the American alligator, the bald eagle, and numerous other iconic species, that legislation required the Secretary of the Interior to maintain a list of species and subspecies facing extinction. It also prohibited any action that might harm any organism on that protected list, which included both native and international forms. The global scope of the problem of wildlife decline gained further recognition with the Convention on Trade in Endangered Species (CITES), which was negotiated in 1973, opened for signature a year later, and entered into force in 1975. By 1980, sixty nations were party to the treaty, a list that has grown to 175 today.

Thomas E. Lovejoy was one of a dozen or so scientists during this period who began issuing frequent public warnings about the threat of extinction. A tropical biologist by training, Lovejoy earned his Ph.D. from Yale, where he studied under the ecologist G. Evelyn Hutchinson. In 1965, he made his first research trip to the Amazon rainforest. Over the next decade he would spend many years in the field there, witnessing firsthand the destruction that was taking place as this vast, biologically rich region was opened up for agricultural development and resource extraction. As early as 1973, he published a paper warning about the environmental consequences of the Trans-Amazonian Highway, one of a series of roads that were beginning to bisect the region. Just as scientists raised concerns about the growing threats the Amazon basin faced, they were also gaining a firmer handle on exactly how diverse and how fragile this unique ecosystem, home to as much as 30 percent of the world's species, really was. At the same time, scientists began speaking about species richness in terms of "biological diversity," a phrase that Lovejoy was one of the first to use and popularize.

Given his credentials and growing prominence, Lovejoy was a natural to turn to for expertise on global extinction rates. In 1977, Donald King, a biologist in the State Department, proposed that President Jimmy Carter include a call for a "one year study of probable changes in the world's population, natural resources, and environment through the end of the century" in an upcoming environmental address. At the urging of his Council on Environmental Quality, Carter took the bait, issuing a directive for what soon became known as the Global 2000 Study. Inspired by the work of the Club of Rome, whose *Limits to Growth* (see Part 2) presented a global model of a rapidly growing human population facing a finite resource base, the director of the Global 2000 Study, Gerald O. Barney, sought to construct a new global model largely using federal data and personnel. When it came to modeling extinction, though, Barney turned to Thomas Lovejoy, who was then serving as director of programs for the World Wildlife Fund—U.S.

Lovejoy's short "Projection of Species Extinction" represented a state-of-the-art attempt to predict global extinction rates. Appearing in a section of the *Global 2000 Report* (1980) that focused on modeling the rates and impacts of deforestation, it relied heavily on the theory of island biogeography to forecast what species loss might look like at different levels of forest destruction across the globe. More than a decade earlier, ecologist Robert H. MacArthur and his colleague E. O. Wilson had published *The Theory of Island Biogeography* (1967), a book that sought to push the science of ecology in more theoretically and mathematically robust directions. More specifically, MacArthur and Wilson sought to explain how migrations to and die-offs on an island eventually reached what they called an equilibrium point, a more or less constant number of species that could be predicted based on the overall size of the island. In subsequent research, Wilson and his collaborators provided experimental confirmation of their theory and sought to apply it to the design of nature reserves. The assumption was that any plot of land surrounded by degraded habitat would resemble an island in terms of the number of species it could support and that fragmentation of that plot would result in a net overall species loss as a new equilibrium point between immigration and species loss was eventually reached. Although Lovejoy's extinction estimates varied widely, even at the low end of the range, they proved quite alarming.

Following publication of the *Global 2000 Report*, the broader environmental movement faced pushback when Ronald Reagan was elected president, while scientists and conservationists continued to develop new responses to the specter of extinction. In the early 1980s, concerned biologists forged an explicitly mission-oriented, crisis-driven, interdisciplinary field known as "conservation biology," which sought to protect, maintain, and restore life on the planet in the face of rising extinction rates. In May 1985, delegates attending the Second International Conference on Conservation Biology voted to create a new organization, the Society for Conservation Biology, to promote and support the new field. A year later, the National Forum on Biodiversity was held in Washington, D.C. The organizer of that internationally broadcast event, Norman Rosen, contracted the phrase *biological diversity* to coin a new term, *biodiversity*, that quickly caught on in conservation circles. Soon biologists were calling attention to the "biodiversity crisis" (see Soulé essay in this part), a threat that they compared to five earlier mass extinction events that had occurred in the earth's history, when as many as 95 percent of the species had been destroyed. Although the decade of the environment had passed, concern about extinction remained strong, while threats to world's flora and fauna continued to grow. As one of the most prominent conservation biologists in the United States, Lovejoy has continued to warn about the dire consequences of mass extinction and to develop concrete strategies, like "debt for nature" swaps, to begin meeting that challenge.

Further Reading

Barney, Gerald O. 1980. Preface to Council on Environmental Quality (Gerald O. Barney [ed.]), *The Global 2000 report to the President of the United States: Entering the 21st century*, vol. 1: *The summary report*. Special edition with the environment projections and the government's global model, vii–xvii. New York: Pergamon Press.

Barrow, Mark V., Jr. 2009. *Nature's ghosts: Confronting extinction from the age of Jefferson to the age of ecology*. Chicago: University of Chicago Press.

Farnham, Timothy J. 2007. *Saving nature's legacy: Origins of the idea of biological diversity*. New Haven: Yale University Press.

Kingsland, Sharon. 2002. "Designing nature reserves: Adapting ecology to real-world problems." *Endeavor* 26(1):9–14.

Lovejoy, Thomas, III. 2000. "Biological diversity." In Heather Newbold (ed.), *Life stories: World-renowned scientists reflect on their lives and the future of life on earth*, 42–54. Berkeley: University of California Press.

MacArthur, R. J., and Wilson, E. O. 1967. *The theory of island biogeography*. Princeton, NJ: Princeton University Press.

What Is Conservation Biology?

MICHAEL E. SOULÉ

[727] Conservation biology, a new stage in the application of science to conservation problems, addresses the biology of species, communities, and ecosystems that are perturbed, either directly or indirectly, by human activities or
other agents. Its goal is to provide principles and tools for preserving biological
diversity. In this article I describe conservation biology, define its fundamental
propositions, and note a few of its contributions. I also point out that ethical
norms are a genuine part of conservation biology, as they are in all mission- or
crisis-oriented disciplines.

Crisis Disciplines

Conservation biology differs from most other biological sciences in one important way: it is often a crisis discipline. Its relation to biology, particularly ecology, is analogous to that of surgery to physiology and war to political science. In
crisis disciplines, one must act before knowing all the facts; crisis disciplines are
thus a mixture of science and art, and their pursuit requires intuition as well as
information. A conservation biologist may have to make decisions or recommendations about design and management before he or she is completely comfortable with the theoretical and empirical bases of the analysis (May 1984, Soulé
and Wilcox 1980, chap. 1). Tolerating uncertainty is often necessary.

Conservation biologists are being asked for advice by government agencies
and private organizations on such problems as the ecological and health consequences of chemical pollution, the introduction of exotic species and artificially
produced strains of existing organisms, the sites and sizes of national parks, the
definition of minimum conditions for viable populations of particular target species, the frequencies and kinds of management practices in existing refuges and

Michael E. Soulé. 1985. Excerpt from "What is conservation biology?" *Bioscience* 35(11):
727–734. [Full list of references included; original page numbers in square brackets.]

managed wildlands, and the ecological effects of development. For political reasons, such decisions must often be made in haste.

For example, the rapidity and irreversibility of logging and human resettlement in Western New Guinea (Irian Jaya) prompted the Indonesian government to establish a system of national parks. Two of the largest areas recommended had never been visited by biologists, but it appeared likely that these areas harbored endemic biotas (Jared M. Diamond *pers. com.*). Reconnaissance later confirmed this. The park boundaries were established in 1981, and subsequent development has already precluded all but minor adjustments. Similar crises are now facing managers of endangered habitats and species in the United States—for example, grizzly bears in the Yellowstone region, black-footed ferrets in Wyoming, old-growth Douglas-fir forests in the Pacific Northwest, red-cockaded woodpeckers in the Southeast, and condors in California.

Other Characteristics of Conservation Biology

[728] Conservation biology shares certain characteristics with other crisis-oriented disciplines (Figure 77.1). For example, cancer biology shares conservation biology's synthetic, eclectic, multidisciplinary structure. Both fields take many of their questions, techniques and methods from a broad range of fields,

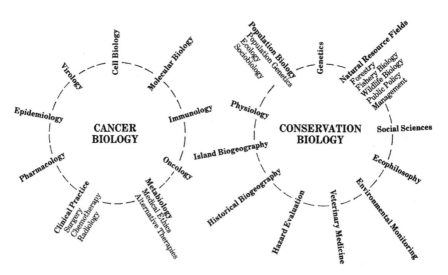

Figure 77.1. Cancer biology and conservation biology are both synthetic and interdisciplinary sciences. The dashed lines indicate the artificial nature of the borders between disciplines and between" basic" and "applied" research. See text

not all biological. This illustration shows the artificiality of the dichotomy between pure and applied disciplines and illustrates the dependence of the biological sciences on social science disciplines. Today, for example, any recommendations about the location and size of national parks should consider the impact of the park on indigenous peoples and their cultures, on the local economy, and on opportunity costs such as forfeited logging profits.

There is much overlap between conservation biology and the natural resource fields, especially fisheries biology, forestry, and wildlife management. Nevertheless, two characteristics of these fields often distinguish them from conservation biology. The first is the dominance in the resource fields of utilitarian, economic objectives. Even though individual wildlife biologists honor Aldo Leopold's land ethic and the intrinsic value of nature, most of the financial resources for management must go to enhancing commercial and recreational values for humans.

The second distinguishing characteristic is the nature of these resources. For the most part, they are a small number of particularly valuable target species (e.g., trees, fishes, deer, and waterfowl)—a tiny fraction of the total biota. This distinction is beginning to disappear, however, as some natural resource agencies become more "ecological" and because conservation biologists frequently focus on individual endangered, critical, or keystone species.

Conservation biology tends to be holistic, in two senses of the word. First, many conservation biologists, including many wildlife specialists, assume that ecological and evolutionary processes must be studied at their own macroscopic levels and that reductionism alone cannot lead to explanations of community and ecosystem processes such as body-size differences among species in guilds (Cody and Diamond 1975), pollinator-plant coevolution (Gilbert and Raven 1975), succession, speciation, and species-area relationships. Even ecological reductionists, however, agree that the proper objective of conservation is the protection and continuity of entire communities and ecosystems. The holistic assumption of conservation biology should not be confused with romantic notions that one can grasp the functional intricacies of complex systems without conducting scientific and technological studies of individual components (Levins and Lewontin 1985, chap. 6). Holism is not mysticism.

The second implication of the term holistic is the assumption that multidisciplinary approaches will ultimately be the most fruitful. Modern biogeographic analysis is now being integrated into the conservation movement (Diamond 1975, Simberloff and Abele 1976, Terborgh 1974, Wilcox 1980). Population genetics, too, is now being applied to the technology of wildlife management (Frankel 1974, Frankel and Soulé 1981, Schonewald-Cox et al. 1983, Soulé and Wilcox 1980). Multidisciplinary research, involving government agencies and wildlife biologists, is also evident in recent efforts to illuminate the question of viable population size (Salwasser et al. 1984).

Another distinguishing characteristic of conservation biology is its time scale. Generally, its practitioners attach less weight to aesthetics, maximum yields, and profitability, and more to the long-range viability of whole systems and species, including their evolutionary potential. Long-term viability of natural communities usually implies the persistence of diversity, with little or no help from humans. But for the foreseeable future, such a passive role for managers is unrealistic, [729] and virtually all conservation programs will need to be buttressed artificially. For example, even the largest nature reserves and national parks are affected by anthropogenic factors in the surrounding area (Janzen 1983, Kushlan 1979), and such refuges are usually too small to contain viable populations of large carnivores (Frankel and Soulé 1981, Shaffer and Samson 1985). In addition, poaching, habitat fragmentation, and the influx of feral animals and exotic plants require extraordinary practices such as culling, eradication, wildlife immunization, habitat protection, and artificial transfers. Until benign neglect is again a possibility, conservation biology can complement natural resource fields in providing some of the theoretical and empirical foundations for coping with such management conundrums.

Postulates of Conservation Biology

Conservation biology, like many of its parent sciences, is very young. Therefore, it is not surprising that its assumptions about the structure and function of natural systems, and about the role of humans in nature, have not been systematized. What are these postulates? I propose two sets: a functional, or mechanistic, set and an ethical, or normative, set.

THE FUNCTIONAL POSTULATES

These are working propositions based partly on evidence, partly on theory, and partly on intuition. In essence, they are a set of fundamental axioms, derived from ecology, biogeography and population genetics, about the maintenance of both the form and function of natural biological systems. They suggest the rules for action. A necessary goal of conservation biology is the elaboration and refinement of such principles.

The first, the evolutionary postulate states: *Many of the species that constitute natural communities are the products of coevolutionary processes.* In most communities, species are a significant part of one another's environment. Therefore, their genetically based physiological and behavioral repertoires have been naturally selected to accommodate the existence and reactions of a particular biota. For example, the responses of prey to a predator's appearance or of a phytophagous insect to potential host plants are continually "tuned" by natural selection. . . .

There are many "corollaries" of this postulate. Strictly speaking, most of them are empirically based generalizations. The following all assume the existence of community processes as well as a coevolutionary component in community structure.

Species are interdependent. Not only have species in communities evolved unique ways of avoiding predators, locating food, and capturing and handling prey, but mutualistic relationships are frequent (Janzen 1975, Seifert and Seifert 1979). This is not to say that every species is essential for community function, but that there is always uncertainty about the interactions of species and about the biological consequences of an extinction. Partly for this reason, Aldo Leopold (1953) admonished conservationists to save all of the parts (species) of a community.

Many species are highly specialized. Perhaps the majority of animal species, including phytophagous insects, parasites, and parasitoids, depend on a particular host (Price 1980). This means that the coattails of endangered host species can be very long, taking with them dozens (Raven 1976) or hundreds (Erwin 1983) of small consumer species when they go.

Extinctions of keystone species can have long-range consequences. The extinction of major predators, large herbivores, or plants that are important as breeding or feeding sites for animals may initiate sequences of causally linked events that ultimately lead to further extinctions (Frankel and Soulé 1981, Gilbert 1980, Terborgh and Winter 1980).

Introductions of generalists may reduce diversity. The introduction of exotic plant and animal species may reduce diversity, especially if they are large or generalist species (Diamond 1984, Elton 1958). Apparently, the larger the land mass, the less the impact of exotics (e.g., Simberloff 1980).

The evolutionary postulate and its corollaries formalize the evidence that natural communities comprise species whose genetic makeups have been mutually affected by their coexistence (Futuyma and Slatkin 1983, Gilbert and Raven 1975). An alternative theory, the null hypothesis that communities are randomly assembled, is usually restricted to "horizontal" subcommunities such as guilds, specific taxa, or trophic levels (e.g., James and Boecklen 1984). In general, this latter thesis lacks empirical support, except that competitive structuring within guilds or trophic levels is often absent or difficult to demonstrate (Strong et al. 1984), and that harsh environments or the vagaries of dispersal may often be more important than biological interactions in determining local community composition (e.g., Underwood and Denley 1984).

The second functional postulate concerns the scale of ecological processes: *Many, if not all, ecological processes have thresholds below and above which they become discontinuous, chaotic, or suspended.* This postulate states that many ecological processes and patterns . . . are interrupted or fail altogether where the system is too small. Smallness and randomness are inseparable.

Nonecological processes may also dominate at the other end of the spatial and temporal scale, in very large or very old systems. In very large systems, such as continents, climatic and physiographic phenomena often determine the major patterns of the landscape, including species distribution. In very old systems, ecological processes give way to [730] geological and historical ones or to infrequent catastrophic events, such as inundation, volcanism, and glaciation. In other words, ecological processes belong to an intermediate scale of physical size and time (MacArthur 1972), and these processes begin to fail or are overwhelmed near the extremities of these ranges.

Two major assumptions, or generalizations, underlie this postulate. First, *the temporal continuity of habitats and successional stages depends on size.* The random disappearance of resources or habitats will occur frequently in small sites but rarely, if ever, in large ones. The reasons include the inherent randomness of such processes as patch dynamics, larval settlement, or catastrophic events, as well as the dynamics of contagious phenomena such as disease, windstorm destruction, and fire. The larger an area, the less likely that all patches of a particular habitat will disappear simultaneously. Species will disappear if their habitats disappear.

Second, *outbursts reduce diversity.* If population densities of ecologically dominant species rise above sustainable levels, they can destroy local prey populations and other species sharing a resource with such species. Outbursts are most probable in small sites that lack a full array of population buffering mechanisms, including habitat sinks for dispersing individuals, sufficient predators, and alternative feeding grounds during inclement weather. The unusually high population densities that often occur in nature reserves can also increase the rate of disease transmission, frequently leading to epidemics that may affect every individual. Taken together, the corollaries of this postulate lead to the conclusion that survival rates of species in reserves are proportional to reserve size. Even though there is now a consensus that several small sites can contain as many species as one large site (when barriers to dispersal are absent), the species extinction rate is generally higher in small sites (Soulé and Simberloff, 1986).

The third functional postulate concerns the scale of population phenomena: *Genetic and demographic processes have thresholds below which nonadaptive, random forces begin to prevail over adaptive, deterministic forces within populations. . . .* The main implication of this postulate for conservation is that the probability of survival of a local population is a positive function of its size. One of the corollaries of this postulate is that below a certain population size (between 10 and 30), the probability of extinction from random demographic events increases steeply (Shaffer 1981).

The next three corollaries are genetic. First, populations of outbreeding organisms will suffer a chronic loss of fitness from inbreeding depression at effective population sizes of less than 50 to 100 (Franklin 1980, Soulé 1980). Second, genetic drift in small populations (less than a few hundred individuals) will cause a progressive loss of genetic variation; in turn, such genetic erosion will reduce immediate fitness because multilocus heterozygosity is generally advantageous in outbreeding species (Beardmore 1983, Soulé 1980 . . .). Finally, natural selection will be less effective in small populations because of genetic drift and the loss of potentially adaptive genetic variation (Franklin 1980).

The fourth functional postulate is that *nature reserves are inherently disequilibrial for large, rare organisms*. There are two reasons for this. First, extinctions are inevitable in habitat islands the size of nature reserves (MacArthur and Wilson 1967); species diversity must be artificially maintained for many taxa because natural colonization (reestablishment) from outside sources is highly unlikely. Second, speciation, the only other nonartificial means of replacing species, will not operate for rare or large organisms in nature reserves because reserves are nearly always too small to keep large or rare organisms isolated within them for long periods, and populations isolated in different reserves will have to be maintained by artificial gene flow if they are to persist. Such gene flow would preclude genetic differentiation among the colonies (Soulé 1980).

THE NORMATIVE POSTULATES

The normative postulates are value statements that make up the basis of an ethic of appropriate attitudes toward other forms of life—an ecosophy (Naess 1973). They provide standards by which our actions can be measured. . . .

Diversity of organisms is good. Such a statement cannot be tested or proven. The mechanisms by which such value judgments arise in consciousness are unknown. The conceptual mind may accept or reject the idea as somehow valid or appropriate. If accepted, the idea becomes part of an individual's philosophy.

We could speculate about the subconscious roots of the norm, "diversity is good." In general, humans enjoy variety. We can never know with certainty whether this is based on avoiding tedium and boredom or something else, but it may be as close to a universal norm as we can come. . . .

A corollary of this postulate is that the untimely extinction of populations and species is bad. Conservation biology does not abhor extinction per se. Natural extinction is thought to be either value free or good because it is part of the process of replacing less well-adapted gene pools with better adapted ones. Ultimately, natural extinction, unless it is catastrophic, does not reduce biological diversity, for it is offset by speciation. Natural extinctions, however, are rare events on a human time scale. Of the hundreds of vertebrate extinctions that have occurred during the last few centuries, few, if any, have been natural (Diamond 1984, Frankel and Soulé 1981), [731] whereas the rate of anthropogenic extinctions appears to be growing exponentially. . . . Although disease and suffering in animals are unpleasant and, perhaps, regrettable, biologists recognize that conservation is engaged in the protection of the integrity and continuity of natural processes, not the welfare of individuals. At the population level, the important processes are ultimately genetic and evolutionary because these maintain the potential for continued existence. Evolution, as it occurs in nature, could not proceed without the suffering inseparable from hunger, disease, and predation.

For this reason, biologists often overcome their emotional identification with individual victims. For example, the biologist sees the abandoned fledgling or the wounded rabbit as part of the process of natural selection and is not deceived that "rescuing" sick, abandoned, or maimed individuals is serving the species or the cause of conservation. . . . Therefore, the ethical imperative to conserve species diversity is distinct from any societal norms about the value or the welfare of individual animals or plants. This does not in any way detract from ethical systems that provide behavioral guidance for humans on appropriate relationships with individuals from other species, especially when the callous behavior of humans causes animals to suffer unnecessarily. Conservation and animal welfare, however, are conceptually distinct, and they should remain politically separate.

Returning to the population issue, we might ask if all populations of a given species have equal value. I think not. The value of a population, I believe, depends on its genetic uniqueness, its ecological position, and the number of extant populations. A large, genetically polymorphic population containing unique alleles or genetic combinations has greater value, for example, than a small, genetically depauperate population of the same species. Also, the fewer the populations that remain, the greater the probability of the simultaneous extinction (random or not) of *all* populations, and thus of the species. Hence, how precious a population is is a function of how many such populations exist.

Ecological complexity is good. This postulate parallels the first one, but assumes the value of habitat diversity and complex ecological processes. . . . Ecological diversity can be enhanced artificially, but the increase in diversity can be more apparent than real (especially if cryptic taxa and associations are considered, such as soil biotas and microbial communities). In addition, humans tend to sacrifice ecological and geographic heterogeneity for an artificially maintained, energy-intensive, local species diversity. Take, for example, the large numbers of plant taxa maintained in the warm-temperate and subtropical cities of the world. Most of these species are horticultural varieties that do well in landscaped gardens and parks. . . . But the roses, citrus, camellias, bougainvilleas, daffodils, eucalyptus, and begonias are everywhere similar. This combination of local variety and geographic homogeneity produces several pleasant benefits for humans. Not only are the exotic species more spectacular, but the world traveler can always feel botanically at home. . . . But these aesthetic benefits are costly. The price is low geographic diversity and ecological complexity. Botanical gardens, zoos, urban parks, and aquaria satisfy, to a degree, my desire to be with other species, but not my need to see wild and free creatures or my craving for solitude or for a variety of landscapes and vistas.

Evolution is good. Implicit in the third and fourth functional postulates is the assumption that the continuity of evolutionary potential is good. Assuming that life itself is good, how can one maintain an ethical neutrality about evolution? Life itself owes its existence and present diversity to the evolutionary process. Evolution is the machine, and life is its product. One possible corollary of this axiom is an ethical imperative to provide for the continuation of evolutionary processes in as many undisturbed natural habitats as possible.

Biotic diversity has intrinsic value, irrespective of its instrumental or utilitarian value. This normative postulate is the most fundamental. In emphasizing the inherent value of nonhuman life, it distinguishes the dualistic, exploitive world view from a more unitary perspective: Species have value in themselves, a value neither conferred nor revocable, but springing from a species' long evolutionary heritage and potential or even from the mere fact of its existence. A large literature exists on this subject (Devall and Sessions 1985, Ehrenfeld 1981, Passmore 1974, Rolston 1985, [732] and the journal *Environmental Ethics*).

Endless scholarly debate will probably take place about the religious, ethical, and scientific sources of this postulate and about its implications for policy and management. For example, does intrinsic value imply egalitarianism and equal rights among species? A more profitable discussion would be about the rules to be used when two or more species have conflicting interests (Naess 1985).

Contributions of Conservation Biology

Recently, rapid progress has been made by zoos and similar institutions in the technology and theory of captive breeding of endangered species. It is becoming apparent that nearly 2000 species of large mammals and birds will have to be maintained artificially if they are to avoid premature extinction (Myers 1984,

Soulé & Simberloff 1986). Eventual advances in technology may enable some, if not most, such species to be kept in a suspended, miniaturized state, such as frozen sperm, ova, and embryos. Meanwhile, however, traditional ways to maintain most of the planet's megafauna must be improved.

. . . Many authors have appealed for larger founder sizes in groups of captively bred animals to minimize inbreeding problems and the loss of genetic variability (Senner 1980, Templeton and Read 1983), but specific guidelines have been lacking. . . .

Conservation biology has also contributed to the design and management of wildland areas. An example is the new field of population viability analysis, whose goal is to estimate the (effective) number of individuals needed to maintain a species' long-term genetic fitness and ensure against extinction from other, nongenetic causes. Several relatively independent pathways of research in population biology, community ecology, and biogeography are being joined in this effort, which I believe will contribute significantly to theoretical population biology. . . .

Genetics is also important in viability analysis. At least in outbreeding species, it appears that relatively heterozygous individuals are frequently more fit than relatively homozygous ones. Many fitness criteria have been studied, including growth rates, over-winter survival, longevity, developmental stability, metabolic efficiency, and scope for growth (Beardmore 1983, Frankel and Soulé 1981, [733] and Mitton and Grant 1984). Russell Lande and George Barrowclough (AMNH *pers. com.*) are proposing that populations must reach effective sizes of several hundred if they are to retain genetic variation for quantitative traits. Larger numbers will be needed for qualitative traits, including genetic polymorphisms. The US Forest Service is already beginning to integrate viability analysis into its planning protocols (Salwasser et al. 1984).

Field work in conservation biology is supported by several agencies and organizations, including the World Wildlife Fund, NSF, the New York Zoological Society, and the Smithsonian Institution. These studies have contributed a great deal to our understanding of diversity. . . .

Conclusions

Conservation biology is a young field, but its roots antedate science itself. Each civilization and each human generation responds differently to the forces that weaken the biological infrastructure on which society depends and from which it derives much of its spiritual, aesthetic, and intellectual life. In the past, the responses to environmental degradation were often literary, as in the Babylonian Talmud (Vol. I, Shabbath 129a, chap. xviii, p. 644), Marsh (1864), Leopold (1966), Carson (1962) and others (see Passmore 1974). More recently, legal and regulatory responses have been noticeable, especially in highly industrialized and democratized societies. Examples include the establishment of national parks and government policies on human population and family planning, pollution, forest management, and trade in endangered species. At this point in history, a major threat to society and nature is technology, so it is appropriate that this generation look to science and technology to complement literary and legislative responses.

Our environmental and ethical problems, however, dwarf those faced by our ancestors. The current frenzy of environmental degradation is unprecedented (Ehrlich and Ehrlich 1981), with deforestation, desertification, and destruction of wetlands and coral reefs occurring at rates rivaling the major catastrophes in the fossil record and threatening to eliminate most tropical forests and millions of species in our lifetimes. The response, therefore, must also be unprecedented. It is fortunate, therefore, that conservation biology, and parallel approaches in the social sciences, provides academics and other professionals with constructive outlets for their concern.

Conservation biology and the conservation movement cannot reverse history and return the biosphere to its prelapsarian majesty. The momentum of the human population explosion, entrenched political and economic behavior, and withering technologies are propelling humankind in the opposite direction. It is, however, within our capacity to modify significantly the *rate* at which biotic diversity is destroyed, and small changes in rates can produce large effects over long periods of time. Biologists can help increase the efficacy of wildland management; biologists can improve the survival odds of species in jeopardy; biologists can help mitigate technological impacts. The intellectual challenges are fascinating, the opportunities plentiful, and the results can be personally gratifying.

References Cited

Beardmore, J. A. 1983. Extinction, survival, and genetic variation, in Schonewald-Cox et al: 125–151.

Benirschke, K. 1983. The impact of research on the propagation of endangered species in zoos, in Schonewald-Cox et al: 402–413.

Bercovitz, A. B., N. M. Dzekala, and B. L. Lasley. 1978. A new method of sex determination in monomorphic birds. J. Zoo Anim. Med. 9: 114–124.

Carson, R. 1962. *Silent Spring*. Houghton Mifflin, Boston.

Cody, M. L., and J. M. Diamond, eds. 1975. *Ecology and Evolution of Communities*. Harvard University Press, Cambridge, MA.

Devall, B., and G. Sessions. 1985. *Deep Ecology: Living as if Nature Mattered*, Peregrine Smith Books, Layton, UT.

Diamond, J. M. 1975. The island dilemma: lessons of modern biogeographic studies for the design of natural reserves. *Biol. Conserv.* 7: 129–146.

———. 1984. Historic extinctions: their mechanisms, and their lessons for understanding prehistoric extinctions, in P. S. Martin and R. Klein, eds. *Quarternary Extinctions*, University of Arizona Press, Tucson: 824–862.

Dubos, R. 1980. *The Wooing of the Earth*. Charles Scribner's Sons, New York.

Ehrenfeld, D. 1981. *The Arrogance of Humanism*. Oxford University Press, London.

Ehrlich, P. R., and A. H. Ehrlich. 1981. *Extinction*. Random House, New York.

Elton, C. S. 1958. *The Ecology of Invasions by Animals and Plants*. Methuen, London.

Erwin, T. L. 1983. Tropical forest canopies: the last biotic frontier. *Bull. Entomol. Soc. Am.* 29: 14–19.

Frankel, O. H. 1974. Genetic conservation: our evolutionary responsibility. *Genetics* 99: 53–65.

Frankel, O. H., and M. E. Soulé. 1981. *Conservation and Evolution*. Cambridge University Press, Cambridge, UK.

Franklin, I. A. 1980. Evolutionary change in small populations, in Soulé & Wilcox: 135–149.

Futuyma, D. J., and M. Slatkin, eds. 1983. *Coevolution*. Sinauer Associates, Sunderland, MA.

Gilbert, L. E. 1980. Food web organization and the conservation of neotropical diversity, in Soulé & Wilcox: 11–33.

Gilbert, L. E., and P. H. Raven, eds. 1975. *Coevolution of Plants and Animals*. University of Texas Press, Austin.

James, F. C., and W. J. Boecklen. 1984. Inter-specific morphological relationships and the densities of birds, in Strong et al: 458–477.

Janzen, D. H. 1975. *Ecology of Plants in the Tropics*. Edward Arnold, London.

———. 1983. No park is an island. *Oikos* 41: 402–410.

Kushlan, J. A. 1979. Design and management of continental wildlife reserves: lessons from the Everglades. *Biol. Conserv.* 15: 281–290.

Leigh, E. G., Jr. 1981. The average lifetime of a population in a varying environment. *J. Theor. Biol.* 90: 213–239.

Leopold, A. 1953. *The Round River*. Oxford University Press, New York.

———. 1966. *A Sand County Almanac*. Oxford University Press, New York.

Levins, R., and R. Lewontin. 1985. *The Dialectical Biologist*. Harvard University Press, Cambridge, MA.

MacArthur, R. H. 1972. *Geographical Ecology*. Harper & Row, New York.

MacArthur, R. H., and E. O. Wilson. 1967. *The Theory of Island Biogeography*. Princeton University Press, Princeton, NJ.

Marsh, G. P. 1864. *Man and Nature*. Scribners, New York.

May, R. M. 1984. An overview: real and apparent patterns in community structure, in Strong et al: 3–16.

Mitton, J. B., and M. C. Grant. 1984. Associations among protein heterozygosity, growth rate, and developmental homeostasis. *Annu. Rev. Ecol. Syst.* 15: 479–499.

Myers, N. 1984. Genetic resources in jeopardy. *Ambio* 13: 171–174.

Naess, A. 1973. The shallow and the deep, long-range ecology movement. *Inquiry* 16: 95–100.

———. 1985. Identification as a source of deep ecological attitudes, in M. Tobias, ed. *Deep Ecology*. Avant Books, San Diego: 256–270.

Orians, G. H. 1980. Habitat selection: general theory and applications to human behaviour, in J. S. Lockard, ed. *The Evolution of Human Social Behavior*. Elsevier North Holland, New York: 49–66.

Passmore, J. 1974. *Man's Responsibility for Nature*. Duckworth, London.

Price, P. W. 1980. *Evolutionary Biology of Parasites*. Princeton University Press, Princeton, NJ.

Ralls, K., and J. Ballou. 1983. Extinction: lessons from zoos, in Schonewald-Cox et al: 164–184.

Raven, P. R. 1976. Ethics and attitudes, in Simmons et al: 155–179.

Rolston, H. 1985. Duties to endangered species. *BioScience* 35: 718–726.

Ryder, O. A., and E. A. Wedemeyer. 1982. A cooperative breeding programme for the mongolian wild horse *Equus przewalskii* in the United States. *Biol. Conserv.* 22: 259–271.

Salwasser, H., S. P. Mealey, and K. Johnson. 1984. Wildlife population viability: a question of risk. *Trans. N. Am. Wildl. Nat. Resource Conf.*: 49.

Schonewald-Cox, C. M., S. M. Chambers, B. MacBryde, and W. L. Thomas, eds. 1983. *Genetics and Conservation*. Benjamin-Cummings Publishing, Menlo Park, CA.

Seifert, R. P., and F. H. Seifert. 1979. A *Heliconia* insect community in a Venezuelan cloud forest. *Ecology* 60: 462–467.

Senner, J. W. 1980. Inbreeding depression and the survival of zoo populations, in Soulé & Wilcox: 209–224.

Shaffer, M. L. 1981. Minimum population sizes for species conservation. *BioScience* 31: 131–134.

Shaffer, M. L., and F. B. Samson. 1985. Population size and extinction. *Am. Nat.* 125: 144–151.

Simberloff, D. S. 1980. Community effects of introduced species in M. H. Nitecki, ed. *Biotic Crises in Ecological and Evolutionary Time*. Academic Press, New York: 53–83.

Simberloff, D. S., and L. G. Abele. 1976. Island biogeography theory and conservation practice. *Science* 191: 285–286.

Soulé, M. E. 1980. Thresholds for survival: maintaining fitness and evolutionary potential. in Soulé & Wilcox: 151–169.

———. 1983. What do we really know about extinction? in Schonewald-Cox et al. 111–125.

Soulé, M. E., M. E. Gilpin, W. Conway, and T. Foose. 1986. The millenium ark. *Zoo Biol.*, 101–113.

Soulé, M. E., and D. S. Simberloff. 1986. What do genetics and ecology tell us about the design of nature reserves? *Biol. Conserv.* 35: 19–40.

Soulé, M. E., and B. A. Wilcox, eds. 1980. *Conservation Biology* Sinauer Associates, Sunderland, MA.

Strong, D. R., Jr., D. S. Simberloff, L. G. Abele, and A. B. Thistle, eds. 1984. *Ecological Communities*. Princeton University Press, Princeton, NJ.

Templeton, A. R., and B. Read. 1983. The elimination of inbreeding depression in a captive herd of Speke's gazelle, in Schonewald-Cox et al: 241–262.

Terborgh, J. 1974. Preservation of natural diversity: the problem of extinction-prone species. *BioScience* 24: 715–722.

Terborgh, J., and B. Winter. 1980. Some causes of extinction. in Soulé & Wilcox: 119–134.

Underwood, A. J., and E. J. Denley. 1984. Paradigms, explanations, and generalizations in models for the structure of intertidal communities on rocky shores, in Strong et al: 151–180,

Wilcox, B. A. 1980. Insular ecology and conservation in Soulé & Wilcox: 95–117.

Wilson, E. O. 1984. *Biophilia*. Harvard University Press, Cambridge, MA.

Commentary

Michael E. Soulé, "What Is Conservation Biology?" (1985)

LIBBY ROBIN

This seminal paper, still often cited today, had three distinct purposes. The first was to define a new "field," conservation biology, which had at its core "the application of science to conservation problems" (727). The second was to quest for a rigorous, quantitative approach to defining "biological diversity," and "principles and tools" that will preserve it. The third was to situate conservation biology in the policy arena, to explore why the world urgently needed conservation biology.

The "field" is Soulé's term; he asserts that conservation biology is much more than a science. It takes its "questions, techniques and methods from a broad range of fields, not all biological," and depends on "social science" (728). Its origins can be traced to an International Conference on Conservation Biology (1978) and an important book by Soulé and Bruce Wilcox, *Conservation Biology: An Evolutionary Ecology* (1980), which informs much of this paper.

Conservation Biology started with a focus on genetic diversity in the late 1970s, and moved to a broader interdisciplinary guide to environmental management by the mid-1980s, when this paper emerged. In 1986, the Society for Conservation Biology was established and a second edition of *Conservation Biology* appeared. This was in fact a very different book. Soulé was its sole editor, and it had a different subtitle: *The Science of Scarcity and Diversity*. The book was 25 percent longer, and 90 percent of its contributing authors were different. If "diversity and rarity are synonyms for 'everything' in ecology," Soulé observed in this book (Soulé 1980: 117), practical conservation biology needs to start with these concepts. And scientists quickly realized the importance of the new disciplinary field. The ranks of the Society for Conservation Biology swelled to five thousand in its first five years.

Principles

Conservation biology is practical, but it also has theory: *functional postulates*, which deal with the way the natural world works in fairly traditional biological or ecological terms, and *normative postulates*, which nest the natural world inside the human world and suggest the principles for managing the natural world. It is a "young" science, like ecology itself, and "many of its parent disciplines," Soulé asserts (394). The drive to articulate a scientific approach is a classic response to a science perceived as "young" and not fully formed, and this is why Soulé spells out his postulates.

Defining Biodiversity

Conservation biology's most important key concept was biodiversity. "Bio-Diversity" emerged in this era as a media-friendly variant of the earlier biological diversity. This was the title of a Smithsonian Institution teleconference (1986) featuring Thomas Lovejoy (another prominent defender of biological diversity), E. O. Wilson, Paul Ehrlich (see Part 1), and others.

Biodiversity provided a framework for understanding both the "phenomenon of life" and the impact of human activity on it. It is more than just "nature": it is a *measure* that can be used to document change. While the term carries a veneer of scientific independence, its emergence as a buzzword in the 1980s was because the idea of loss of natural variety was useful in environmental activism. Biodiversity offered a way to measure what was an essentially moral imperative to take action on environmental crisis.

The Crisis of Mass Extinction and the Language of Triage

Crisis was the key to the new developments in conservation biology. Soulé draws an analogy with medical biology, particularly emergency management, in his choice of the term "triage." In making choices about which species or ecosystems to save, the triage principle focuses on the most serious cases, and the ones which, if treated, will be most likely to recover. The idea of "bang for the buck" is crucial to convincing policy makers to direct funds to particular causes. "Holism is not mysticism," he declares. As in medical biology, "multi-disciplinary approaches will ultimately be the most fruitful" (p. 393).

The crisis that Soulé identified demanded concerted action on more than just local and regional scales: this was emergency management for the planet. The future of the biota depended on politics and people, ecologists realized: they were no longer documenting species in the wild, but rather performing triage for bigger emergencies and on bigger scales than ever before.

Making biological diversity urgent, rather than merely descriptive, directed attention to prediction and sharpened questions about extinctions. What happens to an ecosystem when a species becomes extinct? What are the knock-on effects for other biota? For individual species, extinction is the endpoint. But within the broader ecosystem, one extinction may lead to more, as its functions become increasingly impaired. After one (or more) species becomes extinct, the ecosystem may collapse, or it may adapt. Predicting which outcome is likely is an element of the triage technique.

Extinction has a long history. Fossils in the eighteenth century revealed extinctions that challenged biblical ideas of creation and the balance of nature.

They greatly deepened ideas about time. As Mark Barrow has explored, fossils of extinct creatures were even part of nationalist pride. Thomas Jefferson argued that the New World's "incognitum" mammoth or big buffalo (six times the size of an elephant) disproved the notion put forward by Buffon that the Old World fauna was more vigorous and well developed than that of the New.

In recent years, extinctions are now measured and counted, as they have become a focus of conservation effort. Extinctions of "charismatic" species are signs of a mismanaged world. In this paper, Soulé writes of conservation biology's particular concern, not just for "endangered" (charismatic) species but also for "keystone" species, those that have a disproportionately large effect on their environment and play a critical role in maintaining the structure of an ecological community. "The proper objective of conservation is the protection and continuity of entire communities and ecosystems," so keystone species and "complete" environments, such as national parks and nature reserves, have a special role.

But there is a paradox here: if nature becomes most carefully studied where it is least threatened, in national parks and nature reserves, much important biodiversity is not protected at all. In 2000, the World Conservation Monitoring Centre (part of the United Nations Environment Programme) recognized seventeen megadiverse countries that together harbor more than two-thirds of the planet's biological wealth, in an effort to prioritize spending. The following decade saw an increase in "hot spots" for biological diversity, often beyond parks, and increasingly in nations with emerging economies.

A Conservation Biology Ethic

Soulé's paper sets out a philosophy of conservation biology. He makes it clear that its focus is not just animal welfare, however unpleasant disease and suffering may be. Conservation is "the protection of the integrity and continuity of natural processes, not the welfare of individuals" (p. 398), and its outcomes must be defined in terms of whole populations. The important processes here are genetic and evolutionary, drawing on Soulé's own disciplinary background. He acknowledges the intrinsic value of nature, but in the end realizes that it is people who manage conservation efforts and fund its science. Conservation biology is, therefore, a social science as well as an ecological one. Ecology is a necessary, but no longer sufficient, expertise for biodiversity in a time of planetary crisis.

Further Reading

Barrow, Mark V., Jr. 2009. *Nature's ghosts: Confronting extinction from the age of Jefferson to the age of ecology.* Chicago: Chicago University Press.

Farner, Timothy. 2007. *Saving nature's legacy: The origins of the idea of biodiversity.* New Haven: Yale University Press.

Fazey, I., J. Fischer, and D. B. Lindenmayer. 2005. "What do conservation biologists publish?" *Biological Conservation* 124:63–73.

Robin, Libby. 2011. "The rise of the idea of biodiversity: Crises, responses and expertise," *Quaderni (Journal of l'Institut des Sciences Humaines et Sociales du CNRS)* [Special Issue: Les promesses de la biodiversité] 76(1):25–38.

Soulé, Michael (ed.). 1986. *Conservation biology*, 2nd ed. Sunderland, MA: Sinauer Associates.

Radical American Environmentalism and Wilderness Preservation

A Third World Critique

RAMACHANDRA GUHA

Even God dare not appear to the poor man except in the form of bread.

—Mahatma Gandhi

Introduction

The respected radical journalist Kirkpatrick Sale has celebrated "the passion of a new and growing movement that has become disenchanted with the environmental establishment and has in recent years mounted a serious and sweeping attack on it—style, substance, systems, sensibilities and all" (Sale 1986: 26). The vision of those whom Sale calls the "New Ecologists"—and what I refer to in this chapter as deep ecology—is a compelling one. Decrying the narrowly economic goals of mainstream environmentalism, this new movement aims at nothing less than a philosophical and cultural revolution in human attitudes towards nature. In contrast to the conventional lobbying of environmental professionals based in Washington, it proposes a militant defence of "Mother Earth," an unflinching opposition to human attacks on undisturbed wilderness. With their goals ranging from the spiritual to the political, the adherents of deep ecology span a wide spectrum of the American environmental movement. As Sale correctly notes, this emerging strand has in a matter of a few years made its presence felt in a number

Ramachandra Guha. 1997. "Radical American environmentalism and wilderness preservation: A Third World critique." Chapter 5 in Ramachandra Guha and Joan Martínez Alier (eds.), *Varieties of environmentalism: Essays north and south*, 92–108, 214–216. London: Earthscan.

of fields: from academic philosophy, as in the journal *Environmental Ethics*, to popular environmentalism, for example, the group Earth First!

In this chapter I develop a critique of deep ecology from the perspective of a sympathetic outsider. My treatment of deep ecology is primarily historical and sociological in nature, rather than philosophical. Specifically, I examine the cultural rootedness of a philosophy that likes to present itself in universalistic terms. I make two main arguments: first, that deep ecology is uniquely American, and despite superficial similarities in rhetorical style, the social and political goals of radical environmentalism in other cultural contexts (e.g. Germany and India) are quite different; second, that the social consequences of putting deep ecology into practice on a worldwide basis (what its practitioners are aiming for) are very grave indeed.

The Tenets of Deep Ecology

While I am aware that the term *deep ecology* was coined by the Norwegian philosopher Arne Naess (1973), this chapter refers specifically to the American variant. Adherents of the deep ecological perspective in the United States, while arguing intensely among themselves over its political and philosophical implications, share some fundamental premises about human-nature interactions. As I see it, the defining characteristics of deep ecology are fourfold.

First, deep ecology argues that the environmental movement must shift from an anthropocentric to a biocentric perspective. In many respects, an acceptance of the primacy of this distinction constitutes the litmus test of deep ecology. A considerable effort is expended by deep ecologists in showing that the dominant motif in Western philosophy has been anthropocentric (i.e. the belief that man and his works are the centre of the universe), and conversely in identifying those lonely thinkers (Leopold, Thoreau, Muir, Aldous Huxley, Santayana, etc.) who, in assigning man a more humble place in the natural order, anticipated deep ecological thinking. In the political realm, meanwhile, establishment environmentalism (shallow ecology) is chided for casting its arguments in human-centred terms. Preserving nature, the deep ecologists say, has an intrinsic worth quite apart from any benefits preservation may convey to future human generations. The anthropocentric/biocentric distinction is accepted as axiomatic by deep ecologists, it structures their discourse, and much of the present discussion remains mired within it.

The second characteristic of deep ecology is its focus on the preservation of unspoilt wilderness—and the restoration of degraded areas to a more pristine condition—to the relative, and sometimes absolute, neglect of other issues on the environmental agenda. I later identify the cultural roots and portentous consequences of this obsession with wilderness. For the moment, let me indicate three distinct sources from which it springs. Historically, it represents a playing out of the preservationist (read radical) and utilitarian (read reformist) dichotomy that has plagued American environmentalism since the turn of the century.

Morally, it is an imperative that follows from the biocentric perspective; other species of plants and animals, and nature itself, have an intrinsic right to exist. And finally, the preservation of wilderness also turns on a scientific argument, namely the value of biological diversity in stabilising ecological regimes and in retaining a gene pool for future generations. Truly radical policy proposals have been put forward by deep ecologists on the basis of these arguments. The influential poet Gary Snyder, for example, would like to see a 90 per cent reduction in human populations to allow a restoration of pristine environments, while others have argued forcefully that a large portion of the globe must be immediately cordoned off from human beings.

Third, there is a widespread invocation of Eastern spiritual traditions as forerunners of deep ecology. Deep ecology, it is suggested, was practiced both by major religious traditions and at a more popular level by "primal" peoples in non-Western settings. This complements the search for an authentic lineage in Western thought. At one level, the task is to recover those dissenting voices within the Judeo-Christian tradition; at another, to suggest that religious traditions in other cultures are, in contrast, dominantly if not exclusively "biocentric" in their orientation. This coupling of (ancient) Eastern and (modern) ecological wisdom seemingly helps to consolidate the claim that deep ecology is a philosophy of universal significance.

Fourth, deep ecologists, whatever their internal differences, share the belief that they are the "leading edge" of the environmental movement. As the polarity of the shallow/deep and anthropocentric/biocentric distinctions make clear, they see themselves as the spiritual, philosophical and political vanguard of American and world environmentalism.

Towards a Critique

Although I analyse each of these tenets independently, it is important to recognise, as deep ecologists are fond of remarking in reference to nature, the interconnectedness and unity of these individual themes.

SHIFT TO A BIOCENTRIC PERSPECTIVE

Insofar as it has begun to act as a check on man's arrogance and ecological hubris, the transition from an anthropocentric (human-centred) to a biocentric

(humans as only one element in the ecosystem) view in both religious and scientific traditions is only to be welcomed. What is unacceptable are the radical
conclusions drawn by deep ecology, in particular, that intervention in nature
should be guided primarily by the need to preserve biotic integrity rather than
by the needs of humans. The latter for deep ecologists is anthropocentric, the
former biocentric. This dichotomy is, however, of very little use in understanding
the dynamics of environmental degradation. The two fundamental ecological
problems facing the globe are (1) overconsumption by the industrialised world
and by urban elites in the Third World and (2) growing militarisation, both in a
short-term sense (i.e., ongoing regional wars) and in a long-term sense (i.e. the
arms race and the prospect of nuclear annihilation). Neither of these problems
has any tangible connection to the anthropocentric/biocentric distinction. Indeed, the agents of these processes would barely comprehend this philosophical
dichotomy. The proximate causes of the ecologically wasteful characteristics of
industrial and of militarisation are far more mundane: at an aggregate level, the
dialectic of economic and political structures, and at a microlevel, the lifestyle
choices of individuals. These causes cannot be reduced, whatever the level of
analysis, to a deeper anthropocentric attitude toward nature; on the contrary, by
constituting a grave threat to human survival, the ecological degradation they
cause does not even serve the best interests of human beings. If my identification of the major dangers to the integrity of the natural world is correct, invoking
the bogey of anthropocentricism is at best irrelevant and at worst a dangerous
obfuscation.

FOCUS ON THE PRESERVATION OF WILDERNESS

If the above dichotomy is irrelevant, the emphasis on wilderness is positively
harmful when applied to the Third World. If in the United States the preserva-
tionist/utilitarian division is seen as mirroring the conflict between the "people"
and the "interests," in countries such as India the situation is very nearly the
reverse. Because India is a long-settled and densely populated country in which
agrarian populations have a finely balanced relationship with nature, the setting
aside of wilderness areas has resulted in a direct transfer of resources from the
poor to the rich. Thus Project Tiger, a network of parks hailed by the international community as an outstanding success, puts the interests of the tiger ahead
of those of poor peasants living in and around the reserve. The designation of
tiger reserves was made possible only by the physical displacement of existing
villages and their inhabitants; their management requires the continuing exclu-
sion of peasants and livestock. The initial impetus for setting up parks for tigers
and other mammals such as the rhinoceros and elephant came from two social

groups: (1) a class of ex-hunters turned conservationists belonging mostly to the declining feudal élite, and (2) representatives of international agencies, such as the World Wildlife Fund (WWF) and the International Union for the Conservation of Nature and Natural Resources (IUCN), seeking to transplant the American system of national parks on to Indian soil. In no case have the needs of the local population been taken into account, and as in many parts of Africa, the designated wildlands are managed primarily for the benefit of rich tourists. Until very recently, wildlands preservation has been identified with environmentalism by the state and the conservation élite; in consequence, environmental problems that impinge far more directly on the lives of the poor (e.g. fuel, fodder, water shortages, soil erosion, and air and water pollution) have not been adequately addressed.

Deep ecology provides, perhaps unwittingly, a justification for the continuation of such narrow and inequitable conservation practices under a newly acquired radical guise. Increasingly, the international conservation elite is using the philosophical, moral and scientific arguments used by deep ecologists in advancing their wilderness crusade. A striking but by no means atypical example is the recent plea by a prominent American biologist for the takeover of large portions of the globe by him and his scientific colleagues. Writing in a prestigious scientific forum, the *Annual Review of Ecology and Systematics*, Daniel Janzen argues that only biologists have the competence to decide how the tropical landscape should be used. As "the representatives of the natural world," biologists are "in charge of the future of tropical ecology," and only they have the expertise and mandate to "determine whether the tropical agroscape is to be populated only by humans, their mutualists, commensals, and parasites, or whether it will also contain some islands of the greater nature—the nature that spawned humans, yet has been vanquished by them." Janzen exhorts his colleagues to advance their territorial claims on the tropical world more forcefully, warning that the very existence of these areas is at stake: "if biologists want a tropics in which to biologise, they are going to have to buy it with care, energy, effort, strategy, tactics, time, and cash" (Janzen 1986a: 305–6).

This frankly imperialist manifesto highlights the multiple dangers of the preoccupation with wilderness preservation that is characteristic of deep ecology. As I have suggested, it seriously compounds the neglect by the American movement of far more pressing environmental problems in the Third World. But perhaps more importantly, and in a more insidious fashion, it also provides an impetus to the imperialist yearning of Western biologists and their financial sponsors, organisations such as the WWF and IUCN. The wholesale transfer of a movement culturally rooted in American conservation history can only result in the social uprooting of human populations in other parts of the globe.

EASTERN SPIRITUAL TRADITIONS

I come now to the persistent invocation of Eastern philosophies as antecedent in time but convergent in their structure with deep ecology. Complex and internally differentiated religious traditions—Hinduism, Buddhism and Taoism—are lumped together as holding a view of nature believed to be quintessentially biocentric. Individual philosophers such as the Taoist Lao Tzu are identified as being forerunners of deep ecology. Even an intensely political, pragmatic, and Christian-influenced thinker such as Gandhi has been accorded a wholly undeserved place in the pantheon of deep ecology. Thus the Zen teacher Robert Aitken Roshi makes the strange claim that Gandhi's thought was not human-centred and that he practiced an embryonic form of deep ecology which is "traditionally Eastern and is found with differing emphasis in Hinduism, Taoism and in Theravada and Mahayana Buddhism" (Roshi Appendix, in Devall & Sessions 1985). Moving away from the realm of high philosophy and scriptural religion, deep ecologists make the further claim that at the level of material and spiritual practice "primal" peoples subordinated themselves to the integrity of the biotic universe they inhabited.

I have indicated that this appropriation of Eastern traditions is in part dictated by the need to construct an authentic lineage and in part a desire to present deep ecology as a universalistic philosophy. Indeed, in his substantial yet quixotic biography of John Muir, Michael Cohen goes so far as to suggest that Muir was the "Taoist of the [American] West" (Cohen 1984: 120). This reading of Eastern is selective and does not bother to differentiate between alternative (and changing) religious and cultural traditions; as it stands, it does considerable violence to the historical record. Throughout most recorded history the characteristic form of human activity in the "East" has been a finely tuned but none the less conscious and dynamic manipulation of nature. Although mystics such as Lao Tzu did reflect on the spiritual essence of human relations with nature, it must be recognised that such ascetics and their reflections were supported by a society of cultivators whose relationship with nature was a far more *active* one. Many agricultural communities do have a sophisticated knowledge of the natural environment that may equal (and sometimes surpass) codified "scientific" knowledge, yet the elaboration of such traditional ecological knowledge—in both material and spiritual contexts—can hardly be said to rest on a mystical affinity with nature of a deep ecological kind. Nor is such knowledge infallible; as the archaeological records powerfully suggest, modern Western man has no monopoly on ecological disasters.

In a brilliant article, the Chicago historian Ronald Inden (1986) points out that this romantic and essentially positive view of the East is a mirror image of the scientific and essentially pejorative view normally held by Western scholars of the Orient. In either case, the East constitutes the Other, a body wholly separate and alien from the West; it is defined by a uniquely spiritual and non-rational "essence," even if this essence is valorised quite differently by the two schools. Eastern man exhibits a spiritual dependence with respect to nature—on the one hand, this is sympathetic of his pre-scientific and backward self, on the other, of his ecological wisdom and deep ecological consciousness. Both views are monolithic, simplistic, and have the characteristic effect—intended in one case, perhaps unintended in the other—of denying agency and reason to the East and making it the privileged orbit of Western thinkers.

The two apparently opposed perspectives have then a common underlying structure of discourse in which the East merely serves as a vehicle for Western projections. Varying images of the East are raw material for political and cultural battles being played out in the West; they tell us far more about the Western commentator and his desires than about the "East." Inden's remarks apply not merely to Western scholarship on India, but to Orientalist constructions of China and Japan as well:

> Although these two views appear to be strongly opposed, they often combine together. Both have a similar interest in sustaining the Otherness of India. The holders of the dominant view, best exemplified in the past in imperial administrative discourse (and today probably by that of "developed economies"), would place a traditional, superstition-ridden India in a position of perpetual tutelage to a modern, rational West. The adherents of the romantic view, best exemplified academically in the discourses of the Christian liberalism and analytic psychology, concede the realm of the public and impersonal to the positivist. Taking their foundations and self-help institutes, and from allies in the "consciousness industry," not to mention the important industry of tourism, the romantics insist that India embodies a private realm of the imagination and the religious which modern, western man lacks but needs. They, therefore, like the positivists, but for just the opposite reason, have a vested interest in seeing that the Orientalist view of India as "spiritual," "mysterious" and "exotic" is perpetuated. (Inden 1986: 442)

THE RADICALISM OF DEEP ECOLOGY

How radical, finally, are the deep ecologists? Notwithstanding their self-image and strident rhetoric, even within the American context their radicalism is different and it manifests itself quite differently elsewhere. To my mind, deep ecology is best viewed as a radical trend within the wilderness preservation movement. Although advancing philosophical rather than aesthetic arguments and encouraging political militancy rather than negotiation, its practical emphasis—that is, preservation of unspoilt nature—is virtually identical. For the mainstream movement, the function of wilderness is to provide a temporary antidote to modern civilisation. As a special institution within an industrialised society, the national park "provides an opportunity for respite, contrast, contemplation, and affirmation of values for those who live most of their lives in the workday world" (Sax 1980: 42). Indeed, the rapid increase in visits to the national parks in post-war America is a direct consequence of economic expansion. The emergence of a popular interest in wilderness sites, the historian Samuel Hays points out, was "not a throwback to the primitive, but an integral part of the modern standard of living as people sought to add new 'amenity' and 'aesthetic' goals and desires to their earlier preoccupation with necessities and conveniences" (Hays 1982: 21).

Here the enjoyment of nature is an integral part of the consumer society. The private automobile, and the life style it has spawned, is in many respects the ultimate ecological villain, and an untouched wilderness the prototype of ecological harmony; yet, for most Americans it is perfectly consistent to drive a thousand miles to spend a holiday in a national park. Americans possess a vast, beautiful and sparsely populated continent and are also able to draw on the natural resources of large portions of the globe by virtue of their economic and political dominance. In consequence, America can simultaneously enjoy the material benefits of an expanding economy and the aesthetic benefits of unspoilt nature. The two poles of "wilderness" and "civilisation" mutually coexist in an internally coherent whole, and philosophers of both poles are assigned a prominent place in this culture. Paradoxically as it may seem, it is no accident that Star Wars technology and deep ecology both find their fullest expression in that leading sector of Western civilisation, California.

Deep ecology runs parallel to the consumer society without seriously questioning its ecological and socio-political basis. In its celebration of American wilderness, it also displays an uncomfortable convergence with the prevailing climate of nationalism in the American wilderness movement. For spokesmen such as the historian Roderick Nash (1982), the national park system is America's distinctive cultural contribution to the world, reflective not merely of its economic but also of its philosophical and ecological maturity. In what Henry Luce called the American century, the "American invention of national parks" must be exported worldwide. Betraying an economic determinism that would make even a Marxist shudder, Nash believes that environmental preservation is a "full stomach" phenomenon that is confined to the rich, urban, and sophisticated.

None the less, he hopes that "the less developed nations may eventually evolve economically and intellectually to the point where nature preservation is more than a business."

The error which Nash makes, and which deep ecology in some respects encourages, is to equate environmental protection with the protection of wilderness. This is a distinctively American notion, borne out of a unique social and environmental history. The archetypal concerns of radical environmentalists in other cultural contexts are in fact quite different. The German Greens, for example, have elaborated a devastating critique of industrial society which turns on the acceptance of environmental limits to growth. Pointing to the intimate links between industrialisation, militarisation and conquest, the Greens argue that economic growth in the West has historically rested on the economic and ecological exploitation of the Third World. Rudolf Bahro is characteristically blunt:

> The working class here [in the West] is the richest lower class in the world. And if I look at the problem from the point of view of the whole of humanity, not just from that of Europe, then I must say that the metropolitan working class is the worst exploiting class in history . . . What made poverty bearable in eighteenth-or nineteenth-century Europe was the prospect of escaping it through exploitation of the periphery. But this is no longer a possibility, and continued industrialism in the Third World will mean poverty for whole generations and hunger for millions. (Bahro 1984)

Here the roots of global ecological problems lie in the disproportionate share of resources consumed by the industrialised countries as a whole and the urban élite in the Third World. Since it is impossible to reproduce an industrial monoculture worldwide, the ecological movement in the West must begin by cleaning up its own act. From time to time, American scholars have themselves criticised these imbalances in consumption patterns. In the 1950s, William Vogt [see Part 4] made the charge that the United States, with one-sixteenth of the world's population, was utilizing one-third of the globe's resources. The Greens advocate the creation of a "no-growth" economy, to be achieved by scaling down current, and clearly unsustainable, consumption levels. This radical shift in consumption and production patterns requires the creation of alternate economic and political structures — smaller in scale and more amenable to social participation — but it rests equally on a shift in cultural values. The expansionist character of modern Western man will have to give way to an ethic of renunciation and self-limitation, in which spiritual and communal values play an increasing role in sustaining social life. This revolution in cultural values, however, has as its point of departure an understanding of environmental processes quite different from deep ecology.

Many elements of the Green programme find a strong resonance in countries such as India, where a history of Western colonialism and industrial development has benefited only a tiny elite while exacting tremendous social and environmental costs. The ecological battles presently being fought in India have as their epicentre the conflict over nature between the subsistence and largely rural sector and the vastly more powerful commercial-industrial sector. Perhaps the most celebrated of these battles concerns the Chipko movement, a peasant movement against deforestation in the Himalayan foothills. Chipko is only one of several movements that have sharply questioned the non-sustainable demand being placed on the land and vegetative base by urban centres and industry. These include opposition to large dams by displaced peasants, the conflict between small-scale artisan fishing and large-scale trawler fishing for export, the countrywide movements against commercial forest operations, and opposition to industrial pollution among downstream agricultural and fishing communities (Agawal & Narain 1985).

Two features distinguish these environmental movements from their Western counterparts. First, for the sections of society most critically affected by environmental degradation—poor and landless peasants, women, and tribals—it is a question of sheer survival, not enhancing the quality of life. Second, and as a consequence, the environmental solutions they articulate strongly involve questions of equity as well as economic and political redistribution. Highlighting these differences, a leading Indian environmentalist stressed that "environmental protection *per se* is of least concern to most of these groups. Their main concern is about the use of the environment and who should benefit from it" (Agarwal 1986: 167). The Indian movements seek to wrest control of nature away from the state and the industrial sector and place it in the hands of rural communities who live within that environment but are increasingly denied access to it. These communities have far more basic needs, their demands on the environment are far less intense, and they can draw on a reservoir of co-operative social institutions and local ecological knowledge in managing the "commons"—forests, grasslands and the waters—on a sustainable basis. If colonial and capitalist expansion has both accentuated social inequalities and signalled a precipitous fall in ecological wisdom, an alternative ecology must rest on an alternative society and polity as well.

This brief overview of German and Indian environmentalism has some major implications for deep ecology. Both German and Indian environmental traditions allow for a greater integration of ecological concerns with livelihood and work. They also place a greater emphasis on equity and social justice—both within individual countries and on a global scale—on the grounds that in the absence of social regeneration environmental regeneration has very little chance of succeeding. Finally, and perhaps most significantly, they have escaped the preoccupation with wilderness preservation so characteristic of American cultural and environmental history. One strand in radical American environmentalism, the bioregional movement, by emphasising a greater involvement with the bioregion people inhabit, does indirectly challenge consumerism. However, as yet bioregionalism has hardly raised the questions of equity and social justice (international, intranational, and intergenerational) which I argue must be a central plank of radical environmentalism (see, for example, Sale 1985).

A Homily

In 1958, the economist J. K. Galbraith referred to overconsumption as the unasked question of the American conservation movement. There is a marked selectivity, he wrote, "in the conservationist's approach to materials consumption. If we are concerned about our great appetite for materials, it is plausible to seek to increase the supply, to decrease waste, to make better use of the stocks available, and to develop substitutes. But what of the appetite itself? Surely this is the ultimate source of the problem. If it continues its geometric course, will it not one day have to be restrained? Yet in the literature of the resource problem this is the forbidden question. Over it hangs a nearly total silence" (Galbraith 1958: 91–92).

The consumer economy and society have expanded tremendously in the four decades since Galbraith wrote these words, yet his criticisms are nearly as valid today. I say "nearly," for there are some hopeful signs. Within the environmental movement several dispersed groups are working to develop ecologically benign technologies and to encourage less wasteful lifestyles. Moreover, outside the self-defined boundaries of American environmentalism, opposition to the permanent war economy is being carried on by a peace movement that has a distinguished history and impeccable moral and political credentials.

It is precisely these—to my mind, most hopeful—components of the American social scene that are missing from deep ecology. In their widely read book, Bill Devall and George Sessions (1985) make no mention of militarisation or the movements for peace, while activists whose practical focus is on developing ecologically responsible life styles (e.g., Wendell Berry 1987) are derided as "falling short of deep ecological awareness." A truly radical ecology in the American context ought to work towards a synthesis of the appropriate technology, alternative lifestyles, and peace movements. While Barry Commoner (1987), for example, makes a forceful plea for the convergence of the environmental movement (viewed by him primarily as the opposition to air and water pollution and to the

institutions that generate such pollution) and the peace movement, he significantly does not mention consumption patterns, implying that "limits to growth"
[see Part 2] do not exist. By making the (largely spurious) anthropocentric/bio-
centric distinction central to the debate, deep ecologists may have appropriated
the moral high ground, but they are at the same time doing a serious disservice to
American and global environmentalism.

Postscript: Deep Ecology Revisited

The preceding pages first appeared as an article in *Environmental ethics*,
volume 11, number 1, 1989. They were written at the end of an extended period
of residence in the United States, after several years of research on the origins
of Indian environmentalism. That background might explain the puzzlement
and anger which, in hindsight, appear to mark the chapter. To my surprise, the
article evoked a variety of responses, both pro and con. The veteran Vermont
radical Murray Bookchin, himself engaged in a polemic with American deep
ecologists, offered a short (three-line) letter of congratulation. A longer (30-page)
response came from the Norwegian philosopher Arne Naess (1989), the originator of the term "deep ecology." Naess felt bound to assume responsibility for the
ideas I had challenged, even though I had distinguished between his emphases
and those of his American interpreters. Other correspondents, less known but no
less engaged, wrote privately to praise and to condemn (see also Callicott 1991).
Over the years, the essay has appeared in some half dozen anthologies, as a voice
of the "Third World," the token and disloyal opposition to the reigning orthodox-
ies of environmental ethics (for example, Mappes and Zembaty 1992; Merchant
1994). The article having acquired a life of its own, I felt it prudent to include it
here without any changes. This postscript allows me to look at the issues anew,
to expand and strengthen my case with the aid of a few freshly arrived examples.

Woodrow Wilson once remarked that the United States was the only ideal-
istic nation in the world. It is indeed this idealism which explains the zest, the
zeal, the almost unstoppable force with which Americans have sought to impose
their vision of the good life on the rest of the world. American economists urge on
other nations their brand of energy-intensive, capital-intensive, market-orientated
brand of development. American spiritualists, saving souls, guide pagans to one
or other of their eccentrically fanatical cults, from Southern Baptism to Moral
Rearmament. American advertisers export the ethic of disposable containers—of
all sizes, from coffee cups to automobiles—and Santa Barbara.

Of course, other people have had to pay for the fruits of this idealism. The
consequences of the forward march of American missionaries include the undermining of political independence, the erosion of cultures and the growth of

an ethic of sheer greed. In a dozen parts of the world, those fighting for political, economic or cultural autonomy have collectively raised the question whether the American way of life is not, in fact, the Indian (or Brazilian, or Somalian) way of death.

One kind of US missionary, however, has attracted virtually no critical attention. This is the man who is worried that the rest of the world thinks his country has a dollar sign for a heart. The clothes he wears are also coloured green, but it is the green of the virgin forest. A deeply committed lover of the wild, in his country he has helped put in place a magnificent system of national parks. But he also has money, and will travel. He now wishes to convert other cultures to his gospel, to export the American invention of national parks worldwide.

The essay to which these paragraphs are a coda was one of the first attacks on an imperialism previously reckoned to be largely benign. After all, we are not talking here of the Marines, with their awesome firepower, or even the World Bank, with its money power and the ability to manipulate developing country governments. These are men—and, more rarely, women—who come preaching the equality of all species, who worship all that is good and beautiful in nature. What could be wrong with them?

I had suggested in my essay that the noble, apparently disinterested motives of conservation biologists and deep ecologists fuelled a territorial ambition—the physical control of wilderness in parts of the world other than their own—that led inevitably to the displacement and harsh treatment of their human communities who dwelt in these forests. Consider in this context a recent assessment of global conservation by Michael Soulé [see essay in this part], which complains that the language of policy documents has "become more humanistic in values and more economic in substance, and correspondingly less naturalistic and eco-centric." Soulé seems worried that in theory (though certainly not in practice!) some national governments and international conservation organisations (ICOs) now pay more attention to the rights of human communities. Proof of this shift is the fact that "the top and middle management of most ICOs are economists, lawyers and development specialists, not biologists." This is a sectarian plaint, a trade union approach to the problem spurred by an alleged "takeover of the international conservation movement by social scientists, particularly economists" (Soulé unpub.).

Soulé's work, with its talk of conspiracies and takeover bids, manifests the paranoia of a community of scientists which has a huge influence on conservation policy but yet wants to be the sole dictator. A scholar acclaimed by his peers as the "dean of tropical ecologists" has expressed this ambition more nakedly than most. I have already quoted from a paper published by Daniel Janzen (1986a) in the *Annual review of ecology and systematics*, which urges his fellow biologists to raise cash so as to buy space and species to study. Let me now quote from a report he wrote on a new National Park in Costa Rica, whose tone and thrust perfectly complements the other, ostensibly "scientific" essay. "We have the seed and the biological expertise: we lack the control of the terrain," wrote Janzen in 1986 (1986b). This situation he was able to remedy for himself, by raising enough money to purchase the forest area needed to create the Guanacaste National Park. One can only marvel at Janzen's conviction that he and his fellow biologists know all, and that the inhabitants of the forest know nothing. He justifies the taking over of the forest and the dispossession of the forest farmer by claiming that "Today virtually all of the present-day occupants of the western Mesoamerican pastures, fields and degraded forests are deaf, blind and mute to the fragments of the rich biological and cultural heritage that still occupies shelves of the unused and unappreciated library in which they reside."

This is an ecologically updated version of the White Man's Burden, where the biologist, rather than the civil servant or military official, knows that it is in the native's true interest to abandon his home and hearth and leave the field and forest clear for the new rulers of his domain. In Costa Rica we only have Janzen's word for it, but elsewhere we are better placed to challenge the conservationist's point of view. A remarkable book on African conservation has laid bare the imperialism, unconscious and explicit, of Western wilderness lovers and biologists working on that luckless continent. I cannot here summarise the massive documentation of Raymond Bonner's *At the hand of man*, but will simply quote some of his conclusions:

> Above all, Africans [have been] ignored, overwhelmed, manipulated and outmaneuvered—by a conservation crusade led, orchestrated and dominated by white Westerners.
>
> Livingstone, Stanley and other explorers and missionaries had come to Africa in the nineteenth century to promote the three C's—Christianity, commerce and civilization. Now a fourth was added: conservation. These modern secular missionaries were convinced that without the white man's guidance, the Africans would go astray.
>
> [The criticisms] of egocentricity and neo-colonialism . . . could be levelled fairly at most conservation organisations working in the Third World.
>
> As many Africans see it, white people are making rules to protect animals that white people want to see in parks that white people visit. Why should Africans support these programs? . . . The World Wildlife Fund professed to care about what the Africans wanted, but then tried

to manipulate them into doing what the Westerners wanted: and those Africans who couldn't be brought into line were ignored.

Africans do not use the parks and they do not receive any significant benefits from them. Yet they are paying the costs. There are indirect economic costs—government revenues that go to parks instead of schools. And there are direct personal costs [i.e., of the ban on hunting and fuel-collecting, or of displacement]. (Bonner 1993: 35, 65, 70, 85, 221)

Bonner's book focuses on the elephant, one of the half dozen or so animals that have come to acquire "totemic" status among Western wilderness lovers. Animal totems existed in most pre-modern societies, but as the Norwegian scholar Arne Kalland (1994) points out, in the past the injunction not to kill the totemic species applied only to members of the group. Hindus do not ask others to worship the cow, but those who love and cherish the elephant, seal, whale or tiger try to impose a worldwide ban on its killing. No one, they say, anywhere, anytime, shall be allowed to touch the animal they hold sacred even if—as with the elephant and several species of whale—scientific evidence has established that small-scale hunting will not endanger its viable populations and will, in fact, save human lives put at risk by the expansion, after total protection, of the *Lebensraum* of the totemic animal. The new totemists also insist that their species is the "true, rightful inhabitant" of the ocean or forest, and ask that human beings who have lived in the same terrain, and with the animals, for millennia be sent elsewhere.

To turn, last of all, to an ongoing controversy in my own bailiwick. The Nagarhole National Park in southern Karnataka has an estimated 40 tigers, the species toward whose protection enormous amounts of Indian and foreign money and attention has been directed. But Nagarhole is also home to about 6,000 tribals, who have been in the area longer than anyone can remember, perhaps as long as the tigers themselves. The state Forest Department wants to expel the tribals, claiming they destroy the forest and kill wild game. The tribals answer that their demands are modest, consisting in the main of firewood, fruit, honey and the odd quail or partridge. They do not own guns, although coffee planters living on the edge of the forest do; maybe it is the planters who poach the big game? In any case, they ask the officials, if the forest is only for tigers, why have you invited India's biggest hotel chain to build a hotel inside it while you plan to throw us out?

Into this controversy jumps a "green missionary" passing through Karnataka. Dr John G. Robinson works for the Wildlife Conservation Society in New York, for whom he oversees 160 projects in 44 countries. He conducts a whistle-stop tour of Nagarhole, and before he flies off to the project on his list, hurriedly calls a press conference in the state capital, Bangalore. Throwing the tribals out of the park, he says, is the only means to save the wilderness. This is not a one-off case but a sacred principle, for in Robinson's opinion "relocating tribal or traditional people who live in these protected areas is the single most important step towards conservation." Tribals, he explains, "compulsively hunt for food," and compete with tigers for prey. Deprived of food, tigers cannot survive, and "their extinction means that the balance of the ecosystem is upset and this has a snowballing effect."

One does not know how many tribals Robinson met—none, is the likely answer. Yet the Nagarhole case is hardly typical. All over India, the management of parks has sharply posited the interests of poor tribals who have traditionally lived there against those of wilderness lovers and urban pleasure seekers who wish to keep parks "free of human interference"—that is, free of other humans. These conflicts are being played out in the Rajaji sanctuary in Uttar Pradesh, in Simlipal in Orissa, in Kanha in Madhya Pradesh, and in Melghat in Maharashtra. Everywhere, Indian wildlifers have ganged up behind the Forest Department to evict the tribals and rehabilitate them far outside the forests. In this they have drawn sustenance from American biologists and conservation organisations, who have thrown the prestige of science and the power of the dollar behind the crusade to kick the original owners of the forest out of their home.

Specious nonsense about the equal rights of all species cannot hide the plain fact that green imperialists are possibly as dangerous and certainly more hypocritical than their economic or religious counterparts. For the American advertiser and banker hopes for a world in which everyone, regardless of colour, will be in an economic sense an American—driving a car, drinking Pepsi, owning a fridge and a washing machine. The missionary, having discovered Jesus Christ, wants pagans also to share in the discovery. The conservationist wants to "protect the tiger (or whale) for posterity," yet expects *other* people to make their sacrifice.

Moreover, the processes unleashed by green imperialism are very nearly irreversible. For the consumer titillated into eating Kentucky Fried Chicken can always say "once is enough." The Hindu converted to Baptism can decide later on to revert to his original faith. But the poor tribal, thrown out of his home by the propaganda of the conservationist, is condemned to the life of an ecological refugee in a slum, a fate, for these forest people, which is next only to death.

The illustrations offered above throw serious doubt on Arne Naess' claim that the deep ecology movement is "from the point of view of many people all over the world, the most precious gift from the North American continent in our time" (Naess 1989: 23). For deep ecology's signal contribution has been to invest with privilege, above all other varieties and concerns of environmentalism, the protection of wild species and wild habitats, and to provide high-sounding, self-congratulatory but none the less dubious moral claims for doing so. Treating "biocentric equality" as a moral absolute, tigers, elephants, whale etc. will need more space to flourish and reproduce while humans—poor humans—will be expected to make way for them.

The authors of this book [Guha and Martinez Alier] by no means wish to see a world completely dominated by "human beings, their mutualists, commensals, and parasites." We have time for the tiger and the rainforest, and also wish to protect those islands of nature not yet fully conquered by us. Our plea rather is to put wilderness protection—and its radical edge, deep ecology—in its place, to recognise it as a distinctively North Atlantic brand of environmentalism, whose export and expansion must be done with caution, care, and above all, with humility. For in the poor and heavily populated countries of the South, protected areas cannot be managed with guns and guards but must, rather, take full cognisance of the rights of the people who lived in, and oftentimes cared for, the forest before it became a National Park or a World Heritage Site (Kothari et al. 1995; Gadgil and Rao 1994; Sukumar 1994).

Putting deep ecology in its place is to recognise that trends it derides as "shallow" ecology might in fact be varieties of environmentalism that are more apposite, more representative , and more popular in the countries of the South. When Arne Naess says that conservation biology is the "spearhead of scientifically based environmentalism" (Naess 1990: 45), we wonder why "agroecology," "pollution abatement technology" or "renewable energy studies" cannot become the "spearhead of scientifically based environmentalism." For to the Costa Rican peasant or Ecuadorian fisherman, the Indonesian tribal or slum dweller in Bombay, wilderness preservation can hardly be more "deep" than pollution control, energy conservation, ecological urban planning or sustainable agriculture.

References

Agarwal, A., 1986. "Human-nature interactions in a third world country." *The environmentalist*, 6(2): 167–187.

Agarwal, A., Narain, S., eds., 1985. *India: The state of the environment, 1984–85: A citizen's report*, New Delhi: Centre for Science and Environment.

Bahro, R., 1984. *From red to green*, London: Verso Books.

Berry, W., 1987. "Preserving wildness," *Wilderness*, Spring 1987: 39–40, 50–54.

Bonner, R., 1993. *At the hand of man: Peril and hope for Africa's wildlife*, New York: Alfred A. Knopf.

Callicott, J. B., 1991. "The wilderness idea revisited: The sustainable development alternative," *The environmental professional*, 13(2):235–247.

Cohen, M., 1984. *The pathless way*. Madison: University of Wisconsin Press.

Commoner, B., 1987. "A reporter at large: The environment." *New Yorker*, 15 June.

Devall, B., Sessions, G., 1985. *Deep ecology: Living as if nature mattered*, Salt Lake City: UT: Peregrine Smith Books.

Gadgil, M., Rao, P. R. S., 1994. "A system of positive incentives to conserve biodiversity," *Economic and political weekly*, 6 August.

Galbraith, J. K., 1958. "How much should a country consume?" in Henry Jarett, ed., *Perspectives on conservation*, Baltimore, MD: Johns Hopkins Press.

Hays, S. P., 1982. "From conservation to environment: Environmental politics in the United States since world war two," *Environmental review*, 6: 14–41.

Inden, R., 1986. "Orientalist constructions of India," *Modern Asia studies*, 20: 401–46.

Janzen, D., 1986a. "The future of tropical ecology," *Annual review of ecology and systematics*, 17: 304–24.

Janzen, D. H., 1986b. *Guanacaste National Park: Tropical ecological and cultural restoration*, San Jose: Editorial Universidad Estatal a Distancia.

Kalland, Arne, 1994. "Seals, whales and elephants: Totem animals and the Anti-Use campaigns," in *Proceedings of the conference on responsible wildlife management*, Brussels: European Bureau for Conservation and Development.

Kothari, A., Suri, S., and Singh, N. 1995. "Conservation in India: A new direction," *Economic and political weekly*, 28 October.

Mappes, T. A., Zembaty, J. S., eds., 1992. *Social ethics: morality and public policy*, 4th ed., New York: McGraw-Hill.

Merchant, C., ed., 1994, *Key concepts in critical theory: Ecology*. New Jersey: Humanities Press.

Naess, A., 1973. "The shadow and the deep, long-range ecology movement: A summary," *Inquiry*, 6:96.

Naess, A., 1989. "Comments on the article "Radical American environmentalism and wilderness preservation: A third world critique" by Ramachandra Guha," typescript.

Naess, A., 1990. *Ecology, community and lifestyle*, translated by David Rothenberg, Cambridge: Cambridge University Press.

Nash, R., 1982. *Wilderness and the American mind*, 3rd ed. New Haven, CT: Yale University Press.

Sale, K., 1985. *Dwellers in the land: The bioregional vision*, San Francisco: Sierra Club Books.

Sale, K., 1986. "The forest for the trees: Can today's environmentalists tell the difference?" *Mother Jones*, 11(8):26.

Sax, J., 1980. *Mountains without handrails: Reflections on the National Parks*, Ann Arbor: University of Michigan Press.

Soulé, M., unpub. *The tigress and the little girl* (manuscript of forthcoming book), Chapter 6, "International conservation politics and programs."

Sukumar, R., 1994. "Wildlife-human conflict in India," in R. Guha, ed., *Social ecology*, New Delhi: Oxford University Press.

Commentary

Ramachandra Guha, "Radical American Environmentalism and
Wilderness Preservation" (1997)

ROB NIXON

Environmental movements emanating from the global south have typically
stressed the interdependence of community survival and environmental change
in circumstances where the illusion of a static purity cannot be sustained, far less
exalted. Such movements have also often emphasized how easily outside forces
(including transnational corporations, the International Monetary Fund [IMF],
and the World Bank) and internal authoritarian regimes, often in cahoots with
each other, can rend the delicate, always mutable mesh between cultural survival
and a sustainable environment. Particularly in the global south, but also in indigent communities in the north, the environmentalism of the poor has pushed
back against ideas of natural spaces as set aside for the recreational or spiritual
pursuits of the affluent.

Sociologist and environmental historian Ramachandra Guha has arguably
done more than any intellectual to dispel the myth that environmentalism is a
"full-stomach phenomenon" affordable only to the middle and upper classes of
the world's richest societies. He has drawn on—indeed, drawn out—neglected
strands of American and European environmental thought while refusing them
a global centrality. Guha has insisted that environmentalism be granted a genea-
logical diversity that is historically and materially grounded and more inclusive
in geographical, class, and racial terms.

Guha has underscored the need to keep environmentalism connected to
global questions of distributive justice, connected as well to the unequal burdens
of consumption and militarization imposed on our finite planet by the world's
rich and poor, in their capacity as individuals and as nation-states. While un-
earthing tenacious traditions of environmental thought and activism among the
poor, Guha has resisted sentimentalizing "traditional" cultures as peopled by
"natural" ecologists.

In 1989, he began to question the well-intentioned but ultimately counter-
productive project of deep ecology that, while posing as planetary, was at root
profoundly parochial. In the excerpt above, Guha extends his earlier critique,
insisting that in societies as different as India and Germany, environmental practices and ideals have departed from American preoccupations with wilderness
and deep ecology.

These tenacious preoccupations of U.S. environmentalism, which came to
the fore in the 1970s and 1980s, are not representative of the kinds of environ-

mentalism that can be advanced in most of the world. During this period, as Peter Sauer has noted, the social justice concerns enunciated so powerfully by Rachel Carson began to hive off from America's environmental mainstream. The biocentric focus of the inaugural Earth Day in 1970 exacerbated this split, leaving American environmentalism isolated from trends elsewhere and reducing the nation's incipient environmental justice movement to (in both senses) a minority affair.

Compared with most nation-states, the United States is unusually affluent and lightly populated. Thus to proselytize internationally for wilderness areas that are fenced off from human habitation and human use is impractical. Indeed, a top-down, heavy-handed insistence on such nature–human segregation can tarnish the reputation of environmentalism, reinforcing pervasive suspicions of "green imperialism." How can relatively poor, densely populated societies—like India, China, Nigeria, Indonesia, and Bangladesh, among others—afford to set aside large tracts of unpopulated wilderness areas? At best, such strategies are unviable; at worst, they have resulted in forced removals of those whom Mark Dowie has called conservation refugees. Even in affluent European societies, the pressures of population are such that a purist vision of set-aside wilderness areas has typically achieved little purchase.

To drive humans off their land in the name of preserving animals—or more broadly, in the name of saving wilderness, or indeed, the planet—is to court disaster. The people making the decisions are typically rich and far off, while the people on whom they impose the sacrifices are typically poor and inhabit, in a state of bodily intimacy, diminished prospects of individual and community survival. When their lands (or waters) are turned into green sacrifice zones, the evicted—whether in India or Indonesia, Kenya or Costa Rica—have often fought back in an effort to reclaim their historic access to natural resources. Thus, the environmentalism of the poor gets pitted against the environmentalism of the rich. During such showdowns rich environmentalists have repeatedly dismissed poor environmentalists as anti-environmentalist.

We can see here how the environmentalism of the poor in the global south connects—theoretically, historically, and strategically—with the environmental justice movement among marginalized communities in affluent societies. It was radical African American theologian Benjamin Chavis who in 1982 coined the terms "environmental justice" and "environmental racism" in response to toxic dumping in poor, minority communities in North Carolina. In the United States, civil rights activist and sociologist Robert Bullard did more than anyone to popularize, document, and theorize environmental justice. If much of the impetus for the environmental justice movement in North America and Europe derived from Rachel Carson's exposé of the long-term fallout from toxicity for humans and nonhumans alike, Carson never elaborated on the discriminatory distribution of toxic risk in terms of race and class.

Writers as varied as Guha, Bullard, Vandana Shiva, Wangari Maathai, Ken Saro-Wiwa, and Joan Martinez-Alier have insisted that our planet's future—and the future of environmentalism as a global movement—is inseparable from the plight and priorities of the poor. Impoverished resource rebels can seldom afford to be single-issue activists: their green commitments are seamed through with other economic and cultural causes as they experience environmental threat not as a planetary abstraction but as a set of inhabited risks, some immediate, others long term. Many of these risks arise from what I have called slow violence—that is, unspectacular, attritional, often exponential threats posed by such forces as climate change, toxic seepage, unregulated mining practices, military residues, and deforestation. Those whom Guha has called "ecosystem people" face disproportionate assaults from both slow violence and more immediate violence against the environments they depend on, environments vulnerable to resource capture by transnational corporations; military, civilian, and transnational elites; and international conservation organizations.

The environmentalism of the poor is frequently catalyzed by resource imperialism inflicted on the global south to maintain unsustainable consumer appetites of rich country citizens and, increasingly, the urban middle classes in the global south itself. The outsourcing of environmental crisis, whether through rapid or slow violence, has a particularly profound impact on the world's ecosystem people—those billions who depend for their livelihood on modest resource catchment areas at the opposite extreme from the planetary resource catchment areas plundered by the wealthy—those people whom Guha and his collaborator, Madhav Gadgil, have dubbed "resource omnivores." Crucially, as I have argued, such outsourcing of environmental crisis is not just geographical but temporal, as future generations—above all, the future poor—bear the brunt of rich nation profligacy, especially via the two-hundred-year experiment in carbon extraction and consumption whose historic beneficiaries have been disproportionately white and well off.

The 1997 postscript to Guha's essay ends with a call for more diverse environmental priorities than those promulgated by deep ecologists. In Guha's words: "For the Costa Rican peasant or Ecuadorian fisherman, the Indonesian tribal or slum dweller in Bombay, wilderness preservation can hardly be more 'deep' than pollution control, energy conservation, ecological urban planning or sustainable agriculture." In the intervening years we have witnessed considerable progress toward these ends, as environmental justice movements have become more prolific, more vocal, and more strategic, prompting Geneva-, Washington-, and London-centered green NGOs to become more attentive to the agendas advanced by indigent denizens of embattled environments.

The radical changes at Greenpeace are one measure of this shift. In the late 1980s—just as Guha's original critique of deep ecology was appearing—a Nigerian writer turned activist, Ken Saro-Wiwa, appealed to Greenpeace for assistance in his efforts to expose the despoliation being inflicted on the Niger Delta by Royal Dutch Shell and Chevron in cahoots with the Nigerian dictatorship. The Greenpeace response was one of bewilderment: they said they did not work in Africa and clearly did not see what Saro-Wiwa saw—what he called "genocide by environmental means." Save the whales, yes. But save the Ogoni nation from slow ecological violence? That was outside the Greenpeace idea of green.

However, by the 2011 Climate Summit in Durban, a tectonic shift had occurred at Greenpeace. The organization's energetic international executive director was Kumi Naidoo, a former South African anti-apartheid activist who insisted that an environmentalism that discounts the needs of the planet's indigent majority is an environmentalism without a future. In terms of establishing globally binding emissions agreements, the Durban summit was an abject failure. However, one positive development did emerge, as speaker after speaker from the global south—be it Grenada, Mali, Bangladesh, or South Africa—raised their voices against the slow violence of climate change that threatens the poor disproportionately. Nnimmo Bassey, a Nigerian environmentalist activist, poet, and chair of Friends of the Earth International, put the matter thus: "Delaying real action till 2020 is a crime of global proportions. This means the world is on track for a 4°C temperature rise, a death sentence for Africa, small island states and the poor and vulnerable worldwide. The richest 1% of the world have decided that it is acceptable to sacrifice the 99%" (Oxfam 2011). Oxfam's Celine Charveriat was even blunter: "Negotiators have sent a clear message to the world's hungry. Let them eat carbon" (Vidal and Harvey 2011).

Guha's 1989 essay—and the extended version that appeared in 1997—was prescient on many fronts. He foresaw the parochialism that afflicted deep ecology, its connections to orientalist thinking, and the limitations of the wilderness ideal as an international blueprint. He helped chronicle the long history of environment activism in the global south and anticipated how the environmentalism of the poor would become more central to green activism globally. However, what he could not have foreseen was the dramatic, twenty-first-century convergence between the environmentalism of the poor and the international campaign for climate justice.

Further Reading

Oxfam International. 2011. "Durban platform leaves world sleepwalking towards four degrees of war." December 11. Available at http://www.oxfam.org/en/grow/pressroom/pressrelease/2011–12-11/durban-platform-leaves-world-sleepwalking-towards-four-degrees-war.

Vidal, John, and Fiona Harvey. 2011. "Durban climate deal struck after tense all-night session." *The Guardian*, December 11. Available at http://www.guardian.co.uk/environment/2011/dec/11/durban-climate-deal-struck.

PART 9

Measuring

How Do We Turn the World into Data?

How do we get a conceptual grip on "the environment" and what is happening to it? If we imagine its future is going to be different from its present (or past), how do we present the nature of that difference? Is the aspect we are interested in going to be more, or less (frequent, intense, reliable, and so on)? Does this make it better, or worse? To begin to answer these questions, we need to have some kind of measure, breaking down a continuous world into discrete elements. At the same time—ironically—to understand a phenomenon like global change, we also need to connect the discrete parts back up again, to see how they are linked and how phenomena interact. This demands that we not only turn the world into data, but also that we ensure that the data are commensurable with each other. How can we describe those relationships between very different things (like weather conditions and the prevalence of a species and the role of that species in an ecosystem on which we depend for our economy) in a form that makes the nature of those relationships clear and the consequences of change comprehensible?

This section begins with the writings of James Rennell, a navigator and close observer of the oceans. Rennell's talent for mapping not just currents, but their relations with winds and weather systems; the world's land masses; and the features of the sea bed makes his work an early example of modern "systems thinking." His work is an example of the way in which the knowledge of flows that came to characterize climatology and oceanography had to be painstakingly built from the ground (or water) up (and down). Global change starts with the most basic question: Where are you on the surface of the earth? Rennell recognized the importance of this. In today's world, where we often seemed overwhelmed by data, it is easy to forget how hard it once was to get the most basic information, and how much the practice of science and environmental knowledge was shaped by what could be done at the time. Measurement of basic phenomena— currents or winds—required traveling vast distances, the invention of scales, the development of instruments. But Rennell also required an infrastructure to work in: the opportunities created by British imperial expansion and its need for maritime knowledge. Thus, early environmental knowledge was often a matter of making nature not only legible, but a world to be ruled by imperial power.

The history of environmental understanding has been deeply affected by two processes. First, the opening up of new questions because of the discovery of something that could be measured (perhaps because of new technology, going to a new place, or a new idea) and that produced demand for explanation. Second, the collection and production of data as a response to the demands of a theory, providing the underpinning that would make it convincing or authoritative. Either way, key experts are people who can discover, measure, and interpret metrics. In the case of climate science, early researchers tended to concentrate on what was most manifest, slowly building up a picture of weather systems from temperature, precipitation, and atmospheric pressure readings, having developed the instruments to do so in the eighteenth century. Alternatively, laboratory-based scientists sought to recreate proxies of atmospheric conditions in their laboratories. Gradually putting these two worlds together required new technology and infrastructure, and the funding to back it, often linked to the development of aviation (opening a direct window into the atmosphere) and air warfare (creating the demand for accurate forecasting and prediction).

One of the key players in the increasingly international network of meteorologists after World War II was Carl-Gustaf Rossby, a Swede dividing his time between the United States and Stockholm, where as early as 1954 he stimulated his colleagues to undertake carbon dioxide measurements to find out more about climate change. He was followed in this pursuit by his colleague Charles David Keeling, who, after several alternative locations had been considered, started carbon dioxide measurements on the Mauna Loa mountain in Hawaii in 1958, which turned out to be the most reliable—and now iconic—global time series revealing the steady increase of carbon dioxide concentration in the atmosphere. At the same time, the emergence of carbon dioxide measures made it an ideal parameter to use in the computer simulation of climate, stimulating further work in the area. The measure provided a means by which change could be scaled, and predicted, which equally generated an expectation that good predictions were those that provided clear results in widely accepted measures.

It was no easy work to produce metrics of physical phenomena, requiring instruments, scales, and agreed ways of recording. But what of those that provide no clear physical phenomenon to measure—the interaction of species, the quality of ecosystems, or, indeed, the usefulness of nature to humans? Increasingly, various kinds of environmental science sought metrical analogs that would seem to give precision to qualities they thought important: biodiversity, or rates of extinction.

As environmental data piled up, there also emerged, over a long period of time, a sense that a lack of a measure of value was preventing humans from fully grasping the meaning of environmental change for themselves. One of the solutions to the crisis was to knit together the worlds of ecology and economics—worlds that in fact have much in common in their history and methods. Gretchen Daily's *Nature's Services* (1997) proposed an economic system that considered and accounted for the benefits of nature, such as clean air and water, that are otherwise taken for granted because they are "free." This attempt, which has enjoyed wide policy resonance, shows many of the characteristics that can be found in earlier attempts at integrating different spheres of study into "global change": a preference for numbers and reductive valuations, and reliance on experts to make them. Often these measures have been deliberately formulated to communicate with policy makers, but can end up driving research into natural processes.

Yet the need to integrate measures also goes in the other direction, because as humans play an increasingly prominent role as agents of global change, it becomes desirable to know how changes in human society and economy will affect natural processes—whether through pollution or habitat change—or, more indirectly, how changes in human welfare as measured by something like gross domestic product (GDP) are likely to affect greenhouse gas emissions, climate change, and—of course—future human welfare. The final document, the famous *Stern Review on the Economics of Climate Change* commissioned by the British government, aimed for major policy shifts by indicating the cost of climate change using mainstream economic measures. In doing so, it joined an increasing body of scholarship that seeks to model and to integrate the environmental impact of economic change, and vice versa. This kind of work has demanded aggregating expertise on a new scale, most notably in the vast teams assembled by the International Panel on Climate Change. The Stern report led the way in understanding of how the climate system and the economic system had become enmeshed, moving beyond single factors such as greenhouse gas emissions, deforestation, and land-use change to consider their interactions with each other, and the global economy, and quickly became the best known of several such reports linking national policy, economics, and global environmental outcomes.

Values and measures have helped us comprehend the natural world; indeed, they have strongly shaped our perception of what it is and who is best placed to speak authoritatively about it. Importantly, numbers are not independent of the people and institutions that provide them—and indeed the authority and trust we invest in those people and institutions. Frequently, scientists working on the environment have had their agendas strongly shaped by political and social forces, and their desire to anticipate and predict. Since environmental science is important in policy making, environmental scientists seek to communicate in ways that are useful beyond science. Turning the world into data suitable for comparative analysis and projections of the results of policy is a valuable tool for this. Measuring is a tool that allows us to identify patterns from the world's data and that enables prediction of future change.

An Investigation of the Currents
of the Atlantic Ocean

JAMES RENNELL

General Observations on Winds and Currents

[1–3] Although the currents of the ocean form a most important part of hydrography, yet it is only since the introduction of chronometers, and of celestial observations for the longitude at sea, (that is, not much more than forty years ago), that a competent idea of their direction and force, in any kind of detail, could be obtained. For although the differences in northing and southing, between the dead-reckonings and observations, might be pointed out by the observations of latitude, yet the error of longitude, or of easting and westing, would, of course, escape detection altogether: and it happens that, in the Atlantic Ocean, which forms the scope of the present inquiry, the streams of current, which most materially affect navigation, both in respect of extent and velocity, run more easterly and westerly than otherwise. To the invention of chronometers, then, and to the improved methods of finding the longitude at sea by celestial observations, to check their rates of going, navigators are indebted, as well for expediting their passages as for greater safety in the mean time: and the invention, as it respects currents, is surely in the next degree of importance to that of showing the ship's place, which may be so much affected by them, during the long intervals that occasionally happen between the celestial observations, and for which a delineation of such streams of current as are prevalent, would prepare the navigator.

The progress in the knowledge of this subject has, accordingly, been very great, since the date of the above invention; and, to every person the least informed concerning nautical science and practice, the use of such knowledge

James Rennell. 1832. Excerpts from "An Investigation of the currents of the Atlantic Ocean, and of those which prevail between the Indian Ocean and the Atlantic," ed. J. Purdy. Nabu Public Domain Reprints. Milton Keynes: Lightning Source UK Ltd.

must be obvious. For, in whatsoever line of direction a portion of the ocean may move, it must rarely happen that it is neutral in respect of a ship's course; but that, on the contrary, it will either favor or impede it. A knowledge of the truth, therefore, will enable the navigator so to arrange his course as to render the current the most advantageous in the one case and the least disadvantageous in the other: or, in certain cases, to avoid delay or danger. . . .

[4–5] The author having employed a large portion of his leisure in collecting materials for illustrating and explaining the subject of the currents of the ocean generally, and more particularly for those in the Atlantic and Indian Oceans; and in forming a system conformable to these observations; flatters himself that the work at large will meet the approbation of those who are competent judges of it. His personal knowledge (and perhaps he may be allowed to add some degree of former professional knowledge) of such subjects, has furnished him with the means of appreciating the value of a great proportion of the materials that engaged his notice, as well as of adding to the whole the result of inquiries and observations made on the spot.

The formation of a great number of facts into a system, may, it is presumed, prove of use in impressing those facts on the mind more strongly than if they were left to operate independently of each other. For oftentimes a fact makes less impression, when standing naked and alone, than when it makes part of a system, which operates like a band to keep the parts together in their proper places, when they may happen to explain and illustrate each other. . . .

[6] It appears that scarcely any portion of the surface of the ocean remains still: and doubtless this is amongst the wise dispensations of THE CREATOR, to preserve its purity. . . .

The winds (with very few exceptions) are to be regarded as the prime movers of the currents of the ocean: and of this agency, the trade-winds and monsoons have, by far, the greatest share; not only in operating on the larger half of the whole extent of the circumambient ocean, but as possessing greater power, by their constancy and elevation, to generate and perpetuate currents: and, although the monsoons change half-yearly, yet the interval during which they continue to blow, in each direction, is long enough to produce effects nearly similar to the operation of the constant trade-winds. . . .

An opportunity occurs of determining, generally, to what distance off shore the influence of the monsoon which displaces the S.E. trade on the coast of Brasil extends. Mr. Fitzmaurice (who with so much trouble and difficulty brought home the two ships of Captain Tuckey from the Congo expedition) touched in his way home at St. Salvador, or Bahia, in lat. 13° 1′ S., long. 38° 32′ W., in the month of December, a season which may be reckoned the height of the N.E. monsoon of the coast of Brasil. The N.E. wind compelled him to stand to the

S.E., until he reached longitude 24° in the parallel of 21° 50', which was no less than 940 miles from St. Salvador, and 860 from the nearest shore of Brasil; and here the wind first enabled him to lay the ships' heads to the northward, in his way home. This result is very satisfactory; as being founded on a *series of examples*, and not on *casual* ones; of which we had no other kind before.

It remains to give a word or two concerning the winds beyond the trades; and which, although not operating constantly, yet return so frequently to their accustomed quarter, (the western), that they are reckoned to blow nine days from the western side of the meridian, to about five and a half from the opposite side; but still produce so slow a current, in the North-Atlantic, clear of the Gulf stream, as only to manifest itself generally, on the whole course of a voyage, as from America to Europe. This subject will be discussed at large hereafter, and will appear curious, from the sameness of the results; which arise from experiments, made with floating bodies, either intentionally or accidentally employed.

. . .

Causes and Effects of the Different Currents

[92–95] Iced Water. — There yet remains to be spoken of an effect generally produced on the temperature of a portion of the Gulf-stream, (as well as of the ocean at large), by the presence of islands of ice: and the means which may be employed to derive greater security from accidents in approaching them, in the night, or in thick fogs.

This was exemplified, in a remarkable manner, on board the *Eliza* packet, in her passage from Halifax towards Europe, in 1810: an attention to which may prove of use to all ships navigating such parts of the ocean, as are at any time incommoded with ice. The particulars are amply detailed on Chart II., and the following observations will explain them more fully.

This ship, late in April, crossed the Sable Island Banks, with an ocean-temperature of 40° to 41°, and then came into 62° and 64°, in latitude 42° 15'; longitude about 60° west. Here they considered themselves to be entered within the verge of the Gulf-stream, though beyond the limit assigned to it by Dr. Franklin. The temperature of 64° is certainly about 4° above the ocean temperature at that season; but which was probably not the stream itself, but its overflow.

From thence, in a general easterly course, 34 leagues, with a temperature of 60° to 62°, and still within the supposed limit of the Gulf-stream, the water suddenly fell to 58°, and afterwards gradually to 45°, in the distance of 11½ leagues. They now discovered the cause to be a number of Ice Islands, which they passed at the distance of seven miles, in 45° to 46° of temperature. The group extended about five leagues west to east.

Passing on, the water rose again to 50°, and afterward to 60°: but the 50° was much farther distant from the ice than the 58° in coming towards it. In effect, a space of 36 miles in extent in advancing to, and 57 after leaving, the ice, was cooled from 10 to 16 degrees below the temperature of the surrounding sea; for at seven miles from the ice, the temperature was lowered to 45°, or 16° below the general temperature of the sea. . . .

From hence the temperature rose again to 60°, which continued generally to about 115 leagues, on an E. by N. course, to longitude 41 2/3° W. in latitude 43°, and very near to a situation where Captain Beaufort, in August of the preceding year, had 73° or 74° of temperature, or 6° above the temperature of the ocean water, in that place. But the *Eliza* had generally 60°, the ocean-temperature at that season, (beginning of May), although she had passed through sixty leagues of the space, very near to which Captain Beaufort, in the preceding summer, had 73° to 76°: and indeed, the whole of the *Eliza's* run, from longitude 60°, was through water scarcely above natural temperature, although the whole had been, from thermometrical observations, considered as within the Gulf-stream: and it is certain that part of the tract in which there was Gulf-water, when Captain Beaufort passed eight months before, was now filled with water of the ocean-temperature. Is it then that ice had cooled the water in all that line, since it is found that a few Ice Islands may render cool a circle more than ninety miles diameter? This is a quarter in which much ice is found in spring and summer; some drifting through the Strait of Belle Isle, whilst a much greater quantity is brought by the southerly current that ranges along the eastern coast of Newfoundland. Many large Icebergs have been seen 40 3/4° at Midsummer, but in a state of great decay; having been, of course, a long time subjected to the warmth of Gulf-water.

But however curious this question may be, it is yet far more useful; since it proves that, by an attention to the thermometer, especially in the night, or in foggy weather, the presence of ice may be ascertained, and preparations made accordingly.

. . .

Appendix. Notes on the Currents, &c.

[347–348] THE FLORIDA OR GULF STREAM. —A bottle from H.M. ship *Breton*, Hon. Captain Gordon, 2d February, 1830, in the Mexican Sea, latitude 27° 50′, long. 84° 40′, the Tortugas bearing nearly S.S.E. 215 miles;—found on the 2d of June, 1830, in latitude 25° 52′, longitude 80° 9′, on the south eastern coast of Florida, near White Islet; where there is now a settlement. (*Notice transmitted to the Admiralty.*)

Gulf-stream to the Azores. — A bottle from H.M. ship *Newcastle*, latitude 39°
12′ N., long. 63° 52′, by Mr. James Napier, master, 20th June, 1819; — found on the
shore of St. George, one of the Azores, 20th of May, 1820, in about 38° 40′ N. and
28° W. *(Notice transmitted by Mr. Wm. Parkin, British Vice-Consul.)*

Gulf-stream to Ireland. — Another bottle from H.M. ship *Newcastle*, 20th of
June, 1819, in latitude 38 52′ , longitude 64° 0′; found on shore the Rosses, on the
N.W. of Ireland, near the Isle of Arran, 2d of June, 1820, and attested by Mr. Nas-
sau Forster. *(This appears to have been erroneously attributed to H.M. ship Pique.)*

Gulf-stream to St. George's Channel. — A bottle from the ship *John Esdaile*,
Henry King, commander, cast into the sea 28th of July 1821, in latitude 36° 55′,
long. 71° 50′; picked up on the sand near the mouth of the Ribble, Lancashire,
5th Dec. 1822. (This bottle, therefore, after leaving the stream, was carried to the
E.N.E. towards the Welsh Coast.)

The main-mast of the *Tilbury*, (a sixty-gun ship,) burnt off Hispaniola, in the
seven-years' war, was also brought to our shores, but I cannot give any time. — J.R.

[349] By a comparison of Captain Livingston's examples [of water temperatures in the Gulf Stream] with those of others, and reduced to the same season, it
would seem that, in the part between 15° and 25° of west longitude, (the still part
of the ocean,) Captain Livingston's thermometer was a degree and a half higher
than the others; that is 70° for 68½° in lat. 38°. But in the Gulf-stream, between
Cape Hatteras and longitude 50°, more particularly between that and 60°, he
agrees with Williams, Pell, and Billings.

Commentary

James Rennell, "An Investigation of the Currents
of the Atlantic Ocean" (1832)

SARAH CORNELL

Prediction requires more than just observation. In Rennell's time, mariners were familiar with the trade winds and knew of the major global currents like the Gulf Stream. But using this knowledge as *foreknowledge* in the way Rennell advocated requires a deeper understanding of their variability, which in turn, for complex systems like global circulation patterns, requires an understanding of causality. Rennell not only mapped the currents; he went on to explain their links to winds and weather systems, the world's land masses, and the features of the sea bed. Together, the available observations and his insightful knowledge began to constitute a predictive system.

And comprehension of the system requires stories or pictures—let us call them "models"—that "keep the parts together in their proper places" (5). Rennell describes currents as rivers in the oceans (as did Benjamin Franklin, in his 1768 mapping of the Gulf Stream). It is a robust model, bringing complex patterns of flow including swirls and eddies and side-streams from the global scale to the comprehensible level of everyday experience. Good model-making takes intuition and imagination. Medawar (1969, 51) echoes Rennell, arguing that science comes "not from the apprehension of 'facts,' but from an imaginative preconception of what might be true." At one point, Rennell writes, "Were we to suppose a globe covered with water" (59), going on to describe an idealized global circulation. For science, this kind of imaginative rendering needs to be true, in the sense of being consistent with available observations. Randall and Wielicki (1997) emphasize the interdependency of models and observations, noting that "models by themselves are just 'stories,'" just as data without models are "just numbers." In Rennell's case, he was both observer and modeler, and could check the consistency himself of his idealized model with the available data: "If the reader casts an eye over the index map . . . he will perceive that the motions of the different streams accord with the above principle, where it is left to operate freely" (59). In some fields of science today, however, particularly those dealing with global-scale processes, a gulf exists between the highly technical and specialist fields of modeling and measurement. Where complex models are to be used for prediction of the actual world, this gulf is one of the things that must be addressed.

Rennell's work shows the power of quantification and systematization, giving both a rich description of the ocean's large-scale behavior and weight to his exhortation to apply the knowledge "to avoid delay or danger" (3). Rennell's predictions of maritime currents and their hazards were valued and taken seriously by navigators and scholars alike. By today's standards, though, the evidence base relies on strikingly few numbers, and such precarious ones: the record of ship's logs is patchy ("Sir Philip Broke's description [of the Gulf Stream] is the only one that the Author has seen of the state of it in winter" [180]), supplemented by occasional reports of bottle landings and bits of shipwrecks washing up on beaches (347–349). Yet at the same time, trust in science today has not been assured merely by its data-intensive, technologically sophisticated nature. Quantification and systematization bring predictive power, but their application often entails other kinds of power, too.

The frequent references to individual people are a feature that marks an "investigation" as a historical scientific text—and by contrast, shows how today's scientific writing, having doggedly pursued the ideal of "objectivity," now risks being problematically dehumanized. Almost all of Rennell's data providers are mentioned by name, indicating that the data he relies on are both traceable to their original source and (presumably, since we no longer recognize the individuals named) attributable to reliable sources. Rennell also refers to his own professional and personal experience of the phenomena he has systematized. His explicit recognition of these forms of experience echoes the attention Collins and Evans (2007) give to "contributory" expertise, as he sets out his "locus of legitimacy." His experience not only equipped him with the ability to see beyond the "naked facts" and link them into a whole picture, but conferred credibility on his ideas in his otherwise apparently divergent target audiences of mariners and scholars.

Rennell also takes ownership and responsibility for his ideas in what now looks a rather unmodern way. Much contemporary global change research takes place with a supposed detachment from values and politics, even as it is channelled directly into global policy contexts. Quantitative trajectories of future social-environmental pathways—leading to utopias and dystopias alike—are described as "model output," as if scientific knowledge now arises and resides entirely outside of the mind of the scientist (of course, much of this research involves the collaborative effort of many, many minds, which brings its own challenges). Yet value judgments are still there, arising in the decisions by the modelers of what to quantify and how, and in the selection and application of the model outputs. These values are merely occluded, and the relationship of hidden agendas to power is often complex; it is no surprise that global change research sometimes faces issues with public trust in science. As quantitative global system models become more elaborate, and as their output increasingly informs policy in the actual world, the need to probe, illuminate, and critique them grows rather than vanishes.

Rennell's quantification and systematization produced a robust conceptual representation of the ocean circulation system, many features of which are still valid today. The way that he gathered his inputs and communicated his insights constructed the vital bridges to the lived experience of ocean navigation. As society faces many new global-scale challenges, we would be wise to remember that the measurement and prediction steps are just part of the process. The meaningful use of these numbers and knowledge depends on ensuring that the people whose lived experience will be changed by this use all share in the story that holds the facts together.

Further Reading

Collins, H. M., and R. Evans. 2007. *Rethinking expertise*. Chicago: University of Chicago Press.

Medawar, P. B. 1969. *Induction and intuition in scientific thought*. London: Methuen & Co. Ltd.

Randall, D. A., and B. A. Wielicki. 1997. "Measurements, models, and hypotheses in the atmospheric sciences." *Bulletin of the American Meteorological Society* 78(3):399–406.

Current Problems in Meteorology

CARL-GUSTAF ROSSBY

[10] During the decades which have passed since meteorology first took shape shortly after the beginning of this century, the network of meteorological stations which is at our disposal for the study of the daily changes in state and movements of the atmosphere, has been extended in an impressive way. It is now possible to give a rather satisfying picture of the air movements of the troposphere and the lower stratosphere twice a day over the major part of the northern hemisphere. At the same time our knowledge of the dynamics and physics of the atmosphere has to some degree become more profound. In spite of, or perhaps because of this better knowledge, one finds that certain fundamental postulates, which earlier were regarded as so self-evident that they were not even dealt with in the meteorological textbooks, now must be looked upon as rather uncertain. The heat balance of the atmosphere serves as a good example.

. . .

[12] In the surface layer of the sea the perpetually shifting winds cause mixing and a vertical homogeneous layer of water, the medium depth of which is of the order of magnitude of 50–100 m. If the heat capacity of this layer is taken into account, it is found that a storage of 1 per cent of the effective incoming solar radiation would lead to a mean temperature increase in the entire storage layer (the atmosphere plus the homogeneous surface layer) of only a few tenths of a degree. . . .

The role of the sea as a secular heat reservoir assumes quite a different character at the moment that one takes up the question of secular changes of the total heat balance, taking into account the circulation of the deeper layers.

. . .

Carl-Gustaf Rossby. (1957) 1959. Excerpt from "Current problems in meteorology." In Bert Bolin (ed.), *The atmosphere and the sea in motion: Scientific contributions to the Rossby memorial volume*, 10–16. Original Swedish version printed in 1957. New York: Rockefeller University Press.

[13] . . . one is probably justified in expressing the following two suggestions:

(a)   The assumption that our planet as a whole stands in firm radiation balance with outer space cannot be accepted without reservations, even if periods of several decades are taken into account.

(b)   Anomalies in heat probably can be stored and temporarily isolated in the sea and after periods of the order of a few decades to a few centuries again influence the heat and water-vapour exchange with the atmosphere.

If this latter assumption is correct, it does not seem unlikely that the problem of post-glacial climate fluctuations lasting a few hundred years can take on new aspects. But it must be pointed out that if these anomalies in heat which are stored in the interior of the sea are gradually distributed in greater water masses, they must, when they finally reach the sea surface again, be characterized by very small temperature amplitudes. How such exceedingly small variations in temperature could possibly have a significant influence on the atmosphere is still an unanswered question. It is perhaps more likely that the changes by no means take place at a constant rate but fluctuate so that the contrast in temperature between the surface water and the deep water shows strong variations with time.

Considering what has been said above, it is obvious that measurements or reliable estimates of the heat exchange between our planet and outer space must be looked upon as a major question for meteorologists and oceanographers interested in the global circulation systems of the sea and atmosphere and their fluctuations. . . .

[14–16] B. *Carbon dioxide and its cycle*

The circulation of water between the surface and the deep layers of the sea, and especially its period of circulation, is of fundamental importance when studying another global meteorological problem of great interest to climatology, i.e. the increase of the carbon-dioxide content of the atmosphere. This increase seems to be a result of the steadily increasing consumption of fossil fuel in the last 50 to 100 years. How large this increase really is must to a great degree depend upon whether the sea, particularly the deep layers, is able to absorb slowly or quickly the excess of carbon dioxide constantly supplied to the atmosphere.

It has been pointed out frequently that mankind now is performing a unique experiment of impressive planetary dimensions by now consuming during a few hundred years all the fossil fuel deposited during millions of years. The meteorological consequences of this experiment are as yet by no means clarified, but there is no doubt that an increase of carbon-dioxide content in the atmosphere would lead to an increased absorption of the outgoing infrared radiation from the earth's surface thus causing an increase of the mean temperature of the atmosphere. As we know, SVANTE ARRHENIUS was first to point out that variations in the carbon-dioxide content of the air, resulting from the volcanic activity of the earth, could explain the variations in climate, which characterize the geological history of our planet. Quite recently G. N. PLASS calculated that, assuming all other factors to be constant, a doubling of the carbon dioxide in the atmosphere would lead to a mean air temperature increase of about 3.6°C, while a reduction of the carbon dioxide to half its value would lower the temperature by 3.8°C. It is almost certain that these figures will be subjected to many strong revisions, depending mainly on the fact that those complicated processes which finally determine the mean temperature of the atmosphere, cannot be dealt with as independent, additive phenomena. For instance, a higher mean temperature caused by carbon dioxide must lead to an increase of atmospheric water vapour content and therefore of the infrared absorption by the water vapour but probably also to an increased cloudiness.

Has there really been a considerable increase in the content of carbon dioxide in the air during the very much expanded industrial activity of the last decades? In 1940 G. S. CALLENDAR thought it possible to show that the carbon-dioxide content of the atmosphere had increased by approximately 10 per cent since the beginning of the century. The observational material at his disposal was very extensive but of very uneven quality with a highly unsatisfactory geographical distribution of the observation sites (most of them were situated at places in central Europe, which were highly polluted by industrial activity). Callendar selected the series of observations that he thought were most reliable and representative, but an inspection of the material used with its enormous spread

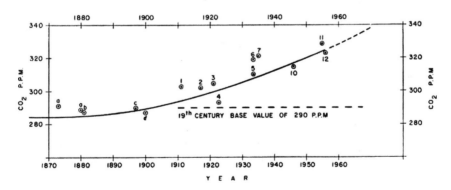

Fig. 1. In this diagram by G. S. Callendar an attempt is made to illustrate the increase, in recent years, of the content of carbon dioxide in the atmosphere by means of observational series which were critically selected. The continuous ascending line represents the theoretically estimated content of carbon dioxide under the assumption that none of the carbon dioxide liberated through combustion is stored in the sea or in increased vegetation.

gives a strong impression of the uncertainty which necessarily characterizes his estimates. In a paper published recently, however, and based on a critical review of older as well as more recent data, Callendar maintains his opinion about the rapid increase of the atmospheric carbon dioxide.

An increase by 10 per cent of the total carbon-dioxide content of the air would, according to Callendar, approximately correspond to the amount of carbon dioxide liberated through the consumption of fossil fuel during the three or four first decades of this century. In order to explain this high value of the increase of carbon-dioxide content, one must assume that only a very small fraction of the amounts released to the atmosphere has been absorbed in the sea in spite of the fact that the capacity of the marine reservoir is about sixty times greater than that of the atmosphere. Thus one is immediately faced with a great number of difficult problems. How should measurements of the total carbon-dioxide content of the atmosphere be conducted in the best way? How should measurements or estimates be made in order to gain increased knowledge of the carbon-dioxide exchange at the sea surface? Finally, how rapid is the exchange between the surface layer and the deep sea?

For almost two years a small group of Scandinavian scientists has maintained a network consisting of 15 stations on the Scandinavian peninsula, Denmark and Finland from which air samples for carbon dioxide analysis have been taken three times every month. The sampling stations and times are chosen to suppress the local sources of error as much as possible. Nevertheless it is found that the carbon-dioxide content varies so much with the origin of the prevailing air masses that it possibly could be used as a diagnostic, synoptic element. It is not unusual to find variations of 10 per cent across a well-developed front. Therefore it seems almost hopeless to arrive at reliable estimates of the atmospheric carbon-dioxide reservoir and its secular changes by such measurements in limited areas.

In order to overcome this difficulty to some extent it has been suggested that regular carbon-dioxide analyses of the air near the surface should be performed in some synoptically inactive parts of the world far from industrial regions, the sea, and densely vegetated regions where also the assimilation could influence locally the values obtained. Carbon-dioxide determinations in the free atmosphere and in desert regions, mainly in the not yet too heavily industrialized southern hemisphere ought therefore to be of special interest, but they must, of course, be made concurrently [*löpande*] and during a great number of years in order to establish secular changes in the total carbon-dioxide content of the atmosphere.

As a contribution to the study of these important problems, a rather extensive observational program will be conducted during the International Geophysical Year. In addition to the rather modest Scandinavian network meteorologists and oceanographers have planned an extended network of synoptic carbon-dioxide stations in North and South America, the Arctic and the Antarctic. Carbon-dioxide determinations will furthermore be made on a great number of islands in the Pacific and the Atlantic, and on mountain stations in North and South America. Furthermore, there will be regular flights along certain meridians in order to determine the carbon-dioxide content in the free atmosphere.

In spite of the rich material which will thus be collected, it is very likely that great difficulties will be encountered in every attempt to compute the content of carbon dioxide in the atmosphere and its secular changes from such scattered observations. For this reason it is of special interest that a new, perhaps more promising method, is being developed based upon comparative determinations of the content of the atmosphere and the biosphere of radioactive carbon (C^{14}). The first attempt to determine the secular change of the carbon-dioxide content of the atmosphere by this method were made by H. SUESS in 1953, and the problem has later been taken up by others.

In principle, this method is based upon the fact that the carbon dioxide, which is brought to the atmosphere by the combustion of fossil fuel, must be free from radioactive carbon, the half-life of which is 5,568 years. By comparison of the C^{14} content of annual tree rings from the middle of the last century with the youngest annual rings in trees recently felled one can thus determine whether the assimilated carbon dioxide originates from the earlier "natural" carbon dioxide reservoir of the atmosphere, in which the C^{14} content represents an equilibrium between the production and decay of radioactive carbon, or from the extra supply of "dead" carbon dioxide which originates from the fossil fuel consumed.

The method has the great advantage that it is very likely to eliminate local synoptic variations in the atmospheric carbon-dioxide content. However, the industrial consumption of industrial fuel shows very great geographical variations with a minimum in the southern hemisphere. Therefore it is obvious that definitive conclusions concerning secular variations can be drawn only when samples from widely separated parts of the world have been analysed.

It should perhaps be stressed that investigations concerning such problems as the total variation of the heat stored in the sea, or of the total content of carbon dioxide in the atmosphere, biosphere or the sea, mean a completely new class of questions in theoretical meteorology and oceanography. In these investigations one is hardly interested in geographical distributions. As a first approximation the problem consequently may be reduced to systems of simultaneous ordinary and usually nonlinear differential equations, which express the interplay between the different reservoirs. Under special conditions thermomechanical systems of this type are able to maintain nonlinear oscillations of finite amplitude, as if their self-regulating properties were defective in some way. E. ERIKSSON AND P. WELANDER have recently suggested that the combined carbon dioxide system should be characterized by such oscillations. Their result depends to a great extent on some much debated assumptions about the interior properties of the system, but it is obvious that the possible existence of such oscillations in the total heat balance system, including the heat storage in the sea, would be of great climatological interest.

Commentary

Carl-Gustaf Rossby, "Current Problems in Meteorology" (1957)

MARIA BOHN AND SVERKER SÖRLIN

Carl-Gustaf Rossby was a Swedish meteorologist who also became an international leader in climate science just as the idea of anthropogenic climate change gained momentum in the 1950s. His long article "Current Problems in Meteorology," written at some point before his premature death in August 1957, is evidence of his interest in the possibility of a human-induced increase of atmospheric carbon and the research he thought most promising in order to gain more knowledge of this phenomenon. "This increase," he said, "seems to be a result of the steadily increasing consumption of fossil fuel in the last 50 to 100 years." He linked the increase, and the ensuing likely climate change, to the burning of fossil fuels: "It has been pointed out frequently that mankind now is performing a unique experiment of impressive planetary dimensions by now consuming during a few hundred years all the fossil fuel deposited during millions of years."

Most of Rossby's climate-related work was conducted at the Institute of Meteorology at Stockholm University, which was established by this charismatic and entrepreneurial scientific leader after his return to his native Sweden in 1947. What was to be known as the Stockholm School of Meteorology grew out of atmospheric physics and applied mathematics, using massive computer-aided research for numerical weather prediction that became available in Stockholm in the early 1950s through the Binary Electronic Sequence Calculator (BESK) at the Royal Institute of Technology.

Rossby's teaching and research style relied on mathematical geophysics, for which he had been trained during the 1920s in Bergen, then the leading center for meteorological research, headed by Vilhelm Bjerknes, whose son Jacob was one of the junior researchers. Rossby was sent by Bjerknes to get further training in Berlin and Leipzig, and he also joined Conrad Holmboe's expedition to East Greenland in 1923, thus acquiring firsthand experience of the Arctic climate. The Bergen school was solidly anchored in physical theories and methods, rather than traditional empirical meteorology "on the ground," a priority that also marked activities at Rossby's institute in Stockholm. Rossby was remembered for his boast that he had "not collected one single empirical observation."

After working in the Swedish Weather Bureau, the SMHA (later SMHI), and with a licentiate in mathematical physics from the Stockholm Högskola, Rossby moved to the United States, where he rose to become the leading meteorological scientist during the 1930s, serving first in the U.S. Weather Bureau and later at the Massachusetts Institute of Technology (MIT) and at the Woods Hole

Oceanographic Institution. He organized comprehensive training programs to support the military weather service during the war, and he headed the leading meteorological graduate school, funded by the military and located at the University of Chicago, where some eight hundred meteorologists trained during the war years. He advised the U.S. president and the highest military leadership on weather issues. His meteorological expertise was also consulted in relation to ecological problems, in particular the Dust Bowl experiences of the 1930s.

"Current Problems" was originally written for the 1956 yearbook of the National Science Research Council of Sweden, printed in 1957. The English version is a translation by members of staff at the International Meteorological Institute in Stockholm, which he directed. It appeared posthumously in a memorial volume dedicated to Rossby published in 1959: *The Atmosphere and the Sea in Motion: Scientific Contributions to the Rossby Memorial Volume*. The memorial volume in itself was a claim about the significance of Rossby and meteorology: five hundred pages in large format, with many of the leading international names of meteorology among the contributors. The declared intent of "Current Problems" was to present state-of-the-art knowledge and research on meteorological problems of a global character concerning the atmosphere and its interaction with the sea, the general circulation of the atmosphere, numerical forecasting, and atmospheric chemistry.

Referring to Arrhenius's 1896 paper, Rossby was confident of the general connection between an increase in the content of carbon dioxide in the atmosphere and an increase of global mean temperature. The extract shows his reasoning—joined to this belief in the possibility of global change—about *how to* find out if the content of carbon dioxide in the air had really increased. Previous series of carbon dioxide data, collected, for example, by Guy Stewart Callendar, were not precise and locations not optimal, Rossby argued. How was one to measure "in the best way"? This was not a trivial question, since at the time, perhaps no one really knew.

Several problems existed. One was that measurements conducted so far suggested major local variations in carbon dioxide content that masked secular changes. A second problem was that very little was known of the carbon dioxide absorption by the sea, which was potentially enormous and which Callendar had disregarded altogether. A third problem was that measurements until then—for example, in the Scandinavian network of fifteen stations in 1954 and 1955—had been undertaken in areas where vegetation was abundant and industrial activities could influence the results. Rossby was hoping instead that measurements to be taken during the upcoming International Geophysical Year 1957–58 in remote locations in Antarctica, on mountaintops in South America, and in the Pacific would eliminate these aberrations.

These assumptions were borne out empirically. In 1956 Rossby, who was chairing the official Swedish Committee for the International Geophysical Year 1957–58 (IGY), was in contact with scientists at the Scripps Institution of Oceanography interested in carbon dioxide, among them Charles David Keeling, who set up measurements both in Antarctica and at a station on top of the Mauna Loa mountain in Hawaii during the IGY. Only a few years later, Keeling could show with new measurements from Antarctica, and later from Mauna Loa, what was going to become the most solid evidence of the increased content of atmospheric carbon dioxide.

Further Reading

Bohn, Maria. 2011. "Concentrating on CO_2: The Scandinavian and Arctic measurements." Special issue, *Revisiting Klima*, ed. James R. Fleming and Vladimir Jankovich. *Osiris* 26:165–179.

Fleming, James R. 1998. *Historical perspectives on climate change*. Oxford: Oxford University Press.

Friedman, Robert Marc. 1989. *Appropriating the weather: Vilhelm Bjerknes and the construction of a modern meteorology*. Ithaca, NY: Cornell University Press.

Keeling, Charles David. 1960. "The concentration and isotopic abundances of carbon dioxide in the atmosphere." *Tellus* 12:200–203.

Nature's Services

Societal Dependence on Natural Ecosystems

GRETCHEN C. DAILY

Introduction: What Are Ecosystem Services?

In the space of a single human lifetime, society finds itself suddenly confronted with a daunting complex of tradeoffs between some of its most important activities and ideals. Recent trends raise disturbing questions about the extent to which today's people may be living at the expense of their descendants, casting doubt upon the cherished goal that each successive generation will have greater prosperity. Technological innovation may temporarily mask a reduction in Earth's potential to sustain human activities; in the long run, however, it is unlikely to compensate for the depletion of fundamental resources, such as productive land, fisheries, old-growth forests, and biodiversity.

On a global scale, different groups of people are now living at one another's expense, as is readily apparent in the disruption and overexploitation of Earth's open-access resources and waste sinks. For example, whereas the levels of disruption caused by energy use were once small, local, and reversible, they have now reached global proportions and carry irreversible consequences. In fueling their industrialization historically and pursuing their activities today, the developed nations appear to have largely used up the atmosphere's capacity to absorb CO_2 and other greenhouse gases without inducing climate change. In the process, they have foreclosed the option of safely using fossil fuels to sustain comparable levels of industrial activity by developing nations.

And, at the local scale, the tradeoffs between competing activities, and between individual and societal interests, are becoming ever more evident. In virtually any community, allocation of land or water to various activities often involves a zero sum game, as is apparent in the widespread loss of farmland and water

Gretchen C. Daily (ed.). 1997. Excerpt from *Nature's services: Societal dependence on natural ecosystems*, 1–19. Washington, DC: Island Press.

to urban and industrial purposes. Thus, constraints on the scale of the human enterprise typically manifest themselves most tangibly not as absolute limits to a particular activity, but rather as tradeoffs, whose resolution is fraught with increasingly difficult practical and ethical considerations.

While civilization is presently careening along on a dangerous course, its fate is not sealed. The close of the 20th century represents a period in history that demands not just a carefully tuned focus on crises of the moment, but also a long-term perspective on challenges to the human future. Second, society is poorly equipped to evaluate environmental tradeoffs, and their continued resolution on the sole basis of the social, economic, and political forces prevailing today threatens environmental, economic, and political security. There is an urgency for developing analytical and institutional frameworks for the informed and wise resolution of these tradeoffs. Third, such decision-making frameworks must ensure the protection of humanity's most fundamental source of well-being: Earth's life support systems. A tremendous amount is known about the importance and value of the natural systems that underpin the human economy, but this information has neither been synthesized nor effectively conveyed to decision-makers or to the general public.

We need to characterize the ways in which Earth's natural ecosystems confer benefits on humanity, to make a preliminary assessment of their value, and to report this in a manner widely accessible to an educated audience. An ecosystem is the set of organisms living in an area, their physical environment, and the interactions between them. Although the distinction between "natural" and "human-dominated" ecosystems is becoming increasingly blurred, our focus is on the natural end of the spectrum, for three related reasons. First, the goods and services flowing from natural ecosystems are greatly undervalued by society. For the most part, the benefits those ecosystems provide are not traded in formal markets and do not send price signals of changes in their supply or condition. This is a major factor driving their conversion to human-dominated systems (e.g., agricultural lands), whose economic value is expressed, at least in part, in standard currency. Second, anthropogenic disruptions of natural ecosystems—such as alteration of the gaseous composition of the atmosphere, introduction and establishment of exotic species, and extinction of native species—are difficult or impossible to reverse on any time scale of relevance to society. Finally, if current trends continue, humanity will dramatically alter or destroy virtually all of Earth's remaining natural ecosystems within a few decades.

What Are Ecosystem Services?

Ecosystem services refer to the conditions and processes through which natural ecosystems, and the species that make them up, sustain and fulfill human life. They maintain biodiversity and the production of *ecosystem goods*, such as seafood, forage, timber, biomass fuels, natural fiber, and many pharmaceuticals, industrial products, and their precursors. The harvest and trade of these goods represent an important and familiar part of the human economy. In addition to the production of goods, ecosystem services are the actual life support functions, such as cleansing, recycling, and renewal, and the conferring of many intangible aesthetic and cultural benefits.

One way to appreciate the nature and value of ecosystem services is to imagine trying to set up a happy, day-to-day life on the moon. Assume for the sake of argument that the moon miraculously already had some of the basic conditions for supporting human life, such as an atmosphere and climate similar to those on Earth. After inviting your best friends and packing your prized possessions, a BBQ grill, and some do-it-yourself books, the big question would be, Which of Earth's millions of species do you need to take with you?

Tackling the problem systematically, you could first choose from among all the species exploited directly for food, drink, spice, fiber and timber, pharmaceuticals, industrial products (such as waxes, lac, rubber, and oils), and so on. Even being selective, this list could amount to hundreds or even several thousand species. The space ship would be filling up before you'd even begun adding the species crucial to *supporting* those at the top of your list. Which are these unsung heroes? No one knows which—nor even approximately how many—species are required to sustain human life. This means that rather than listing species directly, you would have to list instead the life-support functions required by your lunar colony; then you could guess at the types and numbers of species required to perform each. At a bare minimum, other companions on the spaceship would have to include species capable of supplying a whole suite of ecosystem services that Earthlings take for granted.

These services include:

- purification of air and water.
- mitigation of floods and droughts.
- detoxification and decomposition of wastes.
- generation and renewal of soil and soil fertility.
- pollination of crops and natural vegetation.
- control of the vast majority of potential agricultural pests.
- dispersal of seeds necessary for revegetation.
- maintenance of biodiversity, from which humanity has derived key elements of its agricultural, medicinal, and industrial enterprise.
- protection from the sun's harmful ultraviolet rays.
- partial stabilization of climate.

- moderation of temperature extremes and the force of winds and waves.
- support of diverse human cultures.
- imparting of aesthetic beauty and intellectual stimulation that lift the human spirit.

Armed with this preliminary list of services, you could begin to determine which types and numbers of species are required to perform each. This is no simple task! Let's take the soil fertility case as an example. Soil organisms play important and often unique roles in the circulation of matter in every ecosystem on Earth; they are crucial to the chemical conversion and physical transfer of essential nutrients to higher plants and all larger organisms, including humans, depend on them. The abundance of soil organisms is absolutely staggering: under a square-yard of pasture in Denmark, for instance, the soil was found to be inhabited by roughly 50,000 small earthworms and their relatives, 50,000 insects and mites, and nearly 12 million roundworms. And that is not all. A single gram (a pinch) of soil has yielded an estimated 30,000 protozoa, 50,000 algae, 400,000 fungi, and billions of individual bacteria. Which to bring to the moon? Most of these species have never been subjected to even cursory inspection. Yet the sobering fact of the matter is, as Ed Wilson put it: they don't need us, but we need them.

Ecosystem services are generated by a complex of natural cycles, driven by solar energy, that constitute the workings of the biosphere—the thin layer near Earth's surface that contains all known life. The cycles operate on very different scales. Biogeochemical cycles, such as the movement of the element carbon through the living and physical environment, are truly global and reach from the top of the atmosphere to deep into soils and ocean-bottom sediments. Life cycles of bacteria, in contrast, may be completed in an area much smaller than the period at the end of this sentence. The cycles also operate at very different rates. The biogeochemical cycling of carbon, for instance, occurs at rates orders of magnitude faster than that of phosphorus, just as the life cycles of microorganisms may be orders of magnitude faster than those of trees.

All of these cycles are ancient, the product of billions of years of evolution, and have existed in forms very similar to those seen today for at least hundreds of millions of years. They are absolutely pervasive, but unnoticed by most human beings going about their daily lives. Who, for example, gives a thought to the part of the carbon cycle that connects him or her to the plants in the garden outside, to plankton in the Indian Ocean, or to Julius Caesar? Noticed or not, human beings depend utterly on the continuation of natural cycles for their very existence. If the life cycles of predators that naturally control most potential pests of crops were interrupted, it is unlikely that pesticides could satisfactorily take their place. If the life cycles of pollinators of plants of economic importance ceased, society would face serious social and economic consequences. If the carbon cycle were badly disrupted, rapid climatic change could threaten the existence of civilization. In general, human beings lack both the knowledge and the ability to substitute for the functions performed by these and other cycles.

For millennia, humanity has drawn benefits from these cycles without causing global disruption. Yet, today, human influence can be discerned in the most remote reaches of the biosphere: deep below Earth's surface in ancient aquifers, far out to sea on tiny tropical islands, and up in the cold, thin air high above Antarctica. Virtually no place remains untouched—chemically, physically, or biologically—by the curious and determined hand of humanity. Although much more by accident than by design, humanity now controls conditions over the entire biosphere.

Interestingly, the nature and value of Earth's life support systems have been illuminated primarily through their disruption and loss. Thus, for instance, deforestation has revealed the critical role of forests in the hydrological cycle—in particular, in mitigating flood, drought, and the forces of wind and rain that cause erosion. Release of toxic substances, whether accidental or deliberate, has revealed the nature and value of physical and chemical processes, governed in part by a diversity of microorganisms, that disperse and break down hazardous materials. Thinning of the stratospheric ozone layer sharpened awareness of the value of its service in screening out harmful ultraviolet radiation.

A cognizance of ecosystem services, expressed in terms of their loss, dates back at least to Plato and probably much earlier:

> "What now remains of the formerly rich land is like the skeleton of a sick man with all the fat and soft earth having wasted away and only the bare framework remaining. Formerly, many of the mountains were arable. The plains that were full of rich soil are now marshes. Hills that were once covered with forests and produced abundant pasture now produce only food for bees. Once the land was enriched by yearly rains, which were not lost, as they are now, by flowing from the bare land into the sea. The soil was deep, it absorbed and kept the water . . . , and the water that soaked into the hills fed springs and running streams everywhere. Now the abandoned shrines at spots where formerly there were springs attest that our description of the land is true."—Plato

Ecosystem services have also gained recognition and appreciation through efforts to substitute technology for them. The overuse of pesticides, for example, leading to the decimation of natural pest enemies and concomitant promotion of formerly benign species to pest status, has made apparent agriculture's dependence upon natural pest control services. The technical problems and cost of hydroponic systems—often prohibitive even for growing high-priced, specialty produce—underscore human dependence upon ecosystem services supplied by soil. Society is likely to more highly value the services listed above, and to discover (or rediscover) an array of services not listed, as human impacts on the environment intensify and the costs and limits of technological substitution become more apparent.

Valuation of Ecosystem Services

The disparity between actual and perceived value is probably nowhere greater than in the case of ecosystem services. If asked to identify *all* that goes into making a fine cake, a baker willing to share his or her secrets would most likely first identify its ingredients, and the knowledge and skill required to transform them into a culinary work of art. He or she might also describe the type of oven, pan, and various appliances and kitchen gadgets needed. If pressed further, an astute baker might also point out the need for capital infrastructure and human services to process, store, and transport the ingredients. With a helpful hint or two, he or she may even mention the cropland, water, chemical, and energy inputs to the whole process. However, the chances of the baker touching directly upon the natural renewal of soil fertility, the pollination of crops, natural pest control, the role of biodiversity in maintaining crop productivity, clean-up and recycling services outside the kitchen—or, indeed, upon *any* ecosystem service involved—are extremely remote. Ecosystem services are absolutely essential to civilization, but modern urban life obscures their existence.

Once explained, the importance of ecosystem services is typically quickly appreciated, but the actual assigning of value to ecosystem services may arouse great suspicion, and for good reason. Valuation involves resolving fundamental philosophical issues (such as the underlying bases for value), the establishment of context, and the defining of objectives and preferences, all of which are inherently subjective. Even after doing so, one is faced with formidable technical difficulties with interpreting information about the world and transforming it into a quantitative measure of value. Just as it would be absurd to calculate the full value of a human being on the basis of his or her wage-earning power, or the economic value of his or her constituent materials, there exists no absolute value of ecosystem services waiting to be discovered and revealed to the world by a member of the intellectual community. We need to identify and characterize components of ecosystem service value, and to make a preliminary assessment of their magnitude, as a prerequisite to their incorporation into frameworks for decision-making.

We need to begin with use values; aesthetic and spiritual values associated with ecosystem services are only lightly touched upon in this book, having been eloquently described elsewhere. The total value of ecosystem services may be best assessed in terms of physical magnitudes or proportions, such as the amount of human waste processed naturally, the amount of carbon sequestered in soils, the proportion of potential crop pests controlled naturally, and the proportion of pharmaceutical products derived from biodiversity. Where a technological substitute is available for an aspect of an ecosystem service, the market price of the substitute provides a lower-bound index of the value of the service (in terms of avoided costs). As a whole, ecosystem services have infinite use value because human life could not be sustained without them. The evaluation of the trade-offs currently facing society, however, requires estimating the *marginal* value of ecosystem services (the value yielded by an additional unit of the service, all else held constant) to determine the costs of losing—or benefits of preserving—a given amount or quality of services. The information needed to estimate marginal values is difficult to obtain and is presently unavailable for many aspects of the services. Of course, economic indices are likely to underestimate the total value of these systems. Nonetheless, economic markets play a dominant role in patterns of human behavior, and the expression of value—even if imperfect—in a common currency helps to inform the decision-making process. Making economic institutions sensitive and responsive to natural constraints and explicitly dealing with the limitations of such institutions in doing so, are other requisites to effective Earth management.

The Policy Interface

Diverse human societies have now attained the status of ecological super-powers. That is, they have the capacity to seriously impair or destroy essential components of Earth's life support systems; moreover, they are currently using it, almost without restraint. The persistence of all societies ultimately hinges upon those superpowers beginning to wisely coordinate and control the wielding of this power. This will especially be so if the magnitude of human influence continues to expand at unprecedented rates to unprecedented levels, through the momentum and inertia associated with population growth, expanding material desires, and the technical means by which fulfilling the latter are pursued. As the most accessible and suitable resources are sequentially exhausted, each additional person, all else equal, exerts greater per-capita impact in necessarily turning to lower quality resources for the same end.

Historically, human societies have alleviated resource constraints primarily by pushing back intellectual and territorial frontiers. Yet, it would be difficult today for even the most optimistic rates of innovation and of adoption of improved technology (broadly defined) to offset the rates of increase in human disruption caused by rapid population growth and increases in per-capita impacts. Furthermore, opportunities for territorial expansion are now largely foreclosed — or never existed for inherently global impacts, such as those on the composition of the upper atmosphere.

The passage of time leaves in ever sharpening focus a daunting but critical need to tackle social and political frontiers with the same boldness and determination that took the first man to the moon. This will require not only strengthening existing institutions, but also creating entirely new regimes to manage globally human impacts on Earth's life support systems. It will also require an unprecedented level of international cooperation and coordination. It is at these policy frontiers that lie the brightest prospects for resolving the human predicament and converting the world's societies to new and sustainable resource management regimes.

Present scientific understanding of ecosystem services is substantial, wide-reaching, and extremely policy-relevant, and merits urgent attention by decision makers, since current patterns of human activity are unsustainable and threaten to impair critical life support functions. Failure to foster the continued delivery of ecosystem services undermines economic prosperity, forecloses options, and diminishes other aspects of human well-being; it also threatens the very persistence of civilization. While the academic community remains a long way from a fully comprehensive understanding of ecosystem services, the accelerating rate of disruption of the biosphere makes imperative the incorporation of current knowledge into the policy making process.

Commentary

Gretchen Daily, *Nature's Services* (1997)

RICHARD B. NORGAARD

For millennia, people knew they were dependent on nature. Descendants of Abrahamic traditions, however, are no longer shepherds pasturing sheep, dazzled by a starry night sky, nor are descendants of other traditions as conscious of their relation to nature as were their ancestors. Over the past century, a majority of the world's people moved into cities, and, even for those who have not, the cosmos of humanity has become the artifacts and market dynamics of a global economy. People now interact with each other and nature predominantly through markets.

There is a long history of conceptual model sharing between economics and biology and linguistic exchange as well. Both Charles Darwin and Alfred Russel Wallace credit Thomas Malthus for alerting them to the idea of populations pushing up against material constraints. The idea that the most competitive individuals prove evolutionarily fit comes from the ease of explaining evolution using market theory and language, for cooperation occurs in nature just as it does in society. Social Darwinism, in turn, has been misused to justify the gains of the rich and the plight of the poor. Economists were building input-output tables before ecologists were thinking in food webs.

The economic metaphors "environmental services," "public services of global ecosystems," "nature's services," and "ecosystem services" appeared in the biological literature during the 1970s. Walter Westman (1977) wrote the most cited paper of this era, assuaging his discomfort in describing nature in the materialistic terms of economics by prefacing his paper with a quote from William Wordsworth:

> To me the meanest flower that doth blows can give
> Thoughts that do often lie too deep for tears.

By the late 1980s, ecological economists joined conservation biologists in advancing the metaphor of nature as a fixed stock of capital that can sustain a limited flow of ecosystem services. They argued that if society knew of the essential link between ecosystem services and economic well-being, society would protect nature for the well-being of future generations. And one way to ensure recognition of this link is to provide economic incentives, to have users of ecosystem services pay those who manage, or might now take an interest in sustainably managing, the ecosystems that produce the services.

Gretchen Daily, a student of biologist Paul Ehrlich, completed her Ph.D. in biological sciences at Stanford University in 1992. She did a postdoctoral stint at the University of California, Berkeley, to learn how to portray the importance and character of people's relation to nature by using the language and framework of economics. By the late 1990s, she was a central figure in organizing conservation biologists and economists to formalize the concept of ecosystem services and put values on them. The text above by Gretchen Daily is a shortened version of the introduction to the book she edited from a workshop she organized (Daily 1997). Here she lays out the concept and context of ecosystem services, as well as hopes for the metaphor. The authors deliberately wrote in a style to communicate with policy makers to conserve the diversity of life on the planet.

At about the same time, Robert Costanza organized a team of economists and ecologists to dramatically portray the importance of ecosystems to humanity's material well-being. At a time when the global economy was running $18 trillion per year, Costanza's team determined that the world's ecosystem services were worth at least between $16 trillion and $54 trillion per year (Costanza et al. 1997). The sheer boldness of the calculation, let alone the magnitude of the estimate, attracted immediate attention that was slow to wane. Costanza and his colleagues helped direct the spotlights of development and environmental policy prioritizing on the importance of nature. At the same time, the article's many critics documented the difficulties of linking ecology and economics, especially at a global scale.

A search using Google Scholar found twenty references between 1970 and 1980 that used the term "ecosystem services," then 200, 2,670, and 23,100 in the subsequent three decades. By the year 2000, the World Bank and other development agencies were incorporating the concept into their projects. They also promoted payments for ecosystem services to give incentives to sustain the natural capital that produced them. The 2005 U.N. Millennium Ecosystem Assessment incorporated ecosystem services as a central part of its framework for portraying the connection between ecosystem sustenance or degradation in response to human activity and its impacts back on future human options. While many scientists may have assumed that the framework was just for assessing the literature for policy makers, other scientists felt that putting people inside the ecological model was exactly what was needed to think scientifically about the process of environmental change. Soon, linkages were forged between sustaining ecosystem services and the traditional role of economic development: alleviating poverty. The metaphor also affected conservation policy in developed countries, especially in the European Union. An ecosystem service consulting industry emerged to provide advice to land management, conservation, and development agencies as well as landowners. The U.N. sponsored the Economics of Ecosystems and Biodiversity project to elaborate and help implement the ecosystem service concept (Kumar 2010). At this writing, an Intergovernmental Platform on Biodiversity and Ecosystem Services is in the process of being established. There is no doubt that the concept of ecosystem services has successfully transformed development and conservation philosophy and practice, as hoped.

The idea of ecosystem services and their relation to a capital stock of nature is an atomistic-mechanistic framework. Its rise to dominance pushes diverse and better-developed ways of understanding ecosystems—the cycles of population biology, the tight interdependence of coevolving species, or how variations in weather affect primary production and higher-level species through a food web—deeper into our ecological consciousness (Norgaard 2008, 2010). Ecologists regularly warn policy makers and ecosystem service practitioners that the tradeoffs between ecosystem services are poorly understood and that the contexts of particular ecosystems must be well known to correctly design payments for ecosystem services. In short, the concept of ecosystem services has been so successful that many, including Gretchen Daily, are concerned that it has been incorporated into policy and implemented in practice faster than the science has advanced to back it up (Daily and Matson 2008).

Further Reading

Costanza, R., R. D'Arge, R. De Groot, S. Farber, M. Grasso, B. Hannon, et al. 1997. "The value of the world's ecosystem services and natural capital." *Nature* 387(6630):253–260. doi:10.1038/387253a0

Daily, Gretchen C. 1997. *Nature's services: Societal dependence on natural ecosystems.* Washington, DC: Island Press.

Daily, Gretchen C., and Pamela A. Matson. 2008. "Ecosystem services: From theory to implementation." *Proceedings of the National Academy of Sciences of the United States of America* 105(28):9455–9456. doi:10.1073/pnas.0804960105

Kumar, Pushpam. 2010. *The economics of ecosystems and biodiversity: Ecological and economic foundations.* London: UNEP/Earthprint.

Norgaard, Richard B. 2008. "Finding hope in the millennium ecosystem assessment." *Conservation Biology* 22(4):862–869. doi:10.1111/j.1523-1739.2008.00922.x

Norgaard, Richard B. 2010. Ecosystem services: From eye-opening metaphor to complexity blinder. Special issue on payments for ecosystem services. *Ecological Economics* 69(6):1219–1227.

Westman, W. E. 1977. "How much are nature's services worth?" *Science* 197:960–964. doi:10.1126/science.197.4307.960

The Economics of Climate Change

NICHOLAS STERN

The scientific evidence is now overwhelming: climate change presents very serious global risks, and it demands an urgent global response.

This independent Review was commissioned by the Chancellor of the Exchequer, reporting to both the Chancellor and to the Prime Minister, as a contribution to assessing the evidence and building understanding of the economics of climate change.

The Review first examines the evidence on the economic impacts of climate change itself, and explores the economics of stabilising greenhouse gases in the atmosphere. The second half of the Review considers the complex policy challenges involved in managing the transition to a low-carbon economy and in ensuring that societies can adapt to the consequences of climate change that can no longer be avoided.

The Review takes an international perspective. Climate change is global in its causes and consequences, and international collective action will be critical in driving an effective, efficient and equitable response on the scale required. This response will require deeper international co-operation in many areas—most notably in creating price signals and markets for carbon, spurring technology research, development and deployment, and promoting adaptation, particularly for developing countries.

Climate change presents a unique challenge for economics: it is the greatest and widest-ranging market failure ever seen. The economic analysis must therefore be global, deal with long time horizons, have the economics of risk and uncertainty at centre stage, and examine the possibility of major, non-marginal change. To meet these requirements, the Review draws on ideas and techniques from most of the important areas of economics, including many recent advances.

N. H. Stern. 2006. Excerpts from *The economics of climate change: The Stern Review*, 1–2, 25–26, 31–32, 108, 154–157, 230–233. Cambridge: Cambridge University Press.

The Benefits of Strong, Early Action on Climate Change Outweigh the Costs

The effects of our actions now on future changes in the climate have long lead times.

What we do now can have only a limited effect on the climate over the next 40 or 50 years. On the other hand what we do in the next 10 or 20 years can have a profound effect on the climate in the second half of this century and in the next.

No-one can predict the consequences of climate change with complete certainty; but we now know enough to understand the risks. Mitigation—taking strong action to reduce emissions—must be viewed as an investment, a cost incurred now and in the coming few decades to avoid the risks of very severe consequences in the future. If these investments are made wisely, the costs will be manageable, and there will be a wide range of opportunities for growth and development along the way. For this to work well, policy must promote sound market signals, overcome market failures and have equity and risk mitigation at its core. That essentially is the conceptual framework of this Review.

The Review considers the economic costs of the impacts of climate change, and the costs and benefits of action to reduce the emissions of greenhouse gases (GHGs) that cause it, in three different ways:

- Using disaggregated techniques, in other words considering the physical impacts of climate change on the economy, on human life and on the environment, and examining the resource costs of different technologies and strategies to reduce greenhouse gas emissions;
- Using economic models, including integrated assessment models that estimate the economic impacts of climate change, and macroeconomic models that represent the costs and effects of the transition to low-carbon energy systems for the economy as a whole;
- Using comparisons of the current level and future trajectories of the "social cost of carbon" (the cost of impacts associated with an additional unit of greenhouse gas emissions) with the marginal abatement cost (the costs associated with incremental reductions in units of emissions).

. . .

The climate is a public good: those who fail to pay for it cannot be excluded from enjoying its benefits and one person's enjoyment of the climate does not diminish the capacity of others to enjoy it too. Markets do not automatically provide the right type and quantity of public goods, because in the absence of public policy there are limited or no returns to private investors for doing so: in this case, markets for relevant goods and services (energy, land use, innovation, etc.) do not reflect the consequences of different consumption and investment choices for the climate. Thus, climate change is an example of market failure involving externalities and public goods.

. . .

Climate change has special features that, together, pose particular challenges for the standard economic theory of externalities. There are four distinct issues that will be considered in turn in the sections below.

- Climate change is an externality that is global in both its causes and consequences. The incremental impact of a tonne of GHG on climate change is independent of where in the world it is emitted (unlike other negative impacts such as air pollution and its cost to public health), because GHGs diffuse in the atmosphere and because local climatic changes depend on the global climate system. While different countries produce different volumes the marginal damage of an extra unit is independent of whether it comes from the UK or Australia.
- The impacts of climate change are persistent and develop over time. Once in the atmosphere, some GHGs stay there for hundreds of years. Furthermore, the climate system is slow to respond to increases in atmospheric GHG concentrations and there are yet more lags in the environmental, economic and social response to climate change. The effects of GHGs are being experienced now and will continue to work their way through in the very long term.
- The uncertainties are considerable, both about the potential size, type and timing of impacts and about the costs of combating climate change; hence the framework used must be able to handle risk and uncertainty.
- The impacts are likely to have a significant effect on the global economy if action is not taken to prevent climate change, so the analysis has to consider potentially nonmarginal changes to societies, not merely small changes amenable to ordinary project appraisal.

It is common to present policy towards climate change in terms of the social cost of carbon on the margin (SCC) and the marginal abatement (MAC). The former is the total damage from now into the indefinite future of emitting an extra unit of GHGs now—the science says that GHGs (particularly CO_2) stay in the atmosphere for a very long time. Thus, in its simplest form, the nature of the problem is that the stock of gases in the atmosphere increases with the net flow of GHGs emissions in this period, and thus decreases with abatement. Therefore, on the one hand, the SCC curve slopes downwards with increasing abatement in any given period, assuming that the lower the stock at any point in the future, the less the marginal damage. On the other hand, the MAC curve slopes upwards with increasing abatement, if it is more costly on the margin to do more abatement as abatement increases in the given period. The optimum level of abatement must satisfy the condition that MAC equals SCC. If, for example, SCC were bigger than MAC, the social gain from one extra unit of abatement would be less than the cost and it would be better to do a little more. . . .

It should be clear that the SCC curve this period depends on future emissions: if we revised upwards our specified assumptions on future emissions, the whole SCC curve would shift upwards, and so would the optimum abatement level in this period. Thus, if we are thinking about an optimum path over time, rather than simply an optimum emission for this period, we must recognise that the SCC curve for any given period depends on the future stock and thus on the future path of emissions. *We cannot sensibly calculate an SCC without assuming that future emissions and stocks follow some specified path. For different specified paths, the SCC will be different.* For example, it will be much higher on a "business as usual" path (BAU) than it will be on a path that cuts emissions strongly and eventually stabilises concentrations. It is remarkable how often SCC calculations are vague on this crucial point.

. . .

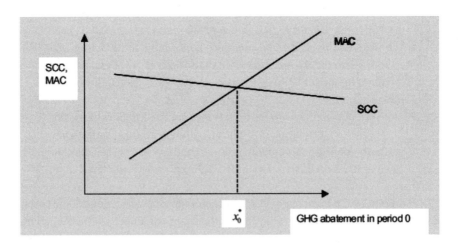

Figure 2.1. The optimum degree of abatement in a given period.

Typically, in the application of the theory of welfare economics to project and policy appraisal, an increment in future consumption is held to be worth less than an increment in present consumption, for two reasons. First, if consumption grows, people are better off in the future than they are now and an extra unit of consumption is generally taken to be worth less, the richer people are. Second, it is sometimes suggested that people prefer to have good things earlier rather than later—"pure time preference"—based presumably in some part on an assessment of the chances of being alive to enjoy consumption later and in some part "impatience."

Yet assessing impacts over a very long time period emphasises the problem that future generations are not fully represented in current discussion. Thus we have to ask how they should be represented in the views and decisions of current generations. This throws the second rationale for "discounting" future consumption mentioned above—pure time preference—into question. We take a simple approach in this Review: if a future generation will be present, we suppose that it has the same claim on our ethical attention as the current one.

. . .

The Stern Review has used the PAGE2002 model (an integrated assessment model that takes account of a wide range of risks and uncertainties) to assess how climate change may affect output and growth in the future. Integrated assessment models can be useful vehicles for exploring the kinds of costs that might follow from climate change. However, these are highly aggregative and simplified models and, as such, the results should be seen as illustrative only.

. . .

The model we use—the PAGE2002 IAM21—can take account of the range of risks by allowing outcomes to vary probabilistically across many model runs, with the probabilities calibrated to the latest scientific quantitative evidence on particular risks.

The first challenge points strongly to the need for a modelling approach based on probabilities (that is, a "stochastic" approach). The PAGE2002 (Policy Analysis of the Greenhouse Effect 2002) IAM [Integrated Assessment Model] meets this requirement by producing estimates based on "Monte Carlo" simulation. This means that it runs each scenario many times (e.g. 1000 times), each time choosing a set of uncertain parameters randomly from pre-determined ranges of possible values. In this way, the model generates a probability distribution of results rather than just a single point estimate. Specifically, it yields a probability distribution of future income under climate change, where climate-driven damage and the cost of adapting to climate change are subtracted from a baseline GDP growth projection. The parameter ranges used as model inputs are calibrated to the scientific and economic literatures on climate change, so that PAGE2002 in effect summarises the range of underlying research studies.

So, for example, the probability distribution for the climate sensitivity parameter — which represents how temperatures will respond in equilibrium to a doubling of atmospheric carbon dioxide concentrations — captures the range of estimates across a number of peer-reviewed scientific studies. Thus, the model has in the past produced mean estimates of the global cost of climate change that are close to the centre of a range of peer-reviewed studies, including other IAMs, while also being capable of incorporating results from a wider range of studies. This is a very valuable feature of the model and a key reason for its use in this study.

PAGE2002 has a number of further desirable features. It is flexible enough to include market impacts (for example, on agriculture, energy and coastal zones) and non-market impacts (direct impacts on the environment and human mortality), as well as the possibility of catastrophic climate impacts. Catastrophic impacts are modelled in a manner similar to the approach used by Nordhaus and Boyer. When global mean temperature rises to high levels (an average of 5°C above pre-industrial levels), the chance of large losses in regional GDP in the range of 5–20% begins to appear. This chance increases by an average of 10% per °C rise in global mean temperature beyond 5°C.

At the same time, PAGE2002 shares many of the limitations of other formal models. It must rely on sparse or non-existent data and understanding at high temperatures and in developing regions, and it faces difficulties in valuing direct impacts on health and the environment. Moreover, . . . the PAGE2002 model does not fully cover the "socially contingent" impacts. As a result, the estimates of catastrophic impacts may be conservative, given the damage likely at temperatures as high as 6–8°C above pre-industrial levels. Thus the results presented . . . should be viewed as indicative only and interpreted with great caution. Given what is excluded, they should be regarded as rather conservative estimates of costs, relative to the ability of these models to produce reliable guidance.

   . . .

Preliminary estimates of average losses in global per-capita GDP in 2200 range from 5.3 to 13.8%, depending on the size of climate-system feedbacks and what estimates of "non-market impacts" are included.

Estimates of losses in per-capita income over time are benchmarked against projected GDP growth in a world without climate change. The baseline-climate/market-impacts scenario generates the smallest losses, where climate change reduces global per-capita GDP by, on average, 2.2% in 2200. However, as discussed in the previous section, the omission of the very real risk of abrupt and large-scale changes at high temperatures creates an unrealistic negative bias in estimates.

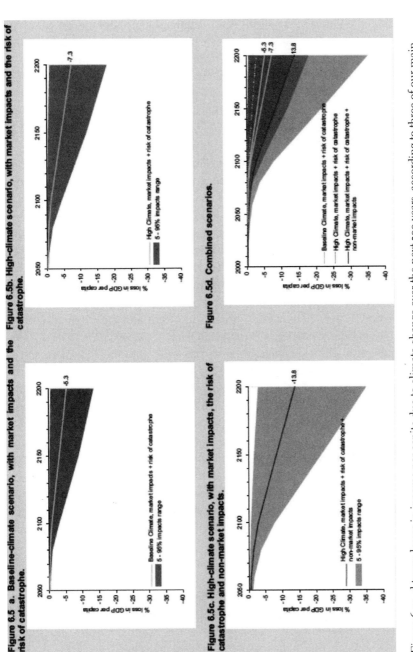

Figure 6.5a–d traces losses in income per capita due to climate change over the next 200 years, according to three of our main scenarios of climate change and economic impacts. The range of estimates from the 5th to the 95th percentile is shaded grey.

Figure 6.5 shows the results of scenarios including a risk of "catastrophe." The lower-bound estimate of the global cost of climate change in Figure 6.5 uses the baseline climate and includes both market impacts and the risk of catastrophic changes to the climate system (Figure 6.5a). In this scenario, the mean loss in global per-capita GDP is 0.2% in 2060. By 2100, it rises to 0.9%, but by 2200 it rises steeply to 5.3%.

There is a substantial dispersion of possible outcomes around the mean and, in particular, a serious risk of very high damage. The grey-shaded areas in Figure 6.5 give the range of estimates in each year taken from the 5th and 95th percentile damage estimates over the 1000 runs of the model. For the lower-bound estimate in 2100, the range is 0.1–3 % loss in global GDP per capita. By 2200, this rises to 0.6–13.4%.

Figures 6.5b to d demonstrate the loss in global GDP per capita when first, the risk of more feedbacks in the climate system is included (the high-climate scenario), and second, estimates of non-market impacts of climate change are included. In the high-climate scenario, the losses in 2100 and 2200 are increased by around 35%. In 2200, the range of losses is increased to between 0.9% and 17.9%.

The inclusion of non-market impacts increases these estimates further still. In this Review, non-market impacts, on health and the environment, are generally considered separately to market impacts. However, if the goal is to compare the cost of climate change in monetary terms with the equivalent cost of mitigation, then excluding non-market costs is misleading. For the high-climate scenario with non-market impacts (Figure 6.5c), the mean total losses are 2.9% in 2100 and 13.8% in 2200. In 2200, the 5th and 95th percentiles increase significantly, to 2.9% to 35.2%.

. . .

For each technology, assumptions are made on plausible rates of uptake over time. It is assumed, for the purposes of simplification, that as the rate of uptake of individual technologies is modest, they will not run into significant problems of increasing marginal cost. . . . Assumptions are also made on the potential for energy efficiency improvements. These assumptions can be used to calculate an average cost of abatement. . . .

An average cost of abatement per tonne of carbon can be constructed by calculating the cost of each technology . . . weighted by the assumed take-up, and comparing this with the emissions reductions achieved by these technologies against fossil-fuel alternatives. This is shown in Figure 9.5, where upper and lower bounds represent best estimates of 90% confidence intervals.

The costs of carbon abatement are expected to decline by half over the next 20 years . . . , and then by a third further by 2050. But the longer-term estimates of

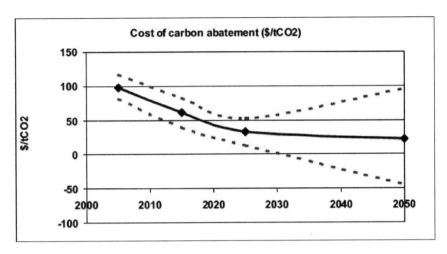

Figure 9.5. Average cost of reducing fossil fuel emissions to 18 GtCO$_2$ in 2050*
* The [broken] lines give uncertainty bounds around the central estimate. These have been calculated using Monte Carlo analysis. For each technology, the full range of possible costs (typically ± 30% for new technologies, ±20% for established ones) is specified. Similarly future oil prices are specified as probability distributions ranging from $20 to over $80 per barrel, as are gas prices (£2–6/GJ), coal prices and future energy demands (to allow for the uncertain rate of uptake of energy efficiency). This produces a probability distribution that is the basis for the ranges given.

shifting to a low-carbon energy system span a very broad range, as indicated in the figure, and may even be broader than indicated here. This reflects the inescapable uncertainties inherent in forecasting over a long time period, as discussed above. It should be noted that, although average costs may fall, marginal costs are likely to be on a rising trajectory through time, in line with the social cost of carbon; this is explained in Box 9.6.

The global cost of reducing total GHG emissions to three quarters of current levels (consistent with 550ppm CO$_2$e stabilisation trajectory) is estimated at around $1 trillion in 2050 or 1% of GDP in that year, with a range of –1.0% to 3.5% depending on the assumptions made.

Anderson's central case estimate of the total cost of reducing fossil fuel emissions to around 18 GtCO$_2$e/year (compared to 24 GtCO$_2$/year in 2002) is estimated at $930 bn, or less than 1% of GDP in 2050 (see table 9.2). In the analysis by Anderson, this is associated with a saving of 43 GtCO$_2$ of fossil fuel emissions relative to baseline, at an average abatement cost of $22/tCO$_2$/year in 2050. However these costs vary according to the underlying assumptions, so these are explored below.

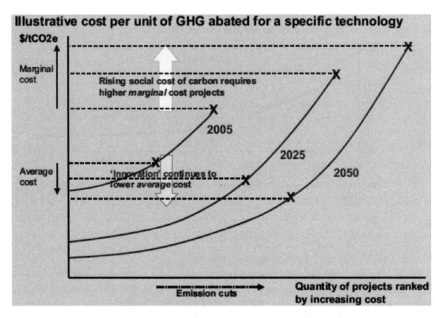

Illustrative cost per unit of GHG abated for a specific technology

$/tCO2e

Marginal cost

Rising social cost of carbon requires higher *marginal* cost projects

2005

2025

2050

Average cost

'Innovation' continues to lower average cost

Emission cuts

Quantity of projects ranked by increasing cost

Box 9.6 The relationship between marginal and average costs over time

It is important not to confuse average costs with marginal costs or the prevailing carbon price. The carbon price should reflect the social cost of carbon and be rising with time, because of increased additional damages per unit of GHG at higher concentrations of gases in the atmosphere. Rising prices should encourage abatement projects with successively higher marginal costs. This does not necessarily mean that the average costs will rise. Indeed, in this analysis, average costs are assumed to fall, quickly at first and then tending to level off (Figure 9.5). At any time, marginal costs will tend to be above average costs as the most costly projects are undertaken last.

At the same time, however, innovation, learning and experience–driven through innovation policy–will lower the cost of producing any given level of output using any specific technology. This is shown in the figure below, which traces the costs of a specific technology through time.

Despite more extensive use of the technology and rising costs on the margin through time (reflecting the rising carbon price), the average cost of the technology may continue to fall. The key point to note is that marginal costs might be rising even where average costs are falling (or at least rising more slowly), as a growing range of technologies are used more and more intensively.

Table 9.2

Annual total costs of reducing fossil fuel emissions to 18 GtCO$_2$ in 2050

	2015	2025	2050
Average cost of abatement, \$/t CO$_2$	61	33	22
Emissions Abated GtCO$_2$	2.2	10.7	42.6
(relative to emissions in BAU)			
Total cost of abatement, \$ billion per year:	134	349	930

Table 9.3

Sensitivity analysis of global costs of cutting fossil fuel emissions to 18 GtCO$_2$ in 2050 (costs expressed as % of world GDP)[a]

Case		2015	2025	2050
(i)	Central Case	0.3	0.7	1.0
(ii)	Pessimistic technology case	0.4	0.9	3.3
(iii)	Optimistic technology case	0.2	0.2	−1.0
(iv)	Low future oil and gas prices	0.4	1.1	2.4
(v)	High future oil and gas prices	0.2	0.5	0.2
(vi)	High costs of carbon capture and storage	0.3	0.8	1.9
(vii)	A lower rate of growth of energy demand	0.3	0.5	0.7
(viii)	A higher rate of growth of energy demand	0.3	0.6	1.0
(ix)	Including incremental vehicle costs[b]			
	• Means	0.4	0.8	1.4
	• Ranges	0.3–0.5	0.5–1.1	−0.6–3.5

[a] The world product in 2005 was approximately \$35 trillion (£22 trillion at the PPP rate of \$1.6/£). It is assumed to rise to \$110 trillion (£70 trillion) by 2050, a growth rate of 2.5% per year, or 1½–2% in the OECD countries and 4–4½% in the developing countries.
[b] Assuming the incremental costs of a hydrogen fuelled vehicle using an internal combustion engine are £2,300 in 2025 and \$1400 in 2050, and for a hydrogen fuelled fuel cell vehicle £5000 in 2025 declining to £1700 by 2050. (Ranges of ~ ±30% are taken about these averages for the fuel cell vehicle.)

Commentary

Nicholas Stern, *The Economics of Climate Change* (2006)

PAUL WARDE

The Stern Review on the Economics of Climate Change was commissioned by the British government in July 2005 and appeared a year later. It was the most prominent of governmental responses to the challenge of turning the increasingly authoritative reports of the Intergovernmental Panel on Climate Change (IPCC) into a domestic policy that carried forward the Kyoto Treaty of 1994, and was mirrored by the subsequent Garnaut Report of 2008 in Australia. Authored by a research team lead by economist Nicholas Stern, the *Review* combined public consultation with commissioned research and reviews of the rapidly developing literature on the impact of climate change, and, crucially, was established by the Treasury. Its remit was to examine the economic cost of global warming. The *Review* also reflected a widespread trend for governments to require policy to be legitimated by being "evidence based," increasingly drawing on independent advice from technical experts. "Evidence," in this case, required provision of numbers as to the potential damage of climate change on the global economy relatively far into the future, and estimates of the costs of mitigation equally far into the future.

The *Review* appeared at a stage where the intertwining of the climate system and the economic system had become an established fact, whether through the influence of greenhouse gas emissions, deforestation, or land-use change on climate, and the feedback effects that altered climates could have on economic development. It seemed increasingly less reasonable to model the dynamics of either the climate or (in admittedly fewer cases) the economic system, especially over a period of decades, without also comprehending the dynamics of the other system. Climate science had identified global warming as a problem primarily from a theoretical and modeling perspective, posing the question of what might happen if the concentration of greenhouse gases in the atmosphere doubled. Increasingly sophisticated computer modeling had allowed the mapping of the probability distributions of possible outcomes. Since the speed of emissions (as well as their scale) had a profound effect upon outcomes, there was a great imperative on accurate modeling of the impact of different economic policies for predictions of climate change.

To meet its remit, the *Review* thus needed to accomplish the herculean task of integrating both climate and (insofar as they existed) long-term models of socio-economic development. It stood in the tradition of mega-audits of the past, such as the *Limits to Growth* (see Part 2) or *Global 2000* (see Part 8) reports. It did so primarily by relying on the work of others, especially the IPCC report of 2001, and the PAGE2002 Integrated Assessment Model. For the costs of climate change, the report examined changes to water availability, food supply, health, erosion, infrastructure, species extinction, impact on development, income falls leading to worse mortality, and likelihood of resource-driven conflict.

Aside from the difficulty of predicting trends and impacts over many decades, the economics of climate change faced a fundamental problem: it sought to equate damages and costs with present market values. But what values would future generations hold? Was there "evidence-based" research to indicate that current generations cared about the future? Economists commonly used a "discount rate" to account for various factors in predicting future values, taking into account the fact that we are likely to be significantly wealthier in a few decades' time, and also a preference for not deferring consumption. These are reasons, of course, for charging interest rates—another kind of discount rate—that reflect the fact that loaned money will not have the same value to us in the future as it does now. Yet the market discount rates commonly used by economists implied that we valued future costs and benefits only to a small degree; indeed, as Ross Garnaut pointed out, they implied that we were indifferent to the extinction of the species within a few decades. In setting the discount rate, Stern thus had to make both a judgment about the likely values of future generations and an ethical decision about how much we, and intervening generations, should be concerned about the future. Many of the results of the *Review* depend on the decision Stern took to place a high present value on costs in the future—a decision that was widely criticized as arbitrary.

Prediction of future costs of both emitted carbon and mitigation were subject to a further major uncertainty; they would depend on how much carbon had been emitted, and how much abatement had been undertaken. They were, in fact, determined by their own history and future. If large amounts of carbon would be emitted, the costs of each individual unit emitted would be much higher over the full time period of its effects. If total emissions turned out to be small, each unit emitted now is not going to be so damaging. So providing a valuation of the costs in today's money of all the future damage caused by current emissions depends on assumptions of what future emissions are going to be.

The consequence of such dilemmas is that the *Stern Review* could do no more than offer a range of possible outcomes and their probability distributions from running a series of models that themselves were based on highly simplifying assumptions. The report itself warned that it could not integrate anything like the complete range of the vast array of data that had been reviewed. Crucial expectations drew on historical trends (albeit from relatively recent decades, and with little detailed historical reconstruction) in the experience of technological change, energy efficiency, and economic growth. Indeed, the implicit assumption that recent trends would continue long into the future—a perspective that subsequent economic developments from 2007 might give pause to—was clear in that both within the review itself, and the responses, the "business as usual" scenario of a continued pattern of fossil fuel–based growth was treated as unproblematic.

In the final estimates stabilizing GHG concentrations at around 550 parts per million (ppm) was estimated to cost around 1 percent of global GDP by 2050 (although the full range of estimates ran from an actual saving to over three times as much). Nicholas Stern has since revised this figure upward. Inevitably, the review garnered both praise, criticism on points of detail, and fierce rebuttal on the grounds that it made too many judgments that were not grounded in empirical data. Politically, the report offered resounding evidence that policies to limit climate change were both necessary, and economically affordable—a highly desirable result from the point of view of the U.K. government. The estimate of costs of mitigation treated as a one-off were in fact considerably less than the global fall in income in the recession of 2008–2009.

Further Reading

Garnaut, Ross. 2008. *The Garnaut climate change review: Final report*. Port Melbourne, Vic.: Cambridge University Press.

Hope, C. W. 2006. "The marginal impact of CO_2 from PAGE2002: An integrated assessment model incorporating the IPCC's five reasons for concern." *Integrated Assessment Journal* 6(1):19–56.

IPCC [Watson, R., et al.]. 2001. *Climate change 2001: A synthesis report*. The Intergovernmental Panel on Climate Change (Approved Wembley September 2001). http://www.ipcc.ch/pdf/climate-changes-2001/synthesis-spm/synthesis-spm-en.pdf.

PART 10

The Anthropocene

How Can We Live in a World Where There Is No Nature Without People?

As the world reached the end of the twentieth century, the environmental news was not good. Climate scientists were close to a consensus that human-forced climate change was happening. Conservation activists and biodiversity experts were talking of a sixth mass extinction. And the world's population was rapidly approaching the 7.8 billion absolute limit put forward by George Knibbs in 1926 (see Part 1). The global view was disturbing, but the world was also becoming hardened to environmental bad news. The increasingly popular slogan of "think global, act local" often just seemed to highlight the gap between aspirations and achievement. The more bad news, it seemed the less that could be done. The world was confronting a set of problems too distant, too multifaceted, and in the end, sometimes too hard to think about.

The idea of the Anthropocene was conceived in this ferment of concern. It had deep roots. Paul Crutzen suggested that the Industrial Revolution was not just a moment of human history, but also of *planetary* history. Crutzen and fellow theoretical chemist E. F. Stoermer identified 1784 as the "onset of the Anthropocene," a new geological era, in the first paper that appears in this part. William Ruddiman (2003) later argued that it was the agricultural revolution five thousand years ago that marked the beginning of humanity's status as a biophysical force. The notion of the Anthropocene as a geological era is still officially "under review" within the geological community, but in the first decade of the twenty-first century, the concept has developed traction as a transdisciplinary tool as humanity struggles to understand its enmeshed role in global futures. Never before have humans witnessed a new geological era, let alone been themselves the physical force that, like asteroids and volcanoes before, abruptly redirected the context for life's evolution.

Millennial thinking takes many forms. The millennial moment in time as defined by a particular calendar is clearly just arbitrary. But the idea that things might change, irreversibly, for ever, is certainly part of the "millennial moment."

As we step over the threshold, we are exhorted to look forward. Whether it is the apocalypse or Y2K, the future is suddenly upon us: it has become our present. How do we live in this new present?

The twentieth century had seen the rise of global institutions for policy making, particularly those connected to the United Nations since the 1940s. In the 1980s they produced landmark reports (the Brundtland Commission of 1987 on sustainable development), held conferences (the Rio Earth summit of 1992), and brokered major international agreements (the Montreal Protocol on ozone depletion of 1987, and the Kyoto protocol on global warming of 1997). But even now, national decision-making remains crucial to global outcomes, whether they are about biodiversity, climate, population, or environmental justice. Ironically, the institutional steps forward of the late twentieth century served only to sharpen the debates about the appropriate measures for the state of the environment, and how responsibility for global impact should be allocated among nations.

The three documents in this final section reflect on our present moment: they are all of the twenty-first century: first, a definition of the Anthropocene idea—a moment where the human and the global are now entwined futures. Second, a paper by Rockström and his many colleagues that looks at nine indicators of planetary health: a stocktake of the planetary issues, and an effort to measure which of them are already beyond our limits. The final paper is a philosophical reflection on how global change opens up new demands on our thinking—are we about to reduce the future of humanity to climate? These three are all documents written by people living with the Anthropocene idea, and trying to make sense of what it means for possible futures, for our children and our grandchildren, and also for ourselves—as our imagined futures shape the present. All are informed by the past, as Mike Hulme's paper suggests. History has a big role to play in future-making, and understanding the history of the concepts that frame global change is important to helping us make sense of our fast-changing world.

Johan Rockström and his colleagues attempts to sketch out "planetary boundaries," the limits to what makes planet Earth a "safe operating space" for humanity. What was novel about this paper, as Susan Owens observes in her commentary, is that the emphasis had shifted to critical earth systems, rather than on nonrenewable resources. Fears of resource scarcity were very old (see the essays by von Carlowitz, Jevons, Ordway, The Club of Rome), but "planetary boundaries" drew attention to the limits of the very functioning of an earth system. This was a problem that could not be resolved by finding new resources or using old ones more efficiently. The economic systems for a world where planetary boundaries were breached would have to prioritize supporting a safe operating space for humanity.

This paper was published only a few years ago, but already Figure 1, "Beyond the Boundary," is reaching iconic status, appearing on the walls of museums, for example. The nine key indicators that are measured by the Rockström team are now being measured in planets, the new global thinking. All the indicators that can be reliably measured subtend a red wedge of danger, some bigger than others. The disk with its exploding wedges (the ones that are already beyond "planetary limits") highlights particularly problems of biodiversity loss and the nitrogen cycle.

We need new ways to think in this global change age. It is too limiting to simply "reduce the future to climate," in the words of Mike Hulme. In our final document and commentary, we draw together many of the threads of this book. These documents provide guidance for how people can continue to live on a planet that is changing—globally, unequally, and probably irreversibly—because of the actions of our own species. Together they describe a new global environmental consciousness that affects not just how major global institutions should move, but informs nations, religions, social movements, and individual, local actions as well.

The "Anthropocene"

PAUL J. CRUTZEN AND EUGENE F. STOERMER

The name Holocene ("Recent Whole") for the post-glacial geological epoch of the past ten to twelve thousand years seems to have been proposed for the first time by Sir Charles Lyell in 1833, and adopted by the International Geological Congress in Bologna in 1885 (1). During the Holocene mankind's activities gradually grew into a significant geological, morphological force, as recognised early on by a number of scientists. Thus, G. P. Marsh already in 1864 published a book with the title "Man and Nature," more recently reprinted as "The Earth as Modified by Human Action" (2). Stoppani in 1873 rated mankind's activities as a "new telluric force which in power and universality may be compared to the greater forces of earth" [quoted from Clark (3)]. Stoppani already spoke of the anthropozoic era. Mankind has now inhabited or visited almost all places on Earth; he has even set foot on the moon.

The great Russian geologist V. I. Vernadsky (4) in 1926 recognized the increasing power of mankind as part of the biosphere with the following excerpt ". . . the direction in which the processes of evolution must proceed, namely towards increasing consciousness and thought, and forms having greater and greater influence on their surroundings." He, the French Jesuit P. Teilhard de Chardin and E. Le Roy in 1924 coined the term "noösphere," the world of thought, to mark the growing role played by mankind's brainpower and technological talents in shaping its own future and environment.

The expansion of mankind, both in numbers and per capita exploitation of Earth's resources has been astounding (5). To give a few examples: During the past 3 centuries human population increased tenfold to 6000 million, accompanied e.g. by a growth in cattle population to 1400 million (6) (about one cow per average size family). Urbanisation has even increased tenfold in the past century. In a few generations mankind is exhausting the fossil fuels that were generated

P. J. Crutzen and E. F. Stoermer. 2000. "The 'Anthropocene.'" *IGBP Newsletter* 41 (May): 17–18.

over several hundred million years. The release of SO_2, globally about 160 Tg/ year to the atmosphere by coal and oil burning, is at least two times larger than the sum of all natural emissions, occurring mainly as marine dimethyl-sulfide from the oceans (7); from Vitousek et al. (8) we learn that 30–50% of the land surface has been transformed by human action; more nitrogen is now fixed synthetically and applied as fertilizers in agriculture than fixed naturally in all terrestrial ecosystems; the escape into the atmosphere of NO from fossil fuel and biomass combustion likewise is larger than the natural inputs, giving rise to photochemical ozone ("smog") formation in extensive regions of the world; more than half of all accessible fresh water is used by mankind; human activity has increased the species extinction rate by thousand to ten thousand fold in the tropical rain forests (9) and several climatically important "greenhouse" gases have substantially increased in the atmosphere: CO_2 by more than 30% and CH_4 by even more than 100%. Furthermore, mankind releases many toxic substances in the environment and even some, the chlorofluorocarbon gases, which are not toxic at all, but which nevertheless have led to the Antarctic "ozone hole" and which would have destroyed much of the ozone layer if no international regulatory measures to end their production had been taken. Coastal wetlands are also affected by humans, having resulted in the loss of 50% of the world's mangroves. Finally, mechanized human predation ("fisheries") removes more than 25% of the primary production of the oceans in the upwelling regions and 35% in the temperate continental shelf regions (10). Anthropogenic effects are also well illustrated by the history of biotic communities that leave remains in lake sediments. The effects documented include modification of the geochemical cycle in large freshwater systems and occur in systems remote from primary sources (11–13).

Considering these and many other major and still growing impacts of human activities on earth and atmosphere, and at all, including global, scales, it seems to us more than appropriate to emphasize the central role of mankind in geology and ecology by proposing to use the term "anthropocene" for the current geological epoch. The impacts of current human activities will continue over long periods. According to a study by Berger and Loutre (14), because of the anthropogenic emissions of CO_2, climate may depart significantly from natural behaviour over the next 50,000 years.

To assign a more specific date to the onset of the "anthropocene" seems somewhat arbitrary, but we propose the latter part of the 18th century, although we are aware that alternative proposals can be made (some may even want to include the entire Holocene). However, we choose this date because, during the past two centuries, the global effects of human activities have become clearly noticeable. This is the period when data retrieved from glacial ice cores show the beginning of a growth in the atmospheric concentrations of several "greenhouse gases," in particular CO_2 and CH_4 (7). Such a starting date also coincides with James Watt's invention of the steam engine in 1784. About at that time, biotic assemblages in most lakes began to show large changes (11–13). Without major catastrophes like an enormous volcanic eruption, an unexpected epidemic, a large-scale nuclear war, an asteroid impact, a new ice age, or continued plundering of Earth's resources by partially still primitive technology (the last four dangers can, however, be prevented in a real functioning noösphere) mankind will remain a major geological force for many millennia, maybe millions of years, to come. To develop a world-wide accepted strategy leading to sustainability of ecosystems against human induced stresses will be one of the great future tasks of mankind, requiring intensive research efforts and wise application of the knowledge thus acquired in the noösphere, better known as knowledge or information society. An exciting, but also difficult and daunting task lies ahead of the global research and engineering community to guide mankind towards global, sustainable, environmental management (15).

References

1. Encyclopaedia Britannica, Micropaedia, IX (1976).

2. G.P. Marsh, *The Earth as Modified by Human Action*, Belknap Press, Harvard University Press, 1965.

3. W. C. Clark, in *Sustainable Development of the Biosphere*, W. C. Clark and R. E. Munn, Eds., (Cambridge University Press, Cambridge, 1986), chapt. 1.

4. V. I. Vernadski, *The Biosphere, translated and annotated version from the original of 1926* (Copernicus, Springer, New York, 1998).

5. B.L. Turner II et al., *The Earth as Transformed by Human Action*, Cambridge University Press, 1990.

6. P. J. Crutzen and T. E. Graedel, in *Sustainable Development of the Biosphere*, W. C. Clark and R. E. Munn, Eds. (Cambridge University Press, Cambridge, 1986). chapt. 9.

7. R. T. Watson, et al., in *Climate Change. The IPCC Scientific Assessment*. J. T. Houghton, G. J. Jenkins and J. J. Ephraums, Eds. (Cambridge University Press, 1990), chapt. 1.

8. P. M. Vitousek et al., *Science*, 277, 494 (1997).

9. E. O. Wilson, *The Diversity of Life*, Penguin Books, 1992.

10. D. Pauly and V. Christensen, *Nature*, 374, 255–257, 1995.

11. E. F. Stoermer and J. P. Smol, Eds. *The Diatoms: Applications for the Environmental and Earth Sciences* (Cambridge University Press, Cambridge, 1999).

12. C. L. Schelske and E. F. Stoermer, *Science*, 173 (1971); D. Verschuren et al. *J. Great Lakes Res.*, 24 (1998).

13. M. S. V. Douglas, J. P. Smol and W. Blake Jr., *Science* 266 (1994).

14. A. Berger and M.-F. Loutre, *C. R. Acad. Sci. Paris*, 323, II A, 1–16, 1996.

15. H. J. Schellnhuber, *Nature*, 402, C19–C23, 1999.

Commentary

Paul J. Crutzen and Eugene F. Stoermer, "The 'Anthropocene'" (2000)

WILL STEFFEN

It is entirely fitting that the first attempt to define the Anthropocene appeared in the newsletter of the global change research program IGBP (International Geosphere-Biosphere Programme) rather than in one of the mainstream scientific journals. The term was introduced in 2000 by Paul Crutzen and Eugene Stoermer in *IGBP Newsletter* 41.

This publication was a crystallization of Paul Crutzen's first use of the term Anthropocene during a discussion at a meeting of the IGBP Scientific Committee in Cuernavaca, Mexico, in February 2000. Scientists from IGBP's palaeo-environment project were reporting on their latest research, often referring to the Holocene, the most recent geological epoch of earth history, to set the context for their work. Paul, a vice-chair of IGBP, was becoming visibly agitated at this usage, and after the term Holocene was mentioned yet again, he interrupted them: "Stop using the word Holocene. We're not in the Holocene any more. We're in the . . . the . . . the . . . (searching for the right word) . . . the Anthropocene!"

The newsletter article built on this impromptu comment and began to flesh out what he actually meant by it, but it was by no means the first suggestion that human activities were beginning to have global impacts. For example, George Perkins Marsh in 1864 published *Man and Nature* and in 1874, *The Earth as Modified by Human Action*, and Italian geologist Stoppani in 1873 compared human activities to the great forces of earth. Twentieth-century commentators on the rising human impact on the global environment included Eduard Suess, Pierre Teilhard de Chardin, and Vladimir Vernadsky (see Part 5). However, the Crutzen/Stoermer article in 2000 made two explicit points that greatly furthered usage of the Anthropocene both as a term and as a concept. First, they proposed "to use the term 'anthropocene' for the current geological epoch" and then later suggested that "to assign a more specific date to the onset of the 'anthropocene' seems somewhat arbitrary, but we propose the latter part of the 18th century." So even at this first usage they proposed that the term Anthropocene be formalized to replace the Holocene as the current geological epoch, and they even suggested a historical start date for the new epoch.

The introduction of the term Anthropocene coincided with the IGBP synthesis project, a synergy that benefited both. Given Paul Crutzen's senior leadership role in the Programme, the concept of the Anthropocene became rapidly and widely used throughout the IGBP as its projects pulled together their main findings. The Anthropocene thus became a powerful concept for framing the ultimate significance of global change. The overall IGBP synthesis volume, which was published in 2004 (Steffen et al. 2004), contained two figures, included here as Figures 1 and 2, which have become widely reproduced as a visual depiction of the Anthropocene. These figures show many shifts in the global environment over the past two centuries away from Holocene patterns and limits. These changes include changes in not only climate, but also in stratospheric ozone, biodiversity, land cover, structure of marine and coastal ecosystems, the water cycle, and the biogeochemical cycles of carbon, nitrogen, phosphorus, and sulfur. Coincident with these global changes in the environment are equally impressive changes in the human enterprise, including population, economic activity, resource use, and connectivity. The evidence for a connection between human activity and change in the global environment is strong. The period from 1950 to 2000 stands out as one of the most remarkable in all of human history for its rapidity and pervasiveness of change, and it is now often called "The Great Acceleration."

Beyond the synthesis project itself, the IGBP research networks and broader community, which numbers thousands of scientists in about seventy countries around the world, played a central role in the spread of the term Anthropocene throughout the global change research community and beyond. Primarily because of its usefulness as a concept in global change research, the term appeared more and more frequently in the scientific literature, albeit still informally, to refer to the post–Industrial Revolution period of intertwined human and environmental change. This rapid spread of the term is in contrast to the 1980s, when Eugene Stoermer began using "anthropocene," but it never caught on then in the wider research community.

Later, in the first decade of the twenty-first century, the Anthropocene concept spread beyond the global change research community and into the mainstream geological community. In 2009 Jan Zalasiewicz and Mark Williams, both stratigraphers based at the University of Leicester, took the lead in the next step—the possible formalization of the term as the next geological epoch in earth history. They formed the Anthropocene Working Group of the Subcommission on Quaternary Stratigraphy (International Commission on Stratigraphy), whose task is to amass the geological (primarily stratigraphic) evidence that the earth has indeed been driven out of the Holocene by human activities and has entered a new geological epoch. The process of formalization is rather tortuous, and the case must be strong enough to convince a series of commissions and subcommissions in the international geological community.

A key event in the formalization process was a workshop in May 2011 at Burlington House in London, organized by the Geological Society of London. The focus of the workshop shifted the emphasis from global change science to the types of evidence in the geological record that would confirm the Anthropocene

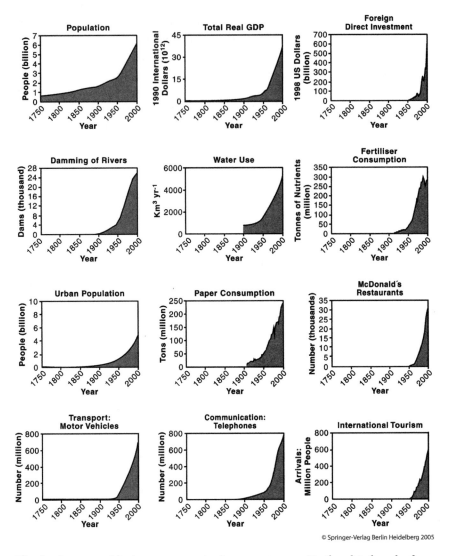

The development of the human enterprise from 1750 to 2000. Further details and references for individual data sets are given in Steffen et al. 2004.

as a new epoch. The participants were mainly from the geological sciences, and the change in the mood of the group as the workshop unfolded was fascinating. Beginning from a position of skepticism, as good scientists should, the participants pointedly questioned whether human activities could really challenge the great forces of nature. As speaker after speaker demonstrated via observations that the

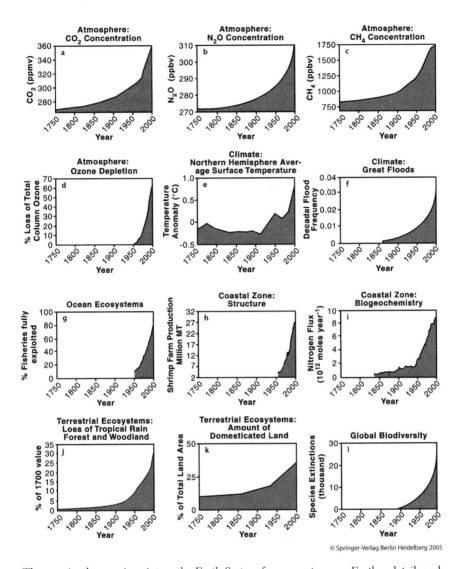

The growing human imprint on the Earth System from 1750 to 2000. Further details and references for individual data sets are given in Steffen et al. 2004.

imprint of the human enterprise was already clear in many stratigraphic features of earth, the mood changed to one of excitement. Participants began to put themselves into the position of geologists centuries or millennia in the future, looking back on this remarkably sharp and profound period of environmental change and speculating on what they would actually find in the stratigraphic record.

When did the Anthropocene start? There are two leading potential dates. The first was proposed by Paul Crutzen, who put the start at the beginning of the Industrial Revolution, near the end of the eighteenth century. In hindsight it is clear that the invention of the steam engine, the access to fossil fuels, and all of the associated developments put humanity on the pathway that has led to the Anthropocene. The other candidate is around 1950, or the end of World War II. This date marks the beginning of the Great Acceleration, when the global environment moved unequivocally out of the Holocene envelope. From a stratigraphic perspective, the end of World War II also produced an unmistakable geological marker, the radioactivity associated with the advent of the nuclear age.

Perhaps the most fascinating question surrounding the concept of the Anthropocene is how humanity will react if the new epoch is formally recognized by the geological community. The concept has already spilled over into the popular press. In 2011 the Anthropocene appeared on the cover of *Le Monde* magazine, was welcomed as the "Age of Man" in *National Geographic* magazine, and was even highlighted as a feature story in *The Economist*. Will humanity charge ahead more deeply and irreversibly into the Anthropocene, perhaps by attempting to geo-engineer its way out of the climate crisis, or will it have the humility (and good sense) to pull away from its present course, redefine its relationship with the rest of nature, and steer back toward a Holocene-like state of the Earth System?

Further Reading

Steffen, W., P. J. Crutzen, and J. R. McNeill. 2007. "The Anthropocene: Are humans now overwhelming the great forces of Nature?" *Ambio* 36(8):614–621.

Steffen, W., J. Grinevald, P. Crutzen, and J. McNeill. 2011. "The Anthropocene: Conceptual and historical perspectives." *Philosophical Transactions of the Royal Society A* 369:842–867.

Steffen, W., A. Sanderson, P. D. Tyson, J. Jäger, P. Matson, B. Moore III, F. Oldfield, K. Richardson, H.-J. Schellnhuber, B. L. Turner II, and R. J. Wasson. 2004. *Global change and the earth system: A planet under pressure.* The IGBP Book Series. Berlin: Springer-Verlag.

A Safe Operating Space for Humanity

JOHAN ROCKSTRÖM, WILL STEFFEN,

KEVIN NOONE, ET AL.

Summary

- New approach proposed for defining preconditions for human development
- Crossing certain biophysical thresholds could have disastrous consequences for humanity
- Three of nine interlinked planetary boundaries have already been overstepped

Although Earth has undergone many periods of significant environmental change, the planet's environment has been unusually stable for the past 10,000 years.[1-3] This period of stability—known to geologists as the Holocene—has seen human civilizations arise, develop and thrive. Such stability may now be under threat. Since the Industrial Revolution, a new era has arisen, the Anthropocene,[4] in which human actions have become the main driver of global environmental change.[5] This could see human activities push the Earth system outside the stable environmental state of the Holocene, with consequences that are detrimental or even catastrophic for large parts of the world.

Johan Rockström, Will Steffen, Kevin Noone, Åsa Persson, F. Stuart Chapin III, Eric F. Lambin, Timothy M. Lenton, Marten Scheffer, Carol Folke, Hans Joachim Schnellhuber, Björn Nykvist, Cynthia A. de Wit, Terry Hughes, Sander van der Leeuw, Hening Rodhe, Sverker Sörlin, Peter K. Snyder, Robert Costanza, Uno Svedin, Malin Falkenmark, Louise Karlberg, Robert W. Corell, Victoria J. Fabry, James Hansen, Brian Walker, Diana Liverman, Katherine Richardson, Paul Crutzen, Jonathan K. Foley. 2009. "A safe operating space for humanity." *Nature* 461:472–475.

During the Holocene, environmental change occurred naturally and Earth's regulatory capacity maintained the conditions that enabled human development. Regular temperatures, freshwater availability and biogeochemical flows all stayed within a relatively narrow range. Now, largely because of a rapidly growing reliance on fossil fuels and industrialized forms of agriculture, human activities have reached a level that could damage the systems that keep Earth in the desirable Holocene state. The result could be irreversible and, in some cases, abrupt environmental change, leading to a state less conducive to human development.[6] Without pressure from humans, the Holocene is expected to continue for at least several thousands of years.[7]

Planetary Boundaries

To meet the challenge of maintaining the Holocene state, we propose a framework based on "planetary boundaries." These boundaries define the safe operating space for humanity with respect to the Earth system and are associated with the planet's biophysical subsystems or processes. Although Earth's complex systems sometimes respond smoothly to changing pressures, it seems that this will prove to be the exception rather than the rule. Many subsystems of Earth react in a nonlinear, often abrupt, way, and are particularly sensitive around threshold levels of certain key variables. If these thresholds are crossed, then important subsystems, such as a monsoon system, could shift into a new state, often with deleterious or potentially even disastrous consequences for humans.[8,9]

Most of these thresholds can be defined by a critical value for one or more control variables, such as carbon dioxide concentration. Not all processes or subsystems on Earth have well-defined thresholds, although human actions that undermine the resilience of such processes or subsystems—for example, land and water degradation—can increase the risk that thresholds will also be crossed in other processes, such as the climate system.

We have tried to identify the Earth-system processes and associated thresholds which, if crossed, could generate unacceptable environmental change. We have found nine such processes for which we believe it is necessary to define planetary boundaries: climate change; rate of biodiversity loss (terrestrial and marine); interference with the nitrogen and phosphorus cycles; stratospheric ozone depletion; ocean acidification; global fresh-water use; change in land use; chemical pollution; and atmospheric aerosol loading (see Fig. 1 and Table).

In general, planetary boundaries are values for control variables that are either at a "safe" distance from thresholds—for processes with evidence of threshold behaviour—or at dangerous levels—for processes without evidence of

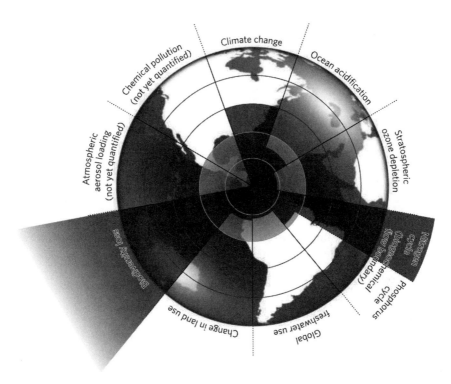

Figure 1. Beyond the boundary. The boundaries in three systems (rate of biodiversity loss, climate change and human interference with the nitrogen cycle) have already been exceeded.

thresholds. Determining a safe distance involves normative judgements of how societies choose to deal with risk and uncertainty. We have taken a conservative, risk-averse approach to quantifying our planetary boundaries, taking into account the large uncertainties that surround the true position of many thresholds. (A detailed description of the boundaries — and the analyses behind them — is given in ref. 10.)

Humanity may soon be approaching the boundaries for global freshwater use, change in land use, ocean acidification and interference with the global phosphorus cycle (see Fig. 1). Our analysis suggests that three of the Earth-system processes — climate change, rate of biodiversity loss and interference with the nitrogen cycle — have already transgressed their boundaries. For the latter two of these, the control variables are the rate of species loss and the rate at which N_2 is removed from the atmosphere and converted to reactive nitrogen for human use,

Planetary Boundaries

Earth-system process	Parameters	Proposed boundary	Current status	Pre-industrial value
Climate change	(i) Atmospheric carbon dioxide concentration (parts per million by volume)	350	387	280
	(ii) Change in radiative forcing (watts per metre squared)	1	1.5	0
Rate of biodiversity loss	Extinction rate (number of species per million species per year)	10	>100	0.1–1
Nitrogen cycle (part of a boundary with the phosphorus cycle)	Amount of N_2 removed from the atmosphere for human use (millions of tonnes per year)	35	121	0
Phosphorus cycle (part of a boundary with the nitrogen cycle)	Quantity of P flowing into the oceans (millions of tonnes per year)	11	8.5–9.5	–1
Stratospheric ozone depletion	Concentration of ozone (Dobson unit)	276	283	290
Ocean acidification	Global mean saturation state of aragonite in surface sea water	2.75	2.90	3.44
Global freshwater use	Consumption of freshwater by humans (km³ per year)	4,000	2,600	415
Change in land use	Percentage of global land cover converted to cropland	15	11.7	Low

Atmospheric aerosol loading	Overall particulate concentration in the atmosphere, on a regional basis	To be determined
Chemical pollution	For example, amount emitted to, or concentration of persistent organic pollutants, plastics, endocrine disrupters, heavy metals and nuclear waste in, the global environment, or the effects on eco-system and function-ing of Earth system thereof	To be determined

Data sources: ref. 10 and supplementary information

respectively. These are rates of change that cannot continue without significantly eroding the resilience of major components of Earth-system functioning. Here we describe these three processes.

Climate Change

Anthropogenic climate change is now beyond dispute, and in the run-up to the climate negotiations in Copenhagen this December, the international discussions on targets for climate mitigation have intensified. There is a growing convergence towards a "2°C guardrail" approach, that is, containing the rise in global mean temperature to no more than 2°C above the pre-industrial level.

Our proposed climate boundary is based on two critical thresholds that separate qualitatively different climate-system states. It has two parameters: atmospheric concentration of carbon dioxide and radiative forcing (the rate of energy change per unit area of the globe as measured at the top of the atmosphere). We propose that human changes to atmospheric CO_2 concentrations should not exceed 350 parts per million by volume, and that radiative forcing should not exceed 1 watt per square metre above pre-industrial levels. Transgressing these boundaries will increase the risk of irreversible climate change, such as the loss of major ice sheets, accelerated sea-level rise and abrupt shifts in forest and agricultural systems. Current CO_2 concentration stands at 387 p.p.m.v. and the change in radiative forcing is 1.5W m^{-2} (ref. 11).

There are at least three reasons for our proposed climate boundary. First, current climate models may significantly underestimate the severity of long-term climate change for a given concentration of greenhouse gases.[12] Most models[11] suggest that a doubling in atmospheric CO_2 concentration will lead to a global temperature rise of about 3°C (with a probable uncertainty range of 2–4.5°C) once the climate has regained equilibrium. But these models do not include long-term reinforcing feedback processes that further warm the climate, such as decreases in the surface area of ice cover or changes in the distribution of vegetation. If these slow feedbacks are included, doubling CO_2 levels gives an eventual temperature increase of 6°C (with a probable uncertainty range of 4–8°C). This would threaten the ecological life-support systems that have developed in the late Quaternary environment, and would severely challenge the viability of contemporary human societies.

The second consideration is the stability of the large polar ice sheets. Palaeoclimate data from the past 100 million years show that CO_2 concentrations were a major factor in the long-term cooling of the past 50 million years. Moreover, the planet was largely ice-free until CO_2 concentrations fell below 450 p.p.m.v. (±100 p.p.m.v.), suggesting that there is a critical threshold between 350 and 550 p.p.m.v. (ref. 12). Our boundary of 350 p.p.m.v. aims to ensure the continued existence of the large polar ice sheets.

Third, we are beginning to see evidence that some of Earth's subsystems are already moving outside their stable Holocene state. This includes the rapid retreat of the summer sea ice in the Arctic ocean,[13] the retreat of mountain glaciers around the world,[11] the loss of mass from the Greenland and West Antarctic ice sheets[14] and the accelerating rates of sea-level rise during the past 10–15 years.[15]

Rate of Biodiversity Loss

Species extinction is a natural process, and would occur without human actions. However, biodiversity loss in the Anthropocene has accelerated massively. Species are becoming extinct at a rate that has not been seen since the last global mass-extinction event.[16]

The fossil record shows that the background extinction rate for marine life is 0.1–1 extinctions per million species per year; for mammals it is 0.2–0.5 extinctions per million species per year.[16] Today, the rate of extinction of species is estimated to be 100 to 1,000 times more than what could be considered natural. As with climate change, human activities are the main cause of the acceleration.

Changes in land use exert the most significant effect. These changes include the conversion of natural ecosystems into agriculture or into urban areas; changes in frequency, duration or magnitude of wildfires and similar disturbances; and the introduction of new species into land and freshwater environments.[17] The speed of climate change will become a more important driver of change in biodiversity this century, leading to an accelerating rate of species loss.[18] Up to 30% of all mammal, bird and amphibian species will be threatened with extinction this century.[19]

Biodiversity loss occurs at the local to regional level, but it can have pervasive effects on how the Earth system functions, and it interacts with several other planetary boundaries. For example, loss of biodiversity can increase the vulnerability of terrestrial and aquatic ecosystems to changes in climate and ocean acidity, thus reducing the safe boundary levels of these processes. There is growing understanding of the importance of functional biodiversity in preventing ecosystems from tipping into undesired states when they are disturbed.[20] This means that apparent redundancy is required to maintain an ecosystem's resilience. Ecosystems that depend on a few or single species for critical functions are vulnerable to disturbances, such as disease, and at a greater risk of tipping into undesired states.[8,21]

From an Earth-system perspective, setting a boundary for biodiversity is difficult. Although it is now accepted that a rich mix of species underpins the resilience of ecosystems,[20,21] little is known quantitatively about how much and what kinds of biodiversity can be lost before this resilience is eroded.[22] This is particularly true at the scale of Earth as a whole, or for major subsystems such as the Borneo rainforests or the Amazon Basin. Ideally, a planetary boundary should capture the role of biodiversity in regulating the resilience of systems on Earth. Because science cannot yet provide such information at an aggregate level, we propose extinction rate as an alternative (but weaker) indicator. As a result, our suggested planetary boundary for biodiversity of ten times the background rates of extinction is only a very preliminary estimate. More research is required to pin down this boundary with greater certainty. However, we can say with some confidence that Earth cannot sustain the current rate of loss without significant erosion of ecosystem resilience.

Nitrogen and Phosphorus Cycles

Modern agriculture is a major cause of environmental pollution, including large-scale nitrogen- and phosphorus-induced environmental change.[23] At the planetary scale, the additional amounts of nitrogen and phosphorus activated by humans are now so large that they significantly perturb the global cycles of these two important elements.[24,25]

Human processes — primarily the manufacture of fertilizer for food production and the cultivation of leguminous crops — convert around 120 million tonnes of N_2 from the atmosphere per year into reactive forms — which is more than the combined effects from all Earth's terrestrial processes. Much of this new reactive nitrogen ends up in the environment, polluting waterways and the coastal zone, accumulating in land systems and adding a number of gases to the atmosphere. It slowly erodes the resilience of important Earth subsystems. Nitrous oxide, for example, is one of the most important non-CO_2 greenhouse gases and thus directly increases radiative forcing.

Anthropogenic distortion of the nitrogen cycle and phosphorus flows has shifted the state of lake systems from clear to turbid water.[26] Marine ecosystems have been subject to similar shifts, for example, during periods of anoxia in the Baltic Sea caused by excessive nutrients.[27] These and other nutrient-generated impacts justify the formulation of a planetary boundary for nitrogen and phosphorus flows, which we propose should be kept together as one boundary given their close interactions with other Earth-system processes.

Setting a planetary boundary for human modification of the nitrogen cycle is not straightforward. We have defined the boundary by considering the human fixation of N_2 from the atmosphere as a giant "valve" that controls a massive flow of new reactive nitrogen into Earth. As a first guess, we suggest that this valve should contain the flow of new reactive nitrogen to 25% of its current value, or about 35 million tonnes of nitrogen per year. Given the implications of trying to reach this target, much more research and synthesis of information is required to determine a more informed boundary.

Unlike nitrogen, phosphorus is a fossil mineral that accumulates as a result of geological processes. It is mined from rock and its uses range from fertilizers to toothpaste. Some 20 million tonnes of phosphorus is mined every year and around 8.5 million–9.5 million tonnes of it finds its way into the oceans.[25,28] This is estimated to be approximately eight times the natural background rate of influx.

Records of Earth history show that large-scale ocean anoxic events occur when critical thresholds of phosphorus inflow to the oceans are crossed. This potentially explains past mass extinctions of marine life. Modelling suggests that a sustained increase of phosphorus flowing into the oceans exceeding 20% of the natural background weathering was enough to induce past ocean anoxic events.[29]

Our tentative modelling estimates suggest that if there is a greater than tenfold increase in phosphorus flowing into the oceans (compared with pre-industrial levels), then anoxic ocean events become more likely within 1,000 years. Despite the large uncertainties involved, the state of current science and the present observations of abrupt phosphorus-induced regional anoxic events indicate that no more than 11 million tonnes of phosphorus per year should be allowed to flow into the oceans — ten times the natural background rate. We estimate that this boundary level will allow humanity to safely steer away from the risk of ocean anoxic events for more than 1,000 years, acknowledging that current levels already exceed critical thresholds for many estuaries and freshwater systems.

Delicate balance

Although the planetary boundaries are described in terms of individual quantities and separate processes, the boundaries are tightly coupled. We do not have the luxury of concentrating our efforts on any one of them in isolation from the others. If one boundary is transgressed, then other boundaries are also under serious risk. For instance, significant land-use changes in the Amazon could influence water resources as far away as Tibet.[30] The climate-change boundary depends on staying on the safe side of the freshwater, land, aerosol, nitrogen-phosphorus, ocean and stratospheric boundaries. Transgressing the nitrogen-phosphorus boundary can erode the resilience of some marine ecosystems, potentially reducing their capacity to absorb CO_2 and thus affecting the climate boundary.

The boundaries we propose represent a new approach to defining biophysical preconditions for human development. For the first time, we are trying to quantify the safe limits outside of which the Earth system cannot continue to function in a stable, Holocene-like state.

The approach rests on three branches of scientific enquiry. The first addresses the scale of human action in relation to the capacity of Earth to sustain it. This is a significant feature of the ecological economics research agenda,[31] drawing on knowledge of the essential role of the life-support properties of the environment for human wellbeing[32,33] and the biophysical constraints for the growth of the economy.[34,35] The second is the work on understanding essential Earth processes[6,36,37] including human actions,[23,38] brought together in the fields of global change research and sustainability science.[39] The third field of enquiry is research into resilience[40–42] and its links to complex dynamics[43,44] and self-regulation of living systems,[45,46] emphasizing thresholds and shifts between states.[8]

Although we present evidence that three boundaries have been overstepped, there remain many gaps in our knowledge. We have tentatively quantified seven boundaries, but some of the figures are merely our first best guesses. Furthermore, because many of the boundaries are linked, exceeding one will have implications for others in ways that we do not as yet completely understand. There is also significant uncertainty over how long it takes to cause dangerous environmental change or to trigger other feedbacks that drastically reduce the ability of the Earth system, or important subsystems, to return to safe levels.

The evidence so far suggests that, as long as the thresholds are not crossed, humanity has the freedom to pursue long-term social and economic development.

Note

1. Dansgaard, W. *et al. Nature* 364, 218–220 (1993).
2. Petit. J. R. *et al. Nature* 399, 429–436 (1999).
3. Rioual, P. *et al. Nature* 413, 293–296 (2001).
4. Crutzen, P. J. *Nature* 415, 23 (2002).
5. Steffen, W., Crutzen, P.J. & McNeill, J. R. *Ambio* 36, 614–621 (2007).
6. Steffen, W. *et al. Global Change and the Earth System: A Planet Under Pressure* (Springer Verlag, 2004).
7. Berger, A. & Loutre, M. F. *Science* 297, 1287–1288 (2002).
8. Scheffer, M., Carpenter, S. R., Foley, J. A., Folke C. & Walker, B. H. *Nature* 413, 591–596 (2001).
9. Lenton, T. M. *et al. Proc. Natl Acad. Sci. USA* 105, 1786–1793 (2008).
10. Rockstrom, J. *et al. Ecol. Soc.* (in the press); available from http://www.stockholm resilience.org/download/18.1fe8f33123572b59ab800012568/pb_longversion_170909.pdf
11. Intergovernmental Panel on Climate Change *Climate Change 2007: The Physical Science Basis. Contribution of Working Group I to the Fourth Assessment Report of the Intergovernmental Panel on Climate Change* (eds Solomon, S. *et al.*) (Cambridge University Press, 2007).
12. Hansen, J. *et al. Open Atmos. Sci. J.* 2, 217–231 (2008).
13. Johannessen, O. M. *Atmos. Oceanic Sci. Lett.* 1, 51–56 (2008).
14. Cazenave, A. *Science* 314, 1250–1252 (2006).
15. Church, J. A. & White, N. J. *Geophys. Res. Lett.* 33, LO1602 (2006).
16. Mace, G. *et al.* Biodiversity in *Ecosystems and Human Wellbeing: Current State and Trends* (eds Hassan, H., Scholes, R. & Ash, N.) Ch. 4, 79–115 (Island Press, 2005).
17. Sala, O. E. *et al. Science* 287, 1770–1776 (2000).
18. Sahney, S. & Benton, M. J. *Proc. R. Soc. Lond. B* 275, 759–765 (2008).
19. Diaz, S. *et al. Biodiversity regulation of ecosystem services in Ecosystems and Human Well-Being: Current State and Trends* (eds Hassan, H., Scholes, R. & Ash, N.) 297–329 (Island Press, 2005).
20. Folke, C. *et al. Annu. Rev. Ecol. Evol. Syst.* 35, 557–581 (2004).
21. Chapin, F. S., Ill *et al. Nature* 405, 234–242 (2000).
22. Purvis, A. & Hector, A. *Nature* 405, 212–219 (2000).
23. Foley, J. A. *et al. Science* 309, 570–574 (2005).
24. Gruber, N. & Galloway, J. N. *Nature* 451, 293–296 (2008).
25. Machenzie F. T., Ver L. M. & Lerman, A. *Chem. Geol.* 190, 13–32 (2002).
26. Carpenter, S. R. Regime shifts in lake ecosystems: pattern and variation, Vol. 15 in *Excellence in Ecology Series* (Ecology Institute, 2003).

27. Zillén, L., Conley, D. J., Andren, T., Andren, E. & Björck, S. *Earth Sci. Rev.* 91(1), 77–92 (2008).

28. Bennett, E. M., Carpenter, S. R. & Caraco, N. E. *BioScience* 51, 227–234 (2001).

29. Handoh, I. C. & Lenton, T. M. *Global Biogeochem. Cycles* 17, 1092 (2003).

30. Snyder, P. K., Foley, J. A., Hitchman, M. H. & Delire, C. J. *Geophys. Res. Atmos.* 109, D21 (2004).

31. Costanza, R. *Struct. Change Econ. Dyn.* 2(2), 335–357 (1991).

32. Odum, E. P. *Ecology and Our Endangered Life-Support Systems* (Sinuaer Associates, 1989).

33. Vitousek, P. M., Mooney, H. A., Lubchenco, J. & Melillo, J. M. *Science* 277, 494–499 (1997).

34. Boulding, K. E. The Economics of the Coming Spaceship Earth in *Environmental Quality Issues in a Growing Economy* (ed. Daly, H. E.) (Johns Hopkins University Press, 1966).

35. Arrow, K. *et al. Science* 268, 520–521 (1995).

36. Bretherton, F. *Earth System Sciences: A Closer View* (Earth System Sciences Committee, NASA, 1988).

37. Schellnhuber, H. J. *Nature* 402, C19–C22 (1999).

38. Turner, B.L. II *et al.* (eds) *The Earth as Transformed by Human Action: Global and Regional Changes in the Biosphere over the Past 300 Years* (Cambridge University Press, 1990).

39. Clark, W. C. & Dickson, N. M. *Proc. Natl Acad. Sci. USA* 100, 8059–8061 (2003).

40. Holling, C. S. The resilience of terrestrial ecosystems: local surprise and global change in *Sustainable Development of the Biosphere* (eds Clark, W. C. & Munn, R. E.) 292–317 (Cambridge University Press, 1986).

41. Walker, B., Holling, C. S., Carpenter, S. R. & Kinzig, A. *Ecol. Soc.* 9, 5 (2004).

42. Folke, C. *Global Environmental Change* 16, 253–267 (2006).

43. Kaufmann, S. A. *Origins of Order* (Oxford University Press, 1993).

44. Holland, J. *Hidden Order: How Adoption Builds Complexity* (Basic Books, 1996).

45. Lovelock, J. *Gaia: A New Look at Life on Earth* (Oxford University Press, 1979).

46. Levin, S. A. *Fragile Dominion: Complexity and the Commons* (Perseus Books, 1999).

Commentary

Johan Rockström, Will Steffen, Kevin Noone, et al.,
"A Safe Operating Space for Humanity" (2009)

SUSAN OWENS

"A Safe Operating Space for Humanity," published as a feature in the journal *Nature* in September 2009, offered a brief but startling analysis of the consequences of human action for global environmental change. Johan Rockström (of the Stockholm Resilience Center) and twenty-eight colleagues, predominantly earth and environmental scientists, argued that the relative environmental stability of the Holocene—the most recent period of the Earth's history, going back about ten thousand years—was increasingly threatened by the destabilizing impacts of multiple human activities. In the absence of such adverse interference, the Holocene might continue for several thousand more years, but if pressures persisted, critical earth systems might shift into states that were much less conducive to human flourishing. The result, the authors suggested, could be "irreversible and, in some cases, abrupt environmental change," with consequences that might be "detrimental or even catastrophic for large parts of the world" (472). Almost four decades earlier, a team at the Massachusetts Institute of Technology had reached broadly similar conclusions, widely publicized in *The Limits to Growth* (see Part 2): they too predicted that business-as-usual would soon breach planetary limits, probably leading to "a rather sudden and uncontrollable decline" in population and industrial capacity (Meadows et al. 1972, 23).

What was different, then, about the Rockström paper in *Nature*, that it should generate renewed attention within academic and political communities? An important part of the novelty lay in the attempt by Rockström et al. to define the "safe operating space for humanity" in terms of "planetary boundaries," each associated with essential biophysical systems and processes. The emphasis was on critical earth systems, rather than on nonrenewable resources; the paper was grounded in extensive research; and it was accompanied by compelling graphics. Eleven processes were identified, and boundaries estimated for nine. These planetary boundaries, often interlinked, were the authors' best estimates of "safe" thresholds beyond which there could be a serious risk of disruptive environmental change. Three of them—related to climate change, the rate of biodiversity loss, and interference with the nitrogen cycle—had already been transgressed.

This was heady stuff, but another important reason for the paper's impact was that it caught the temper of the times. In the early twenty-first century, concern about limits to growth was once again discernible after several decades of being subsumed within the more emollient narratives of ecological modernization and sustainable development. Since the 1980s, much had been made of society's technical and institutional capabilities for doing more with less, while Gro Harlem Brundtland's vision of a sustainable world had been an integrative (and extremely influential) one, in which environment and development were seen not as "separate challenges" but as "inexorably linked" (WCED 1987, 37). In spite of progress in many areas, however, the bigger picture on environmental change, as humanity entered the new century, was not encouraging: societies, it seemed, could become more ecologically modern without becoming environmentally sustainable. Absolute impacts mattered, and the question of limits had never actually gone away. Now it was re-emergent in a new language of tipping points, criticality, and boundaries, and even in warnings of a "perfect storm" for humanity if resource and environmental pressures were not addressed (Beddington 2009). "A safe operating space for humanity" became emblematic of this new environmental discourse.

Although it addressed these re-emergent issues, the planetary boundaries framework might be seen as less deterministic, and less inevitably apocalyptic, than the models underpinning *The Limits to Growth*. The boundaries identified by Rockström and his colleagues—the precautionary thresholds beyond which modification of earth systems should not go—were derived from the authors' assessment of *planetary* limits combined with their judgements about appropriate precaution; the boundaries were not presented as limits to growth per se. In a lively exchange on the *Nature* website (and responding to criticism of his paper's silence on the vexed issue of human numbers), Rockström maintained that the definition of critical earth system processes could be done "*irrespective* of our human impacts on the planet," and that once boundaries were defined, "humanity should be able to thrive within the safe operating space that the boundary levels provide" (emphasis added).

Two arguments follow. One is that if human flourishing involves growth, it must be growth that differs in fundamental respects from the continuous expansion of gross domestic product sought so eagerly by governments worldwide. The idea of growth within boundaries is not new. Attempts to reconcile development and environment, after the polarized and often acrimonious "growth debate" of the 1970s, moved away from the concept of "zero growth" but envisaged development that would be constrained (as well as enabled) by energy, chemicals, and materials cycles (IUCN 1980): in Brundtland's words, patterns of consumption would have to remain "within the bounds of the ecological[ly] possible" (WCED 1987, 44). Other antecedents (such as the work of Kenneth Boulding) were cited by Rockström et al. themselves. What was different about their paper—and this was another reason for its prominence—is that they were bold enough to seek to define what the limits might be.

If boundaries are to be identified a priori, as Rockström suggests, to define the space within which human ends can be pursued, there is a further important implication: a considerable scientific effort would need to be devoted to their delineation. It is not surprising, perhaps, that in initial responses to the paper, particular attention focused on the parameters used by its authors in characterizing the critical thresholds, and on the values that they had attached to the boundaries. The word "arbitrary" recurred frequently in invited commentaries (*Nature* 461), and the authors themselves readily acknowledged that their first estimates were little more than "best guesses" (475). But as some commentators also recognized, the whole framework raised profound questions about scientific authority, social meaning, and political agency. Given that earth system science deals with extreme complexities, and precaution is invariably contested, would it ever be possible to pin down planetary boundaries in some way that could generate consensus? (We can hardly be encouraged by the conflict over atmospheric carbon dioxide concentrations alone.) And can the aggregation of variables at a global scale ever adequately reflect the lived experience of environmental degradation, or provide a basis for political action in particular places at particular times? This important question was discussed by Mike Hulme (2009) in his critical analysis of climate discourse. But perhaps the most challenging issue concerns the very possibility of identifying "the biophysical preconditions for human development" (474) as a purely "scientific" endeavor, *irrespective* of normative judgement and human choice. As Ezrahi (1980, 120) has shown, when scientific uncertainties combine with "unsettled, ambiguous or contradictory" human ends, science and politics "interpenetrate." The project of delineating a "safe operating space for humanity" would seem to fit firmly into this hybrid realm, and that, perhaps, will ensure that it continues to generate intense debate across the natural and the social sciences.

Further Reading

Beddington, J. 2009. Speech by Professor (later Sir) John Beddington, U.K. Government Chief Scientific Advisor, to Sustainable Development UK, 19 March. Available from www.govnet.co.uk/news/govnet/profesr-sir-john-beddington's-speech-at-sduk-09. Accessed 7 February 2012.

Ezrahi, Y. 1980. "Utopian and pragmatic realism: The political context of scientific advice." *Minerva* 18(1):111–131.

Hulme, M. 2009. *Why we disagree about climate change.* Cambridge: Cambridge University Press.

IUCN (International Union for the Conservation of Nature) with United Nations Environment Programme (UNEP) and World Wildlife Fund (WWF). 1980. *World conservation strategy: Living resource conservation for sustainable development.* Gland, Switzerland: IUCN, UNEP, WWF.

Meadows, D. L., et al. 1972. *The limits to growth: A report for the Club of Rome project on the predicament of mankind.* London: Club of Rome.

Rockström, J. 2009. Comment posted 28 September. See http://blogs.nature.com/climate feedback/2009/09/planetary_boundaries_1.html. Accessed 7 February 2012.

WCED (World Commission on Environment and Development) [Brundtland, G. H.]. 1987. *Our common future.* Oxford: Oxford University Press.

Reducing the Future to Climate

A Story of Climate Determinism and Reductionism

MIKE HULME

Introduction

[246] Human beings are always trying to come to terms with the climates
they live with. This is as true for the ways the relationship between society and
climate is theorized as it is for the practical challenges of living fruitfully and
safely with climatic resources and hazards. The story of how the idea of climate
has traveled through the human imagination is well told in Lucian Boia's *The
Weather in the Imagination*, and an exemplary account of how a society seeks
practically to live with its climate is William B. Meyer's *Americans and Their
Weather*. When people reflect on these relationships between society and climate, they frequently adopt two intuitive positions. On the one hand, it is obvious that climates influence and shape human psychological, biological, and
cultural attributes. This is true for individual behaviors, cultural practices, and
environmental resources. Yet it is equally true that an enduring strand of human
encounters with climate seeks both to tame these climatic influences and constraints and to live beyond them. Human beings change microclimates, insulate
themselves against climatic extremes, and adapt technologies and practices for
survival and prosperity.

Attempts to understand and theorize the relationship between climate and
society are therefore prey to two distinct fallacies. The first is that of "climate
determinism," in which climate is elevated to become a—if not *the*—universal
predictor (and cause) of individual physiology and psychology and of collective
social organization and behavior. The second fallacy is that of "climate inde-
terminism," in which climate is relegated to a footnote in human affairs and
stripped of any explanatory power. Geographers have at times been most guilty of
the former fallacy, historians at times most guilty of the latter. Yet not even histori-

Mike Hulme. 2011. Excerpt from "Reducing the future to climate." *Osiris* 26:245–266.
[original page numbers in square brackets]

cal geographers or environmental historians have been always able to hold these two opposing fallacies in adequate and creative tension.

At the beginning of the 20th century, the determinist fallacy achieved considerable salience and popularity in European and, especially, American thought, championed by the likes of the geographers Friedrich Ratzel, Ellen Semple, and Ellsworth Huntington. Climate was viewed as the dominant determinant of racial character, intellectual vigor, moral virtue, and the ranking of civilizations, ideas that had earlier appealed to Greek philosophers and European rationalists alike. However, the ideological wars of the mid-20th century reshaped the political and moral worlds that had nourished such thinking, and determinism became discredited and marginalized within mainstream academic geography.

[247] Now, a hundred years later, and at the beginning of a new century, heightening anxieties about future anthropogenic climate change are fueling—and in turn being fueled by—a new variety of the determinist fallacy. . . . I call this "climate reductionism," a form of analysis and prediction in which climate is first extracted from the matrix of interdependencies that shape human life within the physical world. Once isolated, climate is then elevated to the role of dominant predictor variable. I argue in this article that climate reductionism is a methodology that has become dominant in analyses of present and future environmental change—and that as a methodology it has deficiencies.

This way of thinking and analyzing finds expression in some of the balder (and bolder) claims made by scientists, analysts, and commentators about the future impacts of anthropogenic climate change. . . . [This] reductionism is also contributing to the new discourse about climate change and conflict. For example, climate change is offered by Zhang et al. (2007: 413) as an explanation of cycles of war and conflict in China over the last millennium: "It was the oscillations of agricultural production brought about by long-term climate change that drove China's historical war-peace cycles." The civil war in Darfur is categorized in the media as a harbinger of future climate-driven disputes [and so forth]. . . . [248] In a view related to this belief that climate plays an explanatory role in determining war, climate refugees are seemingly set to threaten global, regional, and national security in an [apparent] rerun of the Mongol invasions of Europe. . . . The term "climate refugees" was invented by Norman Myers in 1993. . . . His estimate of between 150 and 250 million climate refugees by 2050 has been subsequently widely cited.

In this new mood of climate-driven destiny the human hand, as the cause of climate change, has replaced the divine hand of God as being responsible for the collapse of civilizations, for visitations of extreme weather, and for determining the new 21st-century wealth of nations. And to emphasize the message and the mood, the New Economics Foundation and its partners have wound up a climate clock that is now ticking, second by second, until December 1, 2016, when human fate will be handed over to the winds, ocean currents, and drifting ice floes of a destabilized global climate: "We have 100 months to save the planet; when the clock stops ticking we could be beyond the climate's tipping point, the point of no return." Such eschatological rhetoric offers a post-2016 world where human freedom and agency are extinguished by the iron grip of the forces of climate. Such a narrative offers scant [249] chance for humans to escape a climate-shaped destiny. . . .

Such sentiments . . . are enabled by the methodology of climate reductionism (i.e., a form of neoenvironmental determinism). Simulations of future climate from climate models are inappropriately elevated as universal predictors of future social performance and human destiny. . . .

Why should an explanatory logic—if not an ideology—dating from earlier intellectual and imperial eras, a logic subsequently dismissed by many as seriously wanting, have re-emerged in different form in a new century to find new and enthusiastic audiences? . . . I suggest that the hegemony exerted by the predictive natural sciences over human attempts to understand the unfolding future opens up the spaces for climate reductionism to emerge. Because of the epistemological authority over the future claimed, either implicitly or explicitly, by [predictive] modelling activities, climate becomes the one "known" variable in an otherwise unknowable future. The openness, contingency, and multiple possibilities of the future are closed off as these predicted virtual climates assert their influence over everything from future ecology, economic activity, and social mobility to human behavior, cultural evolution, and geosecurity. It is climate reductionism exercised through what I call "epistemological slippage"—a transfer of predictive authority from one domain of knowledge to another without appropriate theoretical or analytical justification. . . .

The Demise of Climate Determinism

[250] The argument of [early twentieth century] determinists [such as Huntington and Griffith Taylor; see Part 3] was that aspects of climate exerted a powerful shaping influence on the physiology and psychology of individuals and races, which in turn shaped decisively the culture, organization, and behaviour of the society formed by those individuals and races. Tropical climates were said to cause laziness and promiscuity, while the frequent variability in the weather of the middle latitudes led to more vigorous and driven work ethics.

This is a determinism that *makes* human and social character. There is also a form of climate determinism that *moves* people. Thus we have accounts that [the Mongols were driven towards Europe], almost involuntarily, by climate variations in search of better pastures, while accounts [like Ian Whyte's 2008 *World Without End?*] suggest Viking arrivals and departures to and from Greenland . . . were driven solely by the oscillations of warmth and cold. . . . Both these manifestations of climate determinism—the making of character and the moving of people—emphasize the agency of climate over the agency of humans. In more extreme articulations of the idea . . . human will becomes hostage to the fortunes of climate, too passive and powerless to respond proactively, or even reactively, to changes in environmental fortune.

The apparent simplicities of climate determinism appealed to philosophers of the Grecian Empire (such as Herodotus and Hippocrates) and to rationalists of the European Enlightenment (such as Montesquieu and Hume). They also appealed to late 19th and early 20th century European and, especially, American geographers. The work of Yale geographer Ellsworth Huntington (1876–1947) best encapsulates [251] the theory's rise during the apogee of modern European and American imperialism. Huntington's major works gained him a contemporary popularity, but among his academic colleagues he generated a range of contrary reactions. . . . For example, anthropologist Franz Boas was consistently irked by Huntington's simplistic statistical methods, which, Boas argued, offered merely a fig leaf of scientific credibility to Huntington's claims.

Based on his belief that there were . . . universally optimal climates for physical and mental activity, Huntington [in *Civilization and Climate* (1915)] drew upon a number of empirical studies of factory workers in America to suggest that 20°C and a 60% humidity maximized productivity. It was a short step from here to postulate that the energy and vigor needed to develop and sustain civilizations was also related to these climatic optima, giving rise to his "mainsprings of civilization" hypothesis. And for Huntington a further short step into the emerging field of genetic selection was to bring him in the 1920s under the influence of the American eugenics movement.

Huntington's determinism was centrally concerned with the tracing of patterns of climate in history, rather than with predicting the future fates of civilizations. The British politician and writer Sydney Markham, however, later developed and applied some of these determinist arguments in a different direction. In *Climate and the Energy of Nations*, published in 1942, Markham argued that climate variations could not only explain the rise and fall of past civilizations, but could also explain and predict the changing geopolitical balance of power in his mid-20th century world. The dependence of contemporary social and economic factors such as trade, wealth creation, and human mortality rates on climate offered Markham a way of interpreting the tumultuous 1940s—and foreseeing the political prospects of nations such as Russia, China, and the United States as well as of Europe.

As with its rise, there is no shortage of accounts of the demise of environmental determinism in geographic thought [by the 1920s or 1930s]. . . . Richard Peet put it . . . bluntly [1985, 327]:

> Environmental determinism became increasingly socially dysfunctional in the 1920s after the main issues of imperialist domination of the world had been settled by World War I.

[252] The strong form of climate determinism was therefore largely discredited and marginalized as the ideological wars of the 20th century reshaped the political and moral worlds that had allowed it to flourish. . . . Vestiges of Huntington's strong determinism nevertheless still lingered among those engaged in talking about and analyzing climatic data in the context of society and behavior. Meyer discusses the persistence of climate determinism in American thought and culture through the middle decades of the 20th century in *Americans and their weather.* . . .

The Rise of Climate Reductionism

[253] Notwithstanding these examples, the fortunes of strong determinism, both as an ideology and as an explanatory framework for climate-society relationships, waned over the 20th century. Yet with the emergence over the last 25 years of anthropogenic climate change as a physical and social phenomenon of worldwide importance, the question of how the challenging relationship between climate and society is conceived has taken on fresh importance. . . .

Reductionism is an approach to understanding the nature of complex entities or relationships by reducing them either to the interactions of their parts or else to simpler or more fundamental entities or relationships. In the case of climate change studies, this means isolating climate as the (primary) determinant of past, present, and future system behavior and response. If crop yield, economic performance, or violent conflict can be related to some combination of climate variables, then knowing the future behavior of these variables offers a way of knowing how future crop yield, economic performance, or violent conflict will unfold. Other factors . . . may be more important than climate (or perhaps less predictable) are ignored or marginalized in the analysis.

The way climate reductionism requires and seeks out simple chains of climatic cause-and-effect is perfectly illustrated in an empirical study of the relationship between climate change and economic growth published by the U.S. National Bureau of Economic Research (Dell *et al.* 2008). The authors recognize that the question of whether climate change has a direct effect on economic development is contentious, but they claim nevertheless that their global analysis using data from over 180 nations reveals [254] a:

> substantial contemporary causal effect of temperature on aggregate [economic] output . . . On average, a 1°C increase in average temperature predicts a fall in *per capita* income by about 8 per cent. (Dell et al., 2008: 4, 6)

Since they find this effect to be asymmetrical between richer and poorer countries, they are then able to extend their analysis to consider the impact of *future* climate change on economic performance. They conclude:

> The negative impacts of climate change on poor countries may be larger than previously thought. Overall, the findings suggest that future climate change may substantially widen income gaps between rich and poor countries. (27–8)

First the complex relationships that exist between climate and economic performance are reduced to a dependent relationship between temperature and GDP *per capita*, and then, using projections of future climate warming, future economic performance is predicted for the 21st century. The many subtleties and contingencies of national and regional economic performance are ignored or suppressed. Climate reductionism opens up the prospect of developing a narrative about future economic growth in which climate change becomes the primary driver of performance.

. . . [This is] a one-eyed view of the future, yet it pervades many recent academic analyses of climate change and social impact and consequently it is an account of the future that enters easily into public perception and discourse. . . . [255] Given the demise of climate determinism described above, . . . how is it possible to have arrived back at an understanding of climate-society relationships that . . . distorts and overemphasizes the causative role of climate in shaping the future prospects of society and the well-being of individuals?

The Hegemony of Model Predictions of the Future

[Andrew] Sluyter . . . suggests that the Enlightenment dichotomy between nature and culture, so pervasive in Western thought and practice, began increasingly to be challenged in the 1980s and 1990s—for example, [he] argues that environmental determinism offered one means for a "quick and dirty integration of the natural and social sciences" (Sluyter 2003: 817). As if the inadvisability of the dualistic thinking pervading Western thought were being belatedly realized, there was a rush to forge a new rapprochement between nature and culture. Determinist thinking was the simplest and most available ideology to hand.

While I think there is some merit in his argument . . . , I wish to suggest a different line of reasoning applies very specifically to the case of climate reductionism. . . . In summary, my argument concerns the hegemony held by the predictive natural and biological sciences over visions of the future. In the case of climate change, this hegemony is rooted in the knowledge claims of climate or Earth system models. In the absence of comparable epistemological reach emerging from the social sciences or humanities, these claims lend disproportionate discursive power to model-based descriptions of putative future climates. It thus becomes tempting to adopt a reductionist methodology when examining possible social futures. . . . The subsequent and derived climate impact modelling then boldly calculates, for example, the billions of people who because of climate change will become starving or thirsty, or the millions who because of climate change will be made destitute or homeless. Climate reductionism is the means by [256] which the knowledge claims of the climate modelers are transferred . . . [to] social, economic, and political analysts.

This transfer of predictive authority, an almost accidental transfer, . . . earlier defined as "epistemological slippage," . . . offers a future written in the unyielding language of mathematics and computer code. These models and calculations allow for little human agency, little recognition of evolving, adapting, and innovating societies, and little endeavor to consider the changing values, cultures, and practices of humanity. The contingencies of the future are whitewashed *out* of the future. Humans are depicted as "dumb farmers," passively awaiting their climate fate. The possibilities of human agency are relegated to footnotes, the changing cultural norms and practices made invisible, the creative potential of the human imagination ignored.

To give some substance to this argument I . . . [examine] the emergence of anthropogenic climate change as a matter of scientific concern in the 1970s and 1980s and as a matter of public policy debate in the 1980s and 1990s. There are three developments that are important for my argument: the retreat of the social sciences, and geography in particular, from working at the nature-culture interface; the emergence of a new epistemic community of global climate modelers; and the asymmetrical incorporation of climate change and social change into envisaged futures.

THE ABSENCE OF THEORY ABOUT
CLIMATE-SOCIETY INTERACTIONS

The . . . academic discipline that had thought the longest and hardest about relationships between climate and society—geography—had by the 1960s become suspicious about grand theories of climate-society interaction, particularly those tinged with any trace of the old determinist ideology. This reaction against the worst excesses of determinism left geographers without a coherent [way of "bridging" between] the social and natural sciences. It also meant that the study of environment-society relationships became a subject without an academic home. . . . [A] small number of historians [e.g., Emmanuel Le Roy Ladurie] and atmospheric scientists [e.g., Reid Bryson, Hubert Lamb] . . . [worked] to reengage substantively with [such] questions. . . . [257] But they did so in the absence of any coherent theoretical framework to explain such interactions and certainly without any basis for prediction.

THE EPISTEMIC COMMUNITY OF GLOBAL CLIMATE MODELING

. . . The 1960s and 1970s had witnessed the development of the first computer-based simulation models of a universal and globally connected climate system. Originally an extension of numerical weather prediction models, these new climate-oriented models allowed experiments with global climate to be performed in virtual reality that were not possible in physical reality. These models were constructed initially by meteorologists and atmospheric scientists in a small number of research centers in the United States, the United Kingdom, and Germany. They were later joined by oceanographers, atmospheric chemists, and biologists as the models extended their representation from simply the climate system (initially the atmosphere) to the deeply coupled components of the Earth system. This move was encapsulated in NASA's 1988 report [258] *Earth system science: A closer view*, the so-called Bretherton report. The . . . goal of this new scientific mission was to obtain a scientific understanding of the entire Earth system on a global scale; predictions were to be secured by using quantitative models . . . to identify and simulate global trends.

In barely 25 years . . . scientific accounts of the causes and properties of climate had become progressively more complex. Climate was now viewed as the outcome of the functioning of an interconnected biogeophysical global system whose past, present, and future behavior could be modeled—and hence "predicted"—using mathematical equations and advanced computing technology. This marked a distinct break from [earlier] more varied conceptions. Clark Miller notes that the *First Annual Conference on Statistical Climatology* was held in 1979, and prior to this time there was no reason to refer to *statistical* climatology because there was no other form of climatology to distinguish it from (Miller 2004). . . . By the 1990s "computer modelling had become *the* central practice for evaluating truth claims" for this community of global climate change scientists (Edwards 2001: 53). Yet as Miller has argued, epistemic communities and the knowledge they produce do not form in isolation from wider social, institutional, and political settings. . . . The development of global climate and . . . questions about [its] future performance . . . occurred against the backdrop of the new environmental geopolitics of . . . the first [259] United Nations Conference on the Human Environment held in Stockholm [in 1972], which presaged a new era of international environmental diplomacy. The World Meteorological Organization's First World Climate Conference, held in 1979, and the 1983 report by the Carbon Dioxide Assessment Committee of the U.S. National Academy of Sciences were also evidence of the growing political saliency of climate change. . . . The early battles about the credibility of anthropogenic climate change in the 1990s were therefore fought largely around the credibility of the models, because [of] the political significance of [such] knowledge claims about the future.

Yet to answer the demanding questions being asked about the significance of anthropogenic climate change for human society required more than mere knowledge of future climate. It demanded some translation of future changes in climate into future impacts for society. The *First Assessment Report* of the IPCC in 1990, for example, was organized into three separate volumes: one on climate science, one on climate impacts, and one on climate policy options. . . . Given the poorly developed and atheoretical understandings of climate-society relationships in the social sciences, how were these demanding questions going to be answered? How did the first IPCC assessments address these relationships?

The Asymmetrical Incorporation of Climate and Social Change into Envisaged Futures

The first studies assessing the consequences of future anthropogenic climate change for society were undertaken in the late 1970s and early 1980s. . . . But methodological challenges these new policy-driven questions [came later], particularly [260] in the International Council of Scientific Unions' Scientific Committee on Problems of the Environment on climate impact assessment *SCOPE* 27 (Kates, Ausubel, and Berberian, 1985) . . . which was a response to the new World Climate Impact Program (WCIP)—whose aim was to advance understanding of the relation between climate and human activities, as agreed at the 1979 First World Climate Conference. The crucial methodological chapter was written by the respected geographer [and lead editor] Robert Kates, who . . . laid out the methodological challenges of performing climate impact assessments, including . . . [the] explicit knowledge hierarchy between the "hard" sciences and the "soft" sciences. As one moves from understanding global heat balances to the impacts of climate change on nutrition, for example, there is "less predictability, more speculation and greater uncertainty" (Kates 1985: 4). Complexity increases, precision decreases, and uncertainties are compounded. . . . [A] second challenge . . . is linking very different methodologies: for example, modeling of global climate with analysis of energy trends or assessment of population dynamics. . . . "As yet there has been no comprehensive study of the problems of integrating such scientific apples and oranges" (Kates 1985: 5). . . . [A] third challenge therefore was how to develop even the most basic of analytical frameworks for performing linked . . . climate impact assessment. Kates offered two models, the "impact model"—where climate change determines the impact directly and the "interaction model" [where] impact is the joint product of the interaction between climate and social change. [Kates argued] that the first . . . was predominant in nearly all attempts at climate impact assessment, which [typically] went "directly from climate events to inferences of higher-order consequences (Kates 1985: 288). This was, he argued, due "partly to disciplinary isolation and partly to [261] the limited effort expended to date on the study of the interaction of climate and society as compared to the study of the dynamics of climate itself" (Kates 1985: 31).

SCOPE 27 therefore reveals, I suggest, how the idea of an explicit knowledge hierarchy, the lack of any theoretical frameworks for integrated analysis, and the preferred linear model of climate response contributed to a climate reductionism at work in impact assessments. At this crucial moment in the 1980s, when climate predictions were asserting their knowledge claims about the future and when policy was demanding knowledge about future consequences of climate change for society, it was easy for simple reductionist accounts of future climate change impacts to emerge.

. . . The development of climate predictions or scenarios, [now a] pivotal component of so many climate impact studies [is covered in a later book]. . . . Published in 1998 under the title Climate Impact and Adaptation Assessment (Parry and Carter eds.) this was a widely read guide to the IPCC approach to assessing climate change impacts and adaptations. It was "a readable guide" to the Technical Guidelines for Assessing Climate Change Impacts and Adaptations [262] (Carter, Parry et al. 1994) . . . that [had become] widely cited and used internationally in the field. In these IPCC assessment guidelines, . . . the default methodological assumptions and practices revealed by Kates in the 1980s were reinforced. [This] was done by privileging predictions of future climate over explorations of how the many other dimensions of cultural, social, and political life may change in the future. Climate reductionism through epistemological slippage was the result. . . .

[264] Reactions against climate reductionism, notably the concepts of vulnerability and resilience, have emerged in the last decade or so [from Janssen and Ostrom (2006) and others]. The origins of these less reductionist conceptual and analytical paradigms are to be found in hazards research and ecology, respectively, and were introduced into climate change research in the late 1990s (vulnerability) and early 2000s (resilience). They offer ways of exploring sensitivities of socioecological systems to climate perturbations—and other environmental and social stresses—without being dependent upon the predictive claims of climate modeling. Although they have gained some visibility in recent climate change research, because vulnerability and resilience approaches to understanding climate-society relationships are less dependent on model-based climate projections, they have been slow to overturn the standard IPCC climate impact methodology.

The combination of these historical developments—the rise of a powerful epistemic community of climate modelers, the asymmetrical incorporation of climate and social change into envisaged futures, and, confounding the whole enterprise, the lack of theory making around climate-society interactions—has allowed a form of climate reductionism to dominate contemporary analysis and thinking about the future. Although it is clear to many social scientists that "the impact of any climatic event depends on the local ecological setting and the organisational complexity, scale, ideology, technology and social values of the local population," current intellectual endeavors in this area unduly privilege climate as the chief determinant of humanity's putative social futures. Quantitative climate predictions for the 2050s, 2080s, or even further ahead continue to be offered by a powerful community of climate modelers, most recently at very high spatial and temporal resolutions. . . . Yet the "complexity, scale, ideology, and social values" of future local populations and communities are for the most part ignored or assumed to be static. . . . Quantified—and often unconditional—predictions of future climate change impacts therefore abound, such knowledge claims drawing power from the epistemic muscle of climate and Earth system models. . . .

And so the future is reduced to climate. By stripping the future of much of its social, [265] cultural, or political dynamism, climate reductionism renders the future free of visions, ideologies, and values. The future thus becomes overdetermined. Yet the future is of course very far from being an ideology-free zone. It is precisely the most important territory over which battles of beliefs, ideologies, and social values have to be fought. And it is these imagined and fought-over visions of the future that—in many indeterminate ways—will shape the impacts of anthropogenic climate change as much as will changes in physical climate alone.

Putting Society Back into the Future

Climate reductionism—a form of neo-environmental determinism—offers a methodology for providing simple answers to complex questions about the relationship between climate, society, and the future. In its crudest form it asserts that if social change is unpredictable and climate change predictable then the future can be made known by elevating climate as the primary driver of change. But such reductionism downgrades human agency and constrains the human imagination.

So, looking back, Jared Diamond claims that "history followed different courses for different peoples because of peoples' environments" (Diamond 1997: 25), while looking forward James Lovelock fears that "despite all our efforts to retreat sustainably, we may be unable to prevent a global decline into a chaotic world ruled by brutal war lords on a devastated Earth" (Lovelock 2006: 198). Although offering accounts of the past and the future that are more popular than academic, both Diamond and Lovelock adopt inadequate and impoverished reductionist frameworks for understanding the past and envisioning the future. Many of the statements concerning the impacts of future climate change emerging from the more analytical research community suffer from the same limitations. The consequence of such reductionism is expressed clearly in Karl Popper's attack from a generation ago on historicism and its deterministic roots: "Every vision of historicism expresses the feeling of being swept into the future by irresistible forces" (Popper, 1957: 160). While Popper, writing in a different era, had historical materialism and the enemies of an open society in mind, his reasoning well applies to climate change today.

The allure of determinist thinking is that it offers the appearance of "naturalistic" explanations . . . or accounts of the future that evacuate it of human agency (as I have contended is the case with climate change today). In contrast to earlier climate determinisms, which flowered in the ascendant and optimistic imperial cultures of classical Greece and of imperialist Europe and a youthful United States, I suggest that the climate reductionism I have described here is nurtured by elements of a Western cultural pessimism that promote the pathologies of vulnerability, fatalism, and fear. It is these dimensions of the contemporary cultural mood that provide the milieu within which this particular form of environmental determinism has re-emerged. By handing the future over to inexorable nonhuman powers, [266] climate reductionism offers a rationalization, even if a poor one, of the West's loss of confidence in the future.

. . . Climate reductionism is a limited and deficient methodology for accessing the future. In his poetic essay "The End of the World," environmental historian Stephen Pyne offers an insight into similar reductionist limitations with regard to the past:

> Reductionism is good for extracting resources and for creating instruments, medicines, gadgets; but it does not—cannot—tell us how to use them or when or why. It cannot convey meaning because meaning requires contrast, connections, context. . . . [Reductionism] cannot tell us what we need to know in order to write genuine history, even when that history involves nature. (Pyne 2007: 650)

If reductionism is a limited form of reasoning for interpreting the past, then climate reductionism is even more inadequate with regard to telling the future. The epistemological pathways offered by climate models and their derived analyses are only one way of believing what the future may hold. They have validity, and they have relevance. But to compensate for the epistemological slippage I have described . . . it is necessary to balance these reductionist pathways to knowing the future with other ways of envisioning the future.

The "contrast, connections, and context" to which Pyne refers must be created by putting society back into the future. Since it is at least possible—if not indeed likely—that human creativity, imagination, and ingenuity will create radically different social, cultural, and political worlds in the future than exist today, greater effort should be made to represent these possibilities in any analysis about the significance of future climate change. Some of these futures may be better; some may be worse. But they will not be determined by climate, certainly not by climate alone, and these worlds will condition—perhaps remarkably, certainly unexpectedly—the consequences of climate change.

Further Reading

Dell, Melissa, *et al.* 2008. "Climate shocks and economic growth" (working paper no. 14132), Cambridge, Mass: National Bureau of Economic Research.

Diamond, Jared 1997. *Guns, germs and steel: The fates of human societies.* New York: Norton.

Edwards, Paul N. 2001. "Representing the Global Atmosphere," in Miller, C., Edwards, P. (eds.) *Changing the Atmosphere,* 31–65.

Lovelock, James 2006. *Revenge of Gaia*. London: Allen Lane.

Miller, Clark. 2004. "Climate Science and the Making of a Global Political Order," in Jasanoff, Sheila (ed.). *States of knowledge: The co-production of science and the social order*, London & New York: Routledge: 46–66.

Myers, Norman. 1993. "Environmental refugees in a globally warmed world," *Bioscience* 43: 752–61.

Popper, Karl. 1957. *The poverty of historicism*. London: Routledge & Kegan Paul.

Sluyter, Andrew. 2003. "Neo-environmental determinism, intellectual damage control, and nature/society science," *Antipode*, 35(4): 813–17.

Zhang, David D., *et al.*, 2007. "Climate change and war frequency in eastern China over the last millennium." *Human ecology* 35: 403–14.

Commentary

Mike Hulme, "Reducing the Future to Climate" (2011)

LIBBY ROBIN, SVERKER SÖRLIN, AND PAUL WARDE

Why have we chosen to end this collection with this salutary paper about "reducing the future to climate"? Because global change science is a relatively new disciplinary area, it can be easily misunderstood. There are many traditional disciplines, including geography and history, as Hulme notes, that might have influenced the Intergovernmental Panel on Climate Change (IPCC). But the actual trajectory of global change ideas, particularly in the past thirty years, has seen a rise in emphasis on earth system science, modeling, and meteorology. Policy makers are increasingly relying on predicted scenarios and forecasting, and it is "climate scenarios" that most often provoke media attention. Modeling and prediction are part of society's "trust in numbers." Decision makers like "hard," impersonal information (numbers are generally regarded as firmer evidence than narratives, for example). They focus on factors that lend themselves to being represented by numbers, rather than factors like ethics and moral attitudes that do not. The "human dimensions" that most easily gain traction among policy makers are increasingly frequently those that can be enumerated: the value of a landscape counted in tourist dollars; the value of impact on a life in terms of health care costs.

Within the global change community, particularly the groups that inform the work of IPCC, there has been a growing awareness of bias to the "biophysical dimensions" over the "human dimensions." Indeed the term "global change science" was adopted because the biophysical scientists wanted to acknowledge and reinstate the very real *human* dimensions of climate change science. For people who work in the humanities, and in practical social situations, no such coy inclusion of "people" is needed: they tend to refer to the subject as *climate* change (because climate is the new factor for them). This is particularly true in the media.

One of the important contributions history can make to public debate is to enable the careful use of words: words that include the human dimensions from the beginning and have an awareness of context. This is very different from adding "people" later to models generated for their numerical usefulness. This does not replace numbers, but adds the additional capacity to open up for discussion the precise and appropriate purposes to which we can or should put numbers, and highlighting why, and for what purposes, numbers were generated. Like Mike Hulme's stimulating paper, this book has reflected on the language, literature, and historical context of global change science. The "human dimensions" are implicated in the very ideas and concepts that frame thinking about climate change, globalization, and prediction, a point well made in this paper. Taken together, the commentaries in this book analyze the way the "climate change experts" have established their authority—and have shown how this has developed historically.

Whatever the scientific or humanistic training of people involved in this field, there is a remarkable consensus that global warming *is* happening, that the activities of people (particularly those living in the developed world) have contributed to this, and that there are not enough planetary resources to support the whole population of the Earth in the lifestyles to which the first world has become accustomed. Mostly, there is agreement that the information about global warming is difficult to translate into fair and transparent action to mitigate damage, since we live in a world where power is organized nationally and locally, rather than globally. So, rather than rehearse the arguments about *whether* climate change is happening at all, or whether it is anthropogenic, this book has accepted this consensus and focused on documents that have dealt historically and in recent times with some dimensions of the question of what should be done about it. Who are the "experts"? How do we, the "global citizens," access such expertise? How can we move beyond the system science, which is a tool for describing a problem, and empower people to act? In seeking better rather than less good outcomes for the planet, including all its humanity, we need to hear many different voices, and we need clear and careful words to frame the problem.

Mike Hulme has been part of our discussions from the beginning of this project, as an attendee at workshops, and as a colleague. We cannot pretend to comment on his document independently. Rather, this commentary is designed as a conclusion to the book as a whole. Trained as a geographer, Hulme has had a distinguished career working in climate science teams. His book *Why We Disagree About Climate Change* (2009) was one of the first to try to disentangle the science from the beliefs of prominent national and global citizens, some of whom have a direct financial interest in denying climate science. Climate change is not a matter of belief: it is documented, and an overwhelming number of people who identify as "scientists" have reached consensus on this point. Most also believe that people are implicated by the changes. But democratic societies choose to debate the ethics of action, and sometimes, when there is confusion about what might constitute a suitable course of action, the response is to disbelieve the models, or to emphasize their uncertainty, rather than to consider the consequences of failing to act.

As the issue of climate change has become more debated in the media (its status is much more debated in the media than in scientific literature), the response of science has been to widen the net and, through the concept of global change, the idea that climate is changing is no longer a single issue, but rather part of a suite of issues—including food security, disease, biodiversity, and urban-ization—that are all (differently) affected by global warming. These issues are much more *social* issues, debatable, ethical, and calling on new, interdisciplin-ary understandings of people and the environment together. In this paper, Mike Hulme explores how the humanities may contribute to climate and environ-mental debates, the big issues of our time. He has been a participant-observer in a broad range of disciplinary forays into these issues, and travels well beyond his own background in geography in this paper.

In "Reducing the Future to Climate," Hulme explores the new climate de-terminism and reductionism that has been aroused by fears of climate change. He traces a historical shift from Ellsworth Huntington–style "determinism" to a new "reductionism," where the future itself is merely *adaptation to climate change*. Only climate futures matter in such scenario-making. Over the past cen-tury, Hulme argues, climate has placed limits on what is possible for humanity in strangely recurrent patterns. In the early twentieth century, Owen Lattimore quipped that we were like the Mongol hordes in the Europe of yesteryear, "er-ratic nomads, ready to start for lost horizons at the joggle of a barometer in search of suddenly vanishing pastures" (quoted in Hulme, 245). But by the 1960s and 1970s, we became increasingly aware of limits: the "Only One Earth" campaign inspired by the images of the lone blue planet beamed back from *Apollo* space expeditions gave Earthlings a new sense of the singularity of our planet. Econ-omist Kenneth Boulding and architect Richard Buckminster Fuller both con-ceived of the planet as "Spaceship Earth," a singular and finite human habitat in a hostile universe (Höhler 2008, 2010; Anker 2010). But since this time, there have been fewer imaginative global concepts, and more scenarios of possible futures, many of which are grim. In the twenty-first century, "the hegemony exercised by the predictive natural sciences over contingent, imaginative, and humanistic ac-counts of social life and visions of the future . . . [has lent] disproportionate power in political and social discourse to model-based descriptions of putative future climates" (Hulme 2011, 245). Our "trust in numbers" is perhaps overwhelming our capacity for *human* relationships with the future.

This paper scrutinizes the idea that what people do is "caused" by climate: this determinist fallacy has a long and now discredited history, particularly be-cause of how it was applied to racial theories in the 1920s and 1930s. It also cri-tiques the idea that climate futures should become the only futures we discuss, as we deal with the issue of global warming. Hulme is careful to differentiate between thinkers concerned with environmental determinism (or patterns of cli-mate in history), the sort of work done by Ellsworth Huntington, and those who argue for particular actions for the future on the basis of predictions about the fates of people under changing climate regimes or scenarios.

The idea of determinism—of "if this, then that"—is attractive in its simplicity but often dangerous in its application, and Hulme's paper is one long argument for maintaining the fullest possible context for all understandings of what future change might mean. Global changes work very differently at different local scales, and global ideas need to be nuanced by local understandings before they will be meaningful to people. We are reminded that all politics, in the end, are local. If understandings are not meaningful to people, useful action is unlikely. Most people live "somewhere," no one lives "everywhere," yet planetary predictions are typically averages over the whole planet. Increasingly we are becoming aware that global change affects the poles more sharply than the tropics, and the oceans differently from the land masses. But there are differences in local impacts, too. For example, if the South Asia Monsoon flips to a dry phase because of global climate shifts (and historically it has flipped from wet to dry in a single year), over a billion people will be caught in the ensuing famines.

Hulme quotes geographer Gordon Manley, who said in 1958: "It is an opportune moment to be reminded that man is still subject to a variety of constraints that may yet be imposed by Nature" (Hulme 2011, 253). Yet the 1950s were, in fact, at the very end of the era of determinism. Manley was on the threshold of the new era of reductionism. A new integrated systems science was already under construction when he wrote, but since the 1960s, enhanced by the revolution in information technology, it has focused increasingly on "drivers." No longer are complex systems fully "determined"; rather, they are "reduced" to key entities or relationships that explain as much of the variance as possible (to put it mathematically). In climate change studies, reductionism demands that climate be considered a *driver* of past, present, and future system behavior and response. Thus, in the mid-century, environmental determinism blurred into climate reductionism, and generated what Hulme calls an understanding that distorted and overemphasized the "causative role of climate in shaping the future prospects of society" (Hulme 2011, 255).

We would argue that the current prominence of climate was made possible by the earlier emergence of "the environment." It was this act of integration that allowed climate to be inserted into models as a "driver," with climate scientists able to deliver powerful and relatively consistent results modeled at a global scale. This expertise had been building over time, bringing together an increasingly interconnected set of "environmental sciences," and eventually climate, which in its modern form was a relative latecomer to the "environmental problem catalogue." Arguably, climate had been much more important as a general explanatory factor in the eighteenth to early twentieth centuries (see Huntington in Part 3). In the 1940s, one might have expected that population would be settled on as the "driver" of environmental change as indeed it was for many authors (e.g., Vogt pp. 187–190), being the most advanced subject in terms of quantified global predictions. Yet population now plays a much reduced role as an explanatory "attractor," ordering other environmental impacts as the outcomes of its own dynamics. This is not because our capacity to predict population change has lessened, but for political and ethical reasons, responses to national population policies, local studies that showed relationships between population and environmental impact to be more complex than expected, and also the fact that the most alarmist predictions turned out to be wrong—usually undershoots, in fact, but without the dismal consequences expected.

Hulme explores why certain kinds of climate models have proven so compelling in controlling the discourse about predicting the future. He identifies three factors:

1. the dearth of theory about how climates and societies interact.
2. the knowledge contexts of the communities doing the modeling (beginning with meteorologists and atmospheric scientists, and later adding oceanographers, atmospheric chemists, and biologists).
3. the asymmetry between the way climate and society were "envisioned" as factors driving change.

Terms like "social-ecological systems" are commonly used, as if they are already integrated entities, thereby getting around the problem of actually *theorizing* the interaction between nature and culture. They are too often statements of what ought to be rather than descriptions of processes we fully grasp. Yet the governance of people and environments is still very separate in most cultures, and the mismatch between the apparently integrated abstract system and the decidedly dis-integrated policy context makes it hard to apply the science to society, to act on information.

Nevertheless, we do not think that the social sciences and humanities have ignored these problems. There is a large literature and body of theory on how communities and households develop strategies to deal with unpredictable climates and environmental risk. An issue here is scale, and whether this was the "relevant" kind of work to be recognized by modelers and institutions operating on a global scale. In fact, the history of environmental prediction shows global science and its precursors appealing to humanity to be different and better very frequently, but with little notion of how this translates between individual acts of will to be less greedy or prolific, and the organization of socio-ecological relations from the household up to the globe. Disciplinary trajectories often leave vacant the middle ground between very localized experiences, which have been extensively researched by the social sciences and development studies, and global processes, which tend to focus on large systems like oceans and the atmosphere. Thus the local becomes positioned as the victim of irresistible forces, merely reactive. When global theorizing is detached from local evidence, there is little possibility for local agency in its scenarios.

This space between the biophysical world and the socio-economic world is where historians, geographers, and others can contribute more—with empirical accounts of spatial and temporal change in the way people and nature (resources) interact. This is by no means new territory for these subjects, but they have to learn how to engage with narratives and scales that work better as part of a collective dialogue. We argue, in fact, that this plays to a strength of these disciplines: their capacity not just to "scale up" from the local to global (as with population studies, or ecology), but integrate different scales in narrative and analysis, a technique at the heart of disciplines that embrace both psychological explanations of individual motivation, and the grand forces of migrating peoples, armies, or political and cultural movements. Geographer Mike Hulme and historian Stephen J. Pyne both seek more emphasis on meaning—on "contrast, connections, context"—a space for the unpredictable human imagination and ingenuity—in facing the next millennium. Climate is certainly changing, but it is not the only thing that will determine the future path of humanity.

Select Bibliography

Adams, W. N. 1996. *Future nature: A vision for conservation*. London: Earthscan.

Adelman, M. A. 1995. *The genie out of the bottle: World oil since 1970*. Cambridge, MA: MIT Press.

Adelson, G., J. Engell, B. Ranalli, and K. Van Anglen (eds.). 2008. *Environment: An interdisciplinary anthology*. New Haven: Yale University Press.

Agarwal, A., and S. Narain (eds.). 1985. *India: The state of the environment 1984–85: A citizen's report*. New Delhi: Centre for Science and Environment.

Andersson, G. 2000. *Kraft och kultur: Kol, petroleum, vattenkraft och vind i människans tjänst*, Geber: Stockholm.

Andersson, J. 2006. "Choosing futures: Alva Myrdal and the construction of Swedish futures studies, 1967–1972." *International Review of Social History* 51:277–295.

Anker, P. 2001. *Imperial ecology: Environmental order in the British empire, 1895–1945*, Cambridge, MA: Harvard University Press.

Anker, P. 2010. *From Bauhaus to Ecohouse: A history of ecological design*. Baton Rouge: Louisiana State University Press.

Arrhenius, S. 1896. "On the influence of carbonic acid in the air upon the temperature of the ground." *Philosophical Magazine and the Journal of Science*, series 5, 41 (April): 237–276.

Bailes, K. E. 1990. *Science and Russian culture in an age of revolutions: V. I. Vernadsky and his scientific school, 1863–1945*. Bloomington: Indiana University Press.

Bailey, R. 2000. "Earth Day, then and now." *Reason* (May). http://reason.com/archives; original.

Barney, Gerald O. 1980. "Preface." In Council of Environmental Quality [Gerald O. Barney (ed.)], *The Global 2000 report to the President of the United States, entering the 21st century*, volume 1: *The summary report*. Special edition with the environment projections and the government's global model, vii–xvii. New York: Pergamon Press.

Barrow, Mark V., Jr. 2009. *Nature's ghosts: Confronting extinction from the age of Jefferson to the age of ecology*. Chicago: University of Chicago Press.

Bashford, A. 2007. "Nation, empire, globe: The spaces of population debate in the interwar years." *Comparative Studies in Society and History* 49(1):170–201.

Bashford, A. 2009. "Energy and population: Global policy in the mid twentieth century." Unpublished paper presented at the workshop "Expertise for the Future: Histories of Predicting Environmental Change." Center for History and Economics & Harvard Center for Sustainability, Harvard University, 17 November 2009.

Bashford, A. 2013. *Geopolitics and the world population problem*. New York: Columbia University Press.

Bashford, A., and C. Strange. 2008. *Griffith Taylor: Visionary, environmentalist, explorer*, Canberra and Toronto: National Library of Australia/University of Toronto Press.

Beck, U. 1992. *Risk society: Towards a New Modernity* (trans. Mark Ritter). London: Sage.

Beck, U. 1997. "Global risk politics." In M. Jacobs (ed.), *Greening the millennium: The new politics of the environment*. Oxford: Wiley, 18–33.

Bell, W. 1997. *Foundations of futures studies*. New Brunswick, NJ: Transaction.

Beveridge, W. H. 1920. "British exports and the barometer." *The Economic Journal* 30: 13–25.

Beveridge, W. H. 1922. "Wheat prices and rainfall in western Europe." *Journal of the Royal Statistical Society* 85(3):412–475.

Blanchard, E. T. Vieille. 2007. "Croissance ou stabilité? L'entreprise du Club de Rome et le débat autour des modèles." In A. Dahan (ed.), *Les modèles du future*. Paris: La Découverte: 19–43.

Bocking, S. 1997. *Ecologists and environmental politics: A history of contemporary ecology*. New Haven: Yale University Press.

Bocking, S. 2004. *Nature's experts: Science, politics, and the environment*. New Brunswick NJ: Rutgers University Press.

Bohn, M. 2011. Concentrating on CO_2: The Scandinavian and Arctic measurements. *Osiris* 26:165–179.

Boia, L. 2005. *The weather in the imagination*. Chicago: University of Chicago Press.

Bonner, R. 1993. *At the hand of man: Peril and hope for Africa's wildlife*. New York: Alfred A. Knopf.

Borgström, G. (1953) 1965. *The hungry planet: The modern world at the edge of famine*. Originally published in Swedish in 1953. New York: MacMillan.

Borgström, G. 1969. *Too many: A study of the Earth's biological limitations*. New York: MacMillan.

Boulding, K. E. 1964. *The meaning of the 20th century*. New York: Harper and Row.

Brand, Stewart. 1968–1972, and intermittently thereafter. *The Whole Earth Catalog*. Menlo Park, CA: Portola Institute.

Bravo, M. T., and S. Sörlin (eds.). 2002. *Narrating the Arctic: A cultural history of Nordic scientific practices*. Cambridge, MA: Science History.

Bretherton, F., et al. 1988. *Earth system science: A closer view*. Washington, DC: NASA.

Broecker, W. S. 1987. "Unpleasant surprises in the greenhouse?" *Nature* 328:123–126.

Broecker, W. S., and R. Kunzig. 2008. *Fixing climate*. New York: Hill and Wang.

Brown, H. 1954. *The challenge of man's future*. New York: Viking.

Brown, J. H., and D. F. Sax. 2004. "An essay on some topics concerning invasive species." *Austral Ecology* 29:530–536.

Bruckmann, G. 1976. *Latin American world model*. Proceedings of the second IIASA symposium on global modelling, 7–10 October 1974.

Buckminster-Fuller, R. 1969. *Operating manual for spaceship Earth*. Carbondale: Southern Illinois University Press.

Buell, F. 2004. *From apocalypse to way of life*. New York: Routledge.

Callendar, G. S. 1938. "The artificial production of carbon dioxide and its influence on temperature." *Quarterly Journal of the Royal Meteorological Society* 64, no. 275 (April): 223–240.

Callicott, J. B. 1991. "The wilderness idea revisited: The sustainable development alternative." *The Environmental Professional* 13(2):235–247.

Carlowitz, H. C. von. 1713. *Sylvicultura oeconomica*. Leipzig: Johann Friedrich Braun.

Carr-Saunders, A. M. 1922. *The population problem: A study in human evolution*. Oxford: Clarendon.

Carson, R. 1951. *The sea around us*. New York: Oxford University Press.

Carson, R. 1962. *Silent spring*. New York: Ballantine.

Carter, T. R., M. L. Parry, H. Harasawa, and S. Nishioka. 1994. *IPCC technical guidelines for assessing climate change impacts and adaptations*. London: Tsukuba.

Chandler, J. 2011. *Feeling the heat*. Carlton: Melbourne University Press.

Chapman, R. N. 1928. "The quantitative analysis of environmental factors." *Ecology* 9:111–122.

Charney, J., et al. 1979. *Carbon dioxide and climate: A scientific assessment*. Washington, DC: National Academy of Sciences.

Christian, David. 2012. "Big history for the era of climate change." *Solutions* 3, no. 2 (March). http://www.thesolutionsjournal.com/node/1066 .

Clements, F. E. 1916. *Plant succession*. Publication 242. Washington, DC: Carnegie Institution.

Cohen, B. R. 2009. *Notes from the ground: Science, soil and society in the American countryside*. New Haven: Yale University Press.

Coleman, D. C. 2010. *Big ecology: The emergence of ecosystem science*. Berkeley: University of California Press.

Collins, H., and Evans, R. 2007. *Rethinking expertise*. Chicago: Chicago University Press.

Connelly, M. 2006. "To inherit the earth: Imagining world population, from the yellow peril to the population bomb." *Journal of Global History* 1:299–319.

Connelly, M. 2008. *Fatal misconception: The struggle to control world population*. Cambridge, MA: Belknap Press.

Connolly, William E. 2011. *A world of becoming*. Durham, NC: Duke University Press.

Costanza, R., et al. 1997. "The value of the world's ecosystem services and natural capital." *Nature* 387(6630):253–260. doi:10.1038/387253a0

Coulson, J., D. Whitfield, and A. Preston (eds.). 2003. *Keeping things whole: Readings in environmental science*. Chicago: Great Books Foundation.

Council of Environmental Quality [Gerald O. Barney (ed.)]. 1980. *The Global 2000 report to the President of the United States: Entering the 21st century*. 3 vols. New York: Pergamon Press.

Crawford, E. 1996. *Arrhenius: From ionic theory to the greenhouse effect*. Cambridge, MA: Science History.

Crosby, Alfred W. (1986) 2004. *Ecological imperialism: The biological expansion of Europe, 900–1900*. Cambridge: Cambridge University Press.

Crowcroft, Peter. 1991. *Elton's ecologists: A history of the Bureau of Animal Population*. Chicago: University of Chicago Press.

Crowley, D., and Pavitt, J. (eds.). 2008. *Cold War Modern: Design, 1945–1970*. London: Victoria & Albert Museum.

Crutzen, P. J., and E. F. Stoermer. 2000. "The 'Anthropocene.'" *IGBP Newsletter* 41 (May): 17–18.

Dai, Q. 1998. *The river dragon has come! The Three Gorges Dam and the fate of China's Yangtze River and its people*. Armonk, NY: M. E. Sharpe.

Dai, Q. 1994. *Yangtze! Yangtze!* London: Earthscan. [Originally published and then banned in 1989.]

Daily, G. (ed.). 1997. *Nature's services: Societal dependence on natural ecosystems*. Washington, DC: Island Press.

Daily, G. 2001. "Ecological forecasts." *Nature* 411(6835):245. doi:10.1038/35077178

Dansgaard, Willi. 2005. *Frozen annals: Greenland ice cap research*. Copenhagen: University of Copenhagen.

Desrochers, P., and C. Hoffbauer. 2009. "The post war intellectual roots of the *Population Bomb*. Fairfield Osborn's 'Our plundered planet' and William Vogt's 'Road to survival' in retrospect." *Electronic Journal of Sustainable Development* 1(3):37–61.

Devall, B., and G. Sessions. 1985. *Deep ecology: Living as if nature mattered*. Salt Lake City: Peregrine Smith Books.

Doel, R. 2003. "Constituting the postwar earth sciences: The military's influence on the environmental sciences in the USA after 1945." *Social Studies of Science* 33(5):635–666.

Driver, F., and L. Martins (eds.). 2005. *Tropical visions in an age of empire*. Chicago: University of Chicago Press.

Dryzek, J., R. Norgaard, and D. Schlosberg (eds.). 2011. *The Oxford handbook of climate change and society*. Oxford: Oxford University Press.

East, G. 1938. *The geography behind history*. London: T. Nelson.

Edwards, P. N. 2010. *A vast machine: Computer models, climate data, and the politics of global warming*. Cambridge, MA: MIT Press.

Egan, M. 2009. *Barry Commoner and the science of survival*. Cambridge, MA: MIT Press.

Ehrlich, P. (1968) 1971. *The population bomb*, revised and expanded ed. New York: Ballantine Books.

Ehrlich, P. 1974. Interview by Mother Earth News. http://www.mnforsustain.org/ehrlich_paul_interview_1974.htm.

Eiseley, L. (1970) 1998. *The invisible pyramid*. Lincoln: University of Nebraska Press.

Ellen, R. 1982. *Environment, subsistence and system*. Cambridge: Cambridge University Press.

Elton, C. S. 1958. *The ecology of invasions by animals and plants*. London: Methuen.

Elton, C. S. 1927. *Animal ecology*. London: Sidgwick & Jackson.

Ezrahi, Y. 1980. "Utopian and pragmatic realism: the political context of scientific advice." *Minerva* 18(1):111–131.

Farnham, Timothy J. 2007. *Saving nature's legacy: The origins of the idea of biological diversity*. New Haven: Yale University Press.

Ferri, Mário Guimarães. 1974. *Ecologia, temas e problemas Brasileiros*. Coleção Reconquista Do Brasil. São Paulo, Belo Horizonte, Brasil: Editora da Universidade de São Paulo; Livraria Itatiaia Editora.

Fleming, J. R. 2005. *Historical perspectives on climate change*. New York: Oxford University Press.

Fleming, J. R. 2007. *The Callendar effect*. Boston: American Meteorological Society.

Fleming, J. R. 2010. *Fixing the sky: The checkered history of weather and climate control*. New York: Columbia University Press.

Forbes, S. A. 1887. "The lake as a microcosm." *Bulletin of the Scientific Association*, 77–87. Reprinted in *Illinois Natural History Survey Bulletin* 15(9):537–550.

Fox, S. R. 1985. *The American conservation movement: John Muir and his legacy*. Madison: University of Wisconsin Press.

Frangsmyr, T., J. L. Heilbron, and R. E. Rider (eds.). 1990. *The quantifying spirit in the eighteenth century*. Berkeley: University of California Press.

Fressoz, J.-B. 2012. *L'Apocalypse joyeuse: Une histoire du risque technologique*. Paris: Le Seuil.

Friman, E. 2002. *No limits: The 20th century discourse of economic growth*. Umeå, Sweden: Department of Historical Studies, Umeå University.

Gadgil, M., and P. R. S. Rao. 1994. "A system of positive incentives to conserve biodiversity." *Economic and Political Weekly*, 6 August.

Galbraith, J. K. 1958. "How much should a country consume?" In Henry Jarett (ed.), *Perspectives on conservation*. Baltimore, MD: Johns Hopkins Press.

Golley, F. B. 1993. *History of the ecosystem concept in ecology: More than the sum of the parts*. New Haven: Yale University Press.

Gould, S. J. 1997. *Questioning the millennium: A rationalist's guide to a precisely arbitrary countdown*. New York: Harmony Books.

Guha, R., and J. Martínez Alier. (eds.). 1997. *Varieties of environmentalism: Essays north and south*. London: Earthscan.

Guha, Ramachandra. 1997. "Radical American environmentalism and wilderness preservation: A third world critique." Chapter 5 in Ramachandra Guha and Joan Martínez Alier (eds.), *Varieties of environmentalism: Essays north and south*, 92–108, 214–16. London: Earthscan.

Hamilton, K. 1994. "Sustainability, the Hartwick rule and optimal growth." *Environmental and Resource Economics* 5:393–411.

Hays, S. P. 1959. *Conservation and the gospel of efficiency*, Cambridge, MA: Harvard University Press.

Hays, S. P. 1982. "From conservation to environment: Environmental politics in the United States since World War Two." *Environmental Review* 6:14–41.

Heymann, M. 2006. "Modelling reality: Practice, knowledge, and uncertainty in atmospheric transport simulation." *Historical Studies of the Physical and Biological Sciences* 37:49–85.

Holling, C. S. 1973. "Resilience and stability of ecological systems." *Annual Review of Ecology and Systematics* 4:1–23.

Huenneke, L., et al. 1988. "SCOPE Program on biological invasions." *Conservation Biology* 2(1):8–10.

Hulme, M. 2009. "On the origin of 'the greenhouse effect': John Tyndall's 1859 interrogation of nature." *Weather* 64(5):122–124.

Hulme, M. 2009. *Why we disagree about climate change*. Cambridge: Cambridge University Press.

Hulme, M. 2011. "Reducing the future to climate: A story of climate determinism and reductionism." *Osiris* 26:245–266.

Humboldt, A. von. 1817. "Sur les lignes isothèrmes." *Annales de chimie et de physique* 5:102–112.

Humboldt, A. von. 1848–1858. *Cosmos, personal narrative, and views of nature*. Trans. E. C. Ottee. London: H. C. Bohn.

Humboldt, A. von, and A. Bonpland. (1807) 2009. *Essay on the geography of plants*. Trans. S. Romanowski, ed. S. T. Jackson, with accompanying essays and supplementary material by S. T. Jackson and S. Romanowski. Chicago: University of Chicago Press.

Huntington, E. 1907. *The pulse of Asia: A journey in central Asia illustrating the geographic basis of history*. Boston: Houghton, Mifflin and Company.

Huntington, E. 1915. *Civilization and climate*. New Haven: Yale University Press.

Hutchinson, G. E. 1948. "Circular causal systems in ecology." *Annals of the New York Academy of Sciences* 50:221–246.

Hutchinson, G. E. 1948. "On living in the biosphere." *Scientific Monthly* 67:303.

Huzar, E. 1855. *La fin du monde par la science*, Paris: Dentu.

Huzar, E. 1857. *L'arbre de la science* [The tree of life]. Paris: Dentu.

Höhler, S. 2008. "'Spaceship Earth': Envisioning human habitats in the environmental age." *GHI Bulletin* 42:65–85.

Höhler, S. 2010. "The environment as a life support system: The case of Biosphere 2." *History and Technology* 26(1):39–58.

Inden, R. 1986. "Orientalist constructions of India." *Modern Asia Studies* 20:401–446.

IUCN (International Union for the Conservation of Nature) with United Nations Environment Programme (UNEP) and World Wildlife Fund (WWF). 1980. *World conservation strategy: Living resource conservation for sustainable development*. Gland, Switzerland: IUCN, UNEP, WWF.

James, P. 2006. *Population Malthus: His life and times*. London: Routledge.

Jankovic, V. 2006. "Change in the weather." *Bookforum*, February/March, 39–40.

Janssen, M. A., and E. Ostrom. 2006. "Resilience, vulnerability and adaptation: A cross-cutting theme of the International Human Dimensions Programme on Global Environmental Change." *Global Environmental Change* 16:237–239.

Janzen, D. 1986. "The future of tropical ecology." *Annual Review of Ecology and Systematics* 17: 304–324.

Jardine, N., J. Secord, and E. Spary (eds.). 1996. *Cultures of natural history*. Cambridge: Cambridge University Press, 287–304.

Jasanoff, S. (ed.). 2004. *States of knowledge: The co-production of science and the social order*, London: Routledge.

Jevons, H. S. 1910. *The sun's heat and trade activity*. London: King and Son.

Jevons, W. S. 1865. *The coal question: An enquiry concerning the progress of the nation, and the probably exhaustion of our coal mines*. London: Macmillan.

Jianqiang, L., and G. Cheng. 2007. "Tiger Leaping Gorge under threat." *Southern Weekly*, September. Available at http://old.fon.org.cn/content.php?aid=8747 .

Kander, A., P. Malanima, and P. Warde. 2013. *Power to the people: Energy in Europe over the last five centuries*. Princeton, NJ: Princeton University Press.

Kates, R. W., J. H. Ausubel, and M. Berberian. 1985. *SCOPE 27: Climate impact assessment*. Chichester: SCOPE.

Keeling, C. D. 1960. "The concentration and isotopic abundances of carbon dioxide in the atmosphere." *Tellus* 12:200–203.

Keeling, C. D. 1978. "The influence of Mauna Loa Observatory on the development of atmospheric CO_2 research." In J. Miller (ed.), *Mauna Loa Observatory: A 20th anniversary report*, 36–54. Boulder, CO: NOAA Environmental Research Laboratories.

Kingsland, S. 2002. "Designing nature reserves: Adapting ecology to real-world problems." *Endeavor* 26(1):9–14.

Kingsland, S. E. 2005. *The evolution of American ecology, 1890–2000*. Baltimore: Johns Hopkins University Press.

Kirk, D. 1996. "Demographic transition theory." *Population Studies* 50:361–387.

Knibbs, G. H. 1928. *The shadow of the world's future, or the population possibilities and the consequences of the present rate of increase of the earth's inhabitants*. London: Ernest Benn.

Kühl, S. 2002. *The Nazi connection: Eugenics, American racism, and German national socialism*, 2nd ed. New York: Oxford University Press.

Ladurie, E. 1972. *Times of feast, times of famine: A history of climate since the year 1000*. Trans. B. Bray. London: Allen & Unwin.

Laplace, P. S. (1814) 1951. *A philosophical essay on probabilities*, 6th ed. Trans. F. W. Truscott and F. L. Emory. New York: Dover.

Lapo, A. V. 2001. "Vladimir I. Vernadsky (1863–1945), founder of the biosphere concept." *International Microbiology* 4(1):47–49.

Latour, B. 1993. *We have never been modern*. Trans. C. Porter. New York: Harvester Wheatsheaf.

Lear, L. J. 1997. *Rachel Carson: Witness for nature*. New York: H. Holt.

Leopold, A. (1941) 1991. *The river of the mother of God and other essays by Aldo Leopold*, Susan L. Flader and J. Baird Callicott (eds.). Madison: University of Wisconsin Press.

Leopold, A. 1949. *A sand county almanac: And sketches here and there*. New York: Oxford University Press.

Levin, S. (ed.). 2012. *The Princeton guide to ecology*. Princeton, NJ: Princeton University Press.

Liebig, J. von. 1842. *Chemistry in its application to agriculture and physiology*. London: Taylor & Walton.

Lightman, B. 2007. *Victorian popularisers of science: Designing nature for new audiences* Chicago: University of Chicago Press.

Linnér, B.-O. 2003. *The return of Malthus: Environmentalism and post-war population-resource crisis*. Isle of Harris: White Horse Press.

Livingstone, D. N. 2002. "Race, space and moral climatology: Notes toward a genealogy." *Journal of Historical Geography* 28:159–180.

Long Now Foundation. 01996. Website: http://longnow.org.

Lopez, Barry (ed.). 2007. *The future of nature: Writing on a human ecology from Orion Magazine*. Minneapolis: Milkweed.

Lotka, A. J. (1925) 1956. *Elements of mathematical biology*. New York: Dover.

Lowe, I. 2012. *Bigger or better?: Australia's population debate*. Brisbane: University of Queensland Press.

Lovejoy, T. 2000. "Biological diversity." In Heather Newbold (ed.), *Life stories: World-renowned scientists reflect on their lives and the future of life on Earth*, 42–54. Berkeley: University of California Press.

Lytle, M. H. 2007. *The gentle subversive: Rachel Carson, Silent Spring, and the rise of the environmental movement*. New York: Oxford University Press.

Maas, H. 2005. *William Stanley Jevons and the making of modern economics*. Cambridge: Cambridge University Press.

Maathai, W. 1985. *The green belt movement: Sharing the approach and the experience*. Nairobi: The Greenbelt Movement (Society).

MacArthur, R. J., and E. O. Wilson. 1967. *The theory of island biogeography*. Princeton, NJ: Princeton University Press.

MacDonald, D., and K. Service (eds.). 2007. *Key topics in conservation biology*. Oxford: Blackwell.

Madureira, N. L. 2012. "The anxiety of abundance: William Stanley Jevons and coal scarcity in the nineteenth century." *Environment and History* 18(3):395–421.

Malthus, T. R. 1798. *An essay on the principle of population*. First printed for J. Johnson, in St. Paul's Church-Yard, London.

Marchetti, C. 1979. "Ten to the twelfth: A check on the Earth-carrying capacity for man." *Energy* 4:1107–1117.

Markham, S. F. 1942. *Climate and the energy of nations*. London: Milford.

Marsh, G. P. 1864. *Man and Nature: or, Physical geography as modified by human action*, London: Murray.

Marshall, A. 1890. *The principles of economics*, 8th ed. London: Macmillan.

Martin, G. J. 1973. *Ellsworth Huntington: His life and thought*. Hamden, CT: Archon Books.

McNeill, J. R. 2001. *Something new under the sun: An environmental history of the twentieth-century world*. New York: W. W. Norton.

Meadows, D. H., J. Randers, and D. L. Meadows for the Club of Rome. 1972. *The limits to growth: A report for the Club of Rome's project on the predicament of mankind*. New York: Universe Books.

Meadows, D., J. Richardson, and G. Bruckmann. 1982. *Groping in the dark: The first decade of global modeling*. Chichester: John Wiley.

Meadows, D. H., J. Randers, and D. L. Meadows. 2004. *The limits to growth: The thirty year update*. London: Routledge.

Merchant, C. (ed.). 1994. *Key concepts in critical theory: Ecology*. Atlantic Highlands, NJ: Humanities Press.

Merchant, C., and T. Paterson (eds.). 2011. *Major problems in American environmental history*, 3rd ed. Belmont, CA: Cengage Learning/Wadsworth.

Meyer, W. B. 2000. *Americans and their weather*. Oxford: Oxford University Press.

Miller, C., and P. Edwards (eds.). 2001. *Changing the atmosphere: Expert knowledge and environmental governance*. Cambridge, MA: MIT Press.

Millikan, R. A. 1930. "Alleged sins of science." *Scribners Magazine* 87(2):119–130.

Mitchell, E. 1946. *Soil and civilization*. Sydney: Angus and Robertson.

Mitchell, T. 2011. *Carbon democracy: Political power in the age of oil*. New York: Verso.

Mittman, G. 1992. *The state of nature: Ecology, community, and American social thought, 1900–1950*. Chicago: University of Chicago Press.

Moon, D. 2005. "The environmental history of the Russian steppes." *Transactions of the Royal Historical Society* 15:149–174.

Muir, C. 2010. "Feeding the world: Our great myth." *Griffith Review* 27:59–73.

Myrdal, A., chair, Royal Ministry for Foreign Affairs in cooperation with the Secretariat for Future Studies. (1972) 1974. *To choose a future: a basis for discussion and deliberations on future studies in Sweden*. Swedish version of the original published as *Att välja framtid*. Stockholm: Swedish Institute.

Naess, A. 1973. "The shadow and the deep, long-range ecology movement: A summary." *Inquiry* 6:96.

Naess, A. 1990. *Ecology, community and lifestyle*. Trans. David Rothenberg. Cambridge: Cambridge University Press.

Nash, R. 1982. *Wilderness and the American mind*, 3rd ed. New Haven: Yale University Press.

Nicolson, M. 1996. "Humboldtian plant geography after Humboldt: The link to ecology." *British Journal for the History of Science* 29:289–310.

Nixon, R. 2011. *Slow violence and the environmentalism of the poor*. Cambridge, MA: Harvard University Press.

Norgaard, R. B. 2010. Ecosystem services: From eye-opening metaphor to complexity blinder. Special issue on payments for ecosystem services. *Ecological Economics* 69(6):1219–1227.

Nyhart, L. K. 2009. *Modern nature: The rise of the biological perspective in Germany*. Chicago: University of Chicago Press.

Odum, E. P. (in collaboration with H. T. Odum). 1953. *Fundamentals of ecology*. Philadelphia: W. S. Saunders.

Oldfield, J., and D. J. B. Shaw. 2006. "V. I. Vernadsky and the noosphere concept: Russian understandings of society-nature interaction." *Geoforum* 37(1):145–154.

Ordway, S. H., Jr. 1953. *Resources and the American dream*. New York: Ronald Press.

Ordway, S. H., Jr. 1956. "Possible limits of raw material consumption." In W. L. Thomas (ed.), *Man's role in changing the face of the earth*, 987–1009. Chicago: University of Chicago Press.

Osborn, F. 1948. *Our plundered planet*. Boston: Little, Brown & Co.

Osborn, F. 1953. *The limits of the Earth*. Boston: Little, Brown & Co.

Owens, S. 2011. "The role of the UK Royal commission on environmental pollution, 1970–2010." In P. Weingart and J. Lentsch (eds.), *The politics of scientific advice*. Cambridge: Cambridge University Press.

Pálsson, Gísli, Sverker Sörlin, Bronislaw Szerzynski, et al. 2012. "Reconceptualizing the 'Anthropos' in the Anthropocene: Integrating the social sciences and humanities in global environmental change research." *Environmental Science and Policy* [online 27 December 2012].

Parry, M. L., and T. R. Carter. 1998. *Climate impact and adaptation assessment: The IPCC method*. London: Earthscan.

Peet, R. 1985. "The social origins of environmental determinism." *Annals of the Association of American Geographers* 75:309–333.

Perkins, J. H., 1997. *Geopolitics and the green revolution: Wheat, genes, and the cold war*. New York: Oxford University Press.

Petersen, W. 1999. *Malthus*, 2nd ed. London: Heinemann.

Petit, J. R., et al. 1999. "Climate and atmospheric history." *Nature* 399:429–436.

Phillips, John. 1934. "Succession, development, the climax and the complex organism: An analysis of concepts." *International Journal of Ecology* 22:554–571.

Political and Economic Planning. 1955. *World population and resources*. London: PEP.

Poovey, M. 1998. *A history of the modern fact: Problems of knowledge in the sciences of wealth and society*. Chicago: University of Chicago Press.

Porter, T. M. 1995. *Trust in numbers: The pursuit of objectivity in science and public life*. Princeton, NJ: Princeton University Press.

Putnam, P. C. 1953. *Energy in the future*. New York: van Nostrand.

Pyne, S. J. 2007. "The end of the world." *Environmental History* 12:649–653.

Radkau, Joachim. 2011. *Wood: A history*. Cambridge: Polity.

Ratcliffe, F. 1938. *Flying fox and drifting sand*. London: Chatto and Windus.

Real, L. A., and J. A. Brown. 1991. *Foundations of ecology: Classic papers with commentary*. Chicago: University of Chicago Press.

Rennell, J. 1832. *Currents of the Atlantic Ocean, and of those which prevail between the Indian Ocean and the Atlantic*. London: J. G. and F. Rivington.

Robertson, T. B. 1959. "A flexible growth function for empirical use." *Journal of Experimental Botany* 10:290–300.

Robertson, T. 2012. "Total war and the total environment: Fairfield Osborn, William Vogt, and the birth of global ecology." *Environmental History* 17:336–364.

Robertson, T. 2012. *The Malthusian moment: Global population growth and the birth of American environmentalism*. New Brunswick, NJ: Rutgers University Press.

Robin, L. 1998. *Defending the Little Desert*. Melbourne: Melbourne University Press.

Robin, L. 2011. "The rise of the idea of biodiversity: crises, responses and expertise." *Quaderni* (Journal of l'Institut des Sciences Humaines et Sociales du CNRS); Special Issue: *Les promesses de la biodiversité* 76(1):25–38.

Robin, L., and W. Steffen. 2007. "History for the Anthropocene." *History Compass* 5(5):1694–1719.

Rockström, J., et al. 2009. "A safe operating space for humanity." *Nature* 461:472–475.

Rossby, C.-G. (1957) 1959. "Current problems in meteorology." In Bert Bolin (ed.), *The atmosphere and the sea in motion: Scientific contributions to the Rossby Memorial Volume*, 9–50. Original Swedish version published in 1957. New York: Rockefeller University Press.

Rostow, W. W. 1960. *The stages of economic growth: A non-communist manifesto.* Cambridge: Cambridge University Press.

Rothschild, E. 1994. "Population and common security." In Laurie Ann Mazur (ed.), *Beyond the numbers: A reader on population, consumption, and the environment.* Washington, DC: Island Press.

Rozwadowski, H. M. 2002. *The sea knows no boundaries: A century of marine science under ICES.* Seattle: University of Washington Press.

Ruddiman, W. F. 2003. "The anthropogenic greenhouse era began thousands of years ago." *Climatic Change* 61(3):261–293.

Sachs, Aaron. 2006. *The Humboldt current: Nineteenth-century exploration and the roots of American environmentalism.* New York: Viking.

Sale, K. 1985. *Dwellers in the land: The bioregional vision.* San Francisco, CA: Sierra Club Books.

Sayre, N. F. 2008. "The genesis, history and limits of carrying capacity." *Annals of the Association of American Geographers* 98:120–134.

Sears, P. B. (1935) 1949. "Deserts in retreat." In *Deserts on the march*, 2nd ed., 165–177. London: Routledge and Kegan Paul.

Shabbas, M. 1990. *A world ruled by number: William Stanley Jevons and the rise of mathematical economics.* Princeton, NJ: Princeton University Press.

Shantz, H. L. 1941. *Conservation of renewable natural resources.* Philadelphia: University of Pennsylvania Press.

Sherratt, T., T. Griffiths, and L. Robin (eds.). 2005. *A change in the weather: Climate and culture in Australia.* Canberra: National Museum of Australia Press.

Shiva, V. 1989. *The violence of the Green Revolution: Ecological degradation and political conflict in Punjab.* Dehra Dun: Research Foundation for Science and Ecology; Natraj Publishers.

Shiva, V. 2000. *Stolen harvest: The hijacking of the global food supply.* Cambridge, MA: South End Press.

Shiva, V. 2005. *Earth democracy: Justice, sustainability, and peace.* Cambridge, MA: South End Press.

Sideris, L. H., and K. D. Moore (eds.). 2008. *Rachel Carson: Legacy and challenge.* Albany: State University of New York Press.

Simberloff, Daniel. 2012. "Charles Elton: Pioneer conservation biologist." *Environment and History* 18(2):183–202.

Simon, J. L., and H. Kahn (eds.). 1984. *The resourceful Earth.* Oxford: Blackwell.

Smil, V. 2005. "Limits to growth: A review essay." *Population and Development Review* 31(1):157–164.

Smil, V. 2008. *Energy in nature and society: General energetics of complex systems.* Cambridge, MA: MIT Press.

Soulé, M. E. 1985. "What is conservation biology?" *Bioscience* 35(11):727–734.

Spate, O. H. K. 1952. "Toynbee and Huntington: A study in determinism." *The Geographical Journal* 118:406–424.

Spencer, H. (1897) 1978. *The principles of ethics*. Indianapolis: Liberty Classics.

Stefansson, V. 1922. *The northward course of empire*. New York: Harcourt, Brace and Company.

Steffen, Alex (ed.). 2006. *Worldchanging: A user's guide for the 21st century*. New York: Abrams.

Steffen, W. 2011. "Global change." In J. Dryzek, R. Norgaard, and D. Schlosberg (eds.), *The Oxford handbook of climate change and society*, 22. Oxford: Oxford University Press.

Steffen, W., et al. 2004. *Global change and the earth system: A planet under pressure*. The IGBP Book Series. Berlin: Springer-Verlag.

Steffen, W., P. J. Crutzen, and J. R. McNeill. 2007. "The Anthropocene: Are humans now overwhelming the great forces of nature?" *Ambio* 36(8):614–621.

Steffen, W., J. Grinevald, P. Crutzen, and J. McNeill. 2011. "The Anthropocene: Conceptual and historical perspectives." *Philosophical Transactions of the Royal Society A* 369:842–867.

Stern, N. H. 2006. *The economics of climate change*. Cambridge: Cambridge University Press.

Stoll. S. 2002. *Larding the lean earth: Soil and society in nineteenth-century America*. New York: Hill and Wang.

Strange, C. 2010. "The personality of environmental prediction: Griffith Taylor as 'latter-day prophet.'" *Historical Records of Australian Science* 21(2):133–148.

Sörlin, S. 2002. "Rituals and resources of natural history: The north and the Arctic in Swedish scientific nationalism." In Michael T. Bravo and Sverker Sörlin (eds.), *Narrating the Arctic: A cultural history of Nordic scientific practices*. Cambridge, MA: Science History.

Sörlin, S., and P. Warde (eds.). 2009. *Nature's end: History and the environment*. London: Palgrave MacMillan.

Tansley, A. G. 1920. "The classification of vegetation and the concept of development." *Journal of Ecology* 8:118–144.

Tansley, A. G. 1922. *The new psychology and its relation to life*. London: Allen & Unwin.

Tansley, A. 1935. "The use and abuse of vegetational concepts and terms." *Ecology* 16:284–307.

Taylor, T. G. 1920. "Nature *versus* the Australian." *Science and Industry* 2(8):1–14.

Thomas, W. L. (ed.), 1956. *Man's role in changing the face of the Earth*. Chicago: University of Chicago Press.

Thompson, W. S. 1929. "Population." *American Journal of Sociology* 34:959–975.

Tyndall, J. 1859. "On the transmission of heat." *Proceedings of the Royal Institution* (London) 3:155–158.

Vogt, W. 1948. *The road to survival*. New York: William Sloane.

Veblen, T. 1899. *The theory of the leisure class: An economic study of institutions*. New York: Macmillan.

Vernadsky, V. 1945. "The biosphere and the noösphere." *American Science* 33:1–12.

Vernadsky, V. (1926/1929) 1998. *La Biosphère* [The Biosphere]. Russian original 1926, French translation 1929; Paris: Félix Alcan. English translation 1998; David B. Langmuir, Mark A. S. McMenamin (eds.). New York: Copernicus.

Walker, B., and D. Salt. 2006. *Resilience thinking: Sustaining ecosystems and people in a changing world*. Washington, DC: Island Press.

Walker, G. T. 1930. "Seasonal foreshadowing." *Quarterly Journal of the Royal Meteorological Society* 56(237):359–364.

Walls, Laura Dassow. 2009. *The passage to cosmos: Alexander von Humboldt and the shaping of America*. Chicago: University of Chicago Press.

Ward, B., and R. Dubos. 1972. *Only one Earth: The care and maintenance of a small planet*. New York: Norton.

Warde, Paul. 2006. "The fear of wood shortage and the reality of the woodlands in Europe, c. 1450–1850." *History Workshop Journal* 62:29–57.

Warde, Paul. 2011. "The invention of sustainability." *Journal of Modern Intellectual History* 8(1):153–170.

WCED (World Commission on Environment and Development) [G. H. Brundtland (ed.)]. 1987. *Our common future*. Oxford: Oxford University Press.

Weart, S. R. 2003. *The discovery of global warming*. Cambridge, MA: Harvard University Press.

White, G. 1789. *The natural history of Selborne*. London: T. Bensley for B. White.

Whyte, I. 2008. *World without end? Environmental disaster and the collapse of empires*, London: I. B. Tauris.

Widmalm, S. 1992. "A commerce of letters: Geodetic networks in the 19th century." *Science Studies* 5(2):43–58.

Williams, M. 2003. *Deforesting the earth: From prehistory to global crisis*. Chicago: Chicago University Press.

Winch, D. (ed.). 1992. *Malthus: An essay on the principle of population*. Cambridge: Cambridge University Press.

Worster, D. (1977) 1991. *Nature's economy: The roots of ecology*, rev. 2nd ed. Cambridge: Cambridge University Press.

Worster, D. 1979. *Dust bowl: The southern plains in the 1930s*. New York: Oxford University Press.

Acknowledgments

We have developed the commentaries and frameworks through a series of workshops in different places. We are grateful to the sponsors of these workshops. The first was held in Norwich at the University of East Anglia (sponsored by the Centre for History and Economics, Cambridge); the second at Harvard (sponsored by the Center for History and Economics, Harvard, and the Harvard University Center for the Environment); the third at the Australian National University, Canberra; and the final one at the Stockholm Resilience Center (supported by the Integrated History and future Of People on Earth IHOPE Programme). We owe much to all the presenters and commentators who participated. The concepts and frameworks for the book were further improved by public forum discussions at several international conferences, including the American Society of Environmental History (Portland) and the European Society of Environmental History (Turku).

Beyond those who have contributed commentaries, we are also indebted to our "virtual community" of readers, including five robust anonymous readers commissioned by Yale University Press, who made excellent suggestions. Valuable comments on the final version also came via Yale from Mary Foskett, Nancy Jacobs, Thomas Lovejoy, Meg Lowman, Edward Melillo, Harriet Ritvo, Jay Turner, and Julianne Lutz Warren. We greatly appreciate the careful readings each offered our project.

For financial and other support we are grateful to: Centre for Environmental History, Australian National University; Centre for History and Economics (Cambridge and Harvard); Division for the History of Science, Technology and Environment, Royal Institute of Technology Stockholm, including (KTH), Environmental Humanities Laboratory; Ely Institute; Fenner School of Environment and Society, Australian National University; Harvard University Center for the Environment; IHOPE Programme (now at Uppsala University); Leverhulme Trust, UK; National Museum of Australia; Stockholm Resilience Center; and the University of East Anglia.

For indexing, technical support and research assistance, we thank Bernadette Hince, Susanna Lidström, and Cameron Muir.

Libby Robin, Sverker Sörlin, and Paul Warde
May 2013

Commentators

BARROW, Mark V., Jr.
Mark V. Barrow Jr. is Professor and Chair of History and an affiliated faculty member with Science and Technology in Society at Virginia Tech. The author of two prize-winning books, *A Passion for Birds: American Ornithology after Audubon* (1998) and *Nature's Ghosts: Confronting Extinction from the Age of Jefferson to the Age of Ecology* (2009), he is currently researching the American alligator, an apex predator that offers a telling window onto our complex perceptions of the natural world.

BASHFORD, Alison
Alison Bashford is Professor of History at the University of Sydney, and has been elected Vere Harmsworth Professor of Imperial and Naval History at the University of Cambridge. She was Chair of Australian Studies at Harvard University in 2009–2010, where she taught in the History of Science Department. Her most recent book is *Global Population: History, Geopolitics and Life on Earth* (2013).

BOCKING, Stephen
Stephen Bocking is Professor of Environmental History and Policy, and Chair of Environmental Studies, Trent University, Canada. His research examines historically the interface between science and environmental policy. His current projects include the environmental history of Arctic science, the transnational science and politics of salmon aquaculture, and local and scientific knowledge in biodiversity conservation. His publications include *Ecologists and Environmental Politics: A History of Contemporary Ecology* (Yale University Press, 1997), and *Nature's Experts: Science, Politics, and the Environment* (Rutgers University Press, 2004).

BOHN, Maria
Maria Bohn is a Ph.D. scholar at the Division of History of Science, Technology and Environment at the Royal Institute of Technology in Stockholm. She is writing a thesis about historical aspects of carbon dioxide measurements. Her empirical focus is the Institute of Meteorology at Stockholm Högskola in the 1950s. She has published on the history of Scandinavian and Arctic carbon dioxide measurements in *Osiris*.

CHU, Pey-Yi

Pey-Yi Chu is Assistant Professor of History at Pomona College. Previously, she was a post-doctoral fellow at the Davis Center for Russian and Eurasian Studies at Harvard University. A historian of Russia and the Soviet Union, she is particularly interested in connections between the history of science and environmental history. She is writing a book about the history of the study of frozen earth and the creation of permafrost science in the Soviet Union.

CORNELL, Sarah

Sarah Cornell works on integrative social-environmental research at the Stockholm Resilience Centre. She is a visiting lecturer at the University of Iceland and a visiting fellow in the Systems Centre, University of Bristol, where from 2004 to 2011 she was the science coordinator for the U.K. Natural Environment Research Council's program for Earth System Science. Her research interests are in anthropogenic changes in global biogeochemistry, and the philosophy and methodologies of interdisciplinarity.

CRUMLEY, Carole

Carole L. Crumley is Professor of Anthropology (emerita), University of North Carolina, Chapel Hill, and Visiting Professor at the Swedish Agricultural University (Uppsala). She is Research Director, Integrated History and future Of People and Earth (IHOPE), Uppsala University, and has been active on other global change core projects (including IGBP, PAGES, AIMES, and DIVERSITAS). She is author of *Historical Ecology: Cultural Knowledge and Changing Landscapes* (1994) and studies long-term landscape change in Burgundy, France.

EGAN, Michael

Michael Egan is an associate professor in the Department of History at McMaster University, where his work concentrates on the histories of science, technology, environment, and the future. He is the author of *Barry Commoner and the Science of Survival: The Remaking of American Environmentalism* (MIT Press, 2007) and co-editor (with Jeff Crane) of *Natural Protest: Essays on the History of American Environmentalism* (Routledge, 2008).

FLEMING, Jim

Jim Fleming is Professor of Science, Technology, and Society at Colby College, Maine. His books include *Historical Perspectives on Climate Change* (1998), *The Callendar Effect* (2007), and *Fixing the Sky* (2010). He is deeply invested in building the community of historians of the geosciences, celebrating and connecting the history of science, technology, and medicine with public policy. New work includes a book on the emergence of atmospheric science and a biography of the carbon dioxide molecule.

FRESSOZ, Jean-Baptiste

Jean-Baptiste Fressoz is a historian of science, technology, and environment at Imperial College London. He is author of *L'Apocalypse joyeuse: Une histoire du risque tech-*

nologique (Paris: Le Seuil, 2012), which studies the history of technological risks, expertise, and regulations in France and Britain in the eighteenth and nineteenth centuries.

GRIFFITHS, Tom
Tom Griffiths is the W. K. Hancock Professor of History at the Australian National University (ANU) and Director of the Centre for Environmental History at ANU. He is the author of *Forests of Ash* (2001), *Hunters and Collectors* (1996), and an international history of Antarctica entitled *Slicing the Silence* (2007), which was a co-winner of the Prime Minister's Prize for Australian History in 2008. In 2002–03 and 2012 he voyaged to Antarctica with the Australian Antarctic Division.

HULME, Mike
Mike Hulme is Professor of Climate Change at the University of East Anglia (UEA), and is in the Science, Society and Sustainability Group. He was Founding Director of the Tyndall Centre for Climate Change Research, UEA (2000–2007). He is the author of *Why We Disagree About Climate Change* (Cambridge University Press, 2009). His current work illuminates the ways in which climate change is deployed in public, political, and scientific discourse, through historical, cultural, and scientific understandings of climate change.

JACKSON, Stephen T.
Stephen T. Jackson is Professor of Botany at the University of Wyoming, where he was founding Director of the Program in Ecology. His research interests center on ecological responses to environmental change, particularly as they play out from decadal to millennial timescales. He uses a variety of ecological and paleoecological approaches to these ends. He is an Aldo Leopold Environmental Leadership Fellow and a Fellow of the American Association for the Advancement of Science.

KAIJSER, Arne
Arne Kaijser is Professor of History of Technology at the Royal Institute of Technology (KTH), Stockholm, and much of his research is on the history of infrastructure. Before entering academia he worked in future studies, energy policy, and development aid. He was senior research fellow at the Dibner Institute, Massachusetts Institute of Technology, in 2001, and at the Netherlands Institute of Advanced Studies in 2010/11. He served as President of the Society for History of Technology (SHOT) in 2009–2010.

LINNÉR, Björn-Ola
Björn-Ola Linnér is Professor in Water and Environmental Studies at the Centre for Climate Science and Policy Research at Linköping University, Sweden, where he was the director in 2006–10. Currently, he is a visiting fellow at the Institute for Science, Innovation and Society (InSIS) at Oxford University. His research focuses on international policy-making on climate change, food security, and sustainable development. His published books include *The Return of Malthus* (2003).

MAUCH, Christof

Christof Mauch is Chair in American history and transatlantic relations, and Director of the Rachel Carson Center for Environment and Society at LMU, Munich. Previously, he was Director of the German Historical Institute in Washington, D.C. He held visiting professorships at Jadavpur University, Kolkata, India; at the University of Alberta, Edmonton, Canada; and at the Center for Environmental History, Vienna, Austria. He is currently President of the European Society for Environmental History (ESEH).

NICHOLLS, Neville

Neville Nicholls has spent forty years in climate research (up to 2005 at the Australian Bureau of Meteorology and subsequently at Monash University), studying the nature, causes, predictability, and impacts of climate variability and change. He has developed methods for predicting seasonal tropical cyclone activity, rainfall, crop yields, and human health impacts (including methods to predict climate-related arbovirus epidemics) and heat wave alert systems. He initiated the production of high-quality data sets for climate monitoring.

NIXON, Rob

Rob Nixon is Rachel Carson Professor of English at the University of Wisconsin and a Senior Fellow at the Institute for Research in the Humanities. He has authored four books, including *Slow Violence and the Environmentalism of the Poor*, which received the 2012 Sprout Prize for the best book in international environmental studies and the 2012 award for the best book in the interdisciplinary humanities. Nixon is a frequent contributor to the *New York Times*.

NORGAARD, Richard

Richard Norgaard is Professor of Energy and Resources, University of California, Berkeley. Trained as an economist, he is among the founders of ecological economics, a critic of the overuse and misuse of economic thinking, and a contributor to a variety of interdisciplinary environmental fields. His philosophical arguments are grounded in professional service and participatory research, currently as a member of the Intergovernmental Panel on Climate Change, UNEP's International Resource Panel, and as Chair of the State of California's Delta Independent Science Board.

OWENS, Susan

Susan Owens is Professor of Environment and Policy and head of the Department of Geography at the University of Cambridge. She has researched and published widely on environmental governance, with particular interests in relations among knowledge, politics, and policy-making. She served on the U.K. Royal Commission on Environmental Pollution 1998–2008. She is a Fellow of the British Academy and in 1998 was appointed an OBE in recognition of her services to sustainable development.

ROBIN, Libby
Libby Robin is Professor of Environmental History at the Australian National University and senior research fellow at the National Museum of Australia, Canberra. She also works in the KTH Environmental Humanities Laboratory at the Royal Institute of Technology, Stockholm. Her books include the prizewinning *How a Continent Created a Nation* (Sydney: University of New South Wales Press, 2007) and *Flight of the Emu* (Melbourne: Melbourne University Press, 2001) and the co-edited collection *Boom and Bust: Bird Stories for a Dry Country* (Melbourne: Commonwealth Scientific and Industrial Research Organisation, 2009). She is vice president of the International Consortium of Environmental History Organizations (ICEHO).

STEFFEN, Will
Will Steffen is Executive Director of the ANU Climate Change Institute at the Australian National University (ANU), Canberra. From 1998 to mid-2004, he served as Executive Director of the International Geosphere-Biosphere Programme, based in Stockholm, Sweden. His research interests span a broad range within the fields of climate and Earth system science, with an emphasis on incorporation of human processes in Earth system modeling and analysis, and on sustainability and climate change.

STRANGE, Carolyn
Carolyn Strange directs the graduate program in history at the Australian National University. She is the author (with Alison Bashford) of *Griffith Taylor: Visionary, Environmentalist, Explorer* (Toronto: University of Toronto Press, 2008). Inspired by Taylor's identity as predictor of environmental limits, she co-convened a symposium at the National Museum of Australia in 2009: *Violent Ends: The Arts of Environmental Anxiety*: http://www.nma .gov.au/history/research/conferences_and_seminars/violent_ends2/home.

SÖRLIN, Sverker
Sverker Sörlin is Professor of Environmental History at the KTH Royal Institute of Technology and a co-founding affiliate of the Stockholm Resilience Center. An adviser to the Swedish government on environment and science policies since the 1990s, his books include a prize-winning two-volume history of European ideas, 1492–1918 (2004), and edited books on northern history and policy: *Science, Geopolitics and Culture in the Polar Region* (London: Ashgate, 2012) and *Northscapes: History, Technology, and the Making of Northern Environments* (with Dolly Jørgensen) (Vancouver: University of British Columbia Press, 2013).

WARDE, Paul
Paul Warde is Reader in Environmental and Economic History at the University of East Anglia, and researches the environmental, economic, and social history of early modern and modern Europe. His publications include *Ecology, Economy, and State Formation in Early Modern Germany* (Cambridge University Press, 2006), and (with Astrid Kander and Paolo Malanima) *Power to the People: energy in Europe over the last five centuries* (Princeton University Press, 2013). In 2008 he was a winner of the Phillip Leverhulme Prize.

Selection Credits

[In order of appearance]

Thomas Malthus. 1798. Excerpts from Chapters 1 and 2 in *An essay on population*. First printed for J. Johnson, in St. Paul's Church-Yard, London.

George Knibbs, 1928. "Conclusions as to population increase." Chapter 11 in *The shadow of the world's future, or, the Earth's population possibilities and the consequences of the present rate of increase of the Earth's inhabitants*. London: Ernest Benn.

George Borgström. (1953) 1965. "Ghost acreage." Chapter 5 in *The hungry planet: The modern world at the edge of famine*. Originally published in Swedish in 1953. English translation, New York: MacMillan. Reprinted with the permission of Scribner, a Division of Simon & Schuster, Inc., from *The hungry planet: The modern world at the edge of famine* by Georg Borgström. Copyright © 1972 by Georg Borgström. Copyright renewed © 2001 by Greta I. Borgström. All rights reserved.

Paul Ehrlich. (1968) 1971. *The Population Bomb*. Excerpts from pp. 3–6, 9–12, and 16–17. From *The population bomb* by Dr. Paul R. Ehrlich, copyright © 1968, 1971 by Paul R. Ehrlich. Used by permission of Ballantine Books, a division of Random House, Inc.

H. C. von Carlowitz. (1713) 2009. *Sylvicultura oeconomica*, 2nd ed., f. 3v–4v, 28–29, 32–33, 68–69. Trans. Paul Warde. Originally published in Leipzig: Johann Friedrich Braun, 1713. Facsimile of 2nd ed., Remagen: Verlag-Kassel, 1732.

W. S. Jevons. (1865) 1866. Chapters 1 and 2 in *The coal question: An enquiry concerning the progress of the nation, and the probable exhaustion of our coal-mines*, 2nd ed., 1–6, 15–17, 28–35. London: Macmillan.

Samuel H. Ordway Jr. 1956. "Possible limits of raw-material consumption." In *Man's role in changing the face of the Earth*, W. L. Thomas (ed.), 987–992. Chicago: University of Chicago Press. Copyright Wenner-Gren Foundation for Anthropological Research and the National Science Foundation by University of Chicago Press.

Donella H. Meadows, Jorgen Randers, and Dennis L. Meadows for the Club of Rome. 1972. *The limits to growth: A report for the Club of Rome's project on the predicament of mankind*, 158–175. New York: Universe Books. Permission to reprint granted with kind permission from Rizzoli International Publications, Inc. and Potomac Associates. From *The limits to growth: A report for the Club Of Rome's project on the predicament of mankind* by Donella H. Meadows. © 1972, Potomac Associates. © 1972, Rizzoli International Publications, Inc. Reproduced with kind permission from Rizzoli International Publications, Inc.

Index

ABOUT THE EDITORS

LIBBY ROBIN is professor of environmental history in the Fenner School of Environment and Society at the Australian National University and a senior research fellow at the National Museum of Australia, Canberra.

SVERKER SÖRLIN is professor of environmental history at the KTH Royal Institute of Technology, Stockholm, and co-founder of the KTH Environmental Humanities Laboratory.

PAUL WARDE is reader in environmental and economic history at the University of East Anglia, an associate lecturer at the University of Cambridge, and associate research fellow at the Centre for History and Economics at Cambridge.

Paul Warde (left), Sverker Sörlin, and Libby Robin.
Photo: George Serras, National Museum of Australia